High-Activity Nuclear Waste Management Policy

High-Activity Nuclear Waste Management Policy: Becoming a Responsible Ancestor describes the technical, social, and political challenges of developing a deep-mined, geologic repository for high-activity nuclear waste. It emphasizes how the technical estimates of repository performance affect non-technical considerations, such as trust and risk perceptions, and how those considerations in turn affect the technical determinations.

Unlike other democracies with mature nuclear power programs, the United States lacks a permanent management plan for its high-activity radioactive waste. This book analyzes how the United States reached this negligent position and examines how eight other countries have addressed five critical waste management activities: establishing and implementing organizations, selecting a repository site, executing phased repository development, licensing a repository, and addressing technical controversies. It concludes with recommendations for reconstituting an effective waste management program in the United States.

The book will interest nuclear waste and nuclear energy and safety researchers, policy makers, and administrators. It will also benefit graduate students taking specialized courses in nuclear waste management, technology ethics, and public/energy policy.

Daniel S. Metlay is an independent researcher in the field of nuclear waste management. He graduated in 1967 from Caltech with a double major in molecular biology and medieval history. He later entered the Ph.D. program for political science at the University of California, Berkeley. While there, he was invited to join a Task Force established by the U.S. Nuclear Regulatory Commission to develop the first set of goals and objectives for nuclear waste management. Dr. Metlay taught organization theory and public policy at Indiana University and MIT for more than a dozen years. He was later appointed to direct a Secretary of Energy Task Force charged with recommending ways to merit and sustain public trust and confidence in DOE's nuclear waste management program. He then joined the staff of U.S. Nuclear Waste Technical Review Board, preparing reports to Congress and serving as the micro-agency's international liaison. Throughout his career, Dr. Metlay has published over 100 articles, technical reports, presentations, and government documents. He retired in 2018 after three decades of public service.

"Daniel Metlay's exquisite sense of nuclear waste history and technical statesmanship combine in producing a path to becoming responsible ancestors by emphasizing social acceptability as a critical dimension in nuclear waste management."

Ronald M. Latanision, *former member of the Nuclear Waste Technical Review Board and Professor Emeritus, Massachusetts Institute of Technology*

High-Activity Nuclear Waste Management Policy

Becoming a Responsible Ancestor

Daniel S. Metlay

Routledge
Taylor & Francis Group
NEW YORK AND LONDON

Designed cover image: Shutterstock

First published 2026
by Routledge
605 Third Avenue, New York, NY 10158

and by Routledge
4 Park Square, Milton Park, Abingdon, Oxon, OX14 4RN

Routledge is an imprint of the Taylor & Francis Group, an informa business

Library of Congress Cataloging-in-Publication Data
Names: Metlay, Daniel S. author
Title: High-activity nuclear waste management policy : becoming a responsible ancestor / Daniel S. Metlay.
Description: New York, NY : Routledge, 2026. | Includes bibliographical references.
Identifiers: LCCN 2025048435 (print) | LCCN 2025048436 (ebook) | ISBN 9781032866093 hardback | ISBN 9781032866086 paperback | ISBN 9781003528302 ebook
Subjects: LCSH: Radioactive waste disposal--Management | Radioactive waste disposal--Government policy--United States
Classification: LCC TD898.14.M35 M48 2026 (print) | LCC TD898.14.M35 (ebook)
LC record available at https://lccn.loc.gov/2025048435
LC ebook record available at https://lccn.loc.gov/2025048436

ISBN: 978-1-032-86609-3 (hbk)
ISBN: 978-1-032-86608-6 (pbk)
ISBN: 978-1-003-52830-2 (ebk)

DOI: 10.1201/9781003528302

Typeset in Times
by KnowledgeWorks Global Ltd.

For my parents, Grischa and Frances, and my sister, Debbie.

Contents

PART III What Can We Learn?

List of Figures

Preface

I have been in the waste-management "business" for a half-a-century as graduate student, professor, and government bureaucrat. I first got involved with the subject because it appeared to have two distinctive and fascinating characteristics. It demanded performance projections of a complex natural and engineered system across millennia. Radioactive waste management also elicited sharp debates and conflicts over values, none of which are easily reconciled.

My first exposure to this issue came when I was just starting my dissertation in the Political Science Department at the University of California, Berkeley. Fortuitously, I was appointed a member of the Nuclear Regulatory Commission's Task Force on Goals and Objectives for managing this material. Subsequently, as a young professor at Indiana University in Bloomington, I served as a full-time consultant to White House Office of Science and Technology Policy. There I drafted a policy analysis for the Interagency Review Group on Nuclear Waste Management. That study became an important input for President Jimmy Carter's decisions to recreate the country's waste-disposition program. I spent the next six years as a junior professor of political science and public policy at MIT.

After leaving university life, I directed a strategic planning group for Energy Secretary James Watkins in the first Bush Administration and for Hazel O'Leary in the Clinton Administration. The task force's remit was to provide advice on how the Department might merit trust and confidence in its efforts to develop a deep-mined, geologic repository for the long-term isolation and containment of high-level radioactive material.

The transition to becoming a member of the senior professional staff of the presidentially appointed Nuclear Waste Technical Review Board proved nearly frictionless. That micro-agency's responsibility was to evaluate the "scientific and technical validity" of the Department of Energy's efforts to implement the Nuclear Waste Policy Amendments Act of 1987. I served for nearly a quarter of a century. In my final dozen years at the Board, I was its international liaison to other nations waste-disposition programs.

So, I have come full circle. What began as an academic pursuit now ends with another one. The subject continues to sustain my interest. It now, however, is tinged with sorrow that after so many decades and considerable investments of time, energy, and resources by hundreds of contributors, the country seems back at the same starting gate it was a half-century ago. It has become an irresponsible ancestor.

Acknowledgments

It should come as no surprise that I am indebted to many individuals and institutions. Professor Todd LaPorte at Berkeley became an early supporter, friend, and mentor. His guidance and interest continues to this day. Without his intellectual encouragement, my own thinking would be even more mundane than it is now. William P. Bishop, the Nuclear Regulatory Commission's first radioactive waste manager, took a risk in asking a close-to-naive and untested student to join his goals and objectives efforts. He continued to stimulate my thoughts when we both worked on Carter's Interagency Review Group. Ted Greenwood directed a key subgroup in that effort. He was also my colleague when we taught at MIT. Not just a sharp mind, he was a man of principle.

Two members of the Nuclear Waste Technical Review Board, Gary Brewer and Warner North, encouraged me to join the Board's staff. Board chairs, Jared Cohon, B. John Garrick, and the Rod Ewing, found value in having a social scientist involved in examining technical issues. All three modeled being outstanding natural scientists and engineers as well as, in their own ways, being sensitive to the societal implications of their vocation. Board members Ali Mosleh and Ronald Latanision also played key roles in my professional growth.

I hesitate to mention the many individuals in other countries with whom I interacted for fear of forgetting some of them. But I must acknowledge those that had a profound impact on my understanding of waste-disposition programs abroad Carl-Reinhold Bråkenhielm, Geráld Ouzounian, Kathryn Shafer, Claudio Pescatore, and Saida Laârouchi Engström. (None, of course, are liable for any remaining errors.)

Allison MacFarlane initially and then Nicholas Vonortas at George Washington University's Institute for International Science and Technology Policy provided a welcoming and intellectually fertile home as a Senior Visiting Scholar (Non-Resident). This volume would not have come to fruition without their generous support. Garrick and Mosleh invited me to become a Senior Fellow at the B. John Garrick Institute for Risk Sciences at the University of California, Los Angeles.

Ewing and MacFarlane, former chairman of the Nuclear Regulatory Commission, are owed a special note of thanks for organizing the *Reset of America's Nuclear Waste Management* study. As a member of the Steering Committee for that effort, I had numerous opportunities to pick their brains and experience. In modified form, the results of some of those interactions reappear in this volume.

I also can't thank enough Mark Holt who read each draft chapter. His understanding of the waste-management program in the United States prevented me from making gross and subtle mistakes. He too is not responsible for those that remain.

My wife, Anita Kay Lampel, an author herself, transformed a long-time bachelor's life. Her love, encouragement, and wisdom made bearable the task of writing almost non-stop for a half-dozen hours a day over eighteen months. With the eye and ear of a *Scientific American* reader, she reviewed each draft chapter. Thus, those who engage with this volume also owe her their gratitude.

Abbreviations and Acronyms

AEC	Atomic Energy Commission
AECL	Atomic Energy of Canada Limited
AkEnd	Committee on a Site Selection Procedure for Disposal Sites (Germany)
Andra	National Waste Management Agency (France)
ASN	Nuclear Safety Authority (France)
ASNR	Nuclear Safety and Radiation Protection Authority (France)
BGE	Federal Company for Radioactive Waste Disposal (Germany)
CEA	Atomic Energy Commission (France)
CEQ	Council on Environmental Quality
CFR	Code of Federal Regulations
CoRWM	Committee on Radioactive Waste Management (United Kingdom)
DOE	Department of Energy
DOI	Department of Interior
DPS	Domestic Policy Council
EKRA	Expert Group on Disposal Concepts for Radioactive Waste (Swiss)
EndKo	Commission on the Disposal of High-Activity Waste (Germany)
ENSI	Federal Nuclear Safety Inspectorate (Switzerland)
EPA	Environmental Protection Agency
ERDA	Energy Research and Development Administration
ESF	Exploratory Studies Facility
FEDCORP	Federally Chartered Corporation
HLW	High-level waste
KTH	Royal Institute of Technology (Sweden)
LBNL	Lawrence Berkeley National Laboratory
LANL	Los Alamos National Laboratory
LILW	Low-level/intermediate level radioactive waste
LLNL	Lawrence Livermore National Laboratory
IFA	Independent Federal Agency
IRG	Interagency Review Group on Nuclear Waste Management
IAEA	International Atomic Energy Agency
ILW	Intermediate-Level Radioactive Waste
INFCE	International Fuel Cycle Evaluation
METI	Ministry of Economy, Trade, and Industry (Japan)
MKG	Environmental Organizations' Nuclear Waste Review Association (Sweden)
MRS	Monitored Retrievable Storage
MUA	Multi-attribute Utility Analysis
Nagra	National Cooperative for the Disposal of Radioactive Waste (Switzerland)
NAS	National Academy of Sciences
NDA	Nuclear Decommissioning Authority (United Kingdom)
NEI	Nuclear Enterprise Institute

NIREX	Nuclear Industry Radioactive Waste Executive (United Kingdom)
NRC	Nuclear Regulatory Commission
NSC	National Security Council
NUCO	Nuclear Utility-Owned, Non-Profit Implementer
NUMO	Nuclear Waste Management Organization (Japan)
NWF	Nuclear Waste Fund
NWMO	Nuclear Waste Management Organization (Canada)
NWPA	Nuclear Waste Policy Act of 1982
NWPAA	Nuclear Waste Policy Amendments Act of 1979
NWTRB	Nuclear Waste Technical Review Board
OCRWM	Office of Civilian Radioactive Waste Management
OMB	Office of Management and Budget
OSTP	Office of Science and Technology Policy
OTA	Office of Technology Assessment
RSSF	Retrievable Surface Storage Facility
SFOE	Federal Office of Energy (Switzerland)
SKB	Swedish Nuclear Fuel and Waste Management Company
SNF	Spent nuclear fuel
SSM	Radiation Protection Authority (Sweden)
STUK	Radiation and Nuclear Safety Authority (Finland)
TBYMS	*Technical Basis of the Yucca Mountain Standard*
TRU	Transuranic-contaminated waste
TSPA	Total System Performance Analysis
TVO	Teollisuuden Voima Oyj (Finland)
USGS	United States Geological Survey
UZ	Unsaturated Zone
WIPP	Waste Isolation Pilot Plant

Source Information for Figures

NWTRB 2015: Figures 1.1, 3.2, 3.3, 3.4, 8.1, 8.2, 8.5

Nuclear Waste Technical Review Board, "Designing a Process for Selecting a Site for a Deep-Mined, Geologic Repository for High-Level Radioactive Waste and Spent Nuclear Fuel," 2015.

Rawpixel: Figure 2.1

https://www.google.com/search?q=rawpixel+public+domain&rlz=1C1ONGR_enUS1101US1101&oq=rawpixc&gs_lcrp=EgZjaHJvbWUqCQgBEAAYChiABDIGCAAQRRg5MgkIARAAGAoYgAQyCQgCEAAYChiABDIJCAMQABgKGIAEMgkIBBAAGAoYgAQyCQgFEAAYChiABDIJCAYQABgKGIAEMgkIBxAAGAoYgAQyCQgIEAAYChiABDIJCAkQABgKGIAE0gEJNjk1N2owajE1qAIIsAIB8QWaaHDWWKPLHvEFmmhw1lijyx4&sourceid=chrome&ie=UTF-8

DOE 2022: Figure 2.2

https://www.energy.gov/em/articles/crews-making-progress-hanfords-former-plutonium-processing-facility

Wikimedia Commons: Figures 2.3, 2.4

https://www.google.com/search?q=wikipedia+public+domain+images&rlz=1C1ONGR_enUS1101US1101&oq=Wikip&gs_lcrp=EgZjaHJvbWUqDAgDEAAYQxiABBiKBTIGCAAQRRg5MgoIARAuGLEDGIAEMgwIAhAAGEMYgAQYigUyDAgDEAAYQxiABBiKBTIMCAQQABhDGIAEGIoFMgwIBRAAGEMYgAQYigUyBggGEEUYPDIGCAcQBRhA0gEIODI2MWowajeoAgiwAgHxBSJlSLiKVRrW&sourceid=chrome&ie=UTF-8

Slovic 1987: Figure 2.7

Slovic, P., "Perception of Risk," *Science*, (236) 1987, 280–285.

NRC: Figure 3.1

https://en.wikipedia.org/wiki/Dry_cask_storage#/media/File:Nuclear_dry_storage.jpg

DOE/RW-0073: Figure 5.1

Department of Energy, "Environmental Assessment Overview Yucca Mountain Site, Nevada Research and Development Area, Nevada," Volume 1, 1986

ORCD-1: Figures 5.3, 6.2

Office of Crystalline Repository Development. "A National Survey of Crystalline Rocks and Recommendations of Regions To Be Explored for High-Level Radioactive Waste Repository Sites," OCRD-1, 1983.

DOE (YM photo): Figure 6.1

Department of Energy, "Viability Assessment of a Repository at Yucca Mountain: Overview," DOE/RW-0508, 1998.

RW-0198: Figure 6.3
Department of Energy, "Site Characterization Plan Overview: Yucca Mountain Site, Nevada Research and Development Area, Nevada," RW-0198, 1988.

Van Luik 1999: Figure 6.4
Van Luik, A., "Overview of Total System Performance Assessment—Viability Assessment," Presentation to NWTRB Winter Meeting, 1999.

Bodvarsson 2004: Figure 6.5
Bodvarsson, B., "Thermal Hydrologic Environment and Thermal Seepage," Presentation to NWTRB Spring Meeting, 2004.

DOE 2008: Figures 6.6, 11.2
https://www.nwtrb.gov/docs/default-source/meetings/2006/september/swift.pdf?sfvrsn=d50ef605_5

Swift 2008: Figure 6.7
Swift, P., "Total System Performance Assessment: Modeling Approach and Overview of Results," Presentation to NWTRB Spring Meeting, 2008.

Thompson 1971: Figure 7.1
Thompson, J., *Organizations in Action: Social Science Bases of Administrative Theory* (New York: McGraw Hill, 1971).

NWTRB 2022: Figures 7.3, 10.2
Nuclear Waste Technical Review Board, "Survey of National Programs for Managing High-Level Radioactive Waste and Spent Nuclear Fuel: 2002 Update," 2002

Nagra 2011: Figure 7.4
Nagra, "Suggestions for the Place of the Site Areas for the Surface Installations of Deep Geological Repositories and Their Development: General Report," Technical Report 11-01, 2011.

SKB 2011: Figures 7.5, 8.12
SKB, "Long-Term Safety for the Final Repository for Spent Nuclear Fuel at Forsmark: Main Report of the SR-Site Project," TR-11-01, 2011.

Flynn et al.: Figure 8.7
Flynn, J. et al., "Trust as a Determinant of Opposition to a High-Level Radioactive Waste Repository: Analysis of a Structural Model," *Risk Analysis*, (12) 417–429, 1992.

Sjöberg 2004: Figure 8.8
Sjöberg, L., "Local Acceptance of a High-Level Nuclear Waste Repository," *Risk Analysis*, (24), 737–749, 2004.

Seidl et al.: Figure 8.9
Seidl, R. et al., "The Role of Trust and Risk Perception in Current German Nuclear Waste Management," *Risk Analysis*, 1–16, 2022.

Posiva 1999: Figure 8.10
Posiva, "Final Disposal Facility for Spent Nuclear Fuel: Environmental Impact Report," 1999.

Andra 2005b: Figure 8.11
Andra, "Dossier 2005 Argile: Evaluation of the Feasibility of a Geological Repository in an Argillaceous Formation," 2005.

NWMO 2023a: Figure 8.13
NWMO, "Confidence in Safety—Revell Site: 2023 Update," NWMO-TR-2023-07, 2023.

NEA 2012: Figures 9.1, 9.2, 10.3
Nuclear Energy Agency, "Reversibility and Retrievability in Planning for Geological Disposal of Radioactive Waste: Proceedings of the 'R&R' International Conference and Dialogue 14-17 December 2010, Reims, France," 2012.

National Council for Nuclear Waste: Figure 10.4
Swedish National Council for Nuclear Waste, "Mechanisms of Copper Corrosion in Aqueous Environments," Report 2009:4e, 2009.

Fabryka-Martin 2006: Figure 11.1
Fabryka-Martin, J. and A. Wolfsberg, "Hydrologic Flow Paths and Rates Inferred from the Distribution of Chlorin-36 in the ESF," Presentation to NWTRB Summer Meeting, 1996.

1 Introduction

The German social philosopher Hans Jonas observed, "The good and evil about which action had to care lay close to the act, either in the praxis itself or its immediate reach. The long-run consequences beyond were left to chance, fate, or providence. A new categorical imperative now requires that the 'future wholeness of Man' be included among the objects of our will."[1] In doing so, he points to a distinction between earlier technologies and a subset of modern ones—those that cause "wicked problems." These conundrums possess substantial technical uncertainties, long time-constants of feedback, significant conflicts over values, and poorly defined decision-making strategies. As Jonas suggests, wicked problems are emerging with increasing frequency. Climate change and artificial intelligence are arguably examples of this class of public-policy challenges. One clear implication of Jonas' analysis is that treating "wicked problems" as if they were "ordinary" is a blueprint for disappointment.

The permanent management of high-activity nuclear waste is another archetypal instance of a wicked problem. Specialists have identified technologies that might isolate this dangerous material from humans and the environment for many millennia. For example, the waste could be rocketed to the sun, disposed of in seabed sediments, placed in deep boreholes, or transmuted into less hazardous or shorter-lived radioisotopes.[2] However, the dominant international consensus is to bury the material in conventionally mined geologic repositories, hundreds of meters below the surface. Figure 1.1 illustrates a generic repository.

To construct a repository, however, scientists, engineers, and other professionals must design a complex system composed of natural barriers, such as a suitable rock formation and hydrogeology, as well as man-made barriers, such as the waste form and package. Although disruptive events, such as earthquakes and volcanoes, need to be evaluated, the primary performance scenario implicates the passage of water through the rock where it then comes in contact with the engineered barriers. Over time, the liquid degrades those obstacles, exposing the waste itself to dissolution and transport outside the bounds of the repository system. The dissolved radionuclides then flow into the biosphere, thereby exposing humans, flora, and fauna to toxic doses. These can result in morbidity, death, and genetic damage.

Selecting a specific site for this facility demands a strong understanding of how the behavior of this intricate arrangement changes over hundreds of thousands of years. Because the repository's performance over time cannot be "proven" in the usual sense of the term, computer models, with all their benefits and drawbacks, must be employed to project the system's effectiveness in isolating radioisotopes. Further, if these projections are seriously flawed, many decades, even centuries, might well pass before the dangers manifest themselves; taking remedial action then is likely to be demanding, if not foreclosed altogether.

DOI: 10.1201/9781003528302-1

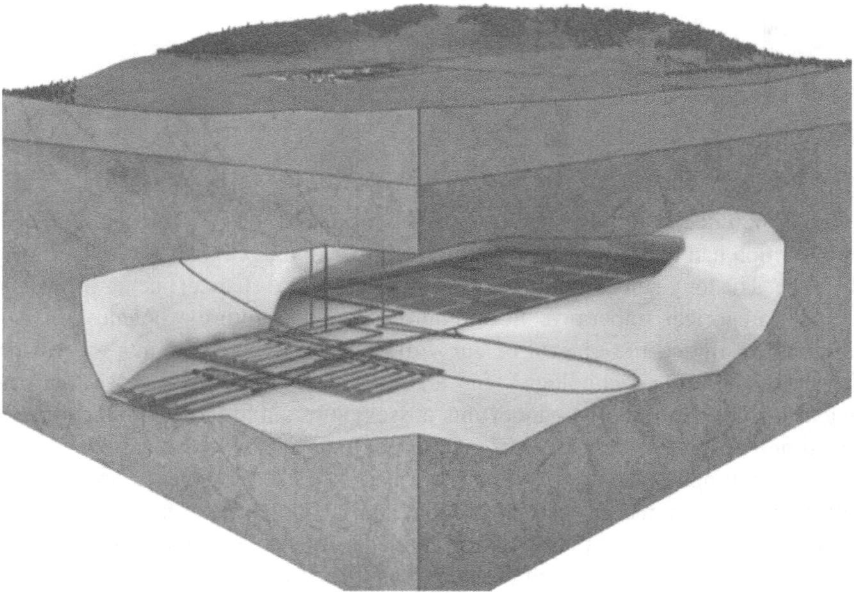

FIGURE 1.1 A layout of a generic repository.

In addition to contested technical challenges, the choice of location almost certainly will evoke public passions, legal challenges, and direct action in opposition. The legitimacy of the decision-making process will almost certainly be questioned. How safe is safe enough? Who should decide? More fundamentally, decision-making theories suggest that our relatively well-understood strategies for choice, such as calculation, bargaining, and trial-and-error, seem ill-matched to the task of managing highly radioactive waste. These political, social, and managerial considerations, it must be emphasized, affect and are affected by the technical analyses and assessments.

MANAGING THE HIGHLY RADIOACTIVE WASTE IN THE UNITED STATES

The United States first created highly radioactive waste material as part of the Manhattan Project's push to fabricate and release the atomic bombs over Hiroshima and Nagasaki. During the next four-score years, the federal government struggled to implement disposal initiative after initiative after initiative without any success. Those efforts intensified when a 1968 fire at a nuclear-weapons fabrication plant in Colorado generated radioactive debris that was eventually sent for storage to a government facility in Idaho.[3] Its governor extracted a commitment from the Atomic Energy Commission (AEC), then the accountable agency, to remove the rubble by the end of the 1970s.

Thus began a fifteen-year process for choosing what hopefully would become a technically suitable and socially acceptable location for a repository, a search that led to the passage of the Nuclear Waste Policy Act (NWPA) in 1982. Adopted overwhelmingly, this legislation embraced four strategic bargains, the sum of which swayed powerful interested parties: (1) geographic equity and geologic diversity would require multiple repositories; (2) a technically driven process would down-select potential repository sites; (3) host-states would be given a meaningful voice over the selection of a site; and (4) nuclear utilities would collect a fee for the wastes' disposal in exchange for a date-certain for the federal government to accept their waste.

Less than five years later, the wickedness of the issue made itself apparent and those agreements began to fall apart. First, geographic equity was, for all practical purposes, abandoned. Then Congress replaced the technically driven site-selection process with an edict that only the Yucca Mountain location in Nevada could be studied. Next, the provision controlling how a state might raise an objection to the selection of a repository site proved to be a fig leaf. Along the way, the Yucca Mountain Project encountered delays, technical and legal issues, political opposition, and funding shortfalls. Finally, a court ordered the federal government to cease collecting the fee—the last bargain still standing. By the end of the 2010s, a bipartisan understanding took hold, conceding that the project was no longer politically viable.

Over three presidential administrations spanning more than two decades, the United States lost the multi-billion-dollar bet it had placed on a potential Yucca Mountain repository. No disposal alternative has since been identified. So, nearly 90,000 metric tonnes of hazardous liquid and vitrified high-level waste (HLW) and spent nuclear fuel (SNF)—collectively high-activity radioactive waste—continue to be stored at more than 100 locations in 39 states.[4] Their security depends on continual attention and maintenance for the indefinite future. By kicking the disposition "can" down the road, the current generation and the next one have become "irresponsible ancestors."

This situation has not gone unnoticed. For example, the former chairman of the U.S. Nuclear Regulatory Commission (NRC) and the former chairman of the U.S. Nuclear Waste Technical Review Board lamented the moral and safety consequences of the current state of affairs, as well as the possible foreclosure of an option to deploy a new generation of reactors that might contribute to combating climate change.[5]

> Without a geologic repository, there is no way forward for the final disposal of this highly radioactive material. Storing it in pools and dry casks at reactor sites is a temporary solution; it is safe for decades, but not the millennia needed to isolate this radioactive material from the environment. The present U.S. policy of indefinite storage at a centralized site is not a viable solution, as it shifts the cost and risks to future generations.

A former attorney at an environmental advocacy organization put it well.[5]

> It's time for a new way forward. After 60 years of failure, some may question whether it's even possible to solve the conundrum of siting a nuclear waste disposal repository. We can *absolutely* solve the problem, but first we need to start with a clear-eyed dismissal of clichéd explanations for why national efforts have failed for so long.

The absence of even the barest plan to dispose of waste material in a repository further exacerbates the ongoing state of affairs. The encroaching risks of climate change impel many to seek out carbon-free energy sources. By some calculations, nuclear energy fits that bill. When incorporating all steps in the fuel cycle, its carbon footprint is substantially lower than fossil fuels. (But its mark is significantly higher than wind, solar, and hydro power.) Nuclear innovators are attempting to bring to market a new wave of so-called small modular and micro-reactors.[7] Amazon, for instance, has signed agreements to support this technology.

Moreover, implementing current and future plans for widening the reach of artificial intelligence will draw prodigious quantities of electricity. So much so that Microsoft contracted with Constellation Energy to purchase electricity from the now-shuttered Three Mile Island nuclear power reactor. The reopened plant will power one of the company's data centers.

In sum, if the United States is to shed its cloak as an irresponsible ancestor and transform itself into a responsible one, federal and local governments, the nuclear industry, and civil society will all have to overcome inertia. As they did in 1982, they will have to reach a new set of strategic bargains. The final chapter of this volume offers some thoughts on how those accords might be structured.

WHAT THE BOOK IS ABOUT

This volume examines the management of HLW that derives from chemically reprocessing defense and commercial SNF, as well as the SNF that is now stored in mounting quantities, mostly at nuclear power plants. Accordingly, it *does not* consider the management of uranium mill tailings, low-level waste, transuranic-contaminated waste (except as it influences the disposal of high-activity material), and mixed radioactive and hazardous waste. Even so limited, there are a variety of subjects and issues that might be considered. Some were omitted from the start. For example, the question of waste transportation, notwithstanding its centrality in the eyes of many members of the public, is not explored. Except as necessary, blow-by-blow descriptions of all the bureaucratic, political, legal, regulatory, and legislative controversies and battles that have been fought, some as many as 80 years ago, are not retold in this book. For those anticipating a tell-all, this is not it. Instead, I focus "narrowly" on the events in the United States and a select number of other nations, leading to the development of deep-mined geologic repositories.

Part I

The remainder of the book contains of three parts. Part I chronicles in five subsequent chapters the evolution of waste-management policymaking in the United States from the mid-1940s to the present. "In the Beginning" discusses the discovery of nuclear fission and how that phenomenon might be tamed to produce both weapons and electricity. It also describes the federal government's early initiatives to store the waste products in vulnerable carbon-steel tanks on the surface *in perpetuity*. After that approach drew strong criticism, the federal government adopted the preferred technology of disposing of the material in deep-mined geologic repositories and set in place preliminary rules about how that might be accomplished. It also includes a discussion of the "many faces of risk."

"Setting the Path" considers initial attempts to find a repository site and how state after state rebuffed federal proposals. After India exploded a nuclear device in 1974, concerns mounted about the dangers of nuclear-weapons proliferation. President Gerald Ford's administration recognized the linkage between the military and civilian applications of fission and took measures to mitigate those fears. These had important implications for the U.S. waste-management program. President Jimmy Carter's administration shut down the domestic "plutonium economy," which anticipated the reprocessing of SNF, recycling of recovered fissile elements into conventional reactors, and widespread, plutonium-fueled fast breeder reactors. Carter also set in motion an interagency group to recommend elements of a national plan for disposing of radioactive wastes, particularly unreprocessed spent fuel.

"Creating the Regulatory and Legal Framework" describes how the Environmental Protection Agency (EPA) and the NRC engaged in a bureaucratic *pas de deux*. The former agency had to establish limits on the radioisotopes that might be released from a repository. The latter had to specify how those limits might be imposed when licensing the construction of a disposition facility. The regulations called for studies of multiple potential repository locations in a variety of hydrogeologic formations. They also mandated that the studies involve *in situ* investigations of the proposed repository horizon. Rather than selecting the "best site," the NRC maintained that the chosen location need only satisfy yet to be specified criteria. Finally, the nuclear regulator instructed the implementor, now the Department of Energy (DOE), how to calculate radioisotope releases from three repository subsystems. This judgment provoked a rain of attacks that would last for more than a decade. The chapter also details the passage of the NWPA in President Ronald Reagan's administration. Starting with the negotiations in the House Interior Committee and concluding with the amended bill's approval by a voice vote in the Senate, it discusses in some detail how the four strategic bargains were struck.

"The Rise of Yucca Mountain Repository" assesses DOE's initial implementation of the NWPA between 1983 and 1987. It describes the department's long-delayed struggles to propose and finalize a set of largely technical guidelines for choosing a repository site. It considers the political controversies attached to winnowing nine potential locations down to five and further narrowing the possibilities from five to three. For the latter decision, the chapter describes the technical controversy surrounding the reliance on Multi-Attribute Utility Analysis methodology. At roughly the same time, Energy Secretary Harrington unilaterally announced that the first strategic bargain, the search for the second repository site, would be "indefinitely postponed." The chapter ends with the passage of the Nuclear Waste Amendments Act of 1987 (NWPAA), which overthrew the technically driven site-selection process—the second strategic bargain.

"The Fall of Yucca Mountain Repository" chronicles DOE's fading Yucca Mountain Project between 1987 and 2020. It considers how the two remaining foundational bargains were broken—Nevada's futile attempt to mount an objection to the choice of the Yucca Mountain site and the collection of a fee in exchange for a date-certain to accept the nuclear utilities' SNF for disposal. It also briefly details the effectiveness of that state's technical, legal, regulatory, and political opposition. The chapter describes how both regulators crafted Yucca Mountain-specific rules and how those protocols were challenged in court. Two related events—Nevada Senator Harry Reid's rise to power and Nevada's early-scheduled presidential primary—placed the Yucca Mountain Project in political peril and ultimately led to its demise.

Part II

Part II of the book probes five overarching issues that played crucial roles in today's irresponsible circumstances. By understanding their evolution in this country and overseas, some clues might be uncovered that suggest how to avoid those difficulties in the future.

"Designing an Implementing Organization" considers a variety of alternatives to affect a nation's nuclear-waste management program. In the United States, the default option has always been an Executive Branch agency. The waste-management torch has passed over the years from the AEC to the Energy Research and Development Administration (ERDA) to DOE. Periodically, however, that organizational design came under fire because it was too political, had transitory leadership, and lacked control over its budget. Indeed, the NWPA authorized a special commission to analyze alternative designs, including an independent federal agency and a federally chartered corporation perhaps like the Post Office and the Tennessee Valley Authority. Although the implementor in France, Germany, and Japan is also the federal government, other organizational forms are more prevalent abroad. In Sweden, Finland, Canada, and Switzerland, the implementor is a non-profit company owned by the countries' waste producers. Their programs are also among the most advanced internationally.

"Selecting a Repository Site" discusses the two prerequisites for choosing a location for a deep-mined geologic repository: technical suitability and social acceptability. It notes that sites can either be selected in a serial or in a parallel process, although the latter is generally the case. Technical suitability was the initial focus in most nations, but over time, social acceptability arose as the most problematic issue. The chapter argues that the two requirements are not necessarily independent of each other but can be reciprocally interdependent. Trust in the implementor, perceived risk, and attitudes toward nuclear power account for most of the variance in attitudes toward siting a repository.

"Phased Repository Development" details the benefits and difficulties of using a staged process to construct a disposition facility. Trivially, all development processes are phased. But staging requires pre-established periods in which the repository design and performance are evaluated and an explicit decision made either to proceed, revise, or stop the effort altogether. Countries such as France, Switzerland, and Canada have adopted a phased approach to repository development.

"Repository Licensing" examines two regulatory determinations: for how long should a repository limit radionuclide releases to the environment and what is an acceptable dose. It also considers the range of methodologies regulators use to demonstrate compliance with those rules. The chapter makes a distinction between establishing performance projections using probabilistic risk assessments and preparing a safety case that includes, but is not limited to, quantitative estimates of dose to an affected population group. The chapter also explores the issue of how prescriptive the regulations should be. It contrasts the rules in the United States with those in Sweden, Switzerland, and Canada.

"Addressing Technical Controversies" details two critical issues affecting the licensing of a disposition facility in the Scandinavian countries, Finland and Sweden, and the United States. In the former case, how likely is the copper canister holding SNF to corrode under anoxic aqueous conditions? In the latter case, what is the magnitude and velocity of the percolation flux at the proposed Yucca Mountain repository? Each issue provides evidence about the obstacles confronting the resolution of technical uncertainty. Both implementors found means to diffuse the controversies, albeit acknowledging that they were not objectively resolved.

PART III

Part III of the book consists of one chapter that advances a set of prescriptions for resurrecting the waste-disposal program in the United States. It offers advice on such issues as standing up a new implementor, siting a deep-mined geologic repository, establishing new regulations, phasing repository development, and addressing technical controversies. These are not items on a *menu* from which to choose the courses of a dinner. Rather, they are a *recipe* where the absence of any one ingredient spoils the dish.

NOTES

1. Hans Jonas, "Technology and Responsibility: Reflections on the New Task of Ethics," *Social Research*, (40) Spring 1973, 35–36.
2. The private company, Deep Isolation, on which I serve as an Advisory Committee member, has initiated the process of listing its shares on a public marketplace. Although the firm is optimistic about its success, it remains uncertain to what extent its horizonal borehole technology will replace deep-mined, geologic repositories worldwide.
3. Storage is distinct from disposal; the latter implies that no plans are put in place to retrieve the waste once emplaced and the repository is sealed.
4. See https://theconversation.com/how-and-where-is-nuclear-waste-stored-in-the-us-252475#:~:text=Around%20the%20U.S.%2C%20about%2090%2C000,all%20to%20one%20secure%20location. Accessed April 16, 2025.
5. Rodney Ewing and Allison MacFarlane, "Nuclear Waste Is Piling Up. Does the United States Have a Plan?" *Scientific American*, (325) 151–152, March 6, 2023.
6. Geoffrey H. Fettus, "Nuclear Waste Is Piling Up. Here's How to Fix the Problem," *Scientific American*, (325) October 18, 2023.
7. A recent, albeit controversial, analysis argues that small modular reactors "exacerbate the challenges of nuclear waste management and disposal." See Lindsay Krall et al., *Proceedings of the National Academy of Sciences*, (119) 1–12. See https://doi.org/10.1073/pnas.2111833119. Accessed April 16, 2025.

Part I

The Nuclear Waste Management Program in the United States

PREFACE

At the urging of President Dwight Eisenhower, Congress authorized the Atomic Energy Act of 1954. In line with the President's Atoms for Peace initiative, the legislation proclaimed that "the development, use, and control of atomic energy shall be directed so as to promote world peace, improve the general welfare, increase the standard of living, and strengthen free competition in private enterprise." The law allowed the private sector to obtain previously restricted data about nuclear energy production and the utilization of fissile materials. It also repealed provisions of an earlier act that forbade the patenting, and thus the ownership, of processes for harnessing nuclear energy and creating elements that might be split to generate electricity.

Not coincidentally, within a week of the new law's passage, Eisenhower broke ground for the nation's first full-scale commercial atomic power plant located on the Ohio River in Shippingport, Pennsylvania. (The reactor actually began its life as the cancelled propulsion unit for an aircraft carrier.) A fleet of more than 100 reactors would ultimately supply one-fifth of the country's electricity. Designing, testing, and selling that novel technology brought together the federal government, an expanding set of national laboratories, contractors, utilities, and entrepreneurs. It was a glamorous and exhilarating period, which attracted bright and pioneering minds.

Yet, so stimulating was the challenge that it completely overshadowed the dull-by-comparison, yet indispensable, component of the enterprise: managing its weapons

DOI: 10.1201/9781003528302-2

and commercial highly radioactive and extremely long-lived waste products. (Of course, further subordinating the gleam of that undertaking was the "successful" storage of the wastes from the weapons-production complex!) A prevailing sense of technological optimism made it easy to trust that whatever might be essential down the long road would somehow materialize.

This part of our chronicle traces the flow and ebb of the federal government's program to store and then permanently dispose of waste from the defense and commercial nuclear activities. It contains five chapters. "In the Beginning" discusses the discovery of nuclear fission and how that phenomenon might be tamed to produce both weapons and electricity. "Setting the Path" considers initial attempts to find a repository site and how state after state rebuffed federal proposals. "Creating the Regulatory and Legal Framework" describes how the Environmental Protection Agency and the Nuclear Regulatory Commission engaged in a bureaucratic *pas de deux*. The chapter also details the passage of the Nuclear Waste Policy Act in President Ronald Reagan's administration. "The Rise of the Yucca Mountain Repository" assesses the Department of Energy's initial implementation of the legislation between 1983 and 1989. "The Fall of the Yucca Mountain Repository" describes how the prospects for that effort faded between 1979 and 2020.

As a consequence of all these events, the United States no longer has even a hypothetical strategy for the long-term, permanent management of its defense and commercial high-level waste and spent nuclear fuel. It lacks the legal authority to identify potentially suitable sites for a disposal facility. It may or may not have the ability to reconstitute a new implementing organization. It is unclear, at the very least, whether funds are available to launch a new programmatic effort. In short, my generation by default has become an ***irresponsible ancestor***.

2 In the Beginning

Looking back at the dawning of the nuclear era at Alamogordo, New Mexico, in 1945, J. Robert Oppenheimer is said to have recalled a stanza from the *Bhagavad Gita*: "Now I am become Death, the destroyer of worlds." A little more than 25 years later, one of the era's foremost contributors, the physicist Alvin Weinberg, reflected, "We nuclear people have made a Faustian bargain with society. On the one hand, we offer…an inexhaustible source of energy…But the price that we demand of society for this magical energy source is both the level of vigilance and of longevity for our social institutions that we are quite unaccustomed to." More recently, a clergyman outside of Washington, D.C., encapsulated the tension between the waste-management now and the later-to-be by urging his congregants to become "responsible ancestors."[1]

MAN CREATES RADIOACTIVE WASTE

It took only a relatively short time between theory development and materialization for "radioactive waste" to join the technical lexicon. In December 1938, the German physical chemists Otto Hahn and Fritz Strassmann noticed that, when natural uranium was irradiated with neutrons, a variety of radionuclides were produced. After chemically separating them, the researchers discovered that some of the elements were like the ones (radium and actinium) that Enrico Fermi had spawned over the course of the previous five years. Hahn and Strassmann's "unexpected observation" was that, in the mix, was barium, an alkaline-earth metal. This finding suggested to them that the heavier uranium atom may have *fissioned* (split) into lighter elements.[2]

They quickly communicated this insight to a co-worker, Lise Meitner, who had fled Nazi Germany for England six months earlier. Over the Christmas holiday in 1938, walking in the snow, Meitner and her physicist-nephew, Otto Frisch, tried to find an explanation for this anomalous result. They speculated that the uranium nucleus might, like a water drop, have stretched when hit by a neutron, collapse in the middle, and form two distinct *nuclei* (drops) that would repulse each other electrically with an energy of roughly 200 MeV. Meitner calculated that the daughter products were less massive than the original uranium atom. Applying Einstein's famous equation, $E=mc^2$, she concluded that the missing mass was approximately 200 MeV! Matter was, in fact, interchangeable with energy.[3]

Nevertheless, it was unclear whether this discovery had practical ramifications, either for the construction of a weapon or as a new energy source. One key to the answer was the discovery in early 1940 by other researchers, including Niels Bohr, that only the ^{235}uranium isotope was fissile. This isotope, however, constitutes just 0.72 percent of natural uranium; the rest is mostly ^{238}uranium. The scarcity of the former isotope meant that somehow it would have to be concentrated for fission—peaceful

DOI: 10.1201/9781003528302-3

or otherwise—to be practicable. Another key was determining whether a split uranium atom would produce at least two free neutrons. Experiments at the University of Chicago under Fermi's leadership began in late 1942. Using a prototypical "atomic pile," those investigators verified that enough neutrons would be released to sustain a *chain reaction*, that is, ^{235}uranium atoms would continue to split.

Several engineers proposed alternatives to separate the fissile isotope from its heavier cousins. Ultimately Manhattan Project leaders decided on an approach in which natural uranium in gaseous form (UF_6) would pass through an extended sequence of semi-permeable membranes. Because of the ^{235}uranium's slightly lower weight, it would move slightly faster through the membranes, and thus the two isotopes could be partitioned. Deploying this process, natural uranium could be *enriched* to the requisite concentrations. By early 1945, high-enriched (greater than 90 percent) ^{235}uranium was produced in Tennessee; it would first be fabricated into the "Little Boy" bomb that was dropped over Hiroshima.[4] Low-enriched (3-5 percent) ^{235}uranium eventually would be assembled into fuel rods that drove civilian reactors and generated electricity. Figure 2.1 shows the large enrichment plant at Oak Ridge, Tennessee, that was operated after the war.

A second nuclear reaction, termed *neutron capture*, can take place in a reactor as well. In this scenario, ^{238}uranium, again loaded into specially designed rods, *absorbs* a neutron to form ^{240}uranium, which quickly emits an electron to transmute into ^{239}neptunium. Neutron absorption and electron emission continue to yield ^{239}plutonium and ^{241}plutonium, which are both fissile, the former nearly twice as likely to split as ^{235}uranium.

The fuel rods are eventually *spent* (depleted).[5] They are then removed and chemically reprocessed. The primary objective here is to extract the fissile plutonium. The

FIGURE 2.1 Oak Ridge enrichment plant (K-25).

FIGURE 2.2 Purex reprocessing plant at Hanford.

first batch was isolated in early 1945 at the Manhattan Project's Hanford Engineering Works in eastern Washington State. It would be used in the "Fat Man" bomb that was dropped on Nagasaki. For our story's purposes, it is important to note that, in addition to plutonium, a witch's brew of other elements, including fission products, actinides, and transuranics, also flowed off separately as *high-level radioactive waste* (HLW).[6] (See Figure 2.2.)

NOW WHAT TO DO WITH THE WASTE: FIRST ITERATION

These waste products are harmful to human health and the biosphere. Although some disagreement persists, both among technical specialists and the general public, about how much, what radioisotopes, and what injuries the material causes, virtually complete agreement now obtains that concentrated HLW and *spent nuclear fuel* (SNF) must be prevented from entering the environment.[7]

The initial approach adopted at Hanford was to dump low concentrations of the HLW into shallow trenches. Higher levels were poured into mammoth tanks. Logistical exigencies brought on by World War II constrained what material could be used to construct the tanks; shortages of acid-resistant stainless steel compelled operators to build them out of vulnerable carbon steel. This practice led over the years to the manufacture of 177 similar tanks that, at one point or another, held 56 million gallons of HLW. (See Figure 2.3.) Although not apparent at that time, within 20 years of the tanks' manufacture, significant quantities of material began to leak.[8]

FIGURE 2.3 Waste Storage Tanks at Hanford.

Starting in the mid-1950s, the successor agency to the Manhattan Project, the Atomic Energy Commission (AEC), held preliminary conversations with several firms to induce them to construct a *commercial* reprocessing facility.[9] The fissile material thereby extracted could be recycled into mixed (uranium and plutonium) oxide fuel that could be used either by the current generation of light-water reactors or stored in anticipation of the next generation breeder reactors. None of the companies were willing to risk the capital required at that time. But in 1962, Nuclear Fuel Services Inc. (NFS) approached the AEC with plans to build such a plant. Located in West Valley, New York, the facility was completed in 1966 at an estimated cost of \$33 million.[10] (See Figure 2.4.)

A desire to reduce operating costs and to proceed expeditiously, rather than waste-management considerations, controlled both the AEC and the NFS's decisions. Most critically, the liquid HLW would be stored in the same type of carbon-steel tanks used at Hanford. To be sure, the AEC did not entirely ignore the implications for waste management of that choice. The facility had to be constructed on land composed of soil that would prevent any migration of radioactive waste should the tanks fail. In addition, the agency required New York State to provide reasonable assurance of care for the tanks by inaugurating surveillance of the wastes and ensuring replacement of the storage facilities in *perpetuity*.

In testimony before the powerful congressional Joint Committee on Atomic Energy (JCAE), Hanford's manager, Herbert Parker, made clear his views about

FIGURE 2.4 Nuclear Fuel Services commercial reprocessing plant.

surveillance and perpetual care. Although not foreclosing better technical advancements that might arise in the future, he asserted[11]:

> This does not mean that it will have to be in [any] particular tank. In other words, if the tanks we have turn out to have a life of 50 years, it will be very simple to be prepared at the right time with an alternative set of tanks and pump the liquid into new tanks. We have extensively moved the liquid from one tank to another and are persuaded we can do this with complete safety.

As importantly, the symbiotic relationship between the federal government and the private sector intensified when SNF moved from weapons-production and electricity-generation sites to West Valley. As part of the AEC's negotiations with NFS, the agency committed to reimburse the company for reprocessing fuel from the *defense* complex. This agreement not only amounted to a subsidy to the firm but also provided a supply of enough SNF to build up a baseload that would tide NFS over until demand grew for commercial plutonium.[12]

Figure 2.5 diagrams the flow of SNF through the West Valley reprocessing plant.[13] The top of the chart presents the stream from the AEC (SNF from Hanford's N-Reactor [534 kg] and SNF from the Navy's nuclear submarine program at Erwin, Tennessee [95 kg]) and from six pioneer power plants. The rest of the chart shows that nearly 80 percent of plutonium ended up in the weapons complex; the remainder, almost 20 percent, found its way into civilian nuclear

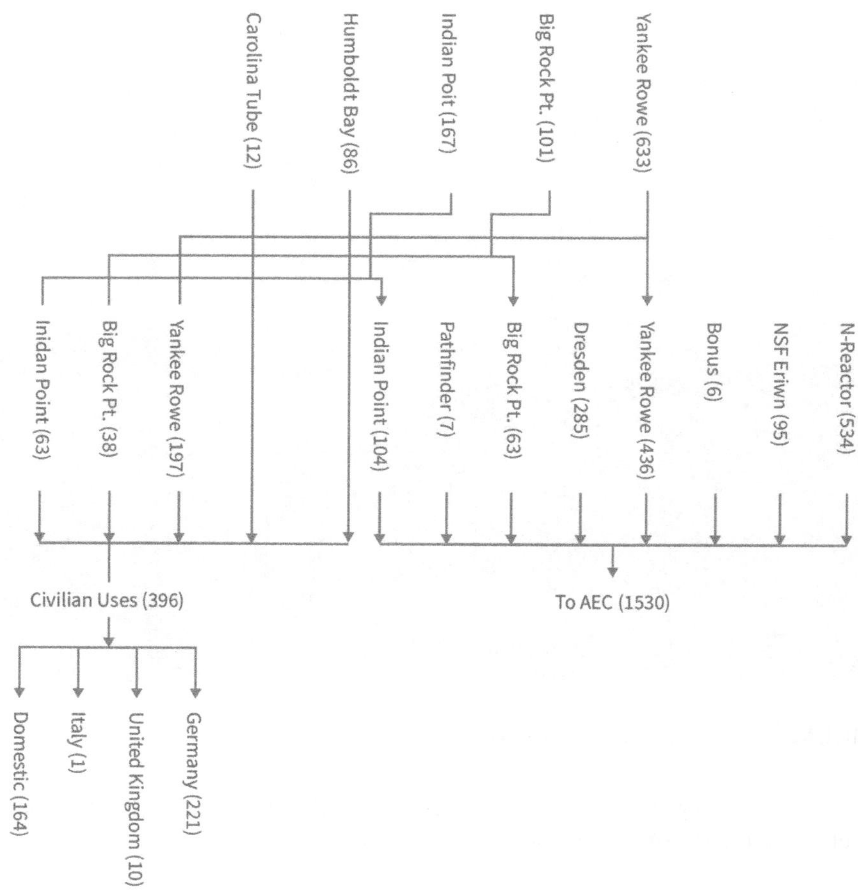

FIGURE 2.5 Kilograms of plutonium recovered at West Valley, New York.

energy programs. As a practical matter then, NFS simply became a substitute for reprocessing (mostly) weapons-dedicated SNF from AEC reservations. Not widely understood today, plutonium from commercial SNF also contributed to the nuclear defense program.

The AEC's venture into commercial reprocessing marked its first direct step in its involvement with the private sector in the arena of radioactive waste management. We shall see that this tight, but in many ways natural, connection will often repeat itself and profoundly influence future waste-management policies.

NOW WHAT TO DO WITH THE WASTE: SECOND ITERATION

Even as the AEC's commitment to commercial reprocessing deepened, questions arose about the soundness of the agency and the industry's strategy for managing high-activity radioactive waste. Notwithstanding Parker's assurances, could institutions behave reliably over many decades to prevent releases of radioisotopes from entering the environment? That concern prompted the U.S. National Academy of Sciences to appoint a panel to examine that and related issues.

The group's 1957 report proposed burying liquid HLW in a conventionally mined repository deep underground. In particular, the panel observed that, because of its plasticity and isolation from flowing water, salt formations might prove to be an appropriate host-rock (media). Moreover, the disposal task would be simplified if the liquid waste was converted into a solid, insoluble form.[14] The study's immediate impact was modest. For instance, efforts expanded to create technologies to solidify the HLW stored in Idaho by the nuclear navy's submarine program.

But it is hard to overestimate the report's long-term effect. On the one hand, the disposal problem appeared to have a scientifically based solution. On the other, the report promoted a sense of compliancy. The AEC could go about its preferred pursuit: building nuclear reactors.

A subsequent National Academy missive did stir matters up.[15]

> No existing AEC installation which generates either high-level or intermediate-level waste appears to have a satisfactory geological location for the safe local disposal of such waste product; neither does any of the present waste-disposal practices that have come to the attention of the committee satisfy its criterion for safe disposal of such wastes.

Although the AEC responded with a 32-page-long report several months later, its reaction was either one of dismissal or misunderstanding. Its most concrete action was to question the need for the Academy's committee.

As qualms grew a decade later, AEC began to reconsider its approach to managing its and the industry's waste. Agency leaders, especially Milton Shaw, now the Director of the Reactor Development and Technology Division, recognized that the Hanford analogy was fundamentally flawed. His arguments found support from Glenn Seaborg, then the Commission's chairman, and James Ramey, the lead Commissioner for the civilian reactor program and former staff director of the JCAE.[16]

Two internal developments ushered the way to policy change. A study completed at the Oak Ridge National Laboratory in 1969 was the first systematic effort to examine the back-end of the nuclear fuel cycle. It concluded that in-tank solidification, as practiced at Hanford, was not an acceptable form of storage on privately owned land because of the obvious difficulties that would be encountered if removal might be required. Further, research at the Idaho National Engineering Station demonstrated that the liquid HLW could be calcined into flakes.[17]

External pressures for change also increased. Not released to the public until four years after it was completed, an influential report from the newly established National Academy of Sciences' Advisory Committee on the Geologic Aspects of Radioactive Waste Disposal strongly made the following point[18]:

> [The AEC's recent efforts] promises relief from the problems of storing and disposing high- level liquid waste where disposal in the ground anywhere near freshwater aquifers would be intolerable. We are especially hopeful about glass or ceramic products because they may be safe from serious leaching and, thus, from release of hazardous radionuclides for periods of many centuries.

The internal and external forces generated enough momentum to overwhelm the inertia surrounding the question "what to do with the waste." In October 1966, Shaw briefed the Commissioners and emphasized three critical ideas[19]:

- Although tank storage was acceptable as a process step, perpetual storage of liquid HLW in surface or near-surface facilities is not an acceptable waste-management practice.
- Ultimate disposal of HLW will require control and possible surveillance of the waste repository in perpetuity.
- Such requirements can best be met by limiting the ownership of the land upon which the repository is situated to the federal government.

Shaw also presented the Commissioners with a draft rule designed to transform those policy views into binding regulations.[20]

Seven months later, the Commission published a proposal in the *Federal Register* that called for public comments on these specific provisions, among others.

- Commercial reprocessing plants can be located on private property.
- Liquid HLW stored at reprocessing plants for more than five years must be solidified.
- These solid wastes must be transferred to a federal repository no later than ten years following the separation of fission products from the irradiated fuel.
- Upon receipt, the federal government will assume physical responsibility for these radioactive waste materials, but the industry will pay all costs of ultimate disposal.
- Ultimate disposal of HLW fission product material will not be permitted on any land other than that owned by the federal government.

Some relatively minor changes were adopted based on comments from interested parties and Commissioner Theos Thompson. In August 1970, over Thompson's persistent dissent, the other Commissioners approved *10 Code of Federal Regulations 50, Appendix F*.[21]

The speedy promulgation of *Appendix F* testified to Shaw's political and bureaucratic skills, not surprising given that he was a protege of Admiral Hyman Rickover, the father of the nuclear Navy. He orchestrated a subtle and delicate internal coalition that included the formidable Production Division, which not only controlled Hanford and other defense complex sites but which also had its own favorite disposal strategy, injection of liquid HLW into bedrocks. He co-opted the nuclear industry by offering side-payments. NFS was grandfathered. Companies, such as Allied Chemical Corporation and General Electric, contemplating entry into the commercial reprocessing business, would still be eligible for licenses without having to design, build, or implement novel solidification technologies beforehand.

The lone voice of internal dissent, albeit muted, belonged to Ernest Tremmel, the Director of the Division of Industrial Participation. He cautioned Shaw that limiting repository development to the federal government and precluding industry's involvement would crimp innovation and stifle competition. Because his unit lacked a significant budget and programmatic responsibility, Tremmel's objections were cast aside in the belief that "long-term considerations should be overriding."[22]

It is almost ironic that the decision whether the federal government or the private sector should be in charge was taken with barely a second serious glance. As we shall see, that fundamental choice, taken *differently* by other nations, will be revisited time and time again in this country even until today.[23]

THE MANY FACES OF RISK

The transition of radioactive waste management from perpetual storage of liquid HLW to permanent disposal of solidified HLW (and later SNF) marked a fundamental shift that has not varied for more than a half-century. Its wisdom rests on two associated pillars: a moral responsibility not to pass current generations' burdens to the future and an environmental ethic not to befoul the earth.

As the policy-significance of the matter gained prominence starting in the 1970s, nuclear critics issued near-apocalyptic assessments of the challenges confronting humanity. For example, Hannes Alfvén, a Nobel Laureate in physics, maintained that[24]:

> The problem is to keep radioactive waste in storage [sic] until it decays after hundreds of thousands of years. The [geologic] deposition must be absolutely reliable as the quantities of poisons are tremendous. It is very difficult to satisfy these requirements for the simple reason that we have had no practical experience with such a long-term project. Moreover, permanently guarded storage [sic] requires a society with unprecedented stability.

More sober assessments noted that waste is not ejected, as in a reactor accident, with the kinetic force of an explosion; rather it is gradually dispersed almost entirely by groundwater moving to the surface.

Nonetheless, any consideration of ancestral responsibility demands context, which brings our story to the subject of risk.

Type I—Calculated Risks: The common quantification of risk holds that the concept is measured by the likelihood that an event will occur multiplied by the impact that the event would have if it did occur, in other words, Risk = Probability x Consequences.[25] This algorithm is near universally employed by natural scientists, engineers, and regulators.

For our purposes, we would like to know what is the risk if the waste is simply left at the storage sites in specially designed steel and concrete containers and how does that risk compare with the risk of disposing the material in a repository. In its *Environmental Impact Statement* for the proposed Yucca Mountain repository, DOE evaluated the consequences of one so-called No-Action Scenario. It assumed that spent nuclear fuel and high-level radioactive waste would remain at the commercial and DOE sites in *perpetuity*, but under institutional control for only about 100 years.[26]

According to DOE's analyses, under this scenario 10 to 100 acres of land would be contaminated at each of the SNF storage sites. As a result, there would be impacts on "biological resources, socioeconomic conditions, cultural resources, and aesthetic resources for long periods… [T]here could be low levels of contamination in the surface watershed and high concentrations of contaminants in the groundwater downstream of the sites for long periods."[27] Without institutional controls, the degraded storage facilities might cause 3,300 latent cancer fatalities over 10,000 years.[28] Conversely, not developing a repository would avoid occupational radiation exposures and fatalities and save the additional costs occasioned in developing and operating a disposal facility. If a repository operated as anticipated, roughly 10^{-5} latent cancer fatalities might arise per year.

The comparison between the risk of the No-Action Scenario and risk of developing a repository, of course, depends on multiple assumptions with regard to the underlying data, models, and physical interactions among the components of the repository and storage systems. But at the very least, DOE seems to have articulated them explicitly, thereby opening its calculations to debate, examination, and criticism. Importantly, DOE's calculations excluded deliberate sabotage and other unseen events. DOE characterized those scenarios as "too speculative" to incorporate. But, of course, those are precisely the reasons for developing a repository in the first place.

Type II—Risk Profiles: William Rowe, who would shortly lead EPA's effort to establish an environmental release standard for a repository (see Chapter 4), is one of a few natural scientists, engineers, and risk regulators to recognize that people hold different risk profiles.[29] These profiles describe how the salience of risk, i.e. its importance or attraction of notice, varies with the level of the risk. *Risk-neutral* individuals present a linear relationship between salience and risk. *Risk-tolerant* individuals exhibit declining salience as the risk rises. *Risk-averse* individuals reveal exponentially increased salience as the risk is amplified. Figure 2.6 depicts the three profiles. (Risk calculations are still subject to the qualifications noted above.)

Type III—*Multi-Dimensionality of Risk*: Cognitive psychologists have brought to the table a more nuanced, albeit a more complex and complicated, notion of risk. The research, sparked by Daniel Kahneman, Amos Tverksy, and Paul Slovic and their colleagues, published an extensive series of papers that argued perceptions of risk were based on eight characteristics including.

- Involuntary
- Unknown to exposed
- Uncontrollable
- New
- Chronic
- Dread
- Fatal
- Unobservable
- Delayed

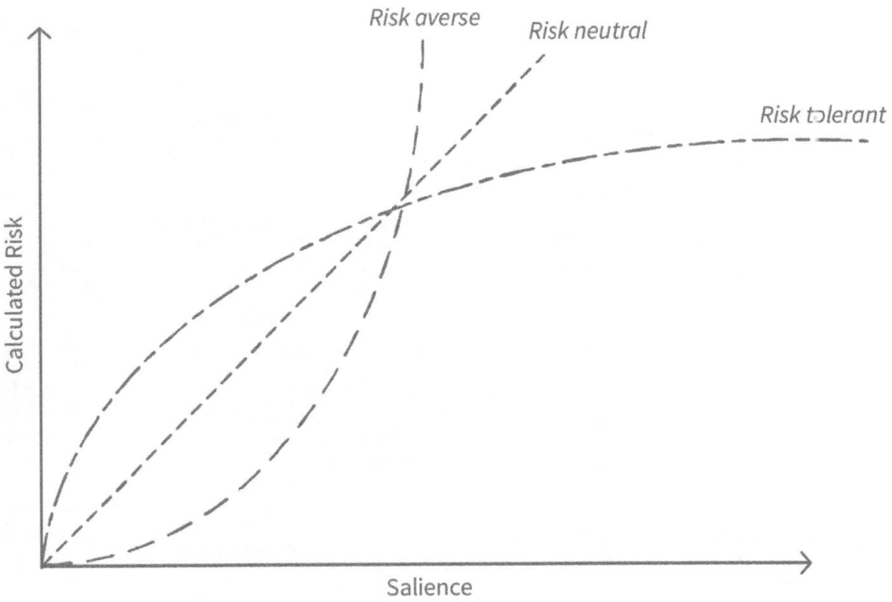

FIGURE 2.6 Relationship between the level of risk and salience of risk.

Using sophisticated statistical techniques, the researchers collapsed the multitude of characteristics into two factors: unknown risks and dread risks. The resulting graph is reproduced in Figure 2.7.[30] For instance, handguns are a bit dreaded but well known; oral contraceptives, however, pose somewhat uncertain risks but are not dreaded. Only two nuclear technology-related actions or events produced more negative reactions along both axes than perceptions of risk concerning radioactive waste.

The implications of these distinctions for becoming responsible ancestors are unmistakable. Policymakers can be comfortable deploying Type I measurements for risk assessment and perhaps risk management purposes. However, calculated risks are very likely to be unpersuasive for members of civil society. It is not that this group is ignorant or uneducated but that it is influenced by risk profiles and the multi-dimensionality of risk perceptions. These risk attributes predispose the general and even elite members of the public, not irrationally, to contest, object, and obstruct efforts to transform irresponsible ancestors into responsible ones. Indeed, the presence of Type II and Type III risk perceptions means that policies for managing high-activity radioactive waste cannot just involve technical debates; rather they are inherently political. As one leader of OCRWM somewhat cynically put it, "Risk perceptions are reality. We just have to deal with it."[31]

In the next chapter, we shall explore how the country's core strategy—geological disposal executed by the federal government—was first implemented beginning in late 1968. There we shall see how the political dimension evolved and how it severely complicated and frustrated the federal government's waste-management program.

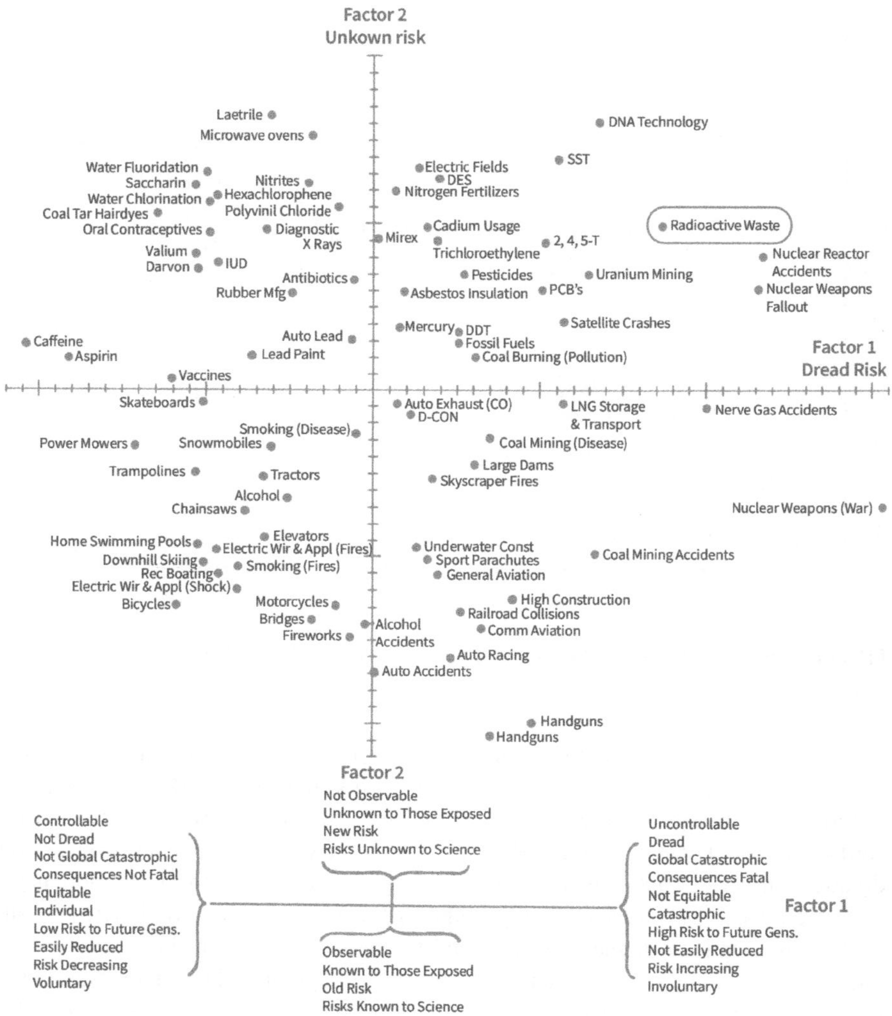

FIGURE 2.7 Perceptions of risk.

NOTES

1. Although some sources suggest that he spoke these words at the time of the explosion, more indicate that his comment at that time was "it works." Ironically, Oppenheimer would later discount the importance for finding a long-term "solution" for managing the waste his bomb produced. Alvin Weinberg, "Social Institutions and Nuclear Energy," *Science*, (177) July 7, 1972, 33. I am grateful to Rabbi Emeritus Rabbi Fred Scherlinder Dobb, of Adat Shalom Synagogue in Bethesda, Maryland, for this formulation. As I will detail in the remainder of the book, behaving as a responsible ancestor differs from but includes concern for intergenerational equity.
2. Otto Hahn and Fritz Strassmann, "The Existence of Alkaline Earth Metals Resulting from Neutron Irradiation of Uranium," *Die Naturwissenschaftten*, (27) 12, 1939.

3. Lise Meitner and Otto Frisch, "Disintegration of Uranium by Neutrons: A New Type of Nuclear Reaction," *Nature*, (143) 1939, 239. Hahn and Frisch were awarded the Nobel Prize in Chemistry in 1944. Meitner, a Jewish woman, was not. See Elisabeth Crawford et al., "A Noble Tale of Postwar Injustice," *Physics Today*, (50) 26–32, 1997. Meitner's escape from Nazi Germany deprived her of recognition for the discovery of fission. See Ruth Lewin Sime, *Lise Meitner: A Life in Physics* (Oakland CA: University of California Press, 1996), 231.
4. The enriched uranium for Little Boy was actually produced by the Y-12 Calutrons using feedstock from the K-25 facility.
5. Spent fuel is no longer able to economically sustain a chain reaction. For weapons production, the fuel rods are not spent but are irradiated for only a few weeks to optimized production of fissionable ^{239}plutonium.
6. In addition, some residual ^{235}uranium is discharged.
7. The U.S. National Academies of Science has published over the years a series of analyses examining this question. See, for instance, its latest report, *Health Risks from Exposure to Low Levels of Ionizing Radiation, BEIR VII*, Phase 2, released in 2006. The International Commission on Radiation Protection has its own, albeit somewhat different, views. See, for instance, its most recent publication, *The 2007 Recommendations of the International Commission on Radiation Protection*, ICPR Publication 103.
8. Legacy waste at Hanford and other weapons material-production sites in Idaho and South Carolina have and will require multiple decades of remediation efforts costing hundreds of billions of dollars. Government Accounting Office, "Issues related to the Closing of the Nuclear Fuel Services, Incorporated, Reprocessing Plant at West Valley, New York," EMD-77-27, March 8, 1977.
9. The AEC's vision at the time was that all SNF from commercial reactors would be reprocessed. The agency felt there would not be enough uranium for 1,000 reactors by the turn of the century. Thus, it was expected that all spent fuel would be moved from reactor sites to reprocessing plants and that only HLW would be disposed of, probably as a liquid in deep repositories.
10. The Joint Committee on Atomic Energy undoubtedly sparked the AEC's interest. See U.S. Congress, Joint Committee on Atomic Energy, Hearings, *Chemical Reprocessing Plant*, 88th Congress, 1st Session (Washington D.C.: Government Printing Office 1963), 56–57.
11. U.S. Congress, Joint Committee on Atomic Energy, Hearings, *Industrial Radioactive Waste Disposal,* 86th Congress, 1st Session (Washington D.C.: Government Printing Office, 1959), 165. That optimistic perspective continued to permeate U.S. waste-management policymaking for decades. In this specific case, more than half a century later, after the expenditure of tens of billions of dollars, the overwhelming majority of the liquid HLW still remains in tanks.
12. AEC-180/88, "Siting of Commercial Fuel Reprocessing Plants and Related Waste-Management Facilities," June 17, 1970, Appendix E-1. The NSF facility proved to have a checkered history. The plant only operated from 1966 until 1972 as stricter regulations forced the plant's shutdown for upgrades. The plant was permanently shut down in 1976 after it was determined that the tighter regulatory requirements could not be economically met. Congress passed the West Valley Demonstration Project Act (Public Law 96-368) in 1980 and directed the Department of Energy to initiate efforts to solidify and remove the accumulated nuclear waste present on the site as well as to decontaminate and decommission the facility and surrounding property. Altogether 278 waste canisters of vitrified HLW have been loaded into 56 vertical storage casks. They await removal to a permanent disposal facility. As of January 2021, the Department has spent $3.1 billion for waste-management activities and to clean up the site. See Government Accountability Office, "Congressional Action Needed to Clarify a Disposal Option at the West Valley Site in New York," GAO-21-115, January 2021.

13. Department of Energy, "Plutonium Recovery from Spent Fuel Reprocessing by Nuclear Fuel Services at West Valley, N.Y. from 1966 to 1972; Department of Energy Press Release, 1996; "DOE Declassifies Location and Forms of Weapons-Grade Plutonium and Highly Enriched Uranium Inventory Excess to National Security Needs," 1996.
14. National Academy of Sciences, "Disposal of Radioactive Waste on Land," 1957.
15. Letter from Harry Hess, Chair of the National Academy of Sciences Committee of Waste Disposal to John McCone, Chairman to the Atomic Energy Commission, June 21, 1960.
16. Much of the information in this section draws heavily on my dissertation, *Error Correction in Bureaucracy*, University of California, Berkeley, Political Science Department, 1978. See Metlay (1978) for a much more detailed discussion of these events.
17. Atomic Energy Commission, AEC-180/47, "Siting of Commercial Reprocessing Plants and Related Waste Management Facilities," October 9, 1968, B-6 – B-7. Frank Pittman, Memorandum to Commissioner Clarence Larson, "High-Level Waste Solidification Program," June 26, 1972.
18. National Academy of Sciences' Advisory Committee on Geology Aspects of Radioactive Waste Disposal, "Report to the Division of Reactor Development and Technology," May 1966, 27.
19. AEC-180/47, 3.
20. Atomic Energy Commission, "Siting of Commercial Fuel Reprocessing Plants and Related Waste Management Facilities: Statement of Proposed Policy," *Federal Register*, 334, June 3, 1969, 8712–8715.
21. With respect to the provisions affecting waste management, the form of the solid was specified in somewhat greater detail. Instead of the federal government taking physical responsibility for the wastes, it would take "permanent custody." Thompson argued for greater flexibility to convert liquid HLW to a solid and was skeptical that salt formations were an appropriate host rock.
22. AEC-180/47, Appendix F, Annex C. Interview with Milton Shaw, March 24, 1976.
23. *Appendix F* stopped short of designating the federal government as the repository's implementor. But for all intents and purposes, it implicitly did so; all interested and affected parties explicitly accepted that interpretation.
24. Quoted in Ralph Nader and John Abbotts, *The Menace of Atomic Energy* (New York: W.W. Norton, 1979), 151.
25. A more sophisticated account of how to quantify risk is the iconic paper, Stanley Kaplan and B. John Garrick, "On the Quantitative Definition of Risk," *Risk Analysis*, (1) 1981, 11–27. See also Chauncey Starr, "Risk Management, Assessment, and Acceptability," *Risk Analysis*, (5) 1985, 97–102.
26. Department of Energy, *Final Environment Impact Statement for a Geologic Repository for the Disposal of Spent Nuclear Fuel and High-Level Radioactive Waste at Yucca Mountain, Nye County, Nevada*, DOE/EIS- 0250, 1–22.
27. DOE/EIS-0250, 2-85–2-86.
28. It is unclear whether this per storage facility or total.
29. William Rowe, *An Anatomy of Risk* (New York: John Wiley & Sons, 1977), 74–75.
30. Paul Slovic, "Perceptions of Risk," *Science*, (236) April 17, 1987, 280–285. See also Spencer Weart, *The Rise of Nuclear Fear* (Cambridge MA.: Harvard University Press, 2012).
31. Comment by Lake Barrett, Acting Director of DOE's Office of Civilian Waste Management, on numerous occasions.

3 Setting the Path

The Atomic Energy Commission's (AEC) adoption in 1970 of *Appendix F* marked the last uncontentious waste-management action for nearly a decade. During that period, non-governmental organizations representing environmental and anti-nuclear advocates surfaced; within a short time, they acquired substantial financing and a reputation for formidable technical, legal, and political talent and expertise. The groups took special advantage of a ground-breaking piece of legislation, the National Environment Policy Act (NEPA), to contest federal decrees and activities.[1] The nuclear-waste honeymoon gave way to an increasingly turbulent organizational milieu as the AEC encountered successive programmatic failures and the presumptions about the back-end of the nuclear fuel cycle were dethroned.

EARLY ATTEMPTS TO FIND A REPOSITORY SITE

On June 12, 1970, after considering potential salt sites in Michigan, and New York, the AEC announced its "tentative" decision to construct a repository near the Carey Salt Mine in Lyons, Kansas. None of the three locations possessed any great geological advantage over the others; each appeared quite technically suitable. Three factors swung the choice[2]:

- Detailed information on the area had been gathered as part of Project Salt Vault.
- Oak Ridge National Laboratories scientists developed a sense of confidence in securing "a favorable reception on the part of local and state officials and private citizens."
- AEC leaders expected that "necessary investigations to prove out the acceptability of [the other] sites would result in considerable delay estimated on the order of two years."

Technical and political difficulties arose almost immediately. Identifying all the boreholes that had been sunk to explore for oil near the Carey Salt Mine proved to be impossible. Moreover, the owners of the adjacent American Salt Mine notified the AEC that it had attempted to extract the salt by solution mining but had no idea where the water went. Galvanized by the actions of William Hambleton, the Director of the Kansas Geologic Survey, Congressman Joe Skubitz, Governor Robert Docking, and Senators Robert Dole and James Pearson began to lobby at the federal level. Ultimately Dole convinced the Senate to pass legislation prohibiting the construction of the repository absent an *imprimatur* from the National Academy of Sciences testifying to its safety. The assurance was not forthcoming. By 1972, the AEC withdrew support for the project.[3]

DOI: 10.1201/9781003528302-4

FIGURE 3.1 Dry SNF storage casks similar to those proposed for RSSF.

Then-AEC Chairman James Schlesinger reacted sharply and shifted his attention from *disposal* to extended *interim storage*. Under his direction and the leadership of a new Director of the Division of Waste Management and Transportation, Frank Pittman, the concept of a Retrievable Surface Storage Facility (RSSF) appeared promising. Mausolea on the surface would hold AEC and commercial spent nuclear fuel (SNF) in concrete storage casks.[4] (See Figure 3.1.) A draft *Environmental Impact Statement* was prepared to support that decision.

The document received the Environmental Protection Agency's (EPA) lowest evaluation.[5]

> A major concern with the employment of the RSSF concept is the possibility that economic factors could later dictate utilization of the facility as a permanent repository, contrary to the stated intent to make the RSSF interim in nature. These economic factors would consist mainly of the fiscal investments attendant to its construction and the activities which arise in the commercial segment of the economy to support its operation.

By then, the AEC too had passed into history. In one of his first actions in January 1975, Robert Seamans Jr., the head of the newly established Energy and Research and Development Administration (ERDA), eliminated funds to build the RSSF. Like the Lyons effort, it was officially dead.[6]

But the new agency could not afford to be seen as having no waste-management policy, notably because its predecessor had promulgated *Appendix F* less than five years earlier. ERDA scrambled in the early to mid-1970s to find other locations where a repository might be located. Not surprisingly, it tightly focused on areas overlaying substantial salt formations.[7] (See Figure 3.2.)

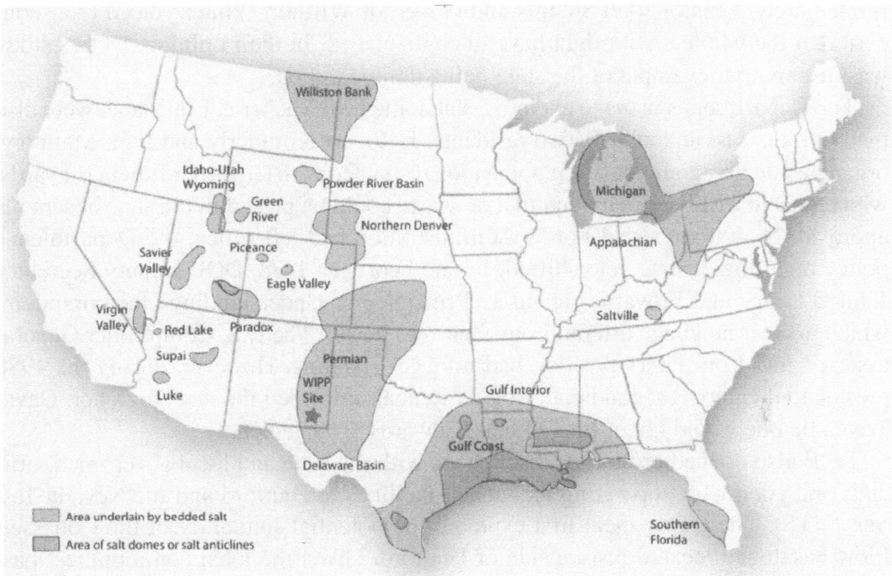

FIGURE 3.2 Potential repository sites in salt formations.

ERDA representatives first arrived in the Permian Basin, which extended from southeast New Mexico through western Texas and Oklahoma into southern Kansas. Although they encountered a generally hospitable political environment in the first state, the agency's leadership decided to reserve that area for the proposed Waste Isolation Pilot Plant (WIPP), a repository for defense-generated transuranic radioactive waste.[8]

Attention then turned to the Salina Basin that swept in an arc through Michigan, Ohio, Pennsylvania, and western New York. Alpena, a small town in Michigan near the shore of Lake Huron, was the initial target. Governor William Milliken announced that there would be no negotiations or commitments until he was assured in writing that a disposal site would only be selected with the state's consent. In what would foreshadow many subsequent actions, Seamans equivocated, arguing that repository development was a federal function, not subject to a state veto. Nonetheless, the question of states' rights in Michigan remained on the table until May 11, 1977, when Milliken wrote to Seamans asking him to terminate the Alpena studies.

After short stops in Ohio (looking at a potential site near Cleveland), in New York (exploring a prospective site near the wine-growing Finger Lakes district), in Texas (considering a possible site overlying the massive Ogallala aquifer), and in Utah (contemplating a promising site near Canyonlands National Park), ERDA's successor bureaucracy, the Department of Energy (DOE), began in 1977 to explore would-be repository sites in the salt formations along the Gulf Coast.

Generic technical questions clouded the suitability of these host rocks, but those uncertainties failed to deter the Department. A leading candidate surfaced in a salt formation near Richton, Mississippi, a tiny community of 1,100 people. Almost immediately, Senator John Stennis and Governor William Winter voiced their concern that the whole town might have to be displaced. In their opinion, DOE's study was just another example of the state being dumped upon.

The Department's plans to explore a salt dome near Vacherie, Louisiana, were also met with distrust and alarm. Two residents, J. Robert Kimmerly and R.H. Manning, organized demonstrations. Their opposition resonated with Governor Edwin Edwards, who repeated Milliken's demand that no waste could be placed in the state absent his approval. Now that Gerald Ford's Administration had left office, the Department's policy regarding a state veto shifted. In late February 1978, DOE Deputy Secretary John O'Leary and Edwards signed a "Principles of Understanding" memorandum, which in clear language offered Louisiana veto power. The U.S. Comptroller General asserted that Louisiana's position had no legal standing. However, during the 1980 presidential campaign, candidate Ronald Reagan informed the new governor, David Treen, that he would honor the O'Leary/Edwards arrangement.

DOE also opened promising discussions with the state and local government officials and contractors operating its nuclear facilities at Hanford and the Nevada Test Site (NTS). The Department first explored the potential suitability of the Cohassett Flow basalt site located just outside of Hanford. Given the local communities' past experiences, enthusiastic encouragement quickly made itself plain, but support never grew in western Washington where the bulk of the population resided. Moreover, DOE's lead contractor, Rockwell Hanford, and independent scientists and engineers, including those from the U.S. Geological Survey (USGS), identified significant technical problems. The layered flow through the volcanic formation would be challenging to model, thereby generating uncertainty about whether radioactive material might reach the Columbia River. Moreover, construction would be problematic because water pressure at the proposed repository's horizon might wash out fractured and brecciated rock zones, creating fast pathways for moving waste to the surface.

DOE began studying Yucca Mountain, a volcanic tuff formation at NTS, in 1979. The area then chosen for investigations was in the saturated zone, below the water table. However, based on a recommendation from USGS scientists, the Department switched in 1982 to the Topopah Spring layer in the unsaturated zone, above the water table. The Survey's hydrologists maintained that the unsaturated zone offered a natural barrier to the migration of any radioisotopes that might diffuse from the buried high-activity waste. Nye County, the host to NTS, saw the advent of a repository as a major economic stimulant. However, the State of Nevada led by its Governor, Richard Bryan, was implacably opposed.[9]

Notwithstanding the obstacles encountered in finding a technically suitable and socially acceptable site during his tenure, Seamans left three other prominent fingerprints on the nation's waste-management efforts. Responding to suggestions from the Joint Committee on Atomic Energy, he sponsored an initiative in early 1976 to prepare a comprehensive study on the technical alternatives available to the waste-disposal program. A five-volume report presented detailed descriptions of the options for treating wastes from power reactors and reprocessing—activities comprising what are collectively known as the "back-end of the nuclear fuel cycle."[10]

At the same time, ERDA staff developed what was billed as "new concept" of placing multiple barriers between the high-activity waste and the environment. As one history of the agency noted, "*both* engineered and natural barriers would be deployed." Further, "acceptable sites would be *simultaneously investigated* in several geographic locations and geologic formations [emphasis added]. The idea of multiple sites for [disposal] would avoid the charge that any one state was being singled out as a 'national waste dump'."[11]

As our story will make plain, claims made about barrier capability, siting strategy, and state veto power continue to be disputed for more than 40 years later.

RETHINKING THE BACK-END OF THE NUCLEAR FUEL CYCLE

The third fingerprint reflected the rising apprehension that a sixth nation (after the United States, the Soviet Union, the United Kingdom, France, and China) might construct a nuclear weapon. The detonation of a small nuclear device on May 18, 1974, in Pokhran, India, introduced a new complication for federal efforts to formulate a viable waste-management program; it also brought home and made manifestly clear the naivety of the vision of "Atoms for Peace" that President Dwight Eisenhower had articulated twenty years earlier. The civilian and the military nuclear domains were not separate but were intractably linked and could not be addressed separately.[12]

In an odd way, the coming into force of the Nuclear Non-Proliferation Treaty in 1970 had further lulled federal policymakers into a false sense of confidence that the spread of nuclear weapons—never their high priority to begin with—had now been contained. Believing that the treaty was both self-implementing and strict, they continued to divorce the military and civilian spheres; policymakers could move on, having checked the box. That complacency was short lived.

Presumptions Have Unexpected Consequences

The trigger for the federal government's newly reignited interest in nuclear nonproliferation was the seemingly disconnected set of decisions and actions taken between 1972 and 1974 to privatize enrichment services. The construction of the K-25 enrichment plant at Oak Ridge by the Manhattan Project was soon followed by the building of two other gaseous diffusion facilities, one in Paducah, Kentucky, and the other in Portsmouth, Ohio. Together they gave the United States a *virtually complete monopoly* over the production of the low-enriched fuel needed by a growing number of domestic and foreign nuclear power plants.[13]

By seeking to privatize enrichment services, President Richard Nixon and officials at the Office of Management and Budget (OMB) saw an opportunity to advance three policy goals:

- Commercialization of the nuclear power industry would be encouraged.
- The federal government would not have to erect new enrichment facilities to meet the increased projected demand for low-enriched uranium; the need for costly upgrades to and maintenance for the operating enrichment plants would also be avoided or deferred.
- The federal government would capture a return on its enrichment investments.

Along with modifications to the enrichment-services' contract, the key to encouraging the private sector to build its own facilities was to increase demand for the product, thereby driving up the price that might be charged. The AEC did so by announcing in June 1974 that it was "closing the book" on new enrichment agreements. At that point, barely anticipated consequences sprung to the fore. Coming just after the 1973 OPEC oil embargo that intensified international worries about energy security, Europeans interpreted the book-closing as evidence that the United States was no longer a reliable trading partner. In response, they saw in the changing situation an economic opportunity to sell their own reactors overseas by coupling those transactions with a promise to transfer the requisite technologies for nuclear autonomy: enrichment and reprocessing facilities. The Germans struck a deal with the Brazilians; the French negotiated with South Korea and Pakistan. Suddenly the old platitude, "the genie is out of the bottle," took on new meaning.

The political environment of the nuclear community grew ever more turbulent as concerns about widespread commercialization of plutonium mounted. Barely ahead of the Executive Branch's learning curve, Congress leaped into action, debating possible legislation to restrict the export of sensitive nuclear technologies. Secretary of State Henry Kissinger, who had previously believed that non-proliferation matters were not worthy of his attention, directed the full machinery of his department to conjure up a remedy. His involvement brought a new political heavyweight into what had been a relatively circumscribed policymaking arena. Presidential candidate Jimmy Carter gave a well-received speech to the United Nations on May 13, 1976. That address firmly positioned him as a foe of domestic reprocessing, thereby injecting the controversy into election-year politics. Thus, in a relatively short time, a pressing and controversial issue had surfaced, and it could no longer be ignored.

To recover from these unexpected chains of fallout meant juggling competing priorities, which previously had never been understood to be in opposition. Not only had the federal government to reassure edgy trading partners, but wide-scale commercialization of separated plutonium was at the core of major AEC plans. By the late 1950s, both the responsible federal agencies and the fission industry envisioned the deployment of the Liquid-Metal Fast-Breeder Reactor (LMFBR). This device would be driven by plutonium extracted from reprocessed SNF discharged from commercial reactors. In principle, it would produce more fuel than it would consume.[14] What should be the fate of this holy grail? Further, just as the Nuclear Fuel Services' facility was closing for good, Allied General Nuclear Services' (AGNS) commitment to operating a large commercial reprocessing plant at Barnwell, South Carolina hung in the balance; the project's economics had turned progressively sour. Should or could the federal government come to its aid? And if so, how? In short, what trade-offs should be struck in the name of reducing the proliferation threat?[15]

Finding a balance among all these considerations preoccupied the final year of Gerald Ford's presidency and the first one of Jimmy Carter's. The former began to lay the preliminary groundwork for slowing the pace of commercial SNF reprocessing in the United States; the latter launched an aggressive campaign to prevent a plutonium economy—reprocessing, recycle, and breeders—from emerging at least in the United States and, aspirationally, in nations overseas as well.[16]

Our chronicle now briefly examines how the two administrations struck those balances. In both, reconstructing the back-end of the nuclear fuel cycle required intensive negotiations with assorted federal bureaucracies and delicate interactions with multinational counterparties and bilateral partners. Our story then notes the two dominant waste-management impacts of the potential policy shift: What technical difference, if any, does it make if SNF is disposed of rather than HLW? What should be done with the accumulating volume of SNF at reactors when a separations facility would no longer accept it?

The Ford Administration Breaks the Ice

In early 1976, Seamans sent a letter to selected White House staff members identifying interlocking issues. He warned that the federal government lacked top-level guidance to face the challenges brought on by the interaction of domestic and foreign nuclear policy, especially with respect to non-proliferation. He also anticipated the emerging crisis affecting the AGNS facility. On the one hand, the Nuclear Regulatory Commission (NRC) was in the midst of preparing its final generic impact statement on the use of recycled plutonium in conventional reactors. On the other, that plutonium could only come from Barnwell, a project now increasingly plagued with technical and legal uncertainties that were causing operating delays and cost overruns. Lurking in the background remained the question of the Nixon/Ford initiative to privatize enrichment. Seaman's message caught the attention of the heads of the Domestic Policy Council, the National Security Council (NSC), and OMB. They urged the president to establish an interagency panel to sort out these issues. Robert Fri, ERDA's Deputy Administrator, very quickly emerged as the consensus choice to lead the group.[17]

The opening page of the Nuclear Policy Review Group's "Memorandum to the President," (generally referred to as the "Fri Memorandum") sharply presented the stakes: "Notwithstanding the difficulty of striking the delicate balance between a "tough" stance domestically [regarding non-proliferation] and international realism, we believe it is crucial to do so." After explaining the considerations noted above, the panel unambiguously informed the president of its foundational premise "We recently judged the spread of nationally controlled reprocessing facilities as undesirable, even if safeguarded...and that such facilities in particularly sensitive countries are intolerable since they will facilitate the availability of weapons-usable plutonium." The committee then teed up four questions for the president's consideration.[18]

- How can we tighten international controls over the storage of sensitive nuclear material, safeguards against diversion of such material, and physical security to prevent its theft?
- How firm should the restraints be that we apply to our nuclear exports, and what sanctions should we invoke if our agreements are violated?
- How aggressively should we apply these restraints to existing agreements, and how, if at all, should the nuclear export licensing processes, as contrasted to diplomatic techniques, be employed for this purpose.?
- What alternatives to national reprocessing should the U.S. offer?

Addressing the first question, the group recommended: "(1) notwithstanding its low 'assurance' in the International Atomic Energy Agency's (IAEA) ability to adequately safeguard the material, excess SNF and separated plutonium should be placed under its control; (2) the U.S. should take the lead in strengthening IAEA's safeguards system; and (3) the U.S. should also take the lead in strengthening physical security arrangements."

Framing the deliberations on the second question was Kissinger's strong opinion that violations of our nuclear agreements should be viewed as "an ominous matter that could lead to a cutoff of nuclear assistance." In particular, the Nuclear Policy Review Group suggested that firmer international restraints and conventions "are the essential vehicles for ensuring proper international controls of sensitive facilities and materials." Further, the committee noted the importance of communicating "more convincing signals to the world community that we would view either a material violation of any nuclear safeguards agreement, or a further spread of nuclear weapons, as an extremely grave development."

The panel evaluated two alternatives for modifying existing agreements: unilaterally insisting on retroactivity and developing a strong, but flexible, initiative on retroactivity. It is noteworthy that the panel took no preferred position.

Finally, the group argued that "the forces favoring reprocessing would lead to widespread national reprocessing—a result we believe is untenable." Moreover, following that path would at best lead to a two percent reduction in the cost of generating nuclear electricity. "Therefore, the U.S. should advocate a policy containing the spread of national reprocessing by, among other things, offering to take back and store SNF from other nations and helping to create alternatives to traditional reprocessing technologies. Our declaratory policy internationally would be to ask all countries to *examine* together [emphasis added] nuclear technologies that avoid weapons-usable material."

At the start, the Nuclear Policy Review Group rejected two potential options. First, a program of direct government assistance to industry to accelerate domestic reprocessing was deemed as extremely costly and most likely premature. Second, to defer reprocessing commitments for the next five to ten years would likely be judged as "temporizing, destructive of investor confidence, and allowing the technical base to atrophy." Instead, the panel offered up two other complementary alternatives: assist industry to gain experience with reprocessing and to develop alternative technologies that might be more proliferation resistant.

But as two staff members from the Domestic Policy Council made clear during the deliberations, attention had to be focused on the capstone for any policy redirection. "If there were no irresistible case to be made on economic and technical grounds, [they] believed that the president should be urged to slow the move to [domestic] reprocessing, and thereby clear the stage for a deliberate review of fresh alternatives."[19]

As the review group's attention turned to crafting a presidential statement, it awaited Ford's decisions. Six weeks would pass between receipt of the "Fri memorandum" and the articulation of a dramatically modified national policy. (The timing of Ford's public statement, perhaps not coincidentally, came just one week before the upcoming election.) For our purposes, the key ideas contained in that document were[20]:

- The world community cannot afford to let potential nuclear weapons material or the technology to produce it proliferate uncontrolled over the globe.
- I have concluded that the reprocessing and recycling of plutonium should not proceed unless there is sound reason to conclude that the world community can effectively overcome the associated risk of proliferation. I believe the avoidance of proliferation must take precedence over economic interests.
- I have decided that the United States should no longer regard reprocessing used nuclear fuel to produce plutonium as a necessary and inevitable step in the nuclear fuel cycle.
- In view of the decisions announced today, I have also directed the Administrator of ERDA to assure that [any] waste repository will be able to handle spent fuel elements as well as the separated and solidified [high-level] waste that would result if we proceeded with nuclear fuel reprocessing.

Left implicit in Ford's pronouncement was the fate of Barnwell and the LMFBR; but the signs were undeniable.[21]

The Carter Administration Delivers the Coup De Grâce to Domestic Reprocessing

Although Ford's policy tip-toed toward a conclusion that reprocessing should be delayed, it left the door at least slightly ajar. A mere two-and-a-half months after his inauguration, Carter firmly shut it.[22]

On January 26, 1977, the NSC sent a memorandum to key agency leaders outlining eight tasks. In a frantic effort to meet the president's short-fused five-week deadline, insiders and outsiders, purists and moderates fashioned a briefing document designed to direct the world's attention to the dangers of separated plutonium; to reconstruct the national and international institutions required to limit the element's spread; and to stop the movement toward commercial reprocessing.

Unlike the Ford pronouncement, Carter understood that achieving those ends would necessitate extended intercessions with the permanent bureaucracy and consultations with other countries, especially France and the Federal Republic of Germany. Given the truncated deadline, the new Nuclear Policy Review Group, which would produce an "options and recommendations" paper for Carter, had little choice but to build on the work of its predecessor.

All the participants accepted the premise that there should be a moratorium of *domestic* reprocessing. Disagreements arose, however, over what stance the president should take on *foreign* reprocessing. A unanimous committee finally recommended that Carter should call for a multilateral forum to undertake an appraisal of the international fuel cycle. One important conclusion emerged from the consultation: disposing of SNF was not appreciably more difficult than disposing of HLW.[23]

In an April 7, 1977, press conference, Carter revealed the core aspects of country's new nuclear course. "We will defer indefinitely the commercial reprocessing and recycling of the plutonium produced in the U.S. nuclear power programs...The plant at Barnwell will receive neither federal encouragement nor funding from us for its completion as a reprocessing facility...We will restructure our own U.S. breeder program to give greater priority to alternative designs other than [those using] plutonium."[24]

Of course, many interagency skirmishes remained, and the heat of the international negotiations grew ever more intense and passionate. Implementing all these objectives also required delicate interactions with the NRC, which, as mentioned earlier, was deciding whether to permit the recycling of mixed-oxide (uranium and plutonium) fuel into light-water reactors. Independent of the Executive Branch, the regulatory agency could approve recycling despite the president's opposition. After much effort, the appointment of new commissioners ensured that NRC's final decision was consistent with Carter's position.[25]

Attracting less notice was another one of the president's comments. "We have to help provide some means for the storage of spent nuclear fuel materials, which are highly explosive [sic], highly radioactive in nature." This offhand remark reflected what appeared at the time to be two developing concerns about the consequences of halting reprocessing. First, unable to move SNF offsite, U.S. reactors would exhaust the space in their spent fuel pools. Second, foreign reactors might be facing the same predicaments and constraints.

DOE announced a new spent fuel storage policy six months later. The federal government would "accept and take title to spent nuclear fuel from domestic utilities. The federal government will also be prepared to accept a limited amount of foreign spent fuel when such action would contribute to meeting non-proliferation goals." The announcement set off a wave of efforts to figure out how to implement it. DOE explored potential storage sites; it estimated the one-time storage charge; contractors prepared draft and final environmental impact statements; and the Department drafted authorizing legislation.[26]

As the matter evolved during the next three years, however, the urgency to develop Away-From-Reactor storage capacity gradually faded. Agreement could not be reached on how to calculate the charge; U.S. utilities reassessed the need, which might be ameliorated by potential technical fixes; foreign utilities expressed no interest; and Congress made it clear that it would not authorize the international transfer of SNF under any circumstances.

But looking back on that period, two outcomes seem evident. First, Carter's efforts created the *necessary*, but perhaps not the *sufficient*, conditions that halted domestic reprocessing for 40 years and counting. Second, although several countries maintain commercial reprocessing programs (France, Russia, China, and India—each already a nuclear-weapons state) and others host clandestine weapons-related reprocessing plants (Israel, Pakistan, and North Korea), the seemingly headlong rush toward a widespread plutonium economy was stymied.

The increased attention given to plutonium, reprocessing, and recycling inevitably renewed efforts to tackle the persistent dilemma of what to do with the waste.

TOWARD A NATIONAL WASTE-MANAGEMENT POLICY

The federal government's early attempts to develop a deep-mined, geologic repository for high-activity radioactive waste suggest an idealized process that seems to reflect both successful and unsuccessful policymaking efforts both in the United States and other pluralistic democracies.

Siting a repository involves the metaphorical passage of specific locations through two "filters:" one for technical suitability and the other for social acceptability.[27] The light areas in Figure 3.3 represent places considered early in the process. The dark areas show a suite of sites investigated later on. In this example, none of the final locations are both technically suitable and socially acceptable. In such a case, development of a repository would be extraordinarily challenging.

The iterative nature of the siting process is depicted in Figure 3.4. Repository implementors, such as DOE, investigate many possible locations.[28] Some are eliminated at the start, either because they are technically unsuitable or politically unacceptable. Those that pass through the first filter are screened by the second. If multiple sites satisfy both filters, they are reassessed until the implementors select only one location.[29] But as Figure 3.3 implies, identifying a potential site is by no means inevitable.

Laws, regulations, and rules adopted by individual nations, it must be noted, determine the order in which sites pass through the two filters. In the United States, both the Executive Branch and Congress presumed that a particular site had to be considered technically suitable *before* its passage through the social acceptability filter could be assessed.[30] As our story continues, we will see that in countries such as Canada, Sweden, and Finland, the order is reversed. Implementors there solicit volunteer communities prior to making suitability determinations. The historical record provides scant evidence *why* the U.S. approach was initially adopted.[31] The passages below as well as the discussion in Chapter 4 detail how policymakers behaved in ways consistent with this metaphor.

On March 3, 1978, DOE's Director of Energy Research John Deutch sent Undersecretary of Energy Dale Myers a memorandum transmitting a draft *Task Force Report for the Review of Nuclear Waste Management*. The report was intended

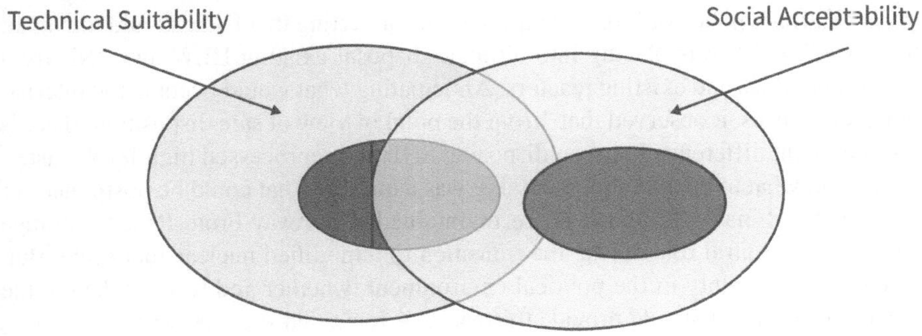

FIGURE 3.3 Prerequisites for choosing a repository site.

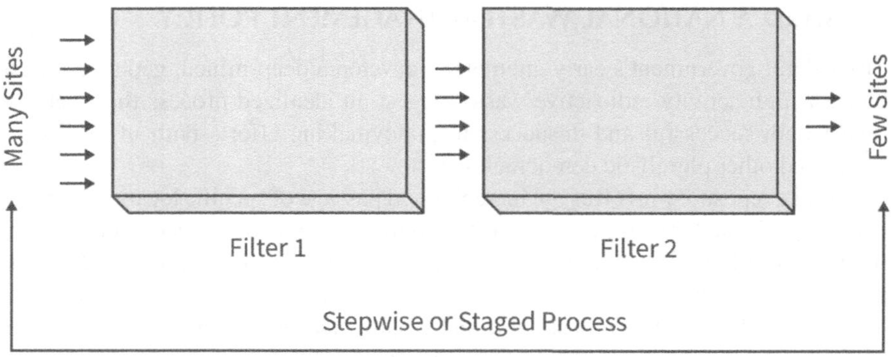

FIGURE 3.4 Winnowing many potential repository sites.

to be the first step in the formulation of an Administration-wide policy. It would, however, also serve as the marker for establishing DOE's position in the interagency process that quickly followed.[32]

Some of the Task Force's views reflected broad agreement within the bureaucracy. Although DOE had not yet released its *Generic Environmental Impact Statement on the Management of Commercial Radioactive Waste*, the Task Force referred to the 20-year collection of National Academy of Sciences' reports as well as research in the Federal Republic of Germany, Sweden, and Canada as evidence of a technical consensus that geologic disposal could safely isolate radioactive wastes for millennia. Nonetheless, it warned that the primary issue related to the adequacy of technical knowledge for the choices of geologic media; that understanding was highly site specific and could not be evaluated in advance. Given this uncertainty, the Task Force argued that careful stepwise implementation, gaining confidence before taking the next step, would be the strategy most likely to yield success over the long run.[33]

ERDA's then-current plan involved looking at potential sites in 36 states, leading to the construction of two repositories each in salt, clay, and granite formations.[34] The *Task Force Report* concluded that meeting that agency's start-of-operations date of 1985 was implausible if not incredible. The report took a pragmatic approach toward Carter's policy of forgoing reprocessing, asserting that DOE should focus "on the prize," which was placing into ultimate disposal existing HLW and SNF from defense facilities and existing reactors. Anticipating what would become the international consensus, it observed that "from the point of view of safe disposition, there is no significant difference between disposing SNF and reprocessed high-level waste." Therefore, what to do about reprocessing was a question that could be postponed for another day. Finally, the Task Force maintained that Away-From-Reactor storage could play a central role during the transition to a modified nuclear fuel cycle. But, given the uncertainty in the political environment, whether and to what degree the federal government should provide this capacity remained speculative.[35]

Once the report was released, however, two other conclusions drew strong criticism from interested and affected parties, both within the federal government and outside of it. First, driven by pressure to demonstrate some movement towards disposal, the

Task Force looked to the proposed Waste Isolation Pilot Plant (WIPP). "In addition to its use for R&D and TRU disposal, WIPP should be considered for a demonstration, on a moderate scale, of the capabilities for ultimate disposal of spent fuel [in salt]. This might involve…using up to one thousand fuel assemblies. Emplacement should be on a technically conservative basis to allow removal of the spent fuel from the demonstration if that were subsequently desired [after] a 15–20-year period."[36]

The second divisive conclusion related to the basic trade-off between *schedule* and *certainty* in identifying a technically suitable and socially acceptable repository site. The report noted that "two different philosophical approaches" were possible.[37]

- Develop a date for permanent repository operation, which would allow time for acquiring more scientific data on performance in salt, a stepped-up program on other [geologic] media, more international interaction, etc. Compare the best salt design with the best design in other media. Then choose the preferred medium and choose between more than one location in that medium.
- Select an earlier operational date than implied by the first approach. Proceed with a repository in salt, which is loaded conservatively from a thermal standpoint. This will provide the potential of retrieval of material emplaced over a period (probably not in excess of 15 years.) Subsequent decisions can then be taken in either direction (i.e., to remove to another location or increase the thermal loading) before the repository is finally sealed.

"The Task Force believes the first approach to be unnecessarily conservative and favors the second."

TRANSFORMING NATIONAL POLICY: INTERAGENCY REVIEW GROUP ON NUCLEAR WASTE MANAGEMENT

In the days that followed the publication of the *Task Force Report*, other parts of the federal government lobbied for a top-level team to wrestle disorder into coordinated action since DOE's marker had yielded no consensus or agreement. Prompted by a recommendation from DOE and support from organizations within the Executive Office of the President, Carter created a 14-member Interagency Review Group for Nuclear Waste Management (IRG). Its remit was to "formulate recommendations to establish an Administration policy with respect to the long-term management of nuclear wastes and supporting programs to implement this policy." Its chair was John Deutch.[38]

The IRG's Executive Committee formed six subgroups, each responsible for assessing specific domains: alternative technology strategies for HLW, SNF, and TRU disposal, spent fuel storage and charges, defense wastes, federal takeover issues (licensing, standards, low-level waste), international issues, and transportation. As one central participant observed, "One important objective of the IRG, understood and shared by [virtually] all members of its Executive Committee, was to move the waste-management issue out of the narrow purview of DOE and into the broader context of the full government and the president."

The core "principals" on the IRG initially included Deutch, Phil Smith, Assistant Director of the Office of Science and Technology Policy (OSTP), Joan Davenport, an Assistant Secretary in the Department of Interior (DOI), home of the USGS, Gus Speth, a member of the Council on Environmental Quality (CEQ), and Jessica Tuchman Mathews from the NSC. These individuals would be responsible for melding the six subgroup products to deliver (1) a draft report to the president; (2) a final report to the president that reflected public input on the draft; and (3) a presidential decision memorandum. (Of course, as in all bureaucracies, others manned the trenches.)[39]

By the time the principals first met in late April 1978, it was apparent that the alternative technology strategies subgroup, chaired by Smith, would become a central locus of the action. Its first assignment was to pinpoint what that idea meant substantively. A draft "Issue Paper" prepared in mid-May 1978 identified two interacting dimensions that had to be addressed: the alternative technologies themselves and the programmatic elements that would best hedge against implementation failures. Those two elements naturally defined the subgroup's efforts. A technical analysis would consider deep-mined, geologic disposal in various environments (hydrology, rock types, and tectonic stability). In addition, more exotic technologies would also be evaluated.

In parallel, the subgroup would prepare a policy analysis that would explore ways to increase the likelihood that at least one acceptable disposal "solution" would surface. The essential concept here was redundancy. Just as an automobile has two independent braking systems to guard against the malfunction of one, multiple tactics for developing a repository system could protect against the breakdown of any single approach. Programmatic redundancy could involve multiple site selections in similar or different hydrogeologic environments or non-conventional methods. Moreover, the redundancy would increase if various alternatives were pursued simultaneously as opposed to if some proceeded at a faster pace than others. (Note here the possible acceptance of the "unnecessarily conservative" philosophy rejected in the *Task Force Report*. Our narrative returns to this matter below.)

Subgroup Report on Alternative Technology Strategies for the Isolation of Nuclear Waste

Technical Evaluation of Disposal Options

Almost from the start, it was evident to many IRG participants that a glaring lacuna pervaded the *Task Force Report*; its technical perspective was caught in the late 1950s when the National Academy of Sciences first endorsed a repository situated in a salt formation. Although acknowledging that other "media," such as granite and clay, might satisfactorily host a repository, the Task Force failed to incorporate the growing wisdom that the hydrogeologic environment also could profoundly affect waste isolation; after all, water was the dominant mechanism for moving waste outside the bounds of the repository.[40]

Two influential contemporaneous studies drove home this point. The first was an analysis undertaken under the auspices of the American Physical Society. Among its principal recommendations was that "criteria for selection of a waste repository should include specifications of appropriate parameters that must be satisfied by present-day hydrogeology and by the projected bounds of the future hydrogeology of any specific site." A report from the USGS reinforced that advice. "Planning a waste repository requires full understanding of the subsurface geology...Finding unexpected groundwater conduits, shear zones, and fault-displacements...could be hazardous and costly. The discovery of such anomalies during the excavation of a nuclear-waste repository could force abandonment of the site."[41]

The subgroup evaluated six different disposal alternatives: geologic repositories, isolation in deep-sea sediments, isolation in geologic formations followed by *in situ* rock melting, deep boreholes, partitioning and transmutation, and isolation in space. The first option, laid out in *Appendix A*, was the only one considered in detail; for the others, the wide uncertainty ranges were simply acknowledged.[42]

Appendix A's value lay in its perspective, which stepped back from the technical weeds. Going beyond the generalities of "Baranowski's Bull's Eye," it explicitly recognized that successful waste isolation depended on three functions: the waste form and how it interacted with the geologic environment (including absorption and adsorption), the impact of the waste's heat on and the mechanical response of the formation, and the transport by water to the accessible ecosystem. This recognition led the subgroup to its most important technical conclusions that corrected the gaps found in the *Task Force Report*.[43]

- Natural barriers [if independent] can provide significant assurance of waste isolation and to some degree can compensate for imperfect predictive capability.[44]
- The waste form and other engineered features...can also provide significant barriers to radionuclide migration.
- Geologic environments and media heretofore not examined may be suitable for repository sites

Thus, the two-decade old, salt-centric waste-management paradigm was overthrown.

Appendix A also emphasized the necessity for site-specific investigations before choosing a location for the repository; generic understanding of potential host-rocks was insufficient. Doing so would require a winnowing process, starting with broad regions and then honing in on narrower settings. Siting criteria would dictate the transition. Once DOE identified a particular locale, site-suitability rules would then determine whether the NRC could issue a construction license for the facility. As noted above, states have historically limited access to particular pieces of real estate. Unless those obstacles can be avoided in the future, any repository-development program will fail.[45]

Central to the winnowing endeavor would be the development of mathematical models for each of the three isolation functions. The developers of *Appendix A* understood that each function did not have to be optimized. Low hydraulic conductivity might not be essential if the leach rate from the waste form was extremely low. Likewise, very long groundwater flow-paths might make one host-rock suitable, whereas the same host-rock in a less isolating hydrogeologic environment might be suitable only if the waste package were virtually invulnerable.

Capturing all the interactions among the large number of parameters influencing waste isolation therefore demanded that a "systems" approach be employed to integrate all the models and calculate a single measure of merit. Risk-assessment methodologies could conceivably be exercised to yield the desired result. However, *Appendix A's* drafters were determined not to oversell this approach.[46]

> Confidence in models can be developed by comparing the output from different models [or natural analogues] that have independently analyzed a hypothetical but realistic situation. The models, of course, can be no better than our understanding of the processes that are intended to represent, and the results will be meaningful to the degree that valid data are utilized and the range of uncertainties in the data are specified.

As our tale moves on, the fate of the systems approach and questions about assessment methodologies will be continually revisited.

Policy Analysis

At the very start of the IRG's activities, the President was warned that "the resolution of institutional issues...is equally important as the resolution of outstanding technical issues and could prove to be more difficult."[47]

Over the course of three months, the subgroup produced 12 drafts of the policy analysis. During that time, informal discussions within the subgroup explored the question of state veto power. Some of the drafters maintained that, given the states' wide authority over matters such as water rights, they already possessed *de facto* veto power; nothing would be sacrificed if they were given it *de jure*. In the end, the subgroup determined that any discussion of this critical policy challenge had to await the report to the President.[48]

The subgroup's drafting took place in an atmosphere largely distinguished by compromise and collegiality, if only because many programmatic battles were just beginning to be fought. Although its final product became more polished and tightly argued, its core structure, logic, and analysis remained fundamentally unchanged. (Deutch did complain to OSTP Director and Presidential Science Advisor Frank Press that the drafts seemed to undermine the siting strategy recommended in his eponymous report. Press supported the two major writers' work but encouraged them to meet with Deutch and to keep his concerns in mind.)

The *Subgroup Report* is noteworthy because it articulated for the first time at the federal level a comprehensive set of technical, implementation, and public involvement objectives for the waste-management system and plans. Among the most novel (at that moment) were these goals[49]:

- The existence of residual technical uncertainties must be recognized and provided for in the program structure.
- The program should include consideration of all aspects of the waste-management system including safety, environmental, organizational, and institutional factors."
- "Budgetary and cost considerations, although important, should not dominate the design of the program or system.

- The system should explicitly include organizational and institutional mechanisms for the detection and correction of errors.
- There should be broadly based involvement of the states and local governments, interest groups, and citizens in the planning and decision-making processes of the program.
- The decision-making process within the program must be reasoned, open, and accessible, including mechanisms for outside review on a continuing basis.

As we shall see, these six goals remain in place today, some fully embraced and others only tangentially. Perhaps not coincidentally, those objectives given only rhetorical adherence and lip-service have created obstacles to becoming responsible ancestors.

NEPA limited what the subgroup and later Carter could decide and implement; the ultimate approach to waste management had to await the completion of a full set of environmental impact statements. "Nevertheless, because near-term waste-management programs must be developed…an interim strategic planning basis is required."[50]

The Subgroup identified four possibilities for disposing HLW and SNF.[51]

- ***Strategy I***: Proceed with a repository in salt. Three to five sites would be investigated intensively before a choice for the first repository were made. Sites for additional repositories would be chosen from the other two to four locations.
- ***Strategy II***: Proceed with the first available geologic environment and gradually include greater diversity. A site for the first repository would be chosen from among whatever types of environments had been adequately characterized (investigated) at the time of choice. Although a system in salt formations would likely be suitable first, over time, a broader variety of hydrogeologic environments and emplacement media would become contenders for subsequent repositories.
- ***Strategy III***: Geologic confinement in mined repositories is selected for the first facility. A range of different geologic environments including potential host rocks and hydrogeologic characteristics would be examined before choosing a site for the first repository.
- ***Strategy IV***: Postpone a decision until a wider range of technical options (both geologic and exotic) can be reasonably assessed. Disposal in deep sea sediments and very deep boreholes would be given research priority.

In true bureaucratic fashion, the strategies were framed so that the outliers were not likely to garner the support of any member agency.

The drafters understood that the policy analysis would be of little value absent an attempt to identify the potential outcomes of each strategy. But they also conceded that "the magnitude of the consequences associated with each strategy cannot be predicted with a high degree of confidence; the evaluation of the strategies must, therefore, be a highly judgmental process."[52]

Of greatest interest to policymakers was the time (including site evaluation, regulatory review, construction, and testing) when the first repository might begin to accept high-activity waste for disposal. For Strategy I and Strategy II, the "idealized" estimated date was 1995; for Strategy III, it was 1998; and for Strategy IV, it was 2009. These target dates may be required to identify programmatic milestones and budget commitments. "However, the Government must refrain from making any firm commitment that target dates will be met since such a commitment cannot be assured, given the technical uncertainties and institutional risks which could conceivably delay any planned accomplishment."[53]

Another salient impact was "assuring technical success." Implementing any of the four strategies would be risky. Sites that appear suitable initially may prove to be unsuitable as characterization intensifies. One means of lowering the risks is to employ a conservative and stepwise approach to repository design, licensing, and operation. An example would be loading the repository with cooler waste and lower thermal output than is considered necessary for safety purposes. Also, repository implementors could experiment, test, and monitor the facility for some time with a small quantity of emplaced waste. Further, multiple sites, environments, and technical options could be investigated simultaneously as a hedge against the possibility that any site might turn out to technically unsatisfactory. In this regard, the subgroup concluded[54]:

> The benefits of sequential examination of several environments (Strategy II) over the lack of any such examination (Strategy I) are very great. The same is true for simultaneous examination of several environments in Strategy III and Strategy IV compared with sequential examination in Strategy II.

Consistent with its overarching philosophy, the subgroup gauged the effect of each strategy on public confidence and creditability. This outcome was especially difficult to evaluate. On the one hand, most segments of the public would be concerned if a site appears to be predetermined without adequate characterization or comparison. If there is no backup plan, unpleasant technical surprises that arise might be ignored or poorly addressed. On the other hand, the more robust the strategy, the more fears might surface that delays will become too costly; delays would also undermine the program's credibility by questioning the federal government's competence and commitment. Strategy I and Strategy II would qualm the doubts of those worried about delays. Strategy III and Strategy IV would appeal to those segments of the public troubled by the consequences of haste.[55]

The subgroup maintained that the cost of disposing defense HLW and SNF should be borne entirely by the taxpayers and that the cost of disposing commercial HLW and SNF should be paid by the generators of the material. Precisely how the total expenditure would be calculated and then split between the two payers was left unspecified. The subgroup also sought to estimate the budgetary impact of each strategy. The best that it could do, however, was to estimate the *programmatic* costs for each federal agency. As an example, DOE would have to spend $1.1 billion for Strategy I, $1.3 billion for Strategy II, $1.8 billion for Strategy III, and $2.1 billion for Strategy IV. These figures did not include outlays for developing the repositories, constructing deep-sea disposal facilities, or space shuttles.[56]

The subgroup briefly investigated the idea of an "intermediate scale facility" as another means to minimize repository-development risks. (The Deutch Task Force, of course, had suggested that WIPP might serve this function for disposing SNF in salt.) To be most worthwhile, the facility should be constructed at a potentially suitable repository site. Putative advantages could include gaining operational experience and exercising the licensing process. What might be learned about the isolation capacity of the site, however, was uncertain as were the timeframes when that information might be gathered.[57]

The subgroup also addressed several cross-cutting issues. (1) It held that developing repositories in more than one region could be attractive but warned that "organizational and political commitments might develop to particular regions or locations to such an extent that less than full attention would be given to safety, environmental, and security considerations." (2) Several site criteria have been proposed but most include the tectonic and volcanic environment, hydrogeologic and hydrologic factors, geometric characteristics, physical and chemical characteristics of rocks, and proximity to possible mineral deposits. "These criteria can be formulated only in the most general way because of large variations within and among geologic formations and environments."

(3) Although the subgroup acknowledged that disposed SNF might contain some residual economic value and that retrievability for economic reasons might be beneficial. "The most important rationale for retrievability would be to maintain a hedge against the possibility that, contrary to the judgment of both developers and regulators, something is belatedly discovered or goes wrong that would jeopardize either the short-term or long-term safety of the disposal system." (4) Public participation was already required under the Administrative Procedures Act and NEPA. But more robust and continual interaction with interested and affected parties must be an integral part of the country's waste-management program. "With extensive and deliberate public input, the greatest [single] public concern—that the federal government, in its eagerness to show progress...might press imprudently with a premature action and thus unintendedly and inadvertently generate a serious public health and safety risk—could be reduced."[58]

Major Conclusions

The report summarized the subgroup's efforts with a set of findings and recommendations; for our story, the most relevant are listed below.[59]

- Increasing the breadth, redundancy, and diversity of the program increases the chances of it achieving technical success.
- There is an urgent need to study specific potential repository sites in different hydrogeologic environments.
- Strategies that look at multiplicity of sites and hydrogeologic environments will more adequately address some public concerns than those that consider only a few locations but may less adequately address others.
- There are sound reasons for incorporating into any programmatic approach that might be adopted the idea that development of a full-scale repository should proceed in a stepwise manner, with licensing or at least regulatory review at each stage.

- The approach of developing more than one repository regionally...is attractive and merits further consideration.
- Programs and activities undertaken prior to full NEPA review must be designed and implemented in a way that does not prejudice the subsequent NEPA process.

Subgroup members and staff completed the final draft of the report on September 29, 1978. It was published three weeks later. But even as the *Subgroup Report* was still being massaged and tweaked, the IRG as a whole had turned in mid-summer to crafting a draft document for public comment. Only then would sharp differences emerge among the members over tone, content, and policy.

The Battle to Prepare a Report to the President

Deutch distributed a DOE-written draft of the *Presidential Report* in late August. Within days, he received several memoranda strongly criticizing the document. Smith from OSTP argued for the inclusion of a separate chapter on financing and a discussion of the costs of the various strategies. He also maintained that the draft failed in its attempt to explain the linkage between the continued production of nuclear electricity and the "resolution of the waste-management issue." Moreover, the discussion of NEPA had to be strengthened. He complained that the tone of the document was overly optimistic. Nuclear-waste management is *not* a routine problem and adjustments *cannot* be done easily. "All is not right in the world." Speth from CEQ conveyed his disagreement with the assertion "there is the capability accurately to predict whether the waste will remain isolated" once disposed. He also challenged the IRG to take a neutral stance toward the future of nuclear power and recast its analyses accordingly. On behalf of DOI, Davenport disputed DOE's continued focus on host-rocks. "We believe the systems approach to be sounder technically and a considerably more versatile search mechanism than the medium [host rock] approach." Given these serious concerns, the IRG decided to replace DOE as the sole drafter and substitute a panel of staff members representing DOE, CEQ, DOI, and OSTP.[60]

A second draft report to Carter, circulated in mid-September, also failed to achieve consensus. For example, Speth claimed that "it widens rather than narrows our differences over the contents and tone of this document...it actually increases its reliance of assertions, conclusions, and lines of argument to which we previously objected." In a similar vein, Smith observed that "unfortunately, the draft was not in as good shape as we had hoped it might be." Now NRC decided to weigh in. Its general comments are noteworthy.[61]

> We have the impression that the report tends to downplay the magnitude of the technical and institutional issues involved in the implementation of acceptable, safe solutions for radioactive waste disposal and contains a number of decisions which are not clearly supported. If this impression is not corrected before the report is made available for public review, there is the risk that the Administration will be unjustifiably criticized for requesting this review of nuclear waste management in order to enhance the viability of the nuclear supply option.

During the last week of September and the first week of October, the panel sought to cobble together additional drafts. Yet deep and seemingly unbridgeable gaps remained between DOE on the one hand and DOI, OSTP, and CEQ on the other. With the internal deadline for releasing a draft for public comment fast approaching, the panel concluded that it had no choice but to return its latest creation back to Deutch for his final "chop." It is fair to say that none of the members were entirely satisfied with what was wrought. But they also understood that a further opportunity to amend the report would present itself when the IRG responded to feedback from interested and affected parties.[62]

The IRG published Deutch's revised version in mid-October. The report adopted the technical perspective advanced in the *Subgroup Report* and much of its policy analysis. One significant departure was the reframing of the Subgroup's four strategies for disposing HLW and SNF. Those descriptions were now viewed as "illustrative," not complete. At the very least, strategies that might comprise "the interim strategic planning basis" had to consider both regional site selection and intermediate scale facilities.[63] Importantly, the IRG declined to enumerate discrete alternative strategies at this juncture.

Beyond the considerations explained in the *Subgroup Report*, the IRG introduced four other ideas that are relevant for this tale. First, it declared that its analysis was "neutral" with respect to the future trajectory of nuclear electricity generation. It held that the existing wastes must be "subjected to the same strict safety criteria" as future wastes. It also asserted that its "conclusions are viable whatever the future course of nuclear power growth." Second, it introduced the idea of "consultation and concurrence." In its view, "a policy preference for either exclusive federal supremacy or state veto" was not appropriate at this time. However, a state would have to agree to every developmental step before the next one could be undertaken. Third, the IRG recommended that creation of an "Executive Planning Council." It would be composed of state and local officials and tribal representatives. The Council would provide a forum to consider every step of the way. Fourth, the panel deliberated on what institution should be given overall authority. It dismissed options for either an *ad hoc* interagency committee or a newly created independent government organization. Instead, it decided that DOE should maintain its current role because it would "avoid disruptive structural changes."[64]

DOE distributed over 15,000 copies of the draft *Presidential Report.* More than 3,300 written comments were received. The panel paid special attention to roughly 1,500 of them. Overwhelmingly, these were received "from government officials, offices and agencies, the private sector, environmental and other special-interest organizations, and technical experts" who communicated detailed views or raised novel issues.[65]

Over the next five months, the IRG released a final *Presidential Report.* In it, the language of the draft was repeated, followed by a summary of comments, and then the IRG's response.[66]

The final *Presidential Report* struggled to reach agreement on how to respond to some of the four ideas. With respect to neutrality, it stressed that the country already had substantial quantities of defense and commercial high-activity waste, which would have to be disposed in any event. It did acknowledge, however, that

IRG members disagreed whether the analyses fully captured the effect of different trajectories on the institutional capacity to adequately manage the wastes. The concept of "consultation and concurrence" drew extensive criticism. Commenters argued it was ill-defined. This ambiguity left open the crucial question of whether a state could refuse, for example, to issue water permits or whether it possessed a *de facto* veto. The closest the IRG came to giving substance to the concept was the following statement: "Under this approach the State effectively has a continuing ability to participate in activities at all points throughout the course of the activity and, if it deems appropriate, to prevent the continuance of Federal activities."

Although most commenters supported the proposal to create an Executive Planning Council, they also complained about the lack of details governing its operations. In its reaction, the IRG provided extensive particulars about its advisory functions. It also recognized that the Council's name was misleading and replaced it with State Planning Council. Opinion was split on what organizational form should be empowered over going forward. Here the IRG recommended that DOE should maintain overall responsibility. "The transfer to a new agency would involve considerable delays...and would not necessarily solve the problems perceived to exist within DOE and is predecessor agencies. The IRG believes that the DOE can and must conduct the waste-management program in a responsible, careful, and open manner and that, over time, DOE can gain public confidence in its ability to do so."

Crafting a Presidential Decision-Making Memorandum and Statement

Nearly six months would pass before the IRG prepared a presidential decision-making document and sent it to Carter. By that time, the IRG had reached agreement about how far NRC's licensing authority should be extended for new waste storage and disposal facilities, on the importance of a systems approach, and on adopting a conservative, step-by-step, and "learn as you go" strategy to selecting a site for a repository. Further, all agreed that alternative geologies should be studied simultaneously, a sharp departure from DOE's earlier preference.

What remained were two interrelated issues that Carter had to resolve. For the first issue—when should the decision be made for the first high-activity waste repository—the IRG presented him with three discrete alternatives.[67]

- **Option 1:** After 2–3 sites with different geologic environments and diverse host-rocks had been [determined to be potentially suitable], a decision would be made to either (1) propose one or more of these sites as repository locations to NRC for licensing or (2) determine that further sites should be qualified prior to such a decision.
- **Option 2:** After 4–5 sites with different hydro geological environments and diverse host-rocks have been qualified, a decision would be made either (1) to propose one or more of the sites as repository locations to NRC or (2) to determine that further sites should be qualified prior to decision.
- **Option 3:** After the completion in 1981 of the comprehensive waste management plan, which would be developed with state participation and would detail processes for decision-making, the schedule for a site comparison leading to site submission to NRC would be determined.

Proponents of Option 1 maintained that a site could conceivably be selected before all three potential locations were investigated, thereby opening a disposal facility at the earliest opportunity. Thus, the recommendation contained in the Deutch *Task Force Report* could be embraced. The option would be viewed as the least technically conservative. It was, however, favored by industry but opposed by the environmental community. DOE, the Arms Control and Disarmament Agency, NSC, and State supported this option, the latter three out of non-proliferation concerns.

Option 2 insisted that 4–5 sites be qualified before any selection is made, thereby pushing back the operating date of a repository by several years. That delay would likely attract "a wider range of support…and therefore is likely to result in greater ease in overcoming institutional barriers to progress." OMB, OSTP, CEQ, EPA, the Department of Commerce, and the Domestic Policy Council embraced this alternative.

Under Option 3, the number of sites that needed to be investigated and the time until repository operation were left ambiguous. The IRG acknowledged that this option represented the greatest departure from past practice and would be seen so by all interested and affected parties. For some, this radical shift was positive; for others, it was distinctly and strongly negative. Only DOI backed this option.

At the bottom of page 12 of the *Decision-Making Memorandum*, Carter, logical as always, scribbled "Why not 3-4?" That query brought the presidential review to a halt, as members fashioned a response. OSTP and CEQ took the lead in framing an answer. For them, the difference between Option 1 and Option 2 was not the number of sites *per se*, but whether the country was "committed to an expanded program of geologic investigation prior to the selection of the first repository site." In particular, the choice of Option 2 made clear that the nation's waste-management policy would "completely fulfill the IRG's recommendation for geologic diversity."[68]

Carter ultimately selected Option 2, but his ambivalence was unmistakable when he wrote on the October 24th memorandum, referring to his chief domestic policy advisor, Stewart Eizenstat, "Stu—Reluctantly agree. You all seem to want to run an experiment instead of finding a waste disposal and use it [them]."

Although presented to Carter separately, Issue 2—what role should WIPP have as a repository for defense TRU waste—carried strong implications for the effort to dispose of HLW and SNF. Recall that the Deutch Task Force advocated that WIPP should become a test-bed, an intermediate scale facility if you will, for disposing SNF in a salt formation. Since then, Congress authorized the facility's construction but prohibited NRC licensing and the disposal of any SNF, positions at odds with the IRG's counsel. The *Decision-Making Memorandum* proffered four alternatives.[69]

The first called for the continuation of the WIPP project only if Congress permitted NRC licensing and disposal of SNF. By now, most of the IRG members had concluded that a TRU-only facility would not be worth the cost. Further, the generic unsuitability of salt continued to be a concern to some members of the technical community while many environmental groups opposed WIPP because its siting failed to consider alternative locations.

The second option was to proceed with WIPP by accepting Congress's conditions. The major argument in favor of this alternative was that some progress would be shown, and the AEC's decade-old commitment to Idaho would be discharged. All the IRG members, except DOE, vigorously opposed this alternative.

Option 3 called for continued planning for a TRU disposal/intermediate scale facility, but for delaying site selection until 2–3 locations (including WIPP) were qualified. This alternative "will be perceived by the environmental community and some members of the public" as being consistent with the IRG's advice to more forward cautiously. This option would likely be understood by many other interested and affected parties as tantamount to inaction by the Administration. All members except DOE, however, voted in favor of Option 3.

Finally, the fourth alternative would also terminate WIPP and focus exclusively on a combined HLW/SNF/TRU facility. The number of sites to be evaluated would be determined by Carter's choice in Issue 1. Although the fourth alternative was "most consistent with the IRG's recommendations," none of its members argued for its selection.

Carter decided to approve Option 3. But in early December, DOE returned to the battle, requesting that Carter rethink his choice. In a memorandum to the president, DOE Secretary Charles W. Duncan, Jr. observed that the Department attempted to alter Congress' position, strongly held by the House and Senate Armed Service Committees, but to no avail. Carter's determination thus placed the Administration and Congress at odds. By deferring to the legislature, DOE could show concrete visible action, deflecting criticism that the new waste-management regime was just as prone to failure as the old one. Strongly written memoranda from OSTP and CEQ rejected DOE's assertions. In the end, the President also stuck to his original judgment. He announced his view in mid-February.[70]

The Carter Administration's formulation of a national waste-management policy took eighteen months. But it was still incomplete. Outside the administration, both the regulators and Congress would still have their say.[71]

NOTES

1. The National Environmental Policy Act imposes strict procedural requirements on the federal government before it can undertake programs that "substantially affect the environment." (Pub. L. 91-190, 42 U.S.C. 4321–4347, January 1, 1970). The defining case was an AEC licensing ruling, *Calvert Cliffs' Coordinating Committee, Inc. v. U.S. Atomic Energy Commission*, 449 F.2d 1109 (D.C. Cir. 1971).
2. AEC-180/87, "Solid Radioactive Wastes: Long Term Storage in Central Kansas Salt Mine."
3. The impetus for moving forward at the Lyons site derived from a pledge that AEC Chairman Glenn Seaborg made to Idaho Senator Frank Church. The AEC promised to remove by the end of the 1970s the radioactive debris from a 1968 fire at the Rocky Flats, Colorado weapons plant then stored at the National Reactor Test Station in Idaho. The definitive account of the rise and fall of the Lyons site can be found in A.A. Albert de la Bruhèze, *Political Construction of Technology: Nuclear Waste Disposal in the United States, 1945–1972* (Netherlands: University of Twente, 1992). See also Luther Carter, *Nuclear Imperatives and Public Trust: Dealing with Radioactive Waste* (Washington DC: Resources for the Future, 1987), and Daniel Metlay, "History and Interpretation of Radioactive Waste Management in the United States," in William Bishop et al. (ed), *Essays on Issues Relevant to the Regulation of Radioactive Waste Management*, NUREG-0412, Nuclear Regulatory Commission, 1978, from which much of the text below heavily draws.
4. Atomic Energy Commission, "Draft Environmental Impact Statement for Management of Commercial High-Level and Transuranic-Contaminated Waste," WASH-1539, November 1974.

5. See Letter from Shelton Meyers to James Liverman, November 21, 1974. The EPA was particularly concerned that "economic factors could later dictate utilization of the facility as a permanent repository contrary to the stated intent to make the RSSF interim in nature." See Luther Carter, 78, for evidence that ERDA had not completely rejected using the RSSF as a repository.
6. For an extended discussion of the rise and fall of the RSSF, see J. Samuel Walker, *The Road to Yucca Mountain* (Oakland: University of California Press, 2009), 79–83.
7. Energy Research and Development Administration, Press Release No. 76-355 (December 2, 1976). See Carter (1987), 143–193. His discussion is summarized here.
8. Because of the decline in the area's potash industry, WIPP was first proposed by the Carlsbad community as a replacement for the Lyons site for HLW.
9. Eugene Roseboom, Jr., "Disposal of High-Level Nuclear Waste Above the Water Table in Arid Regions," *U.S. Geological Survey Circular 903*, 1983. Isaac Winograd, "Radioactive Waste Storage in the Arid Zone," *Eos Transactions*, (55) 1974, 84–94.
10. See U.S. Energy Research and Development Administration, "Alternatives for Managing Wastes from Reactors and Post-Fission Operations in the LWR Fuel Cyle: Alternatives for Waste Isolation and Disposal in the LWR Fuel Cycle," ERDA 76-43, Volume 4. This idea came to be known as the "Baranowski Bull's Eye."
11. Alice Buck, "A History of the Energy Research and Development Administration," Department of Energy, Office of History, March 1982, 13.
12. NRC Commissioner Victor Gilensky would later wonder why the AEC could maintain such an illusion. Of course, the linkage between the two worlds became evident by the early 1960s. See, for example, the discussion surrounding Figure 2.5 above. Its connection was formalized when the Nuclear Non-Proliferation Treaty came into force in 1970.
13. Two European consortia, URENCO and Eurodiff, as well as the Soviet Union, began moving into the enrichment market, but federal control of the intellectual property meant that a U.S. firm could undercut in the near term the European competitors' charges. The end of the U.S. monopoly, however, was not hard to forecast. For an up-to-date review of the low-enriched uranium market, see "On the Verge of a Crisis: The U.S. Nuclear Fuel Gordian Knot," https://www.ans.org/news/article-4909/on-the-verge-of-a-crisis-the-us-nuclear-fuel-gordian-knot. Accessed November 16, 2023.
14. This seemingly unexpected outcome results from the absorption of a neutron by ^{238}uranium to produce fissile ^{239}plutonium. See the discussion of this phenomenon in Chapter 2.
15. A consortium of Gulf Oil, Allied Chemical, and Westinghouse was responsible for constructing the Barnwell plant. Separating plutonium from commercial SNF by chemically reprocessing made possible two technologies for extending uranium supplies. First, it allowed the fabrication of mixed uranium-plutonium fuel assemblies that could be recycled into light water reactors. Second, plutonium could be used directly by breeder reactors in a way that putatively produced more fuel than was consumed. In short, reprocessing, recycle, and breeders were all bound inextricably together; no readily identifiable path could be conceived to advance one development without the others. In addition to Barnwell, General Electric built the small Midwest Fuel Recovery Plant in Morris, Illinois. Results of pre-operational tests with non-radioactive simulated spent fuel revealed significant technical difficulties with remote operations. As a consequence, General Electric decided not to commission the plant to reprocess spent nuclear fuel.
16. The bureaucratic and diplomatic conflicts, alliances, thrusts, and parries that emerged within both administrations provide enthralling and complex tales of their own. I can only touch on some of the highlights that influenced waste-management policymaking. I have been partly guided in that effort by the best discussion by far of the federal government's efforts to redirect nuclear policies to address nuclear proliferation concerns: Michael Brenner, *Nuclear Power and Non-Proliferation: The Remaking of U.S. Policy* (Cambridge: Cambridge University Press, 1981). See also Michael Brenner, "Carter's Bungled Promise," *Foreign Policy*, (36) 89–101, 1979. Certainly, the process

was messy, disjointed, and not entirely logically consistent. Given its long-term impact, however, I believe Brenner's criticism of the Carter Administration's labors is unduly harsh.

Finally, implicit in its bargaining with other nations, but never far from the surface, was the threat of preventing the reprocessing of United States-origin low-enriched uranium. If the U.S. held a virtual monopoly on that supply, the threat was formidable and disruptive. As other countries began to acquire alternative fuel sources, this threat shrank.

17. Brenner provides the details of the group's deliberations. See also Nuclear Regulatory Commission, "Final Generic Environmental Statement on the Use of Recycled Plutonium in Mixed Oxide Fuel in Light-Water Cooled Reactors (GESMO)," NUREG-0002, 1976.
18. The Nuclear Policy Review Group included ERDA, NRC, the Arms Control and Disarmament Agency, various arms of the State Department, not all of whom were reading from the same playbook, OMB, and NSC. The quotations in this paragraph and the next five paragraphs can be found in Memorandum to the President, "Nuclear Policy Review," September 7, 1976.
19. The quotation is from Brenner (1981), 113.
20. President Gerald Ford, "Nuclear Policy Statement," *Public Papers of the President,* 2763–2778, October 28, 1976.
21. General Accounting Office, "Status and Commercial Potential of the Barnwell Nuclear Fuel Plant," GAO/RCED-84-21, March 28, 1984.
22. Here again I found Brenner's discussion quite valuable as is Luther Carter's description of this period. See as well "Presidential Review Memorandum/NSC 15," January 21, 1977. See https://www.jimmycarterlibrary.gov/sites/default/files/pdf_documents/assets/documents/memorandums/prm15.pdf. Accessed September 27, 2024. The "options and recommendations" report remains classified nearly fifty years after its transmission to the president. The Carter Presidential Library informed me that it would take roughly 12 years to get it declassified. Some sense of the deliberations, however, can be found in https://history.state.gov/historicaldocuments/frus1977-80v26, especially documents 322, 325, 328, and 395. Accessed September 27, 2023. These sources provide an imperfect documentary trail of both the interagency conflicts and the difficulties encountered bargaining with other nations.
23. The new Nuclear Policy Review Group consisted of the same agencies and Executive Office of the President bureaucracies as the committee of the same name created by President Ford. However, Carter's Assistant for Energy Policy, James Schlesinger, was now included.

    The International Nuclear Fuel Cycle Evaluation (INFCE) held an organizing meeting in Washington, D.C. on October 19–21, 1977. Fifty-six countries and five organizations participated. See Carter's remarks to the Plenary Session of the Conference, White House Press Release, October 19, 1977. The governments attending agreed that the exercise would be a technical and analytic study and not a negotiation. Eight working groups were formed, covering the full gamut of issues—from reprocessing, plutonium handling, and recycle to fast breeders to SNF management to waste management and disposal. The working groups held 134 meetings, a Technical Coordinating Committee had nine meetings, and three plenary conferences took place, the last one in November 1980. For our purposes, the key product was the *Working Group 7 Report*, International Atomic Energy Agency, Vienna, Austria 1980. The conclusion holding that SNF could be disposed of as readily as HLW can be found in International Nuclear Fuel Cycle Evaluation, *INFCE Summary* (International Atomic Energy Agency: Vienna, Austria, 1980) 16–17, 21. See also R. Skjöldebrand, "The International Nuclear Fuel Cycle Evaluation-INFCE," *International Atomic Energy Agency Bulletin*, (22) 1980. INFCE's impact had an exceedingly short half-life.
24. Carter's statement can be found in *Presidential Documents-Jimmy Carter, 1977,* (13) April 18, 1977. ERDA granted AGNS several, small, short-term contracts to ease the transition. Barnwell shut down completely on December 31, 1983. The LMFBR was finally killed by Congress in 1991.

25. Metlay (1978), "Error Correction in Bureaucracy," provides an extensive analysis of the NRC's licensing activity.
26. *DOE News Release*, R-77-017, October 18, 1977. See, for example, Eric S. Beckjord, Memorandum to Robert D. Thorne, "Expediting Away From Reactor (AFR) Spent fuel Storage," March 3, 1978; Eric Beckjord, Memorandum to George W. Cunningham, "Spend Fuel Storage Options," April 19, 1978; Eric Beckjord, Memorandum to George W. Cunningham, "Information Memorandum from George W. Cunningham to John M. Deutch, Implementing Away-From-Reactor (AFR) Spent Fuel Storage," April 28, 1978; and John Deutch and Worth Bateman, Memorandum to James Schlesinger, "Spent Fuel Storage and Charges," June 12, 1978. Legal support for the spent fuel policy is "Final Environmental Impact Statement: U.S. Spent Fuel Policy," (DOE/EIS-0015), 1979. See also Department of Energy, "Record of Decision—U.S. Spent Fuel Policy," *Federal Register*, (45) 139, July 17, 1980, 47903–47905. Faced with these implementation obstacles, few were surprised when President Ronald Reagan failed to support Carter's spent fuel storage policy. Reagan also overturned his predecessor's ban on domestic reprocessing. See his "Statement on Nuclear Policy Initiatives," https://www.reaganlibrary.gov/archives/speech/statement-announcing-series-policy-initiatives-nuclear-energy. Accessed December 15, 2023.
27. Technical suitability involves a demonstration that a repository system constructed at a particular location complies with relevant performance standards set by independent regulators. Social acceptability is reflected in a variety of ways, such as legislative determinations, referenda, protests, public opinion surveys, and elections. The author formally introduced this conceptual interpretation in a paper entitled "International Experience Developing Deep-Mined Geologic Repositories," delivered at AtmoExpo11, Moscow, Russia, November, 2011.
28. This metaphor allows for either sequential or parallel examination of sites. However, laws such as the National Environmental Policy Act may demand that the performance of repositories located at several sites be compared. This circumstance may disfavor the sequential option.
29. Actual siting processes may depart from the "idealized" one. For example, the progression could stop after only one filter is considered. In such a case, the site selected could either be suitable but unacceptable or acceptable but unsuitable. If either situation arises, development of a repository is very likely to be controversial.
30. This formulation simplifies what is a more complex and complicated dynamic. The filters are not strictly independent. Once the siting processes commences, beliefs about suitability are affected by attitudes about acceptability and vice versa. See the discussion of this circumstance in NWTRB (2015). Chapter 5, "The Rise of Yucca Mountain," and Chapter 6, "The Fall of Yucca Mountain," detail the consequences of the filters' interdependence.
31. My conjecture is that government, military, and industrial policymakers never grasped the institutional and political challenges to developing a repository. They believed that the task was a technically trivial one and that outsiders should defer to the technocrats' judgments. See Daniel Metlay, "A History and Interpretation of Radioactive Waste Management Policy," in W. Bishop, ed., *Essays Relevant to the Regulation of Radioactive Waste Management*, U.S. Nuclear Regulatory Commission, NUREG-0412, 1–18, 1978.
32. John Deutch, Memorandum to Dale D. Myers, "Transmittal of *Report of Task Force for Review of Nuclear* Waste Management," DOE/ER-0004/D, March 3, 1978. It is often referred to as the "Deutch Report."
33. Department of Energy, *Final Environmental Impact Statement: Management of Commercially Generated Radioactive Waste*, DOE/EIS-0046F. This EIS provided the legal support for the federal government's decision to select geologic disposal as the sole technology for the long-term management of HLW and SNF. By doing so, it eliminated the necessity to evaluate other options such as space and deep-borehole disposal.

34. ERDA's then-current planning basis called for site investigations in 36 states. Field studies were to begin by the end of 1977, with the selection of two repository sites scheduled for the end of 1978. ERDA allocated $34 million for all these activities. See Carter (1987), 130–131.
35. Task Force Report, 2.
36. Task Force Report, 3.
37. Task Force Report, 12.
38. Jimmy Carter, "Memorandum to Secretary of State, et al., "Interagency Nuclear Waste Management Task Force," March 13, 1978. Roger LeGassie, "Memorandum to Interagency Task Force on Nuclear Waste Management, Meetings of Interagency Task Force and Working Group," April 11, 1978. The members included: DOE, EPA, NRC, Department of Interior, Department of State, Arms Control and Disarmament Agency, National Aeronautics and Space Administration, Department of Transportation, and the Department of Commerce. In addition, five organizations within the Executive Office of the President were involved: OMB, NSC, Council of Environmental Quality, Office of Science and Technology Policy (OSTP), and Domestic Policy Council. Interagency Review Group on Nuclear Waste Management, "Subgroup Report on Alternative Technology Strategies for the Isolation of Nuclear Waste," TID-28818 (Draft), October 1978. See Ted Greenwood, "Nuclear Waste Management in the United States," in E. William Colglazer, Jr. (ed.), *The Politics of Nuclear Waste* (New York: Pergamon Press, 1982), 23. Greenwood's article offers a unique perspective from an influential insider throughout the IRG process.
39. See Preface to the Subgroup Report for the rest of the IRG's membership.
40. See Draft, "Issue Paper, Subgroup on Alternative Technology Strategies, Interagency Nuclear Waste Management Task Force, Alternatives for Analysis," May 16, 1978.
41. Report of the American Physical Society by the Study Group on Nuclear Fuel Cycles and Waste Management, *Review of Modern Physics*, (5) S1–S183, January 1978; David Bredehoeft et al., "Disposal of High-Level Radioactive Wastes—Earth-Science Perspectives," *U.S. Geological Survey Circular 779*, 1978, 4. It is unclear why these analyses were not incorporated into the *Task Force Report*. One possible explanation is that the studies appeared too close to the deadline for delivering the report. Another plausible explanation is that the DOE Task Force did not include any geologists. This split between the earth scientists and nuclear engineers, it must be noted, would color deliberations that took place over the next 18 months.
42. See "Subgroup Report on Alternative Technology Strategies for the Isolation of Nuclear Waste, Draft," TID-28818D Appendices A-F. Ted Greenwood from OSTP, Peter McGrath, Cyrus Klingsberg, and Jere Nicols from DOE, Dave Stewart and Isaac Winograd from the USGS were the primary contributors to the drafting of the appendices. OSTP and DOE appointed a ten-person Technical Advisory Group to review *Appendix A* of the Subgroup Report. Its members are listed in the Subgroup Report, *Preface* to the *Appendix A*. See, for example, "Minutes of the Meeting of the Nuclear Waste Management Technical Advisory Group," June 15, 1978. Early versions were distributed for comment to interested and affected parties, including city and county government officials, Native American tribes, nuclear industry executives, nuclear utility executives, state legislators, and the National Academy's Committee on Radioactive Waste Management. See Roger LeGassie, Memorandum to IRG Principals and Subgroup Chairman, "Reports of IRG Small Group Meetings Prepared by DOE Waste Management Program Attendees," August 21, 1978. All these comments were read and considered in the redrafting of the appendices.
43. Appendix A, A-25–A-43. Appendix A also discussed seven different host-rocks: Rock salt, anhydrite, granite and other crystalline rocks, shale and other clays, flood basalt, volcanic tuff, and unsaturated rocks.
44. Appendix A, A-iii–A-iv.
45. Appendix A, A-54–A-59.
46. Appendix A, A-18–19.

47. Interagency Review Group for Nuclear Waste Management, "Report to the President by the Interagency Review Group on Nuclear Waste Management," TID-29442, March 1979.
48. Greenwood from OSTP, Colin Heath, Roger LeGassie, Don Vieth, and Peter McGraff from DOE, William Bishop from NRC, and Winograd and Stewart from USGS were key contributors. I served for three months as a full-time consultant to OSTP, drafting the policy analysis beginning in the summer of 1978. See Subgroup Report, 1–89. The report underwent twelve revisions before its release. I commented intermittently on other IRG products until the middle of the following year. Much of what follows relies on discussions with those eight individuals as well as internal documents. I am especially grateful to Greenwood and Stewart for keeping me apprised of events that took place after I left OSTP.
49. Subgroup Report, 13–14. In many respects, the subgroup built on the work of the NRC's first nuclear-waste management report, "Proposed Goals for Radioactive Waste Management," Nuclear Regulatory Commission NUREG-0300.1978.
50. Subgroup Report, 23.
51. See Subgroup Report, 22–29, for a more detailed discussion of the four strategies.
52. Subgroup Report, 41.
53. Subgroup Report, 32–34, 89. The idealized schedule presumed no major technical, regulatory, or political interruptions. It did recognize "normal" contingencies that might be encountered at each stage of the development process.
54. Subgroup Report, 43.
55. Subgroup Report, 48–51.
56. Subgroup Report, 67–71. These figures proved to be optimistic even after adjusting for inflation. The estimated total life-cycle cost of one repository was roughly $20,000 million. By comparison, the 2008 DOE estimate for the proposed Yucca Mountain repository is upwards of $80,000 million. "Analysis of the Total System Life Cycle Cost of the Civilian Radioactive Waste Management Program, Fiscal Year 2007," DOE/RW-0591, 2. The head of the Office of Management and Budget told the IRG Executive Committee that it could not be more specific in its financial recommendations because they might be interpreted as advocating a new tax.
57. Subgroup Report, 37–40.
58. Subgroup Report, 76–89. Quotations are found respectively at 77, 78, 83, and 85.
59. Because the subgroup focused exclusively on disposal strategies, for all practical purposes, it ignored the question of storage in all of its possible forms. In the end, the IRG simply held that storage options would be available if and when needed. Once Congress began considering waste-management legislation, the lacuna took central stage. See the discussion in Chapter 4.
60. Philip Smith, Memorandum to John Deutch and Roger LeGassie, "Comments on the August 21 Draft of the Report to the President," September 1, 1978. Jim Jamieson and Jerry Brubaker, Memorandum to Gus Speth, "Waste Management Issues," August 4, 1978. Gus Speth, Draft Memorandum for John M. Deutch, "CEQ Comments on Draft I of the IRG Report to the President," n.d. Gus Spetch, Memorandum for John Deutch, "IRG and Neutrality Regarding Nuclear Growth," August 29, 1978. Joan Davenport, Memorandum to John Deutch, "Report to the President by the Interagency Review Group on Nuclear Waste Management, Draft of August 21, August 28, 1978. See also, Joan Davenport, Memorandum to Gus Speth, "IRG Executive Group Meeting, Friday, September 8, 1978," September 7, 1978.
61. Gus Speth, Memorandum to John Deutch, "CEQ Comments on Draft 2 of 'Report to the President by the Interagency Review Group on Nuclear Waste Management,'" September 25, 1979. Phil Smith, Memorandum for John Deutch and Roger LeGassie, "OSTP Comments on Draft 2 of the Presidential Paper," September 26, 1978. Clifford V. Smith, Jr., Director, Office of Nuclear Materials Safety and Safeguards, Letter to John Deutch, September 25, 1978.
62. See Michele Zarubica, Memorandum to Joan Davenport, "IRG Report to the President," October 11, 1978.

63. Interagency Review Group, *Report to the President by the Interagency Review Group on Nuclear Waste Management*, TID-28817 (Draft), October 1978, 30–33. An intermediate scale facility, especially one built at a proposed repository site, would hold several hundred spent fuel rods and would be monitored for a period for up to 20 years. The NRC would license it. Intermediate scale facilities would late play a substantial role in Congress' deliberations leading to the passage of the Nuclear Waste Policy Act of 1982. See Chapter 4 for an extended discussion of that debate.
64. TID-28817, iv–v, 52, xv–xvi, 66.
65. Greenwood, 22.
66. Interagency Review Group, *Report to the President by the Interagency Review Group on Nuclear Waste Management*, TID-29442, March 1979. Quotations can be found on 7-8, 93–95, 90, 119 respectively.
67. James T. McIntyre, Jr. and Stuart E. Eizenstat, Memorandum to the President, "Nuclear Waste Management," September 5, 1979, 7–11.
68. Stuart Eizenstat and Kitty Schirmer, Memorandum to the President, "Nuclear Waste," October 24, 1979.
69. McIntyre and Eizenstat, 14–20.
70. See Charles W. Duncan, Jr., Memorandum for the President, "Waste Isolation Plant," December 12, 1979. See also Gus Speth, Memorandum for the President, "DOE's WIPP Project—OMB Issue Paper," December 11, 1979; Stu Eizenstat and Gus Speth, Memorandum for the President, "DOE's WIPP Project—OMB Issue Paper," December 13, 1979. Carter released a statement to Congress about the IRG's efforts on February 12, 1980. See *CQ Almanac*, "Presidential Statement: Carter's Nuclear Waste Program," (38), 1980, E-40.
71. The subsequent regulatory and legislative deliberations took place in an increasingly turbulent political environment. The U.S. Court of Appeals for the District of Columbia Circuit held that reactor licensing must be contingent upon NRC possessing confidence that HLW and SNF can be disposed of in a repository. See *State of Minnesota v. NRC*, 602 F2d 412 (1979). On October 25, 1979, NRC launched a rulemaking to assess the agency's degree of confidence and to determine when a repository might be available. Among its findings: "One or more mined geologic repositories…will be available by the years 2007–2009." See *Federal Register*, (49) August 31, 1984, 34,658–34,696. As DOE's repository program struggled, NRC began a near 30-year retreat from this overly optimistic judgment.

4 Creating a Regulatory and Legal Foundation

PRELUDE

On a bright, sunny day in late September 1979, a convoy of seaplanes touched down on Orcas Island in the Puget Sound just west of Seattle. As the planes taxied to the beach, fresh barbequed salmon and an assortment of vegetables awaited five-dozen participants in a workshop organized by one of the DOE contractors. They included Department and Nuclear Regulatory Commission (NRC) executives, state and local officials, nuclear utility representatives, and university social scientists. The headliners were former Governors Edmund G. Brown, Sr. from California, Jack Campbell from New Mexico, and Mike O'Callaghan from Nevada. Their task was to explore the cryptic meaning of "consultation and concurrence."[1]

All the contributors recognized that the concept was controversial. Some supported the idea as developed by the IRG. Others believed that it was important to use the Group's idea as a basis for proceeding, even if details were later modified. Still others were skeptical of the entire notion from the outset, feeling that ultimate assertion of federal power was necessary to site a repository and that strong state concurrence rights would delay a satisfactory resolution of the nuclear-waste problem. Not surprisingly, the gathering failed to reach a consensus on how to implement the idea. Indeed, it is fair to say that by the end of the workshop, none of the participants had changed their minds.

Several months later, on February 12, 1980, Carter issued Executive Order 12192 and established the State Planning Council, an advisory committee, composed of 14 state, local, and tribal voting officials as well as the non-voting heads of DOE, EPA, and the Departments of Transportation and Interior.[2] Its chair was Richard Riley, the sitting Governor of South Carolina. In its Resolution 3-4, adopted on June 3, 1980, the Council defined consultation as "involving the sharing of information and participation in planned Federal activities and the right of States/Tribes to review, comment, and make recommendations on such activities."

Further, it defined "concurrence" as an incremental process involving the growth of public confidence and the development of a technical consensus that the waste management program can be accomplished with acceptable risk. Importantly, in Resolution 4-10, adopted that same day, the Council emphasized that a "deadlock that cannot be solved through such negotiations should be held open until a point in the repository development process at which decisions can be made through a formal process of petition to the President or the Congress."[3]

DOI: 10.1201/9781003528302-5

TURNING THE BALL OVER

The IRG's efforts ultimately established the Carter Administration's path forward for developing a repository for high-activity waste. Whether that blueprint would be sustainable remained unclear. By early 1980, many politicians, "talking heads," and opinion influencers already were labeling Jimmy Carter a lame duck. What would follow over the next two years would determine whether his nuclear-waste legacy would join the dust bin of other initiatives or would possess sturdier legs. The signs were not auspicious. The terms of nuclear-skeptical NRC Commissioners, Victor Gilinsky and Peter Bradford, were drawing to a close. The landslide election of President Ronald Reagan swept Republicans into office across the nation. Within months, Reagan terminated the spent-fuel policy of his predecessor and reversed Carter's ban on domestic reprocessing. But the new administration's stance on what to do with the waste was publicly cloaked or at best vague.

NRC had kept abreast of the IRG process because of the feedback from its influential staff representative, William Bishop, who also was the chief of the agency's waste management division. Interacting with the executive branch and the IRG only sporadically, Congress was a more distant observer. Its members sponsored dozens of bills between 1978 and 1982; several congressional support offices launched comprehensive studies of the issue. Nonetheless, both the regulators and the lawmakers had to weigh in formally to set a new federal policy in place.

WASTE-MANAGEMENT REGULATIONS

Reorganization Plan No. 3, submitted by President Nixon in 1970, split regulatory authority over the disposal of high-activity waste between EPA and AEC (subsequently NRC). The former, which exercised the power to issue general environmental protection standards outside the boundaries of nuclear facilities, remained a hide-bound bureaucracy. Its inertia was a continuing source of frustration for the IRG's other members, let alone the public. NRC, however, could still advance rules for licensing *specific* disposal facilities, such as repositories, although they might be contingent upon subsequent EPA determinations.

PROCEDURAL RULES

Even before Carter announced his decisions on the issues forwarded by the IRG in February 1980, NRC proposed *procedures* for permitting a repository for HLW and SNF; thirty-four individuals and groups offered comments. The regulator finalized the draft rule a little over a year later. Understanding how NRC addressed five specific issues provides some perspective into the regulators' thinking: (1) characterization of multiple sites, (2) *in situ* testing at depth, (3) whether the site selected by DOE should be the "best," (4) opportunities for State, local, and public participation, and (5) formal public hearings.[4]

NRC lent its voice to those arguing that characterization of multiple sites provided the only "effective means for arriving at a reasoned decision under NEPA." Public comments straddled the issue of how many sites should be mandated. Informed by its understanding of bureaucratic behavior, the Commission maintained that characterization of several sites would prevent "a premature commitment by DOE to a particular site." In a public meeting discussing the final rule, Commissioner Joseph Hendrie noted that he could support a requirement for five sites in three geologic environments. But he feared that such a benchmark "would raise the *ante* for repository [number] one to what might turn out to be a shade too high a threshold." Therefore, NRC just required the Department to investigate at least three sites in two geologic media, one of which would either be a salt bed or a domed salt formation. Nonetheless, it did maintain that it "expected DOE to submit a wider range of alternatives than the minimum."[5]

The conundrum facing any waste-management implementor or regulator was knowing how much information must be gathered to persuasively conclude that a given site was technically suitable. No one contested the importance of surface-based studies, such as drill holes and geophysical techniques. USGS, however, strongly argued that *in situ* geophysical, geochemical, hydrological, and rock mechanics investigations were the only methods that might reveal "the ever-present possibility of lateral changes in the properties of the host rock and the possible presence of inhomogeneities of too small a scale to be detected by remote or borehole techniques." NRC concurred but offered DOE the option of an exemption from this requirement as long as the needed information was secured by other means.[6]

Equally problematic was the classic question "how safe is safe enough." Among the conceivable responses were "the best possible" and, as in other regulatory contexts, and "as low as reasonably achievable." NRC understood that selecting the first reply would send DOE on a fool's errand; other than making Swiss cheese out of the continental United States, there was no rational stopping rule. The second answer was equally questionable; DOE would have to evaluate a large suite of sites before it could conclude whether this standard had been met. So, the regulator simply observed, "It generally has been NRC practice to consider only whether a license application meets prescribed criteria. The Commission perceives no reason to adopt a different philosophy here."[7] (Those targets are described in the next section.)

Grounded on a report to Congress, the draft focused *solely* on possible roles that states might play; seemingly the regulator saw no function that localities, tribal nations, or other interested and affected parties might perform. But even this crabbed interpretation of "participation" was hollow. In the congressional report, the Commission not so subtly revealed its core belief[8]:

> The nature of the Federal system demands that [State] interests be recognized and it is reasonable in the present context to recognize that State agencies could be more responsive to them than Federal decisionmakers would be. But just as the benefits of energy (and weapons) production serve the people of the nation as a whole, so too is there a national need for managing the wastes that result. State actions that inhibit Federal efforts to manage the wastes must be evaluated in terms of their impact upon the achievement of the national objectives.

Given this point of view, it was hardly surprising that the states' role had to be limited. They could assist NRC, in reviewing, in some unclear manner, DOE's site characterization reports, could monitor environmental and radiation releases during the repository's operations, and could temporarily reassign state employees to the Commission.

Unlike the requirements under the Atomic Energy Act of 1954 for licensing reactors, the regulators were not obligated to hold a public hearing about licensing a repository, only to inform state and local officials that one might be held. Also left ambiguous in the proposed rule was the type of hearing. Presumably, if held, it might be a "legislative" hearing, in which the Commissioners or their designees simply listened as the parties presented their case. Alternatively, the regulators could opt for an "adjudicatory" trial on the record, in which parties could discover information from other participants, present evidence, and cross-examine witnesses.

The final rule once again addressed these five points. As implied above, NRC held fast to the core belief expressed in the report to Congress. For all intents, it also reiterated its initial position regarding characterization of multiple sites, *in situ* testing, and licensing criteria. The regulators did expand their understanding of who might participate to include localities, tribal entities, and outside parties. Although now committed to holding a hearing on the repository construction authorization, it never clarified whether it would be legislative or adjudicatory. However, asserting a lack of legal authority, the Commission deflected calls for funding the intervention of any challenger to the DOE's plans and for creating its own version of consultation and concurrence.

Technical Criteria

In developing standards for protecting the general public from being harmed by the migration of radionuclides into the environment, NRC and EPA found themselves entangled in a bureaucratic *pas de deux*. Sideline consultations intensified over time. But for all practical purposes, the former regulator was held hostage to the latter who refused to clarify the uncertainty surrounding its policies. (See Figure 4.1.)[9]

NRC intended in its Advanced Notice of Proposed Rulemaking to advise DOE and members of the interested and affected public of its then-current philosophy for structuring the technical criteria. In the 90 comments received, only one-third

ACTION	DATE
NRC: Advanced Notice of Proposed Rulemaking	May 13, 1980
NRC: Proposed Technical Criteria	July 8, 1981
EPA: Proposed Environmental Standards	December 29, 1982
NRC: Final Technical Criteria	June 21, 1983
EPA: Final Environmental Standards	September 19, 1985

FIGURE 4.1 Timeline for publishing high-level waste-management regulations.

were favorable. "Half of those who disagreed felt we had gone too far in what we were suggesting we wanted to do, and the other half felt we perhaps had not gone far enough."[10] So critical was the feedback that the staff ultimately started from scratch when it released its proposed rule a little more than a year later.

For our purposes, its new approach involved four determinations. First, the timing disconnect with EPA's rulemaking left NRC with no choice but to leave a placeholder for the release limits.[11]

Second, acting on its own, NRC set forth criteria for siting repositories. It catalogued two types—favorable conditions and potentially adverse conditions. Among the former were chemical conditions that promote precipitation or sorption of radionuclides and low population density, remote from population centers. Among the latter were potential for flooding, structural deformations of the host rock, changes in hydrogeologic conditions, and evidence of dissolution or breccia pipes. The regulator emphasized that "a site is not disqualified as a result of the absence of a favorable condition or the presence of a potentially adverse condition."[12]

Third, the Commission determined that release scenarios involving unintended human intrusion had to be incorporated in any application to construct a repository. At the same time, it recognized that speculation on what those scenarios might be was an unrealizable and irrational task. Beyond marking and documenting the repository site, NRC declined to impose any additional requirements.[13]

Fourth, again on its own volition, the Commission sought to guard against the possibility that uncertainties in the modeling might mask projected failures of the repository system. "To assure that radioactivity releases are kept to sufficiently low levels, it concluded that, like the regulation of reactors, safety should depend on "'defense-in-depth'." It, therefore, prescribed *minimum* performance standards for each of the major elements of the geologic repository. This approach gave rise to three subsystem performance standards.[14]

1. Containment of all HLW within the waste packages will be substantially complete for not less than 300 years nor more than 1,000 years after permanent closure of the repository.[15]
2. The [annual] release rate of any radionuclide from the engineered barrier system following the containment period shall not exceed 1 part in 100,000 of the inventory.
3. The repository shall be located so that pre-emplacement groundwater travel time along the fastest path of likely radionuclide travel exceeds 1,000 years.

At a May 7, 1981 meeting of the Commissioners to approve publication of the proposed technical criteria, the subsystem requirements dominated the discussion. The staff detailed its reasoning.

> We have essentially taken action to require engineering solutions to give you a predictable source term out of the repository, one that can be counted on and is not based on a lot of unverifiable assumptions, which of course will allow greater uncertainties in the geological transport problem, which our study shows is going to be tough to narrow the uncertainties much there. I think what we have done is taken the overall system and placed some requirements on subparts of it that we think make it a more tractable problem.

Commissioner Hendrie remained skeptical. In a dialogue that repeated itself several times over, he sought to convince the staff and the other Commissioners that the subsystem constraints were technically redundant and superfluous. "Is it clear that it is either necessary or a good idea to prescribe these sub-elements?" The staff's counter took another tack. In its view, the subsystem conditions were likely to make the difference between a successful showing of compliance and a failed one. So, in the end, the subsystem performance requirements survived in the proposed rule.[16]

This decision unleashed a storm of criticism from parties with a sizeable stake in the proceedings.[17] DOE took NRC to task.

> We believe that the rule should be based on and derived from an overall system performance objective. Instead, the rule centers on the imposition of performance requirements for individual components that are neither derived from nor related to an overall system performance objective.

One professional group, American Nuclear Society, strongly informed NRC that:

> It is our concerted view that overly restrictive and specific performance standards are not necessary, and that such standards in regulation form are likely to add to the overall cost of the waste disposal without achieving any degree of benefit to the public health and safety.

The National Academy of Sciences argued that "NRC has not presented adequate evidence that these numerical criteria can support a finding of 'no unreasonable risk to the health and safety of the public.' Moreover, NRC has not shown that these numerical criteria are either necessary or sufficient to meet the EPA standard."

Even EPA, gently and with some circumspection, relayed its qualms.

> The 1,000-year life requirement for the waste package could necessitate the use of very expensive or exotic materials (such as titanium) for waste canisters. The supporting documentation for the rule does not consider the potential cost of this requirement. In light of the relatively small benefits and possible high cost, we believe the Commission should reexamine this requirement.

The international community also had its concerns. For example, the United Kingdom's Department of the Environment observed:

> This approach to setting technical criteria is incorrect in principle. It leads to criteria which are inflexible because, since they have no clear basis, there can be no basis for changing them. In addition, the approach is very likely to lead to criteria which are too restrictive, thus causing more expenditure on high-level waste disposal than is warranted by radiological protection considerations.

Importantly, several comments suggested that enumerating subsystem performance conditions contradicted one of the IRG's fundamental conclusions: it is the functioning of the system that counts, not the behavior of any single repository element.

The biggest stumbling block to finalizing the rule was how to press ahead with no definitive EPA standard in sight.[18] Indeed, one memorandum to the Commissioners warned that proceeding before its regulatory partner had acted would "would likely subject the agency to considerable criticism, both from the public and the Congress. We would end up taking the blame for EPA's failure to perform. In fact, it would divert much of the attention away from where it belongs—on EPA and OMB." Although sensitive to this concern, NRC felt compelled to move forward.[19]

Compared to the intense internal debate over the proposed rule, especially with regard to the subsystem requirements, the Commission's public deliberations on the final rule were restrained. Commissioners Victor Gilinsky and John Aherne offered minor changes to one definition and a tactic to eliminate what might be perceived as a conflict between EPA's formulation of the ALARA concept and NRC's. (See Figure 4.2 for a summary of the major provisions in the final rule.[20])

As our chronicle continues, the controversy over subsystem performance requirements will play out several years later in the halls of Congress.

WASTE-MANAGEMENT LEGISLATION

In parallel with the IRG's undertaking, the Office of Technology Assessment (OTA), an arm of Congress, began an enquiry into the management of commercial radioactive waste. Two principal staff members kept key committees and lawmakers continuously informed of the study's progress. Many of the review's conclusions

PROVISION	REQUIREMENT
Overall performance standard	Adopt the EPA standard
Waste package integrity	Substantially complete containment for between 300 and 1,000 years
Controlled releases	Not to exceed 1 part in 100,000 per year after containment ends
Groundwater travel time	Greater than 1,000 years on fastest likely path to the accessible environment
Unanticipated (disruptive) events	DOE must calculate the risk by using the human intrusion scenario specified by EPA
Retrievability	The possibility of retrieval should be kept open at least through the emplacement period
Confirmatory testing	Proof of the future performance is not to be had in the ordinary sense of the word. Only reasonable assurance is required.
Geologic testing	Modified to include a repository located in the unsaturated zone

FIGURE 4.2 Major postclosure provisions adopted in the final *10 Code of Federal Regulations 60*.

eventually paralleled those arrived at by the Carter Administration. Thus, a solid policy framework was being erected even before negotiations on the proposed bills had commenced.[21]

But agreement on general principles and philosophy did not necessarily translate into a legislative consensus; both nuanced and stark cleavages began to emerge.[22] By 1980, however, the "order of battle" had unmistakenly jelled. Four parties held significant stakes: DOE, the nuclear industry, environmental and anti-nuclear organizations, and potential repository host states. Five issues dominated the debate.[23]

1. Repository development schedule
2. Need for a "test and evaluation" facility
3. Role of SNF storage: interim or as an alternative to disposal
4. State role in repository siting[24]
5. Relationship between defense and commercial repositories

In order to satisfy those parties, ultimately Congress would pass a bill that contained four strategic bargains related to several of these issues.

A taste of what was to come appeared in the preliminary skirmishes during the 96th Congress (1979-1980). On the Senate side, Bennett Johnston, a Democrat from Louisiana, and Henry Jackson, a Democrat from Washington State, both jurisdictions under consideration as potential sites for a repository, introduced a bill that prioritized storage in casks as an alternative to disposal. Although the underlying concept remained the same, they morphed the Retrievable Surface Storage Facility into a centralized Monitored Retrievable Surface (MRS) facility. DOE and the nuclear industry reluctantly supported the proposal while the non-governmental organizations strongly opposed it.

At the end of 1980, Representative Morris Udall, Democrat from Arizona, offered an alternative. No hearings were held; but under a special procedural rule, the bill passed on the House floor. The bills from each chamber differed in several respects. The Senate bill contained authorization for a MRS and provided for interim away-from-reactor storage, and was more generous to states in its design of a conditional veto. The House bill failed to endorse any form of storage and explicitly focused on developing a repository. Because no agreement could be reached between the two bodies, time ran out before the Senate and House could reconcile their differences.

Congressional Activity Heats Up

The Reagan landslide in the fall of 1980 brought the Republicans into power in the Senate. James McClure, Republican from Idaho, another state storing defense HLW, took over as chair of the Energy and Natural Resources Committee. Although the bill he sponsored early on authorized interim SNF storage and required the DOE Secretary to submit a plan to Congress for a MRS, the proposal sharply broke with the Johnston/Jackson storage-first approach. Away-from-Reactor storage was declared "a last resort." Among other requirements, it obliged the DOE to issue repository site-suitability guidelines, adopt an accelerated schedule for siting a repository, recommend sites for characterization, and transmit to Congress a request to construct at least one test-and-evaluation facility. Moreover, for the first time, a piece of legislation specified exactly how nuclear power plant owners would repay the federal government for the disposal of their waste.

Floor consideration of the McClure bill was delayed until April 28, 1982. Senator Howard Cannon, Democrat from Nevada, tried in vain to strengthen a state's veto power. Senator Johnston's effort to elevate the MRS also failed. The next day the legislation overwhelmingly passed the Senate on a roll-call vote of 69-9.[25] Although the measure pleased DOE and the nuclear industry, environmental and anti-nuclear activists were distressed about the treatment of the away-from-reactor storage and the MRS provisions.[26]

The House Subcommittee on Energy and the Environment of the Committee on Interior and Insular Affairs began three days of hearings on five proposed bills on June 23, 1981.[27] Among them was Udall's, which ultimately evolved into the Nuclear Waste Policy Act of 1982 (NWPA).[28]

The words of the *House Report* on the legislation captured the mood of many lawmakers engaged in these debates.[29]

> An opiate of confidence that the technical issues affecting nuclear waste disposal were easily resolvable for decades rendered Federal officials responsible for providing the facilities apathetic towards addressing those technical issues, and unprepared for the immense social and political problems which would obstruct implementation of a serious repository development program. "Paper" analyses and future plans were accepted as adequate as was assurance that disposal facilities would be available when needed.

Although the Subcommittee's considerations featured passionate statements refighting the reprocessing wars, the deliberations rightly centered on the five waste-management topics itemized above. In order to move the effort forward in an expeditious fashion, Manuel Lujan, Republican from New Mexico, foretelling the future well, asserted that, "where [the wastes] are probably going to go anyway is Nevada or Washington. I think everybody knows that."[30] Yet, David Berick, Washington Representative of the Environmental Policy Center, cautioned[31]:

> There is going to be a direct tradeoff between the speed with which the Department rushes ahead to pick sites and their ability to defend technically and politically those sites...We can't just simply hope that the solution will be quick and expedient, but that we must develop a framework, including things like formal site guidelines, so that the states and the public know what the rules are before we send the Department out with [drilling] rigs to decide that Louisiana or Nevada or Utah is going to be the site.

Colin Heath, the Director of DOE's Office of Waste Isolation and a central participant in the IRG's undertaking, sought to put Berick's concerns to rest.[32]

> If we were in a position where we were working in a State which had no government and no population, and therefore, we were free to proceed purely on a technical basis. I think purely on technical grounds alone you could [select a site] in 4 to 5 years. However, as you well know, using the State of Louisiana as an example, it is not possible to operate that way.

Kenneth Davis, DOE Deputy Secretary, conceded that the primary purpose of a "test and evaluation facility" would be to obtain information on the technology to handle and emplace the waste. But it would also have another rationale.[33]

> The test and evaluation facility is intended to deal with developing ways to handle canisters of actual commercial waste. From the public acceptance point of view, the actual demonstration of the handling of the canisters of waste from commercial plants would seem to me to be a very important aspect of that.

Others offering testimony discounted the value of obtaining operational data, especially given the costs involved. Moreover, whatever public support might be gained would be compromised by DOE's intention not to license the facility.[34]

The Johnston/Jackson bill from the previous year foreshadowed a lengthy debate in 1981 over whether the country should focus on long-term storage or disposal. South Carolina Governor Richard Riley, who also had chaired the State Planning Council, did not equivocate.[35]

> I am strongly opposed to placing primary emphasis on proposals for temporary or long-term storage of high-level wastes as a part of our waste management program. This is, in effect, the program which has existed at the Savannah River plant which is in South Carolina. It has represented our national policy by default. What is needed now is a new policy.

Hugh Parris representing the Tennessee Valley Authority, owner of multiple nuclear power plants, had a shorter-term outlook.[36]

> I think the Nation needs a plan including some demonstrations of knowing exactly what can be done in a final repository. But I do not think there should be a full-scale effort at this juncture. But I think we ought to have the tests and evaluations accomplished. I think we ought to have confidence to do it at the time it was needed to be done. But we firmly believe that the onsite storage is the way that utilities, at least in our case, should proceed.

Although none of hearing's other witnesses disputed the need for some form of interim facility, only Berick came close to labeling an emphasis on long-term storage as a distraction.[37]

Those testifying grappled with the IRG's proposal for "consultation and concurrence." Was concurrence just a more comfortable and pleasing term than veto? Or did it suggest something less rigid? Representative Butler Derrick, Democrat from South Carolina, spoke to the political realities.[38]

> I think it would be unreasonable to expect an absolute veto on the part of the State, because if you do that, quite frankly I think you would get nowhere because there is no State, as a practical consideration or political consideration, that would voluntarily take this waste.

Riley elaborated.[39]

> Adopting the principle of "concurrence" does not mean that we are granting a veto. I have never supported the right of any State or tribe to have an absolute veto on a decision to site a permanent repository, nor does the State Planning Council. What we have proposed is a framework for intergovernmental relations, which will assure that the Federal repository program can be implemented while guaranteeing the legitimate

> right of State and tribal governments to be involved in decisions of vast importance affecting the future of their citizens. We prefer the recommendation that the whole State or tribe have a right at a few key decision points to have its objections considered by the President and the Congress. States and tribes now have the right of meaningful reaction to Federal Government activities within their jurisdiction. We ask for nothing less than to have that right expressed in law: If the legitimate reaction is to object, with cause, to a proposed repository, that is not a veto. It would be a legitimate disagreement that should be resolved by affirmative action of the President and both Houses of Congress.

In the end, Congress morphed "consultation and concurrence" into "consultation and cooperation."[40]

Most of the witnesses adopted a neutral stance regarding whether waste from the defense complex should be disposed in the same repository as commercial HLW and SNF. Derrick's view was typical.[41]

> I have no specific exclusion in my bill for defense waste. It doesn't deal with it one way or the other. I think the site should be developed with the idea that it will handle commercial waste. However, if, at some later date, it becomes necessary or desirable to bury defense waste in this facility, I don't think that it should be excluded, and I think that the NRC has proper procedures for national security licensing.

On behalf of the Administration, Davis took a strong and contrary position.[42]

> We are concerned that several of the bills fail to distinguish between management of defense and commercial waste. We believe, for national security reasons, that any bill that evolves should be primarily directed to the disposal of civilian waste. Defense programs activities should not be subjected to control by an independent regulatory agency.

On December 2, 1982, the House of Representatives passed the NWPA by a vote of 345-6. Maneuvering by the legislators forestalled the need for a conference committee to resolve the differences between the two chambers. The Senate took up Udall's bill on December 20, 1982 and passed it with amendments by a voice vote. The House accepted the Senate amendments the same day.[43] The law came into effect on January 7, 1983, when Reagan signed it.

Strategic Bargains

Like any legislation, continual give-and-take over ancillary matters, such as reprocessing, the role of storage (in its many varieties), the need for a test and evaluation facility, and disposing defense and commercial waste in the same facility, was a feature in Congress' deliberations on the NWPA.[44] But for the primary interested parties—DOE, the nuclear industry, environmental and anti-nuclear organizations, and potential host states—certain provisions were *sine qua non*. Absent those, passage would have been problematic if not impossible. Four strategic bargains ensured that the parties' conditions were satisfied.

Geographic Equity and Geologic Diversity

The NWPA rightfully focused on the development of the first repository because of the strong pressure to advance the program. That concentration, however, discounted three objections. The nine sites DOE initially investigated were selected without any public participation or involvement; they were concentrated either west of the Mississippi River or close to it while the bulk of the reactors were located in the populated East Coast and Midwest[45]; and none of the locations overlay granitic or crystalline rock formations.[46]

To address those concerns the legislation in Section 112(b) almost surreptitiously forced DOE to augment the sites under consideration by adding two more to the three already nominated for characterization. Although the law did not explicitly state that these would be used to construct a second repository, it did declare that one of the two had to possess granite or crystalline formations. The implicit understanding of those involved was that this criteria would force a second repository into the East Coast or Midwest. To emphasize the importance of the second repository, Congress set a limit of 70,000 metric tons of HLW and SNF that could be emplaced in the first repository until the second was operational.

A Technically Driven Process

Lujan was not alone in accurately predicting the final suite of potential repository sites.[47] Although DOE did not acknowledge it publicly, the first repository would almost certainly be located in one of the areas the agency had studied during the1970s and early 1980s. (See Chapter 3.) Fearing that institutional momentum would drive the choice of a site with less-than-full regard to its technical suitability, potential host states as well as environmental and anti-nuclear organizations fought for provisions in the NWPA that might slow the rush to judgment. They hoped to realize their objective by mandating a structured, stepwise course.

The first element of that procedure established the "rules of the game." These rules enumerated properties that DOE had to evaluate as it initiated the siting process.[48] Among the physical features to be assessed were the location's hydrogeology, seismic activity, and potential to retard aqueous transport. In addition, the NWPA restricted DOE from considering sites either in a "metropolitan statistical area, a county urbanized area, or in a place where both the population was [more] than 2,500 individuals and where the population density [exceeded] 1,000 individuals per square mile."[49] Finally, activities at the site should not interfere with military operations include defense undertakings.[50]

The second element was the authorization of a winnowing process, akin to the one depicted in Figure 3.4. DOE would first consider the suite of nine sites that it had previously investigated. These included places along the Gulf Coast, the Permian Basin, Utah, Hanford, and the Nevada Test Site. One site each in Mississippi, Louisiana, Texas, and Utah were eventually eliminated because they ostensibly duplicated the hydrogeologic properties of those retained.[51] The NWPA mandated that DOE further cull potential sites down to three based on its preparation of Environment Assessments.[52] Finally, the Department had to recommend to the President the selection of a single location for the development of a repository. (The law presumed that DOE would not suggest a site that the President would reject.) The President was then obliged to submit the chosen site to Congress.[53] (These subsequent actions are discussed in detail in Chapter 5.)

State Involvement

The NWPA found a simple resolution for defining "concurrence." It transformed the concept into "cooperation," using the model contained in the Waste Isolation Pilot Plant legislation.[54] By doing so, Congress radically altered the political dynamic; it diluted whatever power and influence states might have had as the repository-development process went forward.

Still emplace was the veto. Once the President acted to choose a site, the affected state or Native American tribe could object and petition Congress to resolve the issue. The dialectic that took place within the House Interior and Insular Affairs Subcommittee made clear that an unconditional state veto was political untenable.[55] Instead, the argument revolved around authorizing a "strong" or a "weak" conditional veto. The former would involve one House of Congress sustaining a state's opposition whereas the latter would require both Houses to override a state objection. Udall's bill contained an amendment by Representative James Broyhill, Republican from North Carolina, that endorsed a two-House override. So controversial was his proposal that it passed with a margin of only six votes.

As a practical matter, however, the clash between the two approaches was more symbolic that material. As one OTA staff memorandum put it[56]:

> There may be little if any substantive difference between the one-House-sustain approach and the two-House-override approach in terms of the relative difficulty or ease of overturning a state objection, *so long* [emphasis added] as the procedures for Congressional consideration of the objection ensure that both Houses would ultimately have to vote on a resolution either sustaining or overriding the objection.

Section 115(a) of the NWPA, in fact, details extensive procedures that force both Houses to vote on a state or tribe's objection.[57]

DOE Obligated to Dispose of Waste Upon Payment of a Fee

During the IRG exercise, a clear consensus had already emerged that the generators of commercial HLW and SNF had to pay the full cost of developing and operating a dedicated repository or a proportionate ("ratable") share of the cost for a facility that also accepted defense waste. (The remaining cost would be picked up by the taxpayers.[58])

As the legislative process evolved, however, the necessity of addressing the funding question grew more evident.[59] The McClure bill proposed the authorization of a one mil/kilowatt-hour of electricity generated (the equivalent of $0.001) disposal fee. It did not mandate that DOE enter into a contractual arrangement with the utilities; nor did it depart from the IRG's view that no fixed date be set for accepting the waste.

In commenting on the bill, the OTA staff observed, "It is important to recognize that the commitment in this bill to a schedule for the technical and regulatory steps involved in taking a first candidate repository site through the licensing process will not necessarily assure that an operating repository will in fact be available at the end of the process." For that reason, that there should be "a firm commitment in law to begin accepting waste from utilities on a *conservatively estimated fixed and certain date* [emphasis added] by which we can have great confidence that at least one licensed repository will be available, even allowing for the almost inevitable delays and problems."[60]

Six months later, McClure amended his proposal to instruct DOE to enter into contracts with the utilities. Among the agreement's provisos was a requirement that DOE take title to the HLW and SNF, transport the material to a repository (or MRS), and dispose of it. Still no firm commitment was made to a particular schedule for either the initial operation of a repository (or MRS) or for DOE's acceptance of the high-activity waste.[61]

OTA's later comments to one of the House Committees' deliberating on waste-management legislation noted that[62]:

> Incorporation of such an explicit commitment would represent a *strengthening* of Federal determination, which would give even greater confidence to utilities and the general public that the job will get done within a reasonable time. In effect, it represents an even firmer demonstration of the Federal commitment to predictable resolution of the waste problem.

Possibly because of this observation, the House Environment and Insular Affairs Subcommittee tightened Udall's 1982 legislation. As passed, it contained essentially the same take-title, sign-contract, disposal-fee stipulations as earlier bills. But it added an acceptance-commitment date of January 31, 1998.[63]

Figure 4.3 provides a timeline showing the DOE's legislatively mandated actions to implement the four strategic bargains. The milestones clearly are greatly

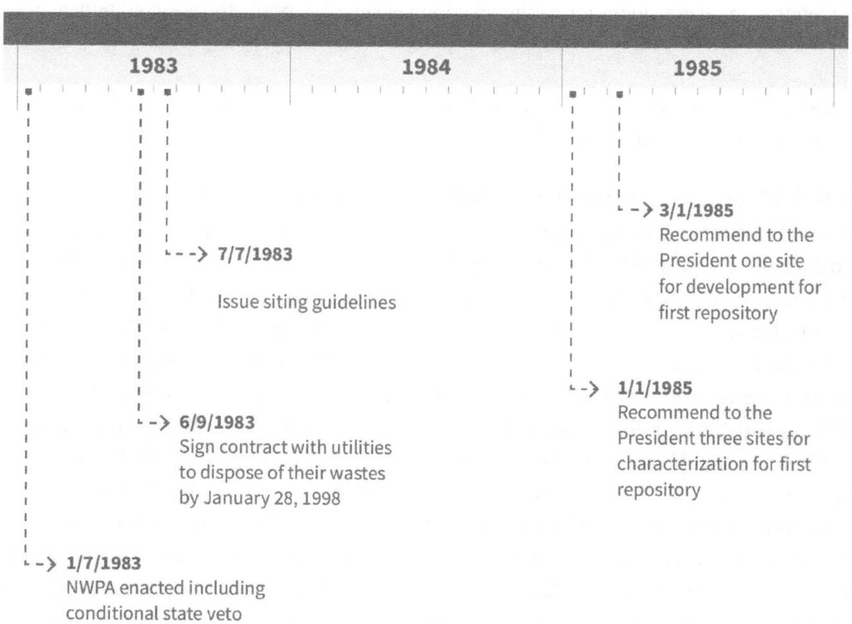

FIGURE 4.3 Nominal timeline for DOE actions under NWPA.

compressed and, even under the best of circumstances, would be organizationally challenging to meet. As our story continues in Chapter 5, we shall see just how demanding were Congress's orders.

NOTES

1. DOE Office of Nuclear Waste Isolation, "Consultation and Concurrence: Workshop Proceedings," ONWI-87, September 23–26, 1979.
2. Carter's Executive Order expired on August 12, 1981. Reagan cancelled it 10 days later.
3. State Planning Council on Radioactive Waste Management, "Memorandum: Resolutions Passed by the State Planning Council," 1981.
4. Nuclear Regulatory Commission, "Disposal of High-Level Radioactive Wastes in Geologic Repositories: Proposed Licensing Procedures," *Federal Register,* (44) December 6, 1979, 70,408–70,421. Nuclear Regulatory Commission, "Disposal of High-Level Radioactive Wastes in Geologic Repositories: Licensing Procedures," *Federal Register,* (48) February 25, 1981, 13,971–13,987. DOE suggested two changes to the final rule: elimination of an NRC requirement for a semi-annual report on the site characterization program and deletion of a requirement for NRC hearings on that project. Neither request was granted. See Letter from Worth Bateman to Chairman John Ahearne, January 6, 1981. My examination of the record indicates that the comments from Governor Richard Riley, the Chairman of the State Planning Council, were among the few taken seriously. See Letter from Richard Riley to John Ahearne, September 17, 1980.
5. The first quote is from *Federal Register* (1981) 13,972. Although this rationale differed from the IRG's reasoning in advocating investigations at multiple sites, NRC noted that its proposed rule was consistent with the IRG's recommendations (*Federal Register* (1980) 70,410). Arguably, in the view of Commissioner Peter Bradford, NRC could have also justified this provision under the 1954 Atomic Energy Act's health and safety mandate but declined to do so. See Memorandum from Peter Bradford to Commissioners Ahearne, Gilinsky, and Hendrie, "Discussion of SECY-80-474: Final Rule, 10 CFR 60," December, 1980, 13–20. The Office of the General Counsel supported reference to the Atomic Energy Act but to no avail. "The use of the Atomic Energy Act would make logical sense to the extent that the safety review of proposed repository sites focuses on maximizing the use of site geology as a barrier to the spread of radionuclides." Memorandum from Leonard Bickwit, Jr. to Commissioners, "Final Rule–10 CFR 60," December 10, 1980, 2.
6. *Federal Register* (1981), 13,972. By *in situ* testing, NRC meant "tests performed from an area at the base of a shaft excavated to the proposed depth of a potential repository." See Transcript, Commission Meeting, May 7, 1981, 21–31.
7. *Federal Register* (1981), 13,973. For a more extensive discussion of the ALARA concept and how it might be applied to repository development, see above and Nuclear Regulatory Commission, "Disposal of High-Level Radioactive Wastes in Geologic Repositories Technical Criteria: Final Rule," *Federal Register,* (48) June 21, 1983, 28,198–28,199.
8. See NRC, "Means for Improving State Participation in the Siting, Licensing, and Development of Federal Nuclear Waste Facilities," NUREG-0539, March 1979, 2. The Commission did, however, urge DOE to involve the states in its siting processes. Early on, NRC engaged DOE informally at high organizational levels on a variety of matters. The Director of the Office of Policy Evaluation cautioned, "In my view, this can give the public the impression that NRC and DOE do not have an adversarial relationship but are rather 'in bed with each other.' Given NRC and DOE's historically close working relationship, their continuing associations during the site review process may taint the present proceedings and cause diminished public confidence in the outcome." Memorandum from Edward J. Hanrahan to Commissioners, "Final Rule–10 CFR

60," January 6, 1981, 2. Perhaps as a result of this communication, meetings between NRC and DOE were opened, a practice that was already in effect for reactor licensing adjudications.

9. NRC, "Technical Criteria for Regulating Geologic Disposal of High-Level Waste—10 CFR 60: Advanced Notice of Rulemaking," *Federal Register*, (45) May 13, 1980, 31,393–31,408. NRC, "Disposal of High-Level Radioactive Waste in Geologic Repositories Technical Criteria: Proposed Rule 10 CFR 60," *Federal Register*, (46) July 8, 1981, 35,280–35,298. NRC, "Disposal of High-Level Radioactive Waste in Geologic Repositories Technical Criteria: Final Rule 10 CFR 60," *Federal Register*, (48) June 21, 1983, 28,194–28,229. EPA, "Environmental Standards for the Management and Disposal of Spent Nuclear Fuel, High-Level and Transuranic Radioactive Wastes: Proposed Rule," *Federal Register*, (47) December 29, 1982, 58,196–58,206. EPA, "Environmental Standards for the Management and Disposal of Spent Nuclear Fuel, High-Level and Transuranic Wastes: 40 CFR 191," *Federal Register*, (50) September 19, 1985, 38,066–38,089. These rules are considered in greater detail in Chapter 10.
10. Transcript, 5.
11. *Federal Register* (1981), 35,289.
12. *Federal Register* (1981), 35,283. Once EPA issued its final rule, NRC had no choice be to adopt that the Agency's reference scenario that specified the number of boreholes and the release consequences that might arise. *Federal Register* (1985), 38,077. *Federal Register* (1983), 28,201, 28,225.
13. 40 *Code of Federal Regulations* [*CFR*] 191, Appendix C.
14. *Federal Register* (1983), 28,196 and 28,224.
15. The "all" was modified in the final rule to "substantially complete."
16. Transcript, 16. Hendrie's skepticism appears in the Transcript, 24–48.
17. These comments are found in Nuclear Regulatory Commission, "Staff Analysis of Public Comments on Proposed Rule 10 CFR Part 60, "Disposal of High-Level Radioactive Wastes in Geologic Repositories," NUREG-0804, December 1983.
18. For members of the general public, EPA eventually limited whole-body radionuclide *releases* to 0.25 millisieverts per year. According to that agency, the risks associated with this overall containment requirement were comparable with "the risks that would have existed if the uranium ore used to create the wastes had not been mined to begin with." Further, EPA concluded that the period over which the implementor had to demonstrate compliance was 10,000 years. In addition to the overall protection standard, EPA issued and NRC later accepted a separate requirement to protect groundwater; it restricted the release to 0.05 millisieverts per year for the first 1,000 years *Federal Register* (1985), 38,066, 38,074.
19. Memorandum from William J. Dircks to Commissioners, "Commission Options on Developing Final Technical Criteria for Disposal of High-Level Wast in Geologic Repositories," SECY-82-427, October 21, 1982. The memorandum suggested three options: Finalize with the exception of the waste package criteria; finalize the rule with the exception of the waste package criteria but request public comments; and finalize the rule with the waste package criteria. The Commission chose the last alternative. As events unfolded, NRC never needed to update its June 1983 rule until it was forced to create "Yucca Mountain-specific" rules, 10 CFR 960, in 2001.
20. See SECY-83-59B, "10 CFR 60—Dispoal of High-Level Radioactive Wastes in Geologic Repositories: Technical Criteria and Conforming Amendments," April 11, 1983.
21. Office of Technology Assessment, "Managing the Nation's Commercial High-Level Radioactive Waste," OTA-O-171, March 1985.
22. Luther Carter, *Nuclear Imperative and Public Trust: Dealing with Radioactive Waste* (Washington DC: Resources for the Future Press, 1987), 195–230, provides the best contemporaneous description of the internecine congressional conflicts.
23. Office of Technology Assessment: Oceans and Environment Program Staff Memorandum, "Summary of Key Issues in Radioactive Waste Management Bills: H.R. 3809, H.R. 5016, H.R. 6598, and S. 1662," September 20, 1982.

24. Even before legislative activities seriously commenced, the Comptroller of the United States asserted that a state could not exercise an absolute veto over the selection of a repository site. See General Accounting Office, "The Nation's Nuclear Waste—Proposal for Organization and Siting," EMD-79-77, June 21, 1977.
25. Several amendments to the bill measured the Senate's policy temperature John Stennis, a Democrat from Mississippi, another potential repository location, failed to require DOE to characterized four sites. Howard Cannon, a Democrat from Nevada, yet another potential repository location, unsuccessfully sought to strengthen to role of states and Native American tribes in the siting of storage facilities. Slade Gorton, a Republican from Washington, convinced his colleagues to limit the disposal of SNF in the first repository to 70,000 MTU. He also succeeded in limiting the amount of SNF both in storage and in the first repository until a second repository began to accept HLW and SNF. Thad Cochran, Republican from Mississippi, won approval for an amendment to require the President to propose legislation to establish an independent entity to manage repositories and storage sites. However, he failed to obtain passage to compel DOE to include a granite formation as a potential repository site.
26. Unbeknownst to his colleagues, McClure had slipped an amendment during the floor debate that would have overridden state laws government nuclear power development and shut down NRC's waste confidence hearing. (See Chapter 3, footnote 61.) Carter (1987), 212.
27. The legislation was also considered by the House Science, Commerce, and Armed Services Committees.
28. *Hearings on Radioactive Waste Legislation*, House of Representatives, Subcommittee on Energy and the Environment, Committee on Interior and Insular Affairs, 97th Congress, 1st Session, Serial Number 97-12, June 23, June 25, July 9, 1981. Nuclear Waste Policy Act (NWPA), Public Law 97-425; 96 Stat. 2201, signed into law on January 7, 1983.
29. *Providing for Repositories...*, House of Representatives, Report 97-491, Part 1, 97th Congress, Second Session, April 27, 1992, 26.
30. Hearings, 201.
31. Hearings, 218–219.
32. Hearings, 283.
33. Hearings, 287.
34. See, for example, Hearings, 195, 202, 223.
35. Hearings, 236.
36. Hearings, 247.
37. Hearings, 217.
38. Hearings, 194.
39. Hearings, 235.
40. NWPA, Section 117(b).
41. Hearings, 196.
42. Hearings, 279.
43. Public Law 97-425, 96 STAT. 2201.
44. See *Providing for Repositories...* for a full discussion of the provisions in the House bill. Briefly, the NWPA authorized a limited amount of interim storage, called upon DOE to present to Congress a plan for a MRS, approved the construction of a test and evaluation facility at an unspecified location, left open the question whether defense and commercial waste should be disposed separately, and adopted a moderately conservative repository-development schedule.
45. Section 112(a) of the guidelines for selecting site states: "Such guidelines shall require the Secretary to consider the various geologic media in which sites for repositories may be located and, to the extent practicable, to recommend sites in different geologic media." See Chapter 5 for an extended discussion of the *Site-Suitability Guidelines.*
46. Paul Witherspoon, an earth scientist from the University of California, Berkeley, made a persuasive case for considering sites in that host rock in testimony before Udall's Subcommittee. See Hearings, 262–274.

47. See, for example, Hearings, 219, 314, 339, and 488.
48. Not surprisingly, these closely followed the "favorable" and "potentially adverse" conditions NRC advanced in its draft technical criteria released 18 months earlier.
49. A strong advocate of this provision was Representative Trent Lott, Republican from Mississippi. He argued, "This is not a parochial question. This square mile area could be a section of the city of Cleveland, Ohio, or a town within the State of Mississippi. My amendment is not intended to make distinctions among regions of the country. What it does is to make a universal statement that the rights of people are primary in deciding where to locate nuclear waste facilities." He also noted, almost in passing, that two of the top three [potential repository sites] in the nation were in his district. *Congressional Record*, November 29, 1982, 27,787.
50. NWPA, Section 112(a).
51. Although the elimination of the only site in Louisiana might be justified on the basis of technical considerations, the influence of Senator Bennett Johnston should not be underestimated.
52. Representative Edward Markey, Democrat from Massachusetts, tried unsuccessfully to convince his colleagues that a full Environment Impact Statement should be required at this stage of the process. "Here is the process that we are now going to engaged in. An Environmental Assessment will be made but the Environmental Assessment is, as we all know, not an Environmental Impact Statement. It is more legalistic; it is more passing of paper between parties. There really is no in-depth scientific analysis as to what the effect of any of these experiments is going to have upon the terrain, upon the environment in a particular State. Most importantly, by reversing what is the present law under NEPA "we take out judicial review with this process so now after, only an Environmental Assessment, after only a perfunctory exchange of 'information between parties, the Federal Government now has the right to come in and drill." Markey's proposal was rejected by a voice vote. *Congressional Record*, December 2, 1982, 28,527.
53. As noted below, the state where the proposed repository is located can object. If Congress sustains the objection, the President has one year to submit another site for approval.
54. NWPA, Section 117(b).
55. This point of view was also emerged in the House floor debate on the passage of the NWPA. Representative Thomas Petri, Democrat Republican from Wisconsin, proposed an amendment during the floor debate to eliminate the provision for congressional override should a state object to the final selection of a site. "In my judgment, the stronger the veto power we give the States, the less likely they will exercise it. Without a strong veto, the public and local officials simply will not trust the Federal Government, and the Department of Energy in particular, not to ram down their throats something potentially injurious to them…Thus an absolute State veto is a positive measure, designed to insure [sic] that the public and local official are really consulted during the entire process of site selection, and that they come to a basic acceptance of an eventual site. Without such local acceptance, any decision will be tied up in court challenges and other obstruction for years." Petri's amendment was defeated in a voice vote. See *Congressional Record*, December 2, 1982, 28,517.
56. See Memorandum from Thomas Cotton to Jack Dooling, "Procedures for Overriding State Objections to Repository Siting," August 26, 1982. See also Letter from Joseph Moakley to Morris Udall, September 21, 1982, which makes the same point.
57. Congress's action on the President's recommendation represents either the passage of the site through the social acceptability filter or its failure to do so.
58. In 1980, Reagan decided that commercial and defense waste would be comingled in a single repository. In 2015, President Barack Obama reversed that decision. See Government Accountability Office, "Nuclear Waste: Benefits and Costs Should Be Better Understood Before DOE Commits to a Separate Repository for Defense Waste," GAO-17-174, January 2017.

59. Much of what follows relies on an essay by OTA's Thomas Cotton, "Chronology of Development of Contractual Commitment in Nuclear Waste Legislation," August 10, 2006. Quotations are from this essay.
60. Testimony of Thomas Cotton before Senate Committee on Energy and Natural Resources, October 6, 1981, 260.
61. Similarly, Udall's 1981 bill contained no fixed date for repository operation, no specification of the fee's size, and no provision for contracts.
62. See undated memorandum for files written by Thomas Cotton of OTA.
63. The contract called for fees to be deposited in a Nuclear Waste Fund. Fees for fuel placed into reactors after July 30, 1983 were to be collected immediately. Utilities had ten years to pay off first-time fees for fuel irradiated before then. Payment of the fee was a prerequisite for holding a NRC reactor operating license. As Cotton's essay asserts, "at this time, this provision was in the repository title of the bill, so it was clear that this subsection referred to disposal in a repository." As this tale notes in Chapter 6, the Federal Court of Appeals for the District of Columbia Circuit held a different view.

5 The Rise of the Yucca Mountain Repository

The leader of the AEC's efforts to dispose of the nation's radioactive material once remarked that "the problem with the waste management program is that it's like the Texan on his wedding night, saying how great it would be but never doing anything about it."[1]

Part of the reason for this inertia was that the waste disposal activity in the AEC and ERDA was plagued by its location buried deep within large bureaucracies, whose priorities were nuclear weapons production, nuclear contamination remediation, and energy research and development of all possible sources. The Nuclear Waste Policy Act (NWPA) sought to remedy that situation by creating the Office of Civilian Radioactive Waste Management (OCRWM).[2] Its Director would have the status of an Assistant Secretary but would report directly to the Secretary of Energy.[3]

Units from the old AEC and ERDA were shuffled into the new Office. Robert Morgan, from the Savannah River Field Office, took over as Acting Director. Amidst the turmoil of trying to do too much with very little, the new organization plunged forward. One of Morgan's first assignments was to begin work on site-suitability guidelines.[4]

SITE-SUITABILITY GUIDELINES

Both the NWPA and the NRC's procedural regulations described, in general terms, the site characteristics that either made a location technically suitable or unsuitable for the construction of a repository. Moreover, the United States, other nations, and international organizations had also undertaken preliminary efforts to delineate features that might be relevant to the suitability of local geologies.[5] What was especially challenging for OCRWM was the obligation to draft *general rules* that might be applicable to the sites conceivably in salt, basalt, tuff (both the saturated and unsaturated layers), clay, and crystalline rock formations.

Guideline Development

On February 7, 1983, OCRWM published its draft guidelines.[6] The purpose of the proposed rule was to provide a basis for the recommendation, characterization, and selection of sites for the disposal of radioactive waste in deep-mined geologic repositories.[7] The broadly stated regulations aimed to protect the health and safety of the public and the environment. In an admirable attempt to counteract the nuclear agencies' reputation for excluding or limiting outside involvement, the Office in March held regional meetings in Chicago, New Orleans, Utah, Washington State,

DOI: 10.1201/9781003528302-6

FIGURE 5.1 Location of nine potential repository sites for the first repository.

and Washington D.C. At roughly the same time, DOE managers met with officials in each of the six states where it was investigating the nine preliminary repository sites for the first repository. (See Figure 5.1.) They connected as well as with governors from the states being considered for the second repository.

Approximately 1,800 comments on the proposed guidelines from other federal agencies, state and local officials, and interested and affected parties overwhelmed the fledgling Office. As one internal memorandum put it: "[A] great many comments indicated that the application of the guidelines in the siting process was neither clear nor specific."[8]

Indeed, no portion of the proposed procedures escaped criticism. The feedback advised OCRWM to rethink its system guidelines, implementation guidelines, and the description of virtually all the favorable, potentially adverse, and disqualifying conditions. An illustrative example and a taste of things to come were the comments from the State of Nevada.[9]

> Nevada believes that the Guidelines are technically, procedurally, and legally flawed. It would be better to take the time now to ensure that the Guidelines are in compliance with the Act and that they can be applied in a meaningful, objective, and understandable manner. This document could go a long ways to begin building public confidence in the ability of government to handle waste disposal safely. Unfortunately, the current version of the Guidelines and the approach to apply them does nothing to build the needed confidence.

Paul Qnirk, president of the consulting firm RE/SPEC, marshaled the DOE community and outside advisors to address the critiques. His first order of business was to organize a workshop in early 1983. Its three-fold purpose was to brief everyone involved, permit project offices to review, ponder, and evaluate comments received to date, and allow opportunities for resolving intradepartmental conflicts. However, he cautioned[10]:

> My overriding concern is that the project managers must fully appreciate the magnitude and significance in the long term of the Guidelines, and at this point the Department and the program cannot afford to tolerate anything less than a unified and intra-supportive position from the top down to the trenches.

By the end of May 1983, the core working team at DOE Headquarters and Field Offices produced and circulated twelve versions of the guidelines. In the hope of a better reception, OCRWM made available for public comment an alternative format.[11] The lead DOE official, Critz George, remarked that "In virtually every case, the comments forthcoming [from the general public] were blatant attempts to doctor the guidelines so as to exclude their states or communities by whatever means could be contrived."[12]

Notwithstanding Qnirk's plea and George's complaint, *intradepartmental* politics continued to intrude on the regulation-writing process.[13] The reason was quite simple: the field offices and associated contractor corps of the sites winnowed down from nine would enjoy decades of generous departmental funding. If any text in the guidelines that might be detrimental to their favored location could be altered or deleted, advantage could be gained.[14]

The Salt Repository Project Office in Chicago, for example, was responsible for overseeing the investigations at the Utah site close to Canyonlands National Park. Its chief, Jeff Neff, sought to remove language that blocked transportation of waste through a national forest.[15] On the one hand, Stephan Whitfield of the Richland Field Office expressed concern about the provision describing a disqualifier that limited groundwater travel time. If adopted, it might have hurt Hanford's chances.[16] On the other, Donald Vieth, Director of the Waste Management Project Office in Nevada, strongly supported the wording on groundwater travel time, believing that water moved very slowly in the unsaturated zone at Yucca Mountain.[17]

Trying to reconcile the comments from inside the organization with the observations elicited from outsiders, OCRWM released yet another revision in August 1983. As NWPA instructed, DOE, on November 22, 1983, forwarded to NRC its latest version and requested the regulator's concurrence. In March of the following year, NRC set forth seven requisites that the Department had to satisfy. In July, the Commissioners unanimously concluded that the Department had met its burden.[18] DOE promulgated the final procedures, 10 CFR 960, on December 6, 1984.[19]

As issued, the Site-Suitability Guidelines adopted the May 1983 alternative format. It consisted of three parts[20]:

- Implementation guidelines covered the need for hydrogeologic and rock type diversity, regionality, and the site-selection process.
- Guidelines after the repository is sealed (postclosure) identified an overall system guideline; it had to be based on a total system analysis to demonstrate that a disposal facility complied with EPA's yet-to-be-issued overall

release parameter. In addition, eight technical guidelines addressed, among other things, geochemistry, erosion, tectonics, climate changes, and human intrusion. In all cases, the guidelines catalogued favorable conditions and potentially adverse conditions. In some cases, the guidelines itemized disqualifying conditions.

- Guidelines before the repository is sealed (preclosure) detailed, for example, requirements dealing with population density, site ownership, meteorology, surface characteristics, socioeconomic impacts, and transportation.

The only major substantive change prompted by the public and NRC's comments was the inclusion of several additional disqualifying conditions. DOE did clarify, however, how the various parts would be weighed: postclosure guidelines would be given substantially greater importance than preclosure standards.[21]

Despite allowances rightly made for the Office's baby legs and the complexity of the tasks that Congress had tendered, its performance two years into the NWPA-era was more than slightly embarrassing. The deadlines shown in Figure 4.3 kept slipping away. In the case of the guidelines, their publication was already 18 months late.

Applying the Guidelines—Draft Environmental Assessments

The development of the Site-Suitability Guidelines was a prerequisite for DOE to advance to the next stage in the repository-selection course: formally sifting the nine potential sites down to the three that the Secretary would nominate to the President as "possibly suitable as candidate sites."[22] A week after OCRWM promulgated the final siting guidelines, it released for public comment nine draft Environmental Assessments to justify its down-selection to potential repository sites.[23] The reports first eliminated formally four sites and then established rankings among the remaining five.[24]

In computing the order, OCRWM applied three simple methodologies—averaging, pair-wise comparison, and utility estimation. DOE then looked only at postclosure standards; subsets of the preclosure parameters; all the preclosure guidelines; and the entire set of rules. Across the three methodologies and the various sets of guidelines, the results were the same. (1) The Lavender Canyon site in Utah, the Swisher Dome in Texas, and the Cypress Dome in Mississippi were eliminated because they duplicated more acceptable adjacent locations. Speculation is probably well-founded that the Vacherie Dome in Louisiana was dropped because of the opposition of Senator Bennett Johnston. (2) The Yucca Mountain, Hanford, and Deaf Smith County were the most favorable, while Richton Dome and Davis Canyon were generally less favorable. These conclusions met with yet another wave of criticism because some locations were clearly targeted while others seemed to have dodged the bullet.[25]

The National Academy's Board of Radioactive Waste Management (NAS) reviewed the common comparative chapter of the five draft Environmental Assessments. It focused sharply on the techniques DOE used to rank the potential sites. In its opinion, of the three methods, averaging and pair-wise comparisons were unsatisfactory. A utility-estimation approach, however, held some promise.[26] That advice triggered slightly over a year of intense action as OCRWM figured out how to respond to the NAS Board while, at the same time, soliciting the panel's further counsel and support.[27]

WINNOWING NINE FIRST-REPOSITORY SITES TO FIVE

DOE's predicament was how to aggregate across the various incommensurate guidelines to derive an ordering of the sites. Two of its initial options, averaging and pair-wise comparison, were unmistakably foreclosed; the Board's not-so-subtle direction to deploy a utility-estimation methodology could not be ignored. Given that constraint, OCRWM set to work during the next three months. By the beginning of July 1985, preliminary calculations indicated that the new approach, at least in principle, could be realized. By the end of the following month, career officials and their contractors prepared to brief the Office's Director, Benard Rusche.[28] Two days afterwards, on August 29, Rusche formally approached the Board once again. "We believe that an independent review of the ranking methodology by an organization such as the NAS Board would be useful in assuring an effective and credible document."[29]

Starting in mid-September, OCRWM madly rushed to re-examine its logic and computations and to transmit them in a draft report for the Board's reconsideration. On October 1, the OCRWM team met with Board members.[30]

Four months after approaching the Board of Radioactive Waste Management, DOE obtained the panel's mixed verdict.[31] Although commending DOE for adopting a "rigorous form" of decision-aiding methodology, the Board cautioned that it did not "consider matters of technical substance" nor did it "examine the specific implementation of the multiattribute utility technique." Thus, it would not go further than complimenting in mostly general terms the appropriateness and value of the methodology.

Beyond that faint praise, the remainder of the four-page letter focused on the pitfalls that could lay ahead as OCRWM sought to apply the technique. It noted that a "scientifically defensible method" demanded a full-blown performance assessment of each of the sites, something that, in OCRWM's view, was clearly premature. The NAS expressed concern that "DOE's use of its own experts…might mask the degree of real uncertainty associated with postclosure issues." A telling observation asserted that the "credibility of the estimates would be substantially enhanced if an independent panel of outside experts were to review the complete analysis" prior to the issuance of the final Environmental Assessments.

Five days later, OCRWM sent out a memorandum setting forth a "management plan" for affecting a formal ranking methodology.[32] Not stated explicitly, but perhaps implied by the communication, thenceforth the Office would prepare two types of narrative: five final Environmental Assessments and a separate report documenting the application of the utility-estimation methodology.[33]

Applying the Guidelines—Final Environmental Assessments

As eventually published in May 1986, the five documents adopted an identical structure.[34] For each location, it laid out the:

- Process for selecting sites for geologic repositories
- Decision process by which the site proposed for nomination was identified
- Site properties
- Expected effects of site characterization activities
- Regional and local effects of locating a repository at the site
- Suitability of the site for site characterization and for development as a repository i.e., the application of the guidelines

The concluding chapter recorded how each of the five sites fared under the multiple provisions of the Site-Suitability Guidelines. Although the final Environmental Assessments avoided any attempt to rank the five remaining sites, Figure 5.2a and Figure 5.2b summarize how OCRWM evaluated the postclosure siting guidelines. Two observations are relevant. Fortuitously, none of the five locations possessed a disqualifying condition. More relevant, all the sites had roughly a half-dozen failed favorable or potential adverse conditions. What should be done to rank them? Here, the role of the utility-estimating technique came to the fore.

Applying the Guidelines—The Multiattribute Utility Analysis (MUA)

OCRWM fully grasped how crucial was the recommendation to the Secretary of which sites to characterize. "In response to numerous comments on the methods used in the draft Environmental Assessments, DOE has undertaken a formal application of a decision-aiding methodology, for the purpose of obtaining a more rigorous evaluation of the nominated sites."[35] The choice of the term "decision-aiding" was itself revealing. It made clear that in the decisive evaluation, *other non-technical* considerations might come into play. As we shall shortly learn, those factors ended up compromising DOE's well-meaning labors to enhance the credibility of its actions.

Internal deliberations on the MUA began seriously in January 1986. They challenged core assumptions and took a "hard look" at the validity of key theories.[36] Among the issues tackled was the overall objective of the analysis. Those involved oscillated between maximizing the likelihood of meeting the EPA standard, minimizing releases, and minimizing health effects.[37] Ultimately, they selected the last option for the postclosure calculations. When it came to the final postclosure ranking, the participants granted that "scoring performance involves technical judgment combined with an appropriate degree of quantitative assessment. The judgment is derived from expert opinion of the group members on all aspects of probable performance of the site for the expected and disruptive scenarios."[38]

The massive MUA is an odd document. On the one hand, it pursued technical exactitude.[39] On the other, its "precision" generated confusion and misinterpretation,

	DC	DS	HN	RD	YM
HYDROGEOLOGY					
Groundwater travel time greater than 10,000 years (FC)[35]	√	√	√	√	√
Hydraulic gradient (FC)	√	√	√	–	√
Nature and rates of long-term processes stable (FC)	√	√	√	√	–
Expected changes will not increase transport (PAC)	√	√	√	√	√
Site has features that can be modeling satisfactory (FC)	√	√	√	√	√
Site has features such as faults and shears (PAC)	√	√	√	√	√
Groundwater sources not fit for crops/humans (PAC)	√	√	√	√	√
GEOCHEMISTRY					
Promote sorption of radionuclides (FC)	√	√	√	√	√
Dissolution of 0.001 less than percent per year (FC)	√	√	√	√	√
Groundwater conditions affect solubility (PAC)	√	√	√	√	√
Geochemical conditions inhibit transport (FC)	√	√	√	√	√
Geochemical conditions not affect isolation (FC)	√	√	√	√	√
Geochemical /retardation process affect isolation (FC)	–	–	√	–	√
Oxidizing conditions (PAC)	√	√	√	√	–
Minerals retard radionuclide transport (FC)	√	√	√	√	√
Geochemical processes degrade rock strength (PAC)	√	√	√	√	√
ROCK CHARACTERISTICS					
High thermal conductivity (FC)	√	√	–	√	√
Phenomena such as fractures affect isolation (PAC)	√	√	–	√	√
Features significantly decrease isolation (PAC)	√	√	√	√	√
Rock features permit conventional mining	√	√	√	√	√
Host rock sufficiently thick to allow flexibility (FC)	√	–	–	√	–
CLIMATE CHANGE					
Surface water not expect to change for 100,000 (FC)	√	√	√	√	√
Climate change over Quaternary Period stable (FC)	×	×	×	×	√
Water table not expected to rise (PAC)	√	√	√	√	√
Climate change raises hydraulic gradient (PAC)					

FIGURE 5.2 **(a, b)** Summary of analysis of postclosure guidelines in the *Environmental Assessments.* *(Continued)*

even as the report conceded that its methodology fundamentally depended on *value judgments* exercised by a small cadre of project participants.[40] Based on separate preclosure and postclosure estimates, the analysis determined that the three salt sites, led by Richton Dome, were superior to either Yucca Mountain or Hanford. By combining the pre- and postclosure stages, however, all the *Sturm und Drang* yielded a surprising conclusion: "The composite overall ranking under a wide range of assumptions is Yucca Mountain, Richton Dome, Deaf Smith, Davis Canyon, and Hanford."[41]

	DC	DS	HN	RD	YM
DISSOLUTION					
Dissolution in Quaternary Period (FC)	√	√	√	√	√
Dissolution degrade isolation (PAC)	√	–	√	–	√
EROSION					
Waste can be placed at least 300m (FC)	√	√	√	√	–
Setting prevents release over 10,000 years (FC)	√	√	√	√	√
Waste not expected exhumed 10^6 years (FC)	√	√	√	√	√
Evidence of extreme erosion Quaternary (PAC)	√	√	√	√	√
Setting does not adversely for 10^6 years (PAC)	√	√	√	√	√
TECTONICS					
Igneous activity/tectonics low over 10^6 years (FC)	√	√	√	√	–
No igneous activity over Quaternary (PAC)	√	√	√	√	√
Historical earthquakes affect isolation (PAC)	√	√	√	√	√
No increase of earthquakes (PAC)	–	√	–	√	–
No increase in earthquake severity (PAC)	√	√	√	√	√
No changes in regional groundwater flow (PAC)	√	√	√	√	√
No potential for tectonic deformation	√	√	√	√	√
NATURAL RESOURCES					
Evidence of subsurface mining (PAC)	√	√	√	√	√
Drilling affects waste isolation (PAC)	√	√	√	√	√
Potential for foreseeable hum activities (PAC)	√	√	–	√	√
No natural resources anticipated (FC)	–	–	–	–	√
Dissolved solids in ground water (FC)	–	–	√	–	√

Legend

√ Passed Guideline × Failed Guideline – Not Applicable

DC = Davis Canyon HN = Hanford YM = Yucca Mt. PAC = Potential Adverse Condition

DS = Deaf Smith RD = Richton Dome FC = Favorable condition

FIGURE 5.2 *(Continued)*

WINNOWING FIVE FIRST-REPOSITORY SITES TO THREE

At this point, the meaning of "decision-aiding" became unmistakably plain. In the same timeframe that the Environmental Assessments and the MUA were released to the public, Secretary of Energy John Herrington recommended to President Ronald Reagan that he approve, in rank order, the Yucca Mountain, Deaf Smith, and Hanford sites for characterization.[42] On May 28th, the President agreed.[43]

The Site-Recommendation

With the publication of the Site-Recommendation Report, DOE had to wrestle with what appeared, on the surface, to be two anomalies. First, Hanford, the poorest postclosure-performing site, was now included. Second, Richton Dome, the best-performing salt site in both the Environmental Assessments and the MUA, was now excluded. Surely, in the minds of many interested and affected parties, something peculiar seemed afoot.

DOE explained Hanford's inclusion using two rationales. The first recalled that the IRG, the NWPA, NRC's regulations, and the Site-Suitability Guidelines all insisted that the three sites nominated for characterization had to sit atop "diverse" hydrogeologies. Conceivably the agency could have claimed that a salt dome, like Richton, was different than a layered-salt site like Deaf Smith or Davis Canyon. (See Figure 3.1.) In that case, both Richton Dome and Deaf Smith might have been included among the three sites designated for characterization. (Technically accurate, this tactic may have been barred by language in the preamble to NRC's rule although not in the regulation itself.[44])

The second line of reasoning was less straightforward. To begin with, DOE held that the overall differences among the sites for postclosure performance were relatively minor—0.23 on a scale of 100. Thus, even the "worst" site, Hanford, met the EPA standard without difficulty. (OCRWM did acknowledge that the projected disparities might be much larger beyond the then-compliance period of 10,000 years.)

In addition, a close inspection of eight preclosure criteria revealed that Richton Dome "would cause the least impacts." However, if transportation and repository costs were excluded from the calculations—"factors that are the least important of all the guidelines"—then the Hanford site would ascend to the top in terms of its preclosure impacts. Justifying this exclusion, DOE cited the Academy Board's thinking,[45]

> The recognition of the heavy dependence on cost reinforces the Board's judgment that the principal usefulness of the multiattribute utility method is to illuminate the factors involved in a decision, rather than to make the decision itself."

Other than by implication, DOE never explained the demotion of the Richton Site. Plausible justifications might include bureaucratic inertia, pressures from the political environment, and local opposition. But it is noteworthy that one analysis employed five different utility-weighting methods, including the one used by OCRWM. In each, the Richton Dome site was superior to the Deaf Smith site.[46]

The criticism of the Department's choices mounted over the coming months. In a letter to DOE Secretary Herrington, Washington's Governor, Booth Gardner, once again called for an "independent" review of the entire endeavor. He warned that without such a reassessment, "the site-selection process is headed for total collapse."[47]

In particular, Gardner cited the views of Detlof von Winterfeldt, a decision analyst who had participated in the NAS Board's evaluation of the MUA: "I believe that the conclusions drawn in the Secretary's report are selective and misleading...Instead I find a convincing analysis that clearly rejects the Hanford site and supports the selection of Richton Dome over the Deaf Smith site."[48]

Two papers written by members of the MUA team only further complicated the matter.[49] The Merkhofer/Keeney paper reported that all five sites "are expected to provide an exceptionally high degree of waste isolation for at least 100,000 years after repository closure." The rankings of the salt sites were tightly clustered and were better than either Yucca Mountain or Hanford. When the analysis combined various weightings for preclosure and postclosure utilities were calculated, Richton Dome was superior to Deaf Smith.

Keeney's own contribution was to shift the focus from site rankings to a portfolio analysis. What counts was which combination of three sites stood the best chance of identifying one that might be successfully developed into a repository. He examined ten combinations. Based on his explicit assumptions, Keeney concluded[50]:

> If it is assumed that three sites must be characterized, it would appear useful to change the sites from those originally recommended by DOE to the portfolio of Yucca Mountain, Richton Dome, and Deaf Smith. This would delete Hanford from the portfolio and add Richton Dome. The portfolio analysis indicates that the value in doing this is between $100 million and $400 million.

But a larger and more unsettling inference might be drawn. Given the opportunity to dice and splice in a multitude of ways the information in the MUA, any decision favored by DOE policymakers could find some quantitative support.

The Aftermath

It would not have been too surprising if many interested and affected parties recalled Shakespeare's words, "full of sound and fury signifying nothing." The absence of an *a priori* implementation protocol as well as an uncertain and mutable technical foundation for choosing sites to characterize allowed a simplistic and convenient, but probably not entirely unwarranted, claim to emerge: once again, politics trumped science.

Although the odds of success were slim, politics became the natural weapon of choice to influence the process.[51]

The most dramatic but short-lived salvo came within weeks of Reagan's decision. For Representatives Jim Weaver, Democrat from Oregon, and Edward Markey, Democrat from Massachusetts, none of OCRWM's rationales rang true.[52] They launched a staff investigation of Herrington's recommendation and, with some flair, announced its results[53]:

- DOE officials systematically deleted passages from the MUA report that were critical of the Hanford site.
- DOE officials systematically deleted passages stating that the Richton Dom site was superior to the Deaf Smith site.
- DOE officials assigned inappropriately low weighting to postclosure safety.
- DOE ignored legal advice that the choice of only one non-salt site would satisfy the legislative and regulatory requirement for geologic diversity.

Reiterating those views, the legislators wrote the Secretary in October 1986[54]:

> The decision to select a permanent waste repository must be based on the soundest scientific and technical judgments possible. Yet we have found conclusive evidence, in many cases supplied by DOE's own internal documents, which leads us to only one possible conclusion: DOE distorted and disregarded its own scientific analyses in order to support selection of the Hanford, Washington site and to avoid selection of the Richton Dome, Mississippi site.

OCRWM's forceful rejoinder came four months later. It is worth quoting at length because it put a political finale—albeit an unsatisfactory one in some minds—to the question of what three sites DOE should characterize to assess their suitability for developing the first repository.[55]

> [Your] statements indicate a belief that the MUA is capable of providing a "scientific" ranking of the five nominated sites—a ranking somehow devoid of *judgment*—which should then be used as the sole basis for selecting three sites for characterization. Without this fundamental premise, there are no logical grounds for criticizing DOE for not selecting the three top-ranked sites identified by the MUA or for inferring that DOE "ignored" the results of the MUA. Indeed, without this premise, there is no incentive for DOE to engage in all of the "manipulation" and "distortions" you believe were undertaken to promote Hanford into the top three sites [emphasis added].

Ultimately, the Weaver/Markey challenge gave way to the legislators' unwillingness to contest an agency's discretion and a fear that, if the siting process were reopened, their states would become the mark. But another political struggle was unfolding simultaneously: what should be the second repository's fate.

SITING THE SECOND REPOSITORY

In the same press release that announced Reagan's decision to characterize three sites, DOE declared that it "has *postponed indefinitely* [emphasis added] site-specific work for a second repository because of the progress in siting the first repository and the uncertainty of when a second repository might be needed."[56] The statement appeared to put an abrupt end to a controversy that had broiled the political landscape for more than three years. But the ramifications of those disputes persisted long afterwards.

As Chapter 4 detailed, attaining geographic equity and geologic diversity surfaced as one of the four strategic bargains that enabled the passage of the NWPA. The siting of the second repository presented an opportunity to fulfill both those goals. The impetus from the start was to investigate sites that overlay crystalline rock formations.[57] These happened to be concentrated along the Eastern Seaboard and in the Upper Midwest, not incidentally where the bulk of the nation's nuclear power plants were located. (See Figure 5.3.)

Beginning slowly in early 1983, OCRWM faced four formidable constraints as it ramped up to launch the second repository project. First, unlike the detailed roadmap contained in the law for siting the first repository, the NWPA was extraordinarily vague about the process to be employed to site the next one. Second, while DOE could deal with potential state jurisdictions on a one-to-one basis for the first repository, it had to deal with as many as 17 state authorities collectively when examining crystalline rock formations. Third, although DOE and its predecessor organizations had explored directly the nine locations under consideration for the first facility, their data cupboards were virtually bare when it came to a new suite of sites. Thus, it had to depend on state geologists for even the most basic information. Fourth, although the Act required DOE to consult with the states and for the states to cooperate with the Department, it was unclear when those activities ought to kick in.

The IRG and the NWPA delivered a robust message to the Department—undertake your responsibilities in a manner that merits public trust and establishes credibility. As

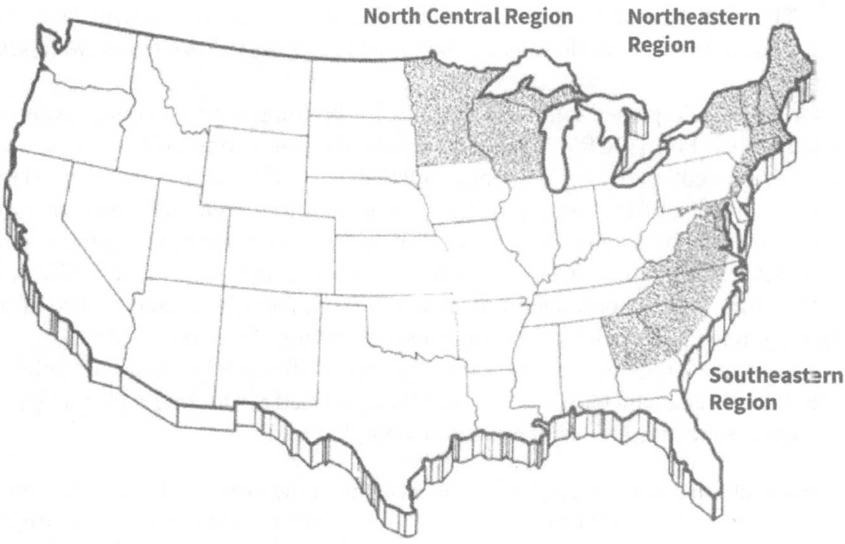

FIGURE 5.3 Regions investigated for the second repository.

OCRWM planned to move forward with the second repository project, it aspired to meet those expectations. Certainly, its design of the process extended many opportunities where states might influence the Office's decisions. (See Figure 5.4.)

OCRWM finalized the Screening Methodology in April 1985. The Regional Geologic Characterization and Environmental Reports identified 235 potential locations that met the requirements of being 1,000 feet below the land surface and having an extent of at least 39 square miles. They were issued in final form in August

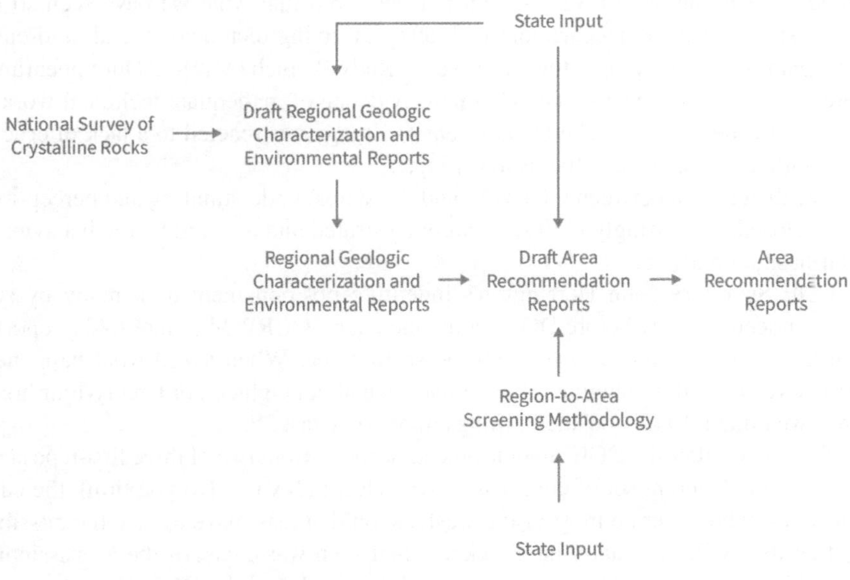

FIGURE 5.4 Siting process for second repository.

1985. The final Region-to-Area Recommendation Reports, released in December 1985, winnowed the possible sites for the second repository down to twelve in seven states.[58]

Not much time passed before aspirations for building trust and confidence confronted reality. From OCRWM's point of view, the states were obstructing the process, either intentionally or by sloppy practice.[59] They failed to appreciate (1) the primary objective of the siting process; (2) the need to screen large portions of the country to identify suitable repository locations; (3) the interim screening was not a final determination that a site might be worthy of selection; (4) that political and social factors were subordinate to technical criteria; and (5) the importance of participating in the process by submitting "unusual excuses" for contributing.

Some state officials saw matters somewhat differently. Marlene Johnson, Minnesota's Lieutenant Governor, for instance, testified before Markey's Energy and Commerce Subcommittee with a litany of complaints.[60]

> First of all, [the process] emphasizes licensability at the expense of safety. Second, the process puts the cart before the horse by beginning a siting process before basic technological questions relating to nuclear waste disposal are answered. Third, it is a mish-mash of arbitrary guidelines, unrealistic schedules, and sloppy research. And, fourth, the selection process is not the open process envisioned by Congress when it enacted the Nuclear Waste Policy Act of 1982. The Federal Government has failed to provide timely replies to requests for information. Its reluctance to share information with affected States and Indian tribes is unconscionable.

Also appearing at Markey's hearing were the governors of five out of the seven states seemingly targeted for hosting the second repository.[61] They recapitulated the same themes. James Martin (NC)—"DOE stacked the deck in the screening process..." Joseph Brennan (ME)—"My message to you is that the process has already failed..." Anthony Earl (WI)—"I am not reassured that what we have seen so far demonstrates that the Department of Energy is giving technical, social, and environmental issues the weight they deserve..." Rudy Perpich (MN)—"Our appearance here is noteworthy because we bring new evidence of inadequate technical work... "Gerald Baliles (VA)—"The Department of Energy has reacted to a lack of crucial data with [a] complacency that is alarming..."

The disjunction between OCRWM and the states' understanding and perceptions never closed; increasingly both sides were frustrated and angered by the behavior of their hesitant partners.

DOE Secretary John Herrington's indefinite postponement took many by surprise. Indeed, the day before DOE's announcement, OCRWM's chief told a reporter that there were no plans to suspend the second site. When asked what happened, Rusche responded, to laughter, that "it has been about eighteen or twenty-four hours since we talked. In this business, things move very fast."[62]

Two weeks after the DOE announcement, Senators from two of three first-repository states had a different set of concerns.[63] Chic Hecht (NV)—"Do you think the eastern states, who generate most of the waste, should at least have to face the possibility that they will eventually have to keep their own waste east of the Mississippi?" Daniel Evans (WA)—"This unfortunate and ill-timed move by DOE does not simply impact the second repository program, it threatens the integrity of the whole high-level waste disposal plan…"

To complicate the situation further, these interactions transpired during the lead-up to elections for Congress in November 1986. Reagan's 1984 landslide victory had produced a three-vote majority in the Republican-controlled Senate. But many of those seats appeared vulnerable, among them was one held by Senator James Broyhill from North Carolina.[64]

Broyhill testified that he had spoken to the Secretary of Energy and the White House, arguing that there was no need for the development of the second repository at this time.[65]

> I think the facts demonstrate [first] that there is no need whatsoever for this second repository and second, above all, there is no need to waste the taxpayers' and ratepayers' money on a second repository program that we may never need and will never need. So that is why I have asked the Department of Energy to suspend all activities on the second repository.

Given all the public criticism of the second repository program, it was hardly surprising that DOE abandoned it.

But there is also solid reason to believe that politics at a much higher level was implicated in the decision. Although he did not testify before the Markey Subcommittee, Governor John Sununu of New Hampshire, a close confidant of Vice President George H.W. Bush, expressed his strong opposition privately.[66] An article in the highly regarded journal *Science* brought some of the behind-the-scenes activity to light.[67] It reported that OCRWM's leaders held 32 meetings or phone conversations with "distraught eastern senators, congressmen, governors, and their staff" during the spring of 1986. For Washington politics, these discussions were again neither startling nor unusual.

Less so, however, was the assertion that one of the opponents to siting the second repository in New Hampshire had conveyed this message to the White House: Unless changes were made, "it would [be] very difficult for any Republican candidate to win the upcoming 1988 presidential primaries. The Bush people saw the handwriting on the wall and saw that something needed to be done." Bush's spokesperson issued a non-denial denial, stating that Bush did not make any recommendations: "The Vice-President's role is not to persuade but to listen and pass along the feelings of people he meets." Yet, so plausible were the reported claims of Bush's intervention in the second-repository decision-making process that they have become entrenched in nuclear-waste lore.

PUTTING THE PIECES BACK TOGETHER

By January 1987, the waste-management program appeared to be in disrepair and disarray. The first-repository states had become progressively more restive. After a short reprieve, the second-repository states again grew nervous when DOE acknowledged that its "indefinite postponement" might, in fact, be illegal. OCRWM struggled to prepare Site Characterization Plans for Yucca Mountain and Hanford. Along the way, their scope swelled so that the costs of studying three sites had ballooned by nearly a factor of twenty in just four years. The Office encountered opposition to its attempt to amend the current Mission Plan.[68] Last but certainly not least, DOE's intention to construct outside Oak Ridge, Tennessee, a Monitored Retrievable Storage (MRS) facility for commercial SNF—a central element in its vision of an integrated waste-management system—garnered only conditional local support but intense resistance from the governor.

For Representative Morris Udall, who had labored mightily to pass the NWPA, the situation was bleak. He was "leaning toward the opinion that the only fair thing to do might be to start all over again…In the coming months, we are all going to have to grapple with this problem and see if we can find a solution that can't be manipulated for expedience or short-sighted political reasons."[69] Udall's notion of "starting over" meant literally starting over. He introduced legislation that would abruptly halt the entire siting and characterization processes and create a Nuclear Waste Policy Commission to chart a new course. Udall told the House of Representatives, "Today, less than five years [after the passage of the NWPA], after billions of dollars of ratepayers' money have been collected and hundreds of millions of dollars spent, the program is in ruins and our goal of siting a repository seems further away than ever."[70]

Senator Bennett Johnston had reached the same conclusion about starting over but with a much different destination in mind.[71] A moratorium, such as the one Udall supported, would grind the program to a halt, from which it might never recover. In Johnston's view, what was essential was an approach that would give the program "the best chance of success."[72] A shrewd tactician, he understood well that the weak and politically isolated could be overpowered to mollify the strong and politically connected.[73] But he also appreciated that if his position was perceived as too heavy-handed, it would kindle resistance from those concerned about procedural fairness.

Johnston's initial proposal contained both a stick and a carrot. DOE would select only a single location to characterize for the first repository. Work on the other two nominated by Reagan would be suspended unless and until disqualifying features were subsequently discovered in a single candidate site. The MRS in Tennessee would be authorized. But the linkage tying its acceptance of an unlimited amount of SNF to progress on development of a repository would be severed. The second repository program would be cancelled.[74] With the money saved by truncating site characterization, a state hosting a repository would receive $50 million upon selection and $100 million annually until closure. A state hosting a storage facility would collect $25 million upon selection and $50 million annually until all the SNF was removed. On the House side, support for Udall's viewpoint was sufficient to block passage of any alternative bill but not enough to authorize the creation of a Nuclear Waste Policy Commission.

Johnston used the mandatory 1987 omnibus budget reconciliation process as the vehicle for advancing the Nuclear Waste Policy Amendments Act (NWPAA).[75] A conference committee considered Johnston's initiative. House conferees rejected his resolution of the MRS dispute but prevailed upon the Speaker to permit a vote on a substitute.[76] This ploy strengthened the House's negotiating power and forced Johnston to accept a tightened linkage between development of a repository and construction of a storage facility. Most importantly, the House conferees, in a bit of a surprise, recommended that DOE investigate only the Yucca Mountain site.[77]

The final legislation passed both chambers without debate three days before Christmas.[78] For the most part, Johnston's fingerprints remained. In addition to addressing the MRS controversy and limiting site characterization to Yucca Mountain, the law preempted arguments that might arise in the future. It eliminated the requirement under the National Environmental Policy Act (NEPA) to consider alternative repository sites and to explore sites in diverse hydrogeologies. To put the second-repository states totally at ease, the legislation barred even the study of crystalline rock formations. The monetary incentives for hosting a repository or a storage facility were cut back substantially and made contingent on a host forgoing its right of disapproval. It prohibited the option of seabed disposal.

The House sponsored two other provisions that would have a lasting impact as our tale continues in the next chapter. The NWPAA created the Office of the Nuclear Waste Negotiator, which was authorized to confer with states and Native American tribes in the hope of finding a willing host for a repository or storage facility. The law also established as an independent agency, the Nuclear Waste Technical Review Board (NWTRB). Composed of 11 members nominated by the NAS for their expertise and appointed by the President, the Board's mission was to "evaluate the scientific and technical validity" of actions taken by DOE to implement the new law.

Of course, where you stood on the NWPAA depended, as they say, on where you sat. No one denied that political pressures confronting OCRWM and Congress. Yet, supporters of the law, such as DOE, the nuclear industry, Texas, Washington, and for all practical purposes, Tennessee, as well as all those possibly hosting a second repository, found the legislation to be both fiscally sensible and technically defensible. After all, the IRG, the U.S. Geological Survey, and the American Physical Society had raised pointed questions about the generic suitability of salt formations to isolate and contain HLW and SNF. Moreover, DOE's Environmental Assessment saddled the Hanford site with several considerable drawbacks. It enjoyed the lowest postclosure performance-ranking in the MUA, faced enormous construction challenges, and, even in the OCRWM's estimation, rested on a problematic groundwater-flow model.

From Nevada's point-of-view, however, the NWPAA will be forever known as the "Screw Nevada" bill.[79] For that state, the passage of the law was an unadulterated exercise of raw political power without any technical merit. At the time, the State of Nevada had a single Member of Congress and a seemingly uninvolved Senator. Arrayed against that impotent representation, the Texas opposition team included Vice-President Bush and the Speaker of the House, Jim Wright, Jr. Washington State had Representative Tom Foley, the Democratic Majority Whip, on its side.

The selection of Yucca Mountain as the nation's sole repository site came at a cost. Two of the four strategic bargains lay in tatters. Geographic equity was discarded. The decision to select sites based on a relatively objective technical comparison of multiple locations was at least compromised if not abandoned altogether.

Yet, in pluralistic democracies, a highly intense minority can often wield its influence over the long term. Further, a skillful and sophisticated state actor can exploit opportunities that arise in what is an inevitably extended and tortuous characterization and licensing process.[80] In the next chapter, our story picks up that theme and explores the fall of proposed the Yucca Mountain repository.

NOTES

1. Interview by the author with Frank Pittman, Director of the AEC's Division of Waste Management and Transportation, January 17, 1976.
2. DOE formally established OCRWM on October 7, 1983.
3. NWPA, Sections 304(a) and Section 304(b). Later on, the OCRWM Director would report either to the Deputy or an Undersecretary of Energy.
4. The definitive, generally reliable, and comprehensive chronology of the waste-management program and especially the Yucca Mountain Project is Michael Voegele and Donald Vieth, *Waste of a Mountain: How Yucca Mountain Was Selected, Studied, and Dumped* (Tonopah NV: Nye County Press, 2016). It is nearly 1,000 pages long across two volumes. Because it contains a much more detailed description of events than does this narrative, the reader would be advised to consult this reference, but regrettably, the tome seems to be out-of-print.
   The authors were two long-time senior managers of the Yucca Mountain Project. The book contains little systematic analyses of why events unfolded and how they were interdependent. Much of what is presented reflects a strong point of view that is reflected in the book's title. My first encounter with Vieth was as a dedicated contributor to the IRG's Alternative Technologies Subgroup Report. Although Voegele was a contractor and an ardent advocate of the Yucca Mountain Repository, my interactions with him over the decades were professional and mostly forthright.
5. See, for example, the following documents. DOE, "Program Objectives, Functional Requirements, and System Performance Criteria," NWTS-33(1), 1982; DOE, "Program Objectives, Functional Requirements, and System Performance Criteria," NWTS-33(2), 1982; U.S. National Research Council-National Academy of Sciences, *Geological Criteria for Repositories for High-Level Radioactive Waste* (Washington D.C.: National Academy of Sciences, 1978); International Atomic Energy Agency, "Site Selection Factors for Repositories of Solid High-Level and Alpha-Bearing Wastes in Geological Formations," Technical Report No. 177, October 1977.
6. DOE, "General Guidelines for Recommendation of Sites for Nuclear Waste Repositories: Proposed Rule," *Federal Register*, (48) February 7, 1983, 5670–5682.
7. NWPA, Section 112(a).
8. DOE, "Record of Responses to Public Comments on Proposed General Guidelines for Recommendation of Sites for Nuclear Waste Repositories," DOE/RW-0001, 1983. See also the cover page of an internal memorandum most likely written by Paul Gnirk, team leader, "Revised Draft Guidelines—(No. 12)," May 26, 1983.
9. Testimony of Robert R. Loux before U.S. House of Representatives Subcommittee on Energy Conservation and Power, February 22, 1984.
10. Memorandum from Paul Qnirk to Carol Hanlon, Critz George, and Thomas Isaacs, "Suggested Purpose and Organizational Features of Project Office Workshops," April 15, 1983.
11. These were never published in the *Federal Register*. The interested reader can find them in Record of Responses, Appendix B.
12. Cited by Voegele and Vieth, 219.
13. The following analysis draws heavily from Daniel Metlay, "From Tin Roof to Torn Wet Blanket: Predicting and Observing Groundwater Movement at a Proposed Nuclear Waste Site," in Daniel Sarewitz et al. (eds), *Prediction: Science, Decision Marking and the Future of Nature* (Washington D.C.: Island Press, 1984).
14. This "conflict-of-interest" would continue for the next few years. See Testimony of Brock Adams, Democratic from Washington, Hearings, Senate Energy and Natural Resources Committee, "S. 839," 100th Congress, 2nd Session, May 7, 1987, 485. "Confidence and

credibility have also suffered from the simple fact that DOE's technical contractors have often had an inherent conflict-of-interest since follow-on contracts depend on finding site suitable."

15. Memorandum from Jeff Neff to B. Hewlett, April 16, 1984.
16. Memorandum from Stephen Winfield to E.S. Burton, March 14, 1984.
17. Memorandum from Donald Vieth to Carol Hanlon, W. Hewlett, and Paul Gnirk, March 10, 1984.
18. Nuclear Regulatory Commission, "Preliminary Decision Related to U.S. Department of Energy's General Guidelines for the Recommendation of Sites for Nuclear Waste Repositories," (49) *Federal Register*, March 14, 1984, 9,650–9,661; Nuclear Regulatory Commission, "Final Decision Related to U.S. Department of Energy's General Guidelines for the Recommendation of Sites for Nuclear Waste Repositories," *Federal Register*, (49) July 10, 1984, 28,130–28,139. See Memorandum from William J. Dircks to Commissioners, "Final Decision on the U.S. Department of Energy's General Guidelines for the Recommendation of Sites for Nuclear Waste Repositories," SECY-84-233, June 11, 1984.
19. Department of Energy, "General Guidelines for the Recommendation of Sites for the Nuclear Waste Repositories: Final Siting Guidelines," *Federal Register*, (49) December 6, 1984, 47,714–47,770. The States of Nevada, Utah, Texas, Minnesota, and Washington challenged the adoption of the Site-Suitability Guidelines. The case was dismissed on the grounds that it was not ripe. See *Nevada v. Herrington*, 777 F.2d, 529 (9th Cir. 1985).
20. *Federal Register*, December 6, 1984, 47,722.
21. 10 Code of Federal Regulations 960.3.
22. NWPA, Section 112(b).
23. As explained in Chapter 4, for this step, the authors of the NWPA made a deliberate choice not to force DOE to prepare full-blown Environmental Impact Statements; instead, it could produce less-demanding Environment Assessments.
24. See DOE/RW-0010 for Davis Canyon, Utah; DOE/RW-0014 for Deaf Smith; DOE/RW-0017 for Hanford; DOE/RW-0011 for Richton Dome; and DOE/RW-0012 for Yucca Mountain.
25. Escape theoretically was only temporary. The discarded sites could be considered as locations for the second repository.
26. Letter from Frank Parker to Benard Rusche, April 26, 1985. Undoubtedly OCRWM had advanced notice of the Board's recommendation.
27. See Letter from Benard Rusche to Peter Myers, August 30, 1985. The Academy's reputation for sound technical studies was strong but not unblemished. See, for example, Phillip Boffey, *The Brain Bank of America* (New York: McGraw-Hill, 1975) as well as Voegele and Vieth, 90–91.
28. By then Qnirk, Tom Longo, and Miley Merkhofer had emerged as the leaders of the effort.
29. Letter from Benard Rusche to Frank Parker, August 29, 1985.
30. Among them were Frank Parker, Susan Wiltshire, William Colglazier, Robert Williams, Chris Whipple, Detlof von Winterfeld, and Kai Lee.
31. Letter from Frank Parker to Benard Rusche, October 10, 1985. A hand-written commentary on the letter of unknown origin or date expressed full agreement with the Board's letter. It even found ways to interpret the criticism graciously!
32. Memorandum from William Purcell to R. Stein, E. Burton, T. Isaacs. W. Hewett, and R. Hilley, "Management Plan for Implementation of Formal Ranking Methodology," October 15, 1985.
33. This bifurcated approach, for the most part, returned the action to OCRWM's staff. It did meet with Board members for an extensive session on March 24, 1986.
34. See the following documents: DOE/RW-0069 for Deaf Smith County, Texas; DOE/RW-0070 for Hanford, Washington; DOE/RW-0071 for Davis Canyon, Utah; DOE/RW-0072 for Richton Dome, Mississippi; and DOE/RW-0073 for Yucca Mountain. These were all released in May 1986.

35. Department of Energy, "A Multi-Attribute Utility Analysis of Sites Nominated the Characterization for the First Radioactive-Waste Repository: A Decision-Aiding Methodology," DOE/RW-0074, May 1986, iii.
36. See, for example, "Minutes of the Nevada Project Office and DOE Headquarters Meeting," January 21, 1986.
37. See Gnirk's handwritten notes of the methodological meeting on January 9, 1986.
38. "Draft Memorandum on Postclosure Meeting: Scoring the Sites," January 29 through February 1, 1986.
39. For instance, see Figure B-2, "Relationships among the factors influencing the numbers of postclosure health effects attributable to the repository," DOE/RW-0074.
40. See Department of Energy, "DOE/OCRWM Management and Their Roles in the Development and Application of the Methodology," RW/DOE-0074, A-12. Given that the Secretary was responsible for making the final decision about three sites, DOE would later assert that it was entirely appropriate to employ project personnel in making these value trade-offs.
41. DOE/RW-0074, 5–16.
42. Department of Energy, "Recommendation by the Secretary of Energy of Candidate Sites for Characterization for the First Radioactive-Waste Repository," DOE/S-0048, May 1, 1986.
43. See Memorandum from John Herrington to Ronald Reagan, May 27, 1986; "DOE News, Statement by John S. Herrington, Secretary of Energy, Nuclear Waste Repository," May 28, 1986.
44. Nuclear Regulatory Commission, "Disposal of High-Level Radioactive Wastes In Geologic Repositories: Licensing Procedures," *Federal Register*, (46) February 25, 1981, 13,972.
45. DOE/S-0048, 5, 6.
46. Katrin Borcherding, et al., "Comparison of Weighting Judgments in Multiattribute Utility Measurement," *Management Science*, (37) December, 1991, 1,603–1,619. See especially Table 11.
47. Letter from Booth Gardner to John Herrington, September 18, 1986.
48. Letter from Detlof von Winterfeldt to Benard Rusche, July 22, 1986.
49. See Miley Merkhofer and Ralph Keeney, "A Multiattribute Utility Analysis of Alternative Sites for the Disposal of Nuclear Waste." *Risk Analysis*, (7) 1987, 173–194 and Ralph L. Keeney, "An Analysis of the Portfolio of Sites to Characterize for Selecting a Nuclear Repository," *Risk Analysis*, (7) 1987, 195–218.
50. Keeney, 217.
51. As a prelude, however, the impacted states sought a remedy from the courts. On the day after the Secretary's decision, the State of Nevada petitioned the Ninth Circuit Court of Appeals to declare the Site-Suitability Guidelines invalid. As matters unfolded at a turtle's pace, Congress rendered that lawsuit moot eighteen months later. The passage of the Nuclear Waste Policy Amendments Act declared Yucca Mountain as the only site that DOE could characterize. *State of Nevada v. Watkins*, 914 F.2d 1545 (9th Cir. 1990). For an excellent commentary of the decision, consult Michael B. Gerrard, "Territoriality, Risk Perception, and Counterproductive Legal Structures: The Case of Waste Facility Siting," *Environmental Law*, (27) 1997, 1,017–1,034. See also the discussion of that legislation later on in this Chapter.
52. Weaver's 4th District was not adjacent to the Columbia River, which flowed through the Hanford site. Markey's involvement, however, was hardly disinterested. Massachusetts was one of the states under consideration for the second repository. Indeed, most of the questions his committee addressed to OCRWM focused on that subject. See the following section of this chapter.
53. Memorandum from Subcommittee staff to Jim Weaver, Chairman, Subcommittee on General Oversight, Northwest Power, and Forest Management, Committee on Interior and Insular Affairs, and Edward Markey, Chairman, Subcommittee on Energy Conservation and Power, Committee on Energy and Commerce, October 20, 1986.

54. Letter from Jim Weaver, Edward Markey, Ron Wyden, and Al Swift to Herrington, October 20, 1986.
55. Letter from Benard Rusche to Edward Markey et al., February 18, 1987.
56. DOE Press Release, May 28, 1986. DOE made the decision notwithstanding explicit language in the NWPA mandating deadlines for siting the second repository. Voegele and Vieth gently label this unlawful act as a "deviation from the law," (273) and "not consistent with the legislation" (274).
57. Crystalline rocks are intrusive (rock forced into another rock bed), igneous, and high-grade metamorphic rocks whose crystals are easily visible with the unaided eye. These rocks include granite and other similar rocks. ERDA had undertaken some cursory crystalline-rock studies in the late 1970s, for example, at the Stripa Mine in Sweden and at the Climax formation at the Nevada Test Site. But until the passage of the NWPA, it had not devoted much time or energy to understanding that hydrogeology.
58. The states chosen were Georgia, Maine, Minnesota, New Hampshire, North Carolina, Virginia, and Wisconsin.
59. Voegele and Vieth, 265.
60. Testimony of Marlene Johnson, Hearings, "DOE Radioactive Waste Repository Program," House Subcommittee on Energy Conservation and Power of the Committee on Energy and Commerce, 99th Congress, 2nd Session, Serial No. 99–102, 21–22.
61. Their testimony can be found in Hearings, "DOE Radioactive Waste Repository Program."
62. Edward Markey, Press Release: "Markey Releases DOE Documents Showing Politics Guided Nuclear Waste Decision," July 31, 1986.
63. Their testimony can be found in Hearing, "Second Waste Repository Site Selection," Senate Subcommittee on Energy Research and Development, Committee on Energy and Natural Resources, 99th Congress Second Session, June 16, 1986.
64. In the 1986 election, the Democrats gained eight seats in the Senate.
65. Hearings, "DOE Radioactive Waste Repository Program, 497–500.
66. Sununu would later lead Bush's campaign for the Republican nomination to succeed Reagan. He also served as Bush's Chief of Staff in the White House.
67. Eliot Marshall, "Nuclear Waste Program Faces Political Burial," *Science*, (233) August 22, 1986, 835–836.
68. According to Rusche, DOE believed it could "legitimatize" the decision to indefinitely postpone the second repository by amending the Office's Mission Plan. That faith proved to be unwarranted.
69. *New York Times*, "Energy Chief Yields to Congress on Waste Storage," February 12, 1987, A24.
70. Introductory Remarks on The Nuclear Waste Policy Commission Act of 1987, July 1, 1987. The bill was H.R. 2888, proposed on that same date.
71. See Hearing, "Nuclear Waste Program, Part 3," Senate Committee on Energy and Natural Resources, 100th Congress, 1st Session, April 29, 1987 and May 7, 1987, and Hearing, "Nuclear Waste Program, Part 4," Senate Committee on Energy and Natural Resources, 100th Congress, First Session, June 29, 1987.
72. Voegele and Vieth, 293.
73. E.W. Colglazier and R.B. Langum, "Policy Conflicts in the Process for Siting Nuclear Waste Repositories," *Annual Review of Energy*, (13), 1988, 347–350.
74. Hearings, "Nuclear Waste Program, Part 4," 4–15.
75. Public Law 100-203, 101 STAT.1330. NRC never took an official position on the NWPAA's merits. An internal staff memorandum, however, seemed to suggest some sympathy toward narrowing characterization of one site. See Memorandum from Victor Stello, Jr. to Commissioners, "Senate Bill 839—Nuclear Waste Policy Act Amendments of 1987," SECY-87-175, July 17, 1987.
76. Colglazier and Langum, 350.

77. OCRWM would likely choose Yucca Mountain if left to its own volition. Perhaps, the family connection between Johnston and SAIC, the Yucca Mountain contractor, would have tipped the decision in that direction.
78. *Congressional Record*, (135), 100th Congress, 1st Session, December 22, 1987, 11,964.
79. Nevada Attorney-General Marta Adams, "Yucca Mountain—Nevada's Perspective," *Idaho Law Review*, (46) 2010, 1–26.
80. As the former OCRWM Acting Director, Lake Barrett, put it on many occasions, "It takes many yeses for the program to move forward but only one no to stop it."

6 The Fall of the Yucca Mountain Repository

The organization theorist Russell Ackoff could have been thinking about the Yucca Mountain Project when he wrote, "Managers are not confronted with problems that are independent of each other, but with dynamic situations that consist of complex systems of changing problems that interact with each other."[1] As our narrative so far described, that effort spawned interdependence and turbulence. If anything, that interdependence and turbulence intensified over the next quarter century.

Yucca Mountain lies about 100 miles northwest of Las Vegas in Nye County. It sits on three federal properties—the Nevada Test and Training Range (formerly Nellis Air Force Range), the Nevada National Security Site (formerly the Nevada Test Site), and terrain controlled by the Bureau of Land Management. (See Figures 6.1 and 6.2.)

The Solitario Canyon fault on the west and the Ghost Dance fault on the east bound the proposed repository site.[2] The rock units at Yucca Mountain consist of successive layers of fine-grained volcanic welded and non-welded tuff. The underground facility is positioned in the unsaturated zone between 200 and 500 meters below the ground surface and approximately 300 meters above the water table. The proposed repository itself rests in the Topopah Spring welded unit.

Annual present-day and projected future precipitation rates range from 130-485 mm/year, 80 percent of which evaporates, transpires through vegetation, or runs off. Thus, only 26-83 mm/year percolates down to the repository horizon.[3] A dozen small, inactive volcanoes lie within 20 km of Yucca Mountain; all of them are southwest of the site in Crater Flat.[4]

VOICES OF DISSENT AND SKEPTICISM

DOE's task to investigate the site encountered many challenges. Some were political. If there were any lingering doubts about the State of Nevada's stance toward the Nuclear Waste Policy Amendments Act (NWPAA), they were unambiguously and immediately clarified. Its legislature passed two resolutions in December 1989. The first held that the risks posed by the repository were so long term and dangerous, they created a palpable hazard to the population. The second asserted that the federal law represented an overreach on the use of its lands, a threat to the state's tourist industry, and a slap in the face to overwhelming public opposition to the repository.

In addition, relying on a previously passed state law prohibiting the construction of a repository, Nevada returned DOE's applications for the air, water, and underground injection permits needed to characterize the site.[5] These clashes foreshadowed Nevada's unrelenting opposition—to some, obstruction—for almost a quarter of a century. Nonetheless, DOE would have to soldier on.

DOI: 10.1201/9781003528302-7

FIGURE 6.1 Yucca Mountain.

Still other challenges stemmed from the underlying structure of the entire process. In July 1990, the National Academy of Sciences (NAS) Board of Radioactive Waste Management published its Rethinking High-Level Radioactive Waste Disposal report. The position statement's key finding was[6]:

> The current approach is poorly matched to the technical task at hand. It assumes that the properties and future behavior of a geological repository can be determined and specified with a very high degree of certainty. In reality, however, the inherent variability of the geological environment will necessitate frequent changes in the specifications, with resultant delays, frustration, and loss of public confidence. The current program is not sufficiently flexible or exploratory to accommodate such changes.

SITE CHARACTERIZATION BEGINS

Under the original Nuclear Waste Policy Act (NWPA), DOE's Office of Civilian Radioactive Waste Management (OCRWM) had to submit a Site Characterization Plan for NRC's approval. After many internal revisions and in response to external comments, the Office did so on December 28, 1988.[7] NRC staff issued its conditional consent seven months later.[8]

OCRWM'S Initial Conceptual Model for Waste Isolation and Containment

OCRWM's initial conceptual model for waste isolation and containment under anticipated scenarios relied on several premises.[9] The Site Characterization Plan would

FIGURE 6.2 Location of Yucca Mountain.

test the validity of those presumptions. If they held, then the repository system—in DOE's view—could be shown to comply with regulatory directives. (See Figure 6.3.)

- The infiltration of groundwater into the unsaturated zone would be very low.
- Any infiltration would percolate very slowly through the rock matrix rather than through fractures.
- Very little water would move past the air gap (disturbed zone) between the rock mass and the engineered barrier system composed of SNF, vitrified HLW, and waste packages.
- The engineered barrier system would be primarily designed to thwart the release to the surface of the gaseous radioactive ^{14}carbon dioxide entrapped in the SNF.
- Only extremely small quantities of radionuclides would be discharged from the engineered barrier system first into the unsaturated zone and then into the saturated zone, where the material would flow to the accessible environment.

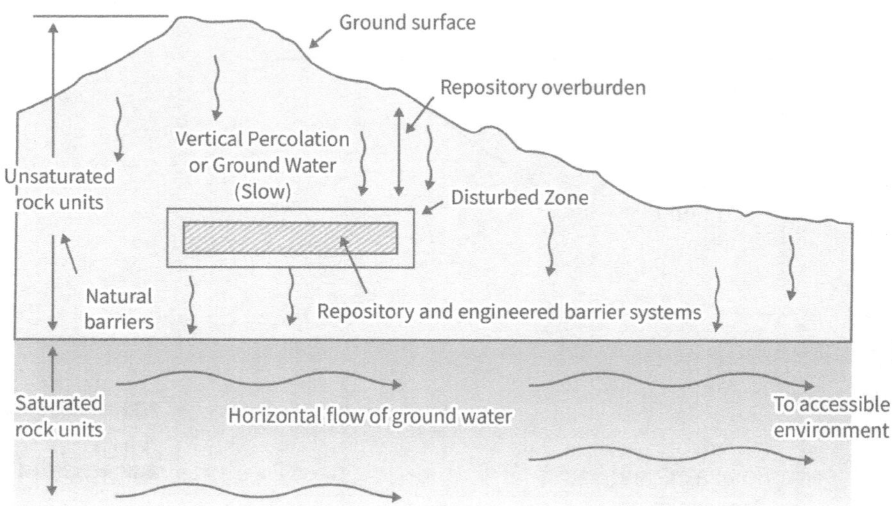

FIGURE 6.3 Initial conceptual model for waste isolation and containment under anticipated scenarios.

Site Investigations Get Off to a Rough Start But Gain Momentum

A call from Nevada's Senator Harry Reid stimulated the publication of an article in the *New York Times*.[10] It brought national attention to an emerging controversy at the Yucca Mountain Project.

During development of the Site Characterization Plan, a report by a DOE staff scientist, Jerry Szymanski, suggested that the groundwater flow in the saturated zone in the geologically recent past had periodically risen well above the level proposed for the repository horizon. He further claimed that such an event could happen again. If the groundwater did in fact well up and flood the facility, DOE's initial conceptual model would fall apart, and Yucca Mountain would have to be disqualified as a disposal site.

Once again, OCRWM called upon the NAS, this time to resolve differences of opinion among members of an external peer review group. The Academy panel's strong overall conclusion was that none of the evidence cited as proof of groundwater upwelling in and around Yucca Mountain could be reasonably attributed to that process.[11] Nonetheless, this episode illustrated how intermittently vulnerable the project had become, even if the critical technical claims were incorrect.

NRC's licensing regulations anticipated that DOE would sink a shaft to access the unsaturated zone and the repository horizon. Bureaucratic inertia kept that alternative stuck in the minds of OCRWM's management. In its first report, however, the newly established Nuclear Waste Technical Review Board (NWTRB) urged the Office to take a different course. Strongly influenced by its Chairman, Don Deere, who was the chief geotechnical consultant for the construction of the "Chunnel" between England and France, the Board counseled: “Consideration should be given to replacing one of the two exploratory shafts with an inclined tunnel (or ramp) driven from the surface into the east side of the Yucca Mountain geologic block.”[12] After weighing thirty-four options, OCRWM selected a design for the Exploratory Studies Facility that was largely consistent with NWTRB’s advice.[13]

Starting in July 1991, two and a half years after the passage of the NWPAA, experiments began.[14] One of them employed more than 100 boreholes to measure present-day infiltration processes and net infiltration rates. Project scientists also elaborated three-dimensional flow-and-transport models for the unsaturated zone. These included a variety of studies to better understand both matrix and fracture flow as well as radionuclide solubility constants.

Perhaps in an effort to show the Project was on track and in an attempt to satisfy the Energy Secretary’s pledge to deliver a timely assessment of the Yucca Mountain site, OCRWM’s contractors prepared the Early Site-Suitability Evaluation.[15] The report walked a fine line between acknowledging the dearth of data and cautious optimism.[16]

> The Core Team concludes that…the current information supports at least lower-level suitability findings for the disqualifying and qualifying conditions for all Postclosure Guidelines. Higher-level suitability findings can be supported for eight out of sixteen conditions.

Although OCRWM never sanctioned this conclusion, it did little to dispel it.

A year later, as internal and external pressures on the program intensified, OCRWM’s management tried a different tack. It prepared a topical report documenting its assessment of what it considered the “easiest” siting guideline: demonstrating that extreme erosion had never occurred during the Quaternary period.[17] Reviewing erosion measures around the globe but concentrating on the Southwest, the Project claimed that the rate at Yucca Mountain was 0.19 cm per thousand years, hardly enough to threaten the repository. (One confidence-building data point was that boulders sitting precariously on the mountain’s edge had remained there for hundreds of thousands of years.)

OCRWM then submitted the topical report to the NAS and the NRC for their evaluation. The National Academy panel considered the statement too narrowly focused. It suggested that OCRWM should address conflicting findings by Nevada’s investigators. It also recommended that different techniques be adopted to add confidence to the project’s findings.[18] The regulatory staff believed that sufficient information was available to support a finding that extreme erosion did not occur at Yucca Mountain. However, the document itself failed to make that case.[19] So, in the end, months of concerted and well-publicized work yielded very little.

By early 1993, the studies to characterize Yucca Mountain took off. Investigations included geologic mapping, tests for perched water, and the Fran Ridge Large Block Test to understand the effect of heat on the rock in the unsaturated zone. Most importantly, excavations started to construct the Exploratory Studies Facility (ESF), which would be used as a launching pad for eventual *in situ* studies at the repository horizon.[20]

OCRWM's POLITICAL ENVIRONMENT GROWS MORE TURBULENT

The fragility of the Yucca Mountain Project, however, continued to manifest itself even as it made incremental technical progress. Significant clashes emerged on three fronts. Congress passed the Energy Policy Act of 1992, which upset the regulatory regime. Owners of nuclear power plants in 1994 sued DOE for failure to fulfill its contractual obligations to accept SNF by January 31, 1998. Finally, Congress cold-bloodedly cut OCRWM's budget.

A Regulatory Upheaval Is Unleashed

A second challenge to DOE's initial conceptual model surfaced when OCRWM and others revisited the issue of how much radioactive ^{14}carbon dioxide in the SNF would, in fact, reach the assessable environment. Counterintuitively, it would be perfectly acceptable to position SNF at the crest of Yucca Mountain and have the gas vent to the atmosphere. However, emplacing the material within the repository and having the SNF then release the gas to the environment through the unsaturated zone might violate EPA's standard.[21]

To address that potentially disqualifying event, Congress passed the comprehensive Energy Policy Act on October 24, 1992.[22] Section 801 of the law charged EPA to write "public health and safety" standards that were *specific* to Yucca Mountain. Those rules were to be "based on and consistent" with the findings and recommendations of an EPA-commissioned NAS report, which was completed in August 1994.

For our purposes, the Technical Bases for Yucca Mountain Standards (TYMS) report endorsed three fundamental modifications to EPA's standards.[23] First, the new standard should "set a limit on the risk to individuals of adverse health effects from releases from the repository."[24] Further, the panel also considered who would be protected. The report recommended that the *critical group* concept be applied here. That concept points to a small group of persons at the highest risk as defined by cautious but reasonable assumptions and present knowledge.

Second, the compliance with the standard should be evaluated at the time of peak dose rather than at 10,000 years. What was troublesome to the panel was that a draft performance assessment provided by DOE indicated that the dose would start rising at 12,000 years.[25] For Yucca Mountain, whose hydrogeologic environment displayed very long-term stability, the peak dose might well occur one million years after repository closure.[26]

Third, the committee rejected the convoluted methodology for evaluating the behavior of the total system.[27] Having concluded that it was feasible, the group instead recommended that DOE undertake a quantitative, probabilistic performance assessment. EPA was generally comfortable with this methodology. The Commission, which typically adopted a classical defense-in-depth philosophy to license reactors, was less convinced.[28]

By compelling DOE to adopt a performance assessment methodology, EPA could gauge whether the individual health effects at the time of projected peak dose met the new standard. Moreover, in the panel's view, the adoption of a performance assessment methodology would obviate the need for the three controversial subsystem requirements.[29] Indeed, as one panel member remembered, "I told an NRC official that 'those constraints were about as useful as a rule to paint the waste canisters blue.'"[30]

Nuclear Utilities Sue DOE For Breach of Contract

The sluggish pace of the investigatory process, coupled with DOE's decision to develop a "multipurpose canister-based system,"[31] forced the Department in April 1995 to recognize that the contractual date of January 31, 1998 for waste acceptance from utilities would be missed.[32] Nonetheless, "DOE has concluded that it does not have an unconditional statutory or contractual obligation to accept HLW and SNF… in the absence of a repository or interim storage facility constructed under NWPA." DOE acknowledged that the legislation did not define "dispose" but held that its meaning should be inferred from "disposal," which referred to "emplacement in a repository."

In response, 27 states and 14 nuclear utilities filed suit in the Court of Appeals for the District of Columbia Circuit asking for damages for DOE's failure to live up to the contract's terms.[33] The Court chose to reject DOE's inference. Instead, it concluded that "dispose" should take on the "common" meaning of the term, which made no reference to the availability of a repository. Thereby, the fourth strategic bargains in the NWPA started to collapse.

Congress Forces OCRWM to Adopt a New Program Approach

The most visible sign of the waste-management program's precarious state was that Congress slashed its Fiscal Year 1996 budget request by 37 percent. This represented an appropriation's cut of nearly 25 percent from the prior year.

The precipitous funding drop forced OCRWM to fundamentally restructure its tasks. Transportation activities, such as route planning, rolling stock design, and interactions with interested and affected parties, were mostly eliminated Also terminated was the multipurpose canister development effort.

Research to determine the suitability of the Yucca Mountain site did continue largely unabated. In particular, design work on the critical elements of the repository system and the waste package progressed. Further, implementation of the Total System Performance Assessment (TSPA) methodology accelerated.[34] All these undertakings would culminate in the preparation of a Viability Assessment to be released by the end of 1998.

FROM VIABILITY ASSESSMENT TO SITE RECOMMENDATION

The period between 1996 and 2001 saw OCRWM recover its equilibrium and, for the most part, tame the challenges of its turbulent political environment. During that time, the Office completed excavation of the 25-kilometer U-shaped tunnel for the ESF, which ran north-south parallel to the Mountain's ridge at a depth of roughly 350 meters. In addition, after considerable controversy with the NWTRB, OCRWM constructed an east-west-running smaller tunnel through the geologic unit that would actually hold the repository.[35] Without much affection and with more than a touch of sarcasm, project personnel would refer to that passageway as the "Board's tunnel."

Is the Yucca Mountain Project Sustainable?

Congressional frustration with the program was not only reflected in the budget cuts. Starting in 1995, the House and the Senate considered legislation that de-emphasized the repository project and called for DOE to construct a large SNF storage facility at Yucca Mountain. In 1997, President William Clinton's Administration let it be known that he would veto any bill that might pass. Nonetheless, advocates feared that waste disposal would no longer be the nation's priority. In a very real sense, the program's hopes rested heavily on the reception the Viability Assessment would garner.[36]

In carrying out that analysis, OCRWM revised in important respects the preliminary conceptual model for waste isolation and containment. Whereas the Office asserted earlier that performance depended on slow percolation through the matrix in the unsaturated zone, it now recognized that water could infiltrate much faster through fractures to the repository horizon.[37] In such a case, rather than being just an insubstantial impediment to radionuclide release, the engineered components needed to be transformed into a robust barrier.

To do so, the repository's man-made elements were radically altered to include nearly four-inch-thick layered steel and nickel-alloy corrosion-resistant waste packages, a concrete invert and pier to retard the transport of any radionuclide releases, and a gantry to support the package to avoid exposure to pooled water. The new conceptual model featured the following claims[38]:

- Limited water would contact the waste packages.
- The waste packages would last for a long time.
- Breached waste packages would release radionuclides at a very low rate.
- Concentration of radionuclides would be reduced as they were transported away from the waste packages through the unsaturated zone and then the saturated zone.

The project carried out two total system performance assessments (TSPA) of the anticipated scenario, one for a deterministic base case and one for a probabilistic base case. A presentation to the NWTRB underscored that the TSPA-VA could not be used to assess compliance with a regulatory standard, show defense-in-depth of the repository, or determine system suitability. Despite those caveats the TSPA-VA maintained that the repository system would comfortably meet the current requirement that individuals be exposed to a radiation dose of less than 0.15 millisieverts/year for 10,000 years.

OCRWM, however, also provided information about a variety of scenarios that might arise afterward.[39] Figure 6.4 illustrates one of those cases: how the repository system might perform at 1,000,000 years, a compliance period more than hinted at in the Academy's TYMS report. At that point, the revised conceptual model begins to broadly break down. Percolating water would have corroded roughly 30 percent of the waste packages. It would turn the SNF contained therein into rubble. And it

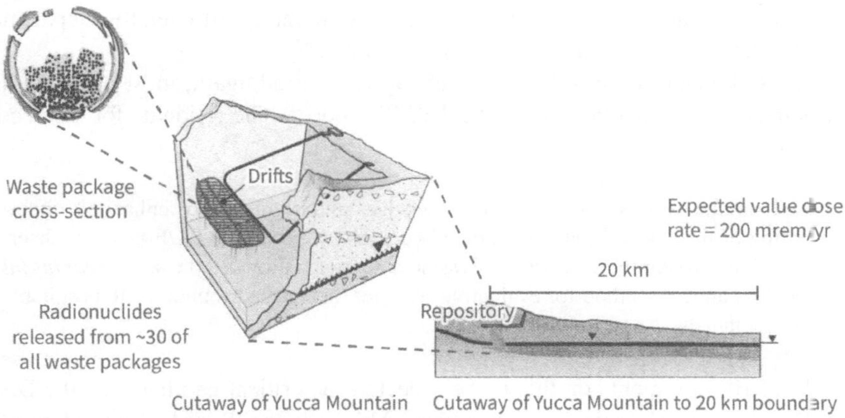

FIGURE 6.4 Expected dose rate at the accessible environment at one million years.

would transport the radionuclides outside the engineered system into the unsaturated zone and subsequently into saturated zone, where the waste material would be carried into the accessible environment to be ingested by a critical group of people.[40]

If OCRWM was looking for external support to revive and sustain the repository program, the NWTRB delivered it (mostly) in the form of an April 1999 report, Moving Beyond the Yucca Mountain Viability Assessment.[41] "Many parts of the VA present cutting-edge scientific analyses in a comprehensible format…The Board concludes that Yucca Mountain continues to merit study as a candidate site for a permanent geologic repository…"

At the same time, the Board reminded DOE of an unsettled technical concern it had raised in its First Report, nearly a decade previously.

> In particular, the VA highlighted the close connections between the repository design and the priority list of key uncertainties about the natural system. For example, such site characteristics as the movement of water in liquid and vapor forms at temperatures above boiling and the effects of high temperatures on rock stability are important *only because* [emphasis added] of the VA's high-temperature repository design. In a low-temperature design, uncertainties about these phenomena would be less significant and might not need resolving before making a suitability determination.

Furthermore, such a design "could increase confidence in the analyses of repository performance required for a site-suitability determination."

Over the coming years, OCRWM attempted, but never quite succeeded, to address the Board's concerns. That failure took on added significance as DOE moved to recommend the Yucca Mountain site for development as a repository.

Changing the Rules of the Game

In the spring of 1994, DOE began contemplating whether its site-suitability guidelines, 10 CFR 960, should be revised in light of the passage of the Nuclear Waste Amendments Act (NWPAA) of 1987. That law, as the reader will recall, eliminated the comparative assessment of multiple sites. A year and a half later, the Department formally determined that no change was warranted.[42]

Yet, by the end of 1996, DOE had changed its mind again, most probably as a consequence of the publication of the TYMS report.[43] The rationale for the second shift was:

> Based on the DOE's accumulated knowledge, and significantly enhanced understanding of the Yucca Mountain site and geologic disposal, the DOE has now determined that a system-performance-assessment approach provides the *most meaningful* [emphasis added] method for evaluating whether the Yucca Mountain site is suitable for development as a repository.

NWTRB's comments on this move reflected the critical gap between the Board and DOE's methodological philosophies. Although the Board observed that by "linking suitability directly and unambiguously to system performance, the proposed revised guidelines seem to be a sound approach." It also held that "uncertainties will be large, and they will become larger as the time horizon for performance projections reaches further into the future." Thus, it expressed doubt whether "relying solely on performance assessment to demonstrate repository safety will ever be possible." Instead, NWTRB recommended that DOE augment its use of TSPA by, in effect, creating a "safety case." Such an account, which would rely on multiple, independent lines of evidence, that explicitly makes apparent the engineering concepts of redundancy and "defense-in-depth."[44]

It must be noted that among the four nuclear agencies involved with the repository development process, the Board was a conspicuous, but persistent, outlier. EPA, NRC, and DOE would soon maintain that it was not necessary to supplement TSPA with other independent analyses, not only as NWTRB advised, but as other nations routinely did.

Environmental Protection Standard and Licensing Rules

The passage of the Energy Policy Act of 1992 and the publication of the TYMS report compelled EPA and NRC to write Yucca Mountain-specific regulations.[45] For our purposes, they touched on five principal areas: the dose limit, the compliance period, who should be protected, the level of confidence in the assessment of protection, and the compliance methodology.[46] As with the regulation-writing effort in the early and mid-1980s, NRC went first.[47] EPA followed. Once its standard was final, NRC conformed its proposal to recognize EPA's ultimate authority.

The first question was what should be the dose limit that DOE had to meet. NRC proposed a whole-body dose limit of 0.25 millisieverts/year. The constraint was outside the range advanced by the NAS study—0.02-0.20 millisieverts/year. In contrast, EPA set forth a 0.15 millisieverts/year standard. The two regulators went back and forth for more than two years before the nuclear one conceded that it had to defer to the environmental one. "The EPA has established a dose limit for individual protection…The NRC is incorporating this limit as required by law."[48]

Departing from TYMS analysis, the Agency and the Commission proposed the same compliance period stipulated in its generic rule: 10,000 years rather than the peak dose during the period of geologic stability, which at Yucca Mountain was roughly one million years.

As a threshold question, EPA asked, "whether it was bound to follow the NAS's determinations without exception or whether it has discretionary decision-making authority." Relying on its interpretation of the direction set forth by Congress, it concluded it was not.[49]

> The Conferees [did] not intend for the National Academy of Sciences, in making its recommendations, to establish specific standards for protection of the public but rather to provide expert scientific guidance on the issues involved in establishing those standard…The provisions of Section 801 are not intended to limit the Administrator's discretion in the exercise of his authority related to public health and safety issues.

"In summary, we do not believe we must, in this rulemaking, adopt all of the positions advanced by the NAS."[50]

Having answered the threshold question, EPA felt it could disregard the advice it was given. The Agency asserted that "*despite NAS's recommendation* [emphasis added], we conclude that there is still considerable uncertainty as to whether current modeling capability allows development of computer models that will provide sufficiently meaningful and reliable projections over a time from up to tens of thousands to hundreds of thousands of years."[51]

EPA defended its deviation from the NAS's advice on three grounds.[52] First, the hazard posed by HLW and SNF decreases rapidly and significantly during the initial 10,000 years.[53] Thus, that original compliance period corresponded to the time when the waste is inherently most dangerous. Second, the compliance period is sufficiently long, such that a wide range of conditions will challenge the natural and engineered barriers, providing a reasonable evaluation of the robustness of the geologic repository. Third, NRC maintained that the compliance period is consistent with other regulations involving geologic disposition of long-lived hazardous materials, including radionuclides disposed at the Waste Isolation Pilot Plant.

On the issue of who should be protected, the two regulators offered different—and arcane—provisions. EPA recommended using the "reasonably maximally exposed individual" and NRC adopted the NAS position of the "critical group." As with the dose limit, NRC by statute had to accept EPA's determination.

The two regulatory agencies answered the fourth question somewhat differently, which sparked another prolonged debate. NRC held fast to its criterion of "reasonable assurance," while EPA championed "reasonable expectation." Ultimately it proved to be a difference without a distinction when NRC conceded EPA's point-of-view.

The most fundamental change in the generic standards and rules came with the adoption by both regulators of TSPA as the compliance-determining methodology for the Yucca Mountain-specific directives. NRC, in particular, explained in its proposed rules why it was abandoning the deterministic engineering concepts of redundancy and defense-in-depth. In its view, those notions might well be appropriate when probabilistic quantitative techniques and analyses were unavailable or unreliable. Given both the regulators and implementor's experience over the previous 15 years and taking into account the findings and recommendation of the TYMS report, NRC concluded that performance assessment should be sole quantitative measure used to demonstrate compliance with the postclosure dose limit.[54]

The Commission, however, did not discard entirely the notion of multiple barriers. It required DOE to evaluate the behavior of structures important to waste isolation and containment. The Department would have to identify which engineered and natural features were, in fact, barriers, describe the capabilities of those features, and provide a technical basis for its claim of capability. Nonetheless, it would be perfectly acceptable if DOE showed that the natural system provided 99 percent of the capability and the engineered system only one percent, and *vice versa*.

Department of Energy Site-Suitability Guidelines

Once EPA and NRC settled upon their Yucca Mountain-specific standards and rules, DOE needed to adjust its site-suitability regulations accordingly. In late November 1999, the Department proposed to amend 10 CFR 960, its generic site-suitability guidelines.[55] It decided to align its rule with the EPA's proposed standard that had been released three months earlier.

Barred by its lack of authority, NWTRB could not reasonably comment on the regulators' proposed rules. But it took the opportunity to make observations that could just as easily pertain to EPA and NRC's anticipated regulations. Of special note, the Board reminded the Department of its previous comment that it doubted that "relying solely on [TSPA] to demonstrate repository safety will ever be possible." Instead, the Board recommended that additional [independent] lines of evidence, such as natural analogs, be used to overcome those limitations. Although the Department acknowledged that it would demonstrate waste isolation by a number of approaches, its proposed guidelines focused only on performance assessment, "leaving unclear how these additional approaches will in fact be used in the context of site-suitability."[56]

DOE finalized its Yucca-Mountain specific rule in November 2001.[57] For all practical purposes, DOE did not alter its previous determinations. What is striking, however, is the way that the Department addressed the NWTRB's concerns. It began by stating that it "agrees with much of the NWTRB's comments and recommendations." It believed that the Board's interest in additional lines of evidence would be accommodated by requirements for conducting the performance assessment. With this sleight-of-hand, DOE rejected the most critical of the Board's recommendations.[58]

The Board immediately recognized and understood that rebuff. It authorized a staff member to meet with OCRWM managers and General Counsel to attempt to remediate the situation. That conference proved unavailing. So, the Board seriously contemplated writing a follow-up letter expressing its strong disagreement with what DOE had put forth. Eventually, NWTRB decided not to pursue the matter further. But it would keep its powder dry.

The Secretary of Energy Recommends the Yucca Mountain Site for Development

Once the three regulations were finalized, DOE prepared to assemble documents that would support the Energy Secretary's site-suitability determination for the Nevada location. In August 2001, DOE issued its Yucca Mountain Preliminary

Site-Suitability Evaluation report.[59] In February 2002, Energy Secretary Spencer Abraham affirmed that "after careful evaluation, I am convinced that the product of over 20 years, millions of hours, and four billion dollars of this research provides a sound scientific basis for concluding that the [Yucca Mountain] site can perform safely during both the pre- and postclosure periods, and that it is indeed scientifically and technically suitable for development as a repository."[60] By mid-month, President George H. Bush transmitted the recommendation to Congress.[61] On April 8, 2002, the Governor of the State of Nevada Kenny Guinn submitted to the Speaker of the House of Representatives a notice of disapproval and a statement of reasons why he disapproved the recommended repository site.

The House of Representatives voted 306-217 to reject Nevada's objection. The action then moved to the Senate, where it placed the Democratic Majority Whip, Harry Reid, in a bind. His entire career had been based on opposition to Yucca Mountain. Notwithstanding the power of Reid's position and the support it received from the Majority Leader Thomas Daschle, the body voted 60-39 to reject the objection.[62] Bush signed a Joint Resolution of Congress formally approving the Yucca Mountain repository site on July 23, 2002.[63] By doing so, it became clear that the third of the four strategic bargains underlying the NWPA—a meaningful voice for the host state—was a mere fig-leaf.[64]

OCRWM's POLITICAL ENVIRONMENT BECOMES TURBULENT AGAIN

As 2002 began, DOE's political environment once again turned chaotic. NWTRB widely distributed a letter widely informing the Department of serious flaws in its latest performance assessment. The State of Nevada and the Natural Resources Defense Council (NRDC) sued EPA, NRC, and DOE to overturn their Yucca Mountain-specific regulations. Together those actions would delay the submission of a license application for several years.

NWTRB Expresses "Weak to Moderate" Confidence in DOE's Performance Estimates

NWTRB used the *Preliminary Site-Suitability Evaluation* to begin an intensive analysis of the status of OCRWM's Project. In a January 24, 2002, letter to the Speaker of the House, the President *Pro Tempore* of the Senate, and the Secretary of Energy, the Board returned to many of the themes it had raised in the past. Because of the disruptive effect the letter had, quoting its main points in full is warranted.[65]

- In evaluating the DOE's technical and scientific work related to individual natural and engineered components of the proposed repository system, the Board finds varying degrees of strength and weakness. Such variability is not surprising, given that the Yucca Mountain Project is in many respects a first-of-a-kind, complex undertaking. When the DOE's technical and scientific work is taken as a whole, the Board's view is that the technical basis for the DOE's repository performance estimates is weak to moderate at this time.

- The DOE uses a complex integrated performance assessment model to project repository system performance. Performance assessment is a useful tool because it assesses how well the repository system as a whole, not just the site or the engineered components, might perform. However, gaps in data and basic understanding cause important uncertainties in the concepts and assumptions on which the DOE's performance estimates are now based. Because of these uncertainties, the Board has limited confidence in current performance estimates generated by the DOE's performance assessment model.
- Confidence in waste package and repository performance potentially could increase if the DOE adopts a low-temperature repository design.
- This limited confidence is not an assessment of the Board's level of confidence in the Yucca Mountain site. At this point, no individual technical or scientific factor has been identified that would automatically eliminate Yucca Mountain from consideration as the site of a permanent repository. DOE can increase confidence in its performance estimates by, among other things, developing multiple lines of evidence and strengthening its arguments about defense-in-depth.

The Court of Appeals Overturns EPA's 10,000-Year Compliance Period

On July 9, 2004, the Court of Appeals for the District of Columbia Circuit rejected most of NRDC and Nevada's claims but agreed that the compliance period was unlawful.

The Court's analysis began with a discussion of the *Chevron* ruling.[66] It established that, notwithstanding EPA's interpretation of the legislative history behind the Energy Policy Act, Congress did not unambiguously speak to the length of the compliance period. Thus, while EPA was legally entitled to use its discretion, there were strong constraints on what it might do.[67]

> It would have been one thing had EPA taken the Academy's recommendations into account and then tailored a standard that accommodated the agency's policy concerns. But that is not what EPA did. Instead, it unabashedly rejected NAS's findings, and then went on to promulgate a dramatically different standard, one that the Academy had expressly rejected. Although Section 801's "based upon and consistent with" standard does not require EPA to walk in lock-step with the Academy, we think it entirely unreasonable for EPA to have acted *inconsistently* with NAS's findings and recommendations.

Thus, by acceding to both plaintiff's challenge to the compliance period, the Court sent EPA, NRC, and DOE back to the drawing boards, where they would stay for the next four years.

PREPARING THE GROUND TO SUBMIT A LICENSE APPLICATION

In addition to its ongoing investigations in Nevada, two other activities preoccupied OCRWM between 2002 and 2008. First, it devoted substantial time and effort attempting to address NWTRB's criticisms that led to the "weak to moderate" evaluation. Second, it engaged EPA and NRC in an attempt to reconstruct the Yucca Mountain-specific standards and rules.

Interactions with the Nuclear Waste Technical Review Board

NWTRB's concerns with a high-temperature design were two-fold.[68] First, the strategy introduced greater uncertainties into the performance assessment. Initially, the heat would drive the humid air within the repository drifts into the rock that lay above. As the repository cooled, the water would re-enter the excavated openings. Modeling such phenomenon would be more difficult than for a low-temperature design, and thus the TSPA would be less definitive.

OCRWM did, in fact, make several changes to its high-temperature design. For instance, it expanded the distance between the emplacement drifts to increase the likelihood that the returning water would flow through the pillars rather than seep into the drift. Further, it argued that, because of the effectiveness of capillary and vaporization barriers, the Board was mistaken in its concern about significant seepage as the temperature waned.[69] (See Figure 6.5.)

The Board's second technical argument was that deliquescence-induced corrosion caused by calcium chloride would corrode the surface of the Alloy-22 waste package. OCRWM maintained that two chemical conditions had to be satisfied for calcium chloride brines to be formed. In its view, neither condition held. The Department's position in this matter received strong support from the NRC and the industry's Edison Electric Institute. Faced with these counterarguments, NWTRB subsequently retreated, although it never explicitly acknowledged its error.

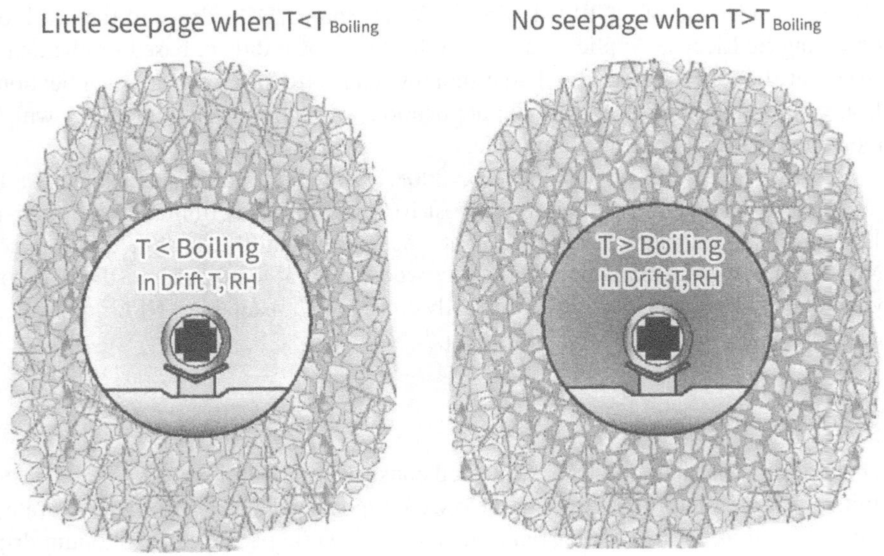

FIGURE 6.5 Seepage into the emplacement drifts.

Rewriting EPA, NRC, and DOE Rules Yet Again

The opinion of the D.C. Court of Appeals vacating EPA's individual-protection-10,000-year compliance period forced the Agency to rethink how it would set a compliance period at the time of peak dose, roughly one million years for Yucca Mountain. In August 2005, it decided to propose a bifurcated rule. For the first 10,000 years, it retained the 0.15 millisieverts/year dose limit specified in its 2001 rule. But for the next 990,000 years it proposed a 3.50 millisieverts/year constraint. For the latter, it did offer alternatives of 0.15, 1.00, and 2.00 millisieverts/year. Although the TYMS report called for using the mean of the TSPA calculations, EPA initially sought to adopt the median because it would not be influenced by extremes in the iterations, and thus, would be more easily met. Two weeks later, NRC proposed its conforming rule.[70]

For almost five years, the regulators and DOE went back and forth on how to proceed. Their disagreements centered on how to make the new rules "implementable." In addition, the three agencies tinkered with the details of how compliance would be actually measured at one million years. In the end, at an interagency meeting hosted by the Office of Management and Budget (OMB), they agreed on details such as the location of the accessible environment, the composition of the "reasonably maximally exposed individual," and exclusion probabilities for features, events, and processes. As a former OCRWM Acting Director later told an NWTRB meeting, he would inform Congress that the regulators had imposed "unimplementable" rules if the Department did not get its way.

DOE Submits Its License Application to NRC

By 2008, time was running out on the George W. Bush Administration. Under the effective leadership of OCRWM Director Ward Sproat, DOE drove to the brink of finalizing its License Application, and in the summer, it did so, based on the deals struck at the OMB gathering. Three months later, NRC docketed the application. It announced in September that the application was substantially complete, which meant that its review could commence.

Shortly thereafter, on October 15, 2008, EPA did finalize its standards. It retreated from the 3.5 millisieverts post-10,000-year dose limit and adopted a 1.00 millisieverts/year. It clarified that "mean" meant the arithmetic average. Notably, the Agency maintained that "sites whose natural features alone did not provide total containment were not necessarily considered unsuitable." NRC conformed its implementation rules five months later.[71]

TSPA Calculations

The repository system that DOE evaluated consisted of natural barriers (the surface, units of unsaturated tuff above and below the repository horizon, and the saturated zone) as well as engineered barriers (the Alloy-22 waste package, the titanium drip shield, the invert, and the platform upon which the invert rested). Figure 6.6 illustrates the entire arrangement.

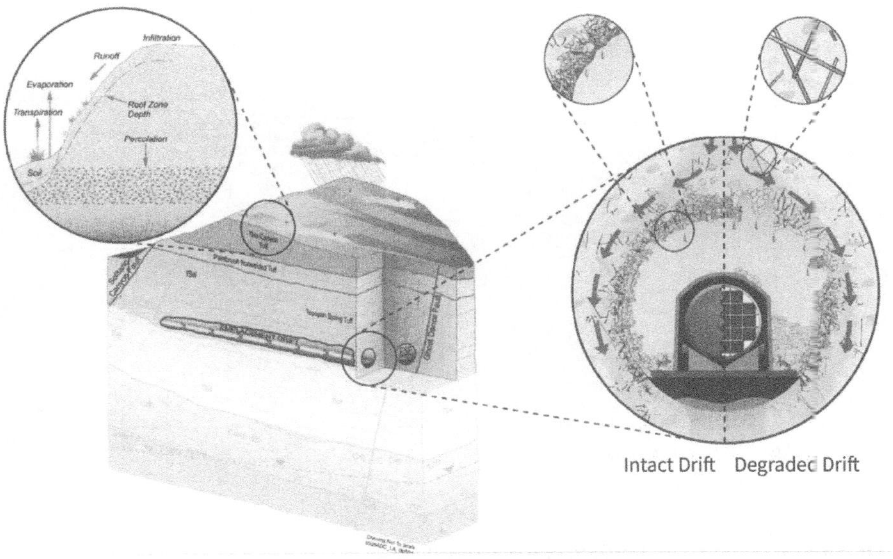

FIGURE 6.6 DOE's conception of the entire repository system.

At the core of the license application were the TSPA results for the total dose received by the reasonable maximally exposed individual at one million years. Each iteration of the performance model yielded a single outcome. Multiple reiterations generated the so-called "horse-tail" diagram. The chart included in the submission is shown in Figure 6.7.

DOE asserted that its analysis was "conservative," that is, it most likely overestimated the projected doses. Even so, the mean was only one percent of the limit and the 95th percentile of the iterations was just ten percent of the constraint. The Department also broke down the total into doses for six projected scenarios. Early drip shield failures were the least consequential followed by volcanic eruptions, early waste package failures, seismic fault displacement, seismic ground motion, and igneous intrusion. DOE therefore advanced its obvious conclusion: the Yucca Mountain repository system was more than acceptably safe.

NRC's Review Process

By the time that DOE forwarded its license application, however, external events yet again intruded. Reid had become Democratic Majority Leader the previous year. The Democratic Party had moved the date of the Nevada's presidential primary so it became the fourth in the nation and an early indicator of the strength of the three major contenders, Barack Obama, Hillary Clinton, and James Edwards. All of them understood the extent of Reid's power as the titular party leader in the state and his likely sway if a Democrat were elected in 2008. Hardly surprisingly, the three pledged fealty and, among other things, vowed to oppose the Yucca Mountain Project as President.

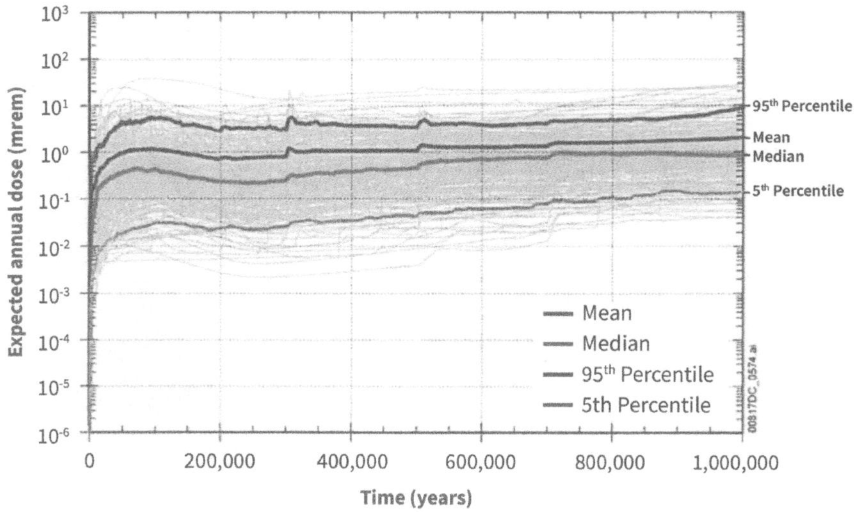

FIGURE 6.7 Horse-tail results of the TSPA iterations.

DOE had expected that the review might be completed within the two-year window (September 2010) stipulated in the NWPA. But normal bureaucratic routines caused delays until the results of the November election showed Obama to be the winner. His administration soon fulfilled its promises to the Nevada Senator. Secretary of Energy, Steven Chu, announced that the Yucca Mountain Project was "unworkable" and shut down OCRWM in early 2010.[72] OMB slashed its Fiscal Year 2011 budget request for waste disposal by 60 percent. In May, DOE ordered its main contractor, USA Repository Services, to cease all work.

DOE then petitioned the NRC's Atomic Safety and Licensing Board to withdraw the License Application.[73] The State of Washington, home for Hanford, and the State of South Carolina and Aiken, South Carolina, hosts to the Savannah River Plant, sought to intervene to oppose DOE's action. Litigation on this matter continued for several years. The outcome of those battles was that the NRC Commissioners prevented DOE from withdrawing the application. The Court of Appeals for the District of Columbia Circuit issued a *Writ of Mandamus* compelling DOE to continue the licensing process until all appropriated funds ran out.

AFTERMATH

Obama asked Chu in January 2010 to form a Blue-Ribbon Commission on America's Nuclear Future. Among its charges was "to conduct a comprehensive review of policies for managing the back end of the nuclear fuel cycle, including all alternatives for the storage, processing, and disposal of civilian and defense spent nuclear fuel and nuclear waste." The remit made clear that the Commission "was not to serve as a siting body." In particular, "it should not evaluate Yucca Mountain or any other location as a potential site for the storage of spent nuclear fuel or disposal of other high-activity radioactive waste."[74] Former Representative Lee Hamilton and former General Brent

Scowcroft chaired the group, which included a future Secretary of Energy, Ernest Moniz, and a future Chairman of the Nuclear Regulatory Commission, Allison MacFarlane. Its central recommendation was that any subsequent siting process for storage or disposal of high-activity waste should be "consent-based."

Moniz made a strategic decision not to push forward with the Blue Ribbon's Commission's advice once he became Energy Secretary in early 2013. He waited two years before instigating any sustained effort in this regard. In 2016, DOE issued a draft document detailing its current view, but without offering any specifics on how to move forward. Once President Donald Trump was elected, his Administration perfunctorily requested appropriations for the next three years to restart the Yucca Mountain licensing proceeding. By 2020, recognizing Nevada's importance in the upcoming primary, he disavowed the Yucca Mountain Project.

Responding to the Court's *Writ of Mandamus* order, NRC in December 2014 released Volume 3 of its Safety Evaluation Report: Repository Safety after Permanent Closure. In it, NRC staff concluded that DOE had complied with the Commission's regulatory rules for postclosure repository safety.[75] Of course, by then, licensing efforts had ceased. But the Report did serve the function of providing talking points for the project's advocates for the next several years.

As this book is being written, no one has a strong stake in promoting a repository-development program. The nuclear reactor owners continue to receive payments—nearly eleven billion dollars and counting—from the Judgment Fund to compensate them for DOE's failure to perform on its contract. President Joseph Biden's Administration launched an initiative, through which potential host states are avoiding being targeted once again for a repository. Communities have been offered the possibility of a federal consolidated interim storage facility on the basis of the consent-based siting process. With the election of President Donald Trump, whether that initiative will continue remains unclear.

Although some diehard defenders of the Yucca Mountain Project remain, that option is rapidly receding into the past. Lacking any semblance of a repository-development program, by drift and benign neglect, our generation is now composed of ***irresponsible ancestors***.

NOTES

1. Russell Ackoff, "The Future of Operational Research is Past," *Journal of Operational Research Society*, (39) 93–104.
2. Henceforth for convenience, the repository itself and its features and structures will not use the label "proposed."
3. DOE/RW-0573 Rev. 1, Yucca Mountain Repository License Application: General Information, Docket No. 63-001, November 2008, 1–3, 5–40.
4. Rob Rechard et al., "Hazards and Scenarios Examine for the Yucca Mountain Disposal System for Spent Nuclear Fuel and High-Level Radioactive Waste," *Reliability Engineering and System Safety*, (122) 2014, 74–95.
5. The early denial of water permits proved to be frustrating. To conduct the investigations OCRWM had to truck in water from Death Valley Junction, which was 45 miles away from the site. Eventually Nevada would allow OCRWM to withdraw a limited amount of water.
6. National Research Council, "Rethinking High-Level Radioactive Disposal: A Position Statement of the Board on Radioactive Waste Management," Washington, D.C., July 1990.

7. Department of Energy, "Site Characterization Plan Yucca Mountain Site, Nevada Research and Development Area," RW-0198, December 1988. The plan was 6,300 pages long. Part of the length was due to the approach that prompted criticism in the *Rethinking* report. Part of the length was caused by project officials and contractors desiring larger budgets.
8. Nuclear Regulatory Commission, "NRC Staff Site Characterization Analysis of the Department of Energy's Site Characterization Plan, Yucca Mountain Site, Nevada," NUREG-1347, August 1989.
9. EPA's standards and NRC's licensing regulations also expected DOE to assess unanticipated scenarios such as earthquake and igneous activity. Lacking space here, these evaluations are only considered when discussing the Department's License Application. But see Chapter 4 for how the regulators explained those expectations.
10. William Broad, "A Mountain of Trouble," *New York Times*, November 1990, Section 6, 37.
11. National Research Council, "Ground Water at Yucca Mountain: How High Can It Rise?" 1992. Six years later, Nevada unsuccessfully tried to revive the controversy by asking the NWTRB to review new information.
12. Nuclear Waste Technical Review Board, "First Report to the U.S. Congress and Secretary of Energy," March 1990, 13. In several conversations I had with OCRWM's senior managers, they maintained that the Project had considered deploying a ramp even before the Board made its recommendation. In addition, the First Report laid down the Board's long-lasting mark in several areas: the need for an east/west tunnel at the repository horizon; studying the presence of radioactive hydrogen and chlorine to better understand recharge rates in the unsaturated zone; developing more robust waste packages; and thermal loading of the repository.
13. The highest ranked option included extending the ramp into the Calico Hills below the repository horizon. Although it highlighted the importance of that rock unit because it impeded the flow to the saturated zone, the Board did not recommend this specific alternative.
14. Information in this section is derived from Michael Voegele and Donald Vieth, *Waste of a Mountain: How Yucca Mountain Was Selected, Studied, and Dumped* (Tonopah NV: Nye County Press, 2016), Chapter 9.
15. Science Applications International Corporation (SAIC), "Report of Early Site-Suitability Evaluation of the Potential Repository Site Yucca Mountain," SAIC-91-8000, January, 1992.
16. SAIC, 4-1.
17. The Quaternary period extended back from the present for approximately 2.6 million years. Yucca Mountain Site Characterization Project, "Topical Report: Evaluation of the Potentially Adverse Condition Evidence of Extreme Erosion During the Quaternary Period at Yucca Mountain, Nevada," YMP/92-41-TPR, March 1993.
18. National Research Council, "Review of U.S. Department of Energy Technical Basis Report for Surface Characteristics, Preclosure Hydrology, and Erosion," 1995.
19. Letter from Michael Bell to Ronald Milner, "Issue Resolution Report on the Potentially Adverse Condition: Evidence of Extreme Erosion During the Quaternary Period at Yucca Mountain," February 29, 1996.
20. NWTRB sharply criticized the decision to employ a large-diameter (25 feet) tunnel boring machine to construct the Exploratory Studies Facility. In the Board's view, such an apparatus would leave the impression that OCRWM was, in fact, building the repository, not studying a potential site. See Nuclear Waste Technical Review Board, "Underground Exploration and Testing at Yucca Mountain: A Report to Congress and the Secretary of Energy," October 1993, 23.
21. Calculations indicated that ^{14}carbon dioxide would likely discharge roughly 8,000 curies of radioactivity to the surface or about 14 percent more than the Agency's current rule would permit.
22. Public Law 102-486, 106 STAT. 2776.

23. National Research Council, "Technical Bases for Yucca Mountain Standards," 1995. The report also considered how human intrusion scenarios might be evaluated, whether the standard provided adequate protection for the general population, and whether a technology-based "as low as reasonably achievable" principle should be incorporated into the standard.
24. Technical Bases..., 2. The report recommended an individual risk standard, which could be straight forwardly converted into an individual dose standard. EPA ultimately chose to use dose rather than risk in promulgating its Yucca Mountain rule.
25. Performance assessments attempt to quantify the risk or dose to a set of individuals that might be expected to arise after a repository is sealed.
26. Testimony of Robert Fri before the Nuclear Waste Technical Review Board, Spring Meeting, Arlington, Virginia, October 17, 1995, 124.
27. 40 CFR 197, Appendix A.
28. See Chapter 10 for a more detailed discussion of this point.
29. See the discussion of that rule in Chapter 4.
30. Interview with Chris Whipple, July 6, 2024.
31. The multipurpose system would reduce handling requirements for SNF. The canister would be designed so that, filled, it could be transferred from storage and transportation casks into disposal packages without directly handling the SNF.
32. Department of Energy, "Final Interpretation of Nuclear Waste Acceptance Issues," *Federal Register*, (60) 21,793, May 3, 1995.
33. *Indiana Michigan Power Co. v. Department of Energy*, 88 F.3d 1272 (D.C. Circuit 1996).
34. This method relies upon sophisticated computer modeling of the potential processes, features, and events leading to releases of radionuclides from the disposal system, subsequent radionuclide transport, and consequent health impacts.
35. NWTRB and OCRWM tangled for more than a year over whether the tunnel should intersect the lower lithophysal rock unit where the repository would actually be constructed. It took a meeting between the responsible Political Associate Director at OMB and a small group of NWTRB members and staff to convince DOE to accept the Board's advice.
36. OCRWM's then Director, Daniel Dreyfus, told a meeting of the NAS Board of Radioactive Waste Management that "you may not like the VA, but to object too strongly would kill the disposal program." For a more tactful version of his comments, see Daniel Metlay, "From Tin Roof to Torn Wet Blanket: Predicting and Observing Groundwater Movement at a Proposed Nuclear Waste Site," in *Prediction: Science, Decision Making and the Future of Nature*, Daniel Sarewitz, Roger Pielke, Jr., and Radford Byerly, Jr. (eds) (Washington, D.C.: Island Press, 2000), 217–218.
37. The possibility of fracture flow became clear from experiments using radioactive ^{36}chlorine tracers. How contradictory data from those studies was interpreted is the subject of Chapter 11.
38. Voegele and Vieth, 475–477.
39. Abraham Van Luik, "Overview of Total System Performance Assessment—Viability Assessment," Presentation to the Nuclear Waste Technical Review Board's Winter 1999 Meeting, January 26–27, 1999.
40. As the discussion below in this chapter details, the TYMS report and later court cases substantially tightened the dose limit and compliance period.
41. Nuclear Waste Technical Review Board, "Moving Beyond the Yucca Mountain Viability Assessment," Washington D.C., April 1999. Quotations are found at 2 and 7 respectively.
42. See the following notices. Department of Energy, "Process for Evaluating the Suitability of the Yucca Mountain Site for Development of a Repository for High-Level Radioactive Waste and Spent Nuclear Fuel," *Federal Register*, (59) 19,680, April 25, 1994, and Department of Energy, "Use of 10 CFR 960 Siting Guidelines in Evaluating the Suitability of the Yucca Mountain Site," *Federal Register*, (60) 47,737, September 14, 1995.

43. Department of Energy, "General Guidelines for the Recommendation of Sites for Nuclear Waste Repositories." *Federal Register*, (61) 66,157–66,169, December 16, 1996. The quotation is found at 66,160.
44. Letter from Jarod Cohon, Board Chair, to April Gil, April 15, 1997. In layman's terms, redundancy and defense-in-depth translated into a "belt and suspenders" approach to design. It is why a car has two independent brake systems.
45. Undoubtedly, EPA and NRC political and career leaders engaged in detailed conversations about provisions in their respective Yucca Mountain-specific regulations. Therefore, their similarities are hardly surprising. What is more remarkable are some the differences between EPA and NRC's proposed rules.
46. Nuclear Regulatory Commission, "Disposal of High-Level Radioactive Waste in a Proposed Geologic Repository at Yucca Mountain, NV: Proposed Rule," *Federal Register*, (64) 8,640–8,679, February 22, 1999; Environmental Protection Agency, "Environmental Radiation Protection Standards for Yucca Mountain, Nevada: Proposed Rule," *Federal Register*, (64) 46,976–47,016, August 27, 1999; Environmental Protection Agency, "Public Health and Environmental Radiation Protection Standards for Yucca Mountain, NV: Final Rule, *Federal Register* (66) 32,074–32,135, June 13, 2001; Nuclear Regulatory Commission, *Federal Register*, "Disposal of High-Level Radioactive Waste in a Proposed Geologic Repository at Yucca Mountain, NV: Final Rule," (66)55,732–55,816, November 2, 2001.
47. Rather wait until EPA promulgated its Yucca Mountain-specific environmental standard, the nuclear regulators chose to push ahead. "[T]he time schedule for development of the Yucca Mountain repository is aggressive, and DOE has stated that it needs to have implementing regulations in place by 2000." *Federal Register* (1999), 8,643.
48. NRC *Federal Register* (2001), 55,750.
49. *Conference Report*, "Energy Policy Act of 1992," No. 1,018, 102nd Congress, 2nd Session, 1992, 391.
50. *Federal Register* (1999), 46,982.
51. *Federal Register* (2001), 32,097.
52. *Federal Register* (1999), 8,647.
53. See Figure 10.1 below.
54. *Federal Register* (1999), 8,647–8,650. The subsystem requirements in 10 CFR 60, of course, were explicit acknowledgment of the need for defense-in-depth. It is unclear how much concern there was that the subsystem requirements might "impose significant additional expenditures without producing any commensurate increase in protection." Quotation is at 8,649.
55. Department of Energy, "General Guidelines for the Recommendation of Sites for Nuclear Waste Repositories; Yucca Mountain Site-Suitability Guidelines: Proposed Rule," *Federal Register*, (64), 67,054–67089, November 30, 1999.
56. Letter from Jared Cohon, Board Chairman, to Ivan Itkin, OCRWM Director, March 20, 2000.
57. Department of Energy, "General Guidelines for the Recommendation of Sites for Nuclear Waste Repositories; Yucca Mountain Site-Suitability Guidelines: Final Rule," *Federal Register*, (66) 37,298–37,340, November 14, 2001.
58. DOE *Federal Register* (2001), 37,322–37,323.
59. Department of Energy, "Yucca Mountain Site-Suitability Evaluation," DOE/RW-0549, 2001.
60. Spenser Abraham, "Recommendation by the Secretary of Energy Regarding the Suitability of the Yucca Mountain Site for a Repository Under the Nuclear Waste Policy Act of 1982," February 2002, 45.
61. In early February, Dr. John Marburger, President Bush's science advisor, held a series of meetings about the Secretary's recommendation, during which he was informed of about the NWTRB's critical January 24, 2002 letter. (See discussion of that communication below.) Marburger met with Board members and staff on February 13, 2002 to explain their concerns. Abraham released his recommendation the next day. See Voegele and Vieth, 624–632, for an overview of those exchanges.

62. In their written dissent, six Congressmen cited several of the NWTRB's concerns.
63. U.S. Congress, Joint Resolution, Public Law 107–200.
64. The legislators' calculations were transparent. To have sustained the objection would have reopened the site-selection process. When three sites were on the table, those calculations would have to incorporate more technical considerations.
65. See also, Nuclear Waste Technical Review Board, "Report to U.S. Congress and the Secretary of Energy: January 1, 2001 to January 31, 2002," Washington D.C., 2002.
66. *Chevron U.S.A. Inc. v. Natural Resources Defense Council, Inc.*, 467 U.S. 2,778, (1984).
67. *Nuclear Energy Institute, Inc. v. Environmental Protection Agency*, 373 F.3rd 1251, 1,270. Noteworthy, the Court rejected Nevada's contention that NRC violated the NWPA's requirement that a repository incorporate a "multiple barriers" system by failing to include "any specific requirement for any barrier to provide any degree of protection that is substantially independent of the others."
68. Letter from Michael Corradini, Board Chairman, to Margaret Chu, OCRWM Director, and the attachment "An Evaluation of Key Elements in the U.S. Department of Energy's Proposed System for Isolating and Containing Radioactive Waste," November 25, 2003.
69. Bo Bodvarsson, Presentation to the Nuclear Waste Technical Review Board, "Thermal Hydrologic Environment and Thermal Seepage," May 19, 2004.
70. Environmental Protection Agency, "Public Health and Environmental Protection Standards for Yucca Mountain: Proposed Rule," *Federal Register*, (70) 49.014-49,065, August 22, 2005; Nuclear Regulatory Commission, "Implementation of a Dose Standard After 10,000 Years: Proposed Rule," *Federal Register*, (70) 53,313–53,320, September 8, 2005.
71. Environmental Protection Agency, "Public Health and Environmental Protection Standards for Yucca Mountain: Final Rule," *Federal Register*, (73) 61,256–61,289; October 15, 2008; Nuclear Regulatory Commission, "Implementation of a Dose Standard After 10,000 Years: Final Rule," *Federal Register*, (74) 10,811–10,830, March 13, 2009. The State of Nevada sued to overturn both regulations. (*State of Nevada v. Environmental Protection Agency*, Court of Appeals for the District of Columbia Circuit, October 10, 2008.) However, with the eventual demise of the Yucca Mountain Project, that litigation has been suspended.
72. In August 2008, ten Directors of DOE's National Laboratories, including Chu, noted the importance of Yucca Mountain as a national resource. Of significance, "unworkable" did not mean technically unsuitable.
73. The Atomic Safety and Licensing Board is an administrative forum that is located within but independent of the NRC's management. As its name implies, it handles reactor and repository matters, among other things. The Commissioners may ultimately review its decisions. Nuclear Regulatory Commission, "Department of Energy's Motion to Withdraw," Atomic Safety and Licensing Board, March 3, 2010.
74. Introductory Letter to Steven Chu, Blue Ribbon Commission on America's Nuclear Future, *Report to the Secretary of Energy*, January 2012.
75. Importantly, Nevada never had the opportunity to challenge the staff's findings in an adjudicatory hearing.

Part II

Lessons From Other Countries

PREFACE

The United States finds itself in an ironic situation. As recently as a dozen years ago, it was among the world's leading nations to pursue a conscientious course for coping with its large and growing inventory of radioactive material. Now it is chasing unconvincing policies while failing to keep its "eye on the prize." The absence of even a conceptual strategy for the long-term management of high-level waste and spent nuclear fuel unescapably incriminates this generation as irresponsible ancestors. What is to be done?

By no means are there heavy villains or bumbling incompetents controlling this story. Nor does our current dilemma lie in the intrusion of unanticipated contingencies. Rather our circumstance arises from a flawed appreciation of the nature of the task.[1] Policymakers addressed a sociotechnical mission by overwhelmingly subordinating the social to the technical. Up until two years ago, all the federal waste-management agencies directly employed a single graduate-trained social scientist. In many other nations, that gap has been filled for decades. Perhaps by turning to their experiences, we can derive some insights that inform how to resurrect our own country's disposal program.[2]

Part II explores five prerequisites for revitalizing our disposition strategy. "Designing an Implementing Organization" considers alternatives to carry out a nation's nuclear-waste management program. In the United States, the default option has always been an Executive Branch agency, lately the Department of Energy. Is that choice still appropriate? In many other countries, the implementor is a non-profit organization owned by the waste generators. Perhaps that institutional form might be more effective than the *status quo*. "Selecting a Repository Site" examines

DOI: 10.1201/9781003528302-8

efforts in the United States and eight other countries to choose a location to construct a deep-mined, geologic repository for high-activity radioactive waste. It considers nearly two dozen attempts, some of which focused initially on demonstrating technical suitability, others of which focused initially on establishing social acceptability, including meriting and sustaining trust. Both requirements, however, must be met if site selection is to be successful.

"Phased Repository Development" considers the efficacy of an incremental strategy to construct and operate a repository. Almost by definition, all national programs feature a phased approach. Analysts have proposed processes that demand a more elaborate set of steps in which explicit decisions are taken frequently before moving to the next developmental stage. Yet implementing such an approach is more problematic than typically presumed. "Repository Licensing" considers how the United States and five other nations have designed procedures that permit the construction and operation of a disposition facility. Of particular importance are determining whether the repository can limit releases to pre-established levels and last for pre-established time, perhaps as long as one million years. "Addressing Technical Uncertainties" goes to the heart of developing a repository. What happens when traditional science is open to interpretative flexibility? What if creditable specialists disagree? Those questions are explored in two case studies. The first considers the issue of whether the copper canister that holds the waste in Finland and Sweden will corrode in oxygen-free groundwater. The second explores how much and how fast water travels from the surface of Yucca Mountain to the repository horizon.

As Part I makes evident, these five matters singly and in combination have been root causes of the existing state of affairs in the United States. Unless better means can be identified to reset the enterprise, we are likely to be ensnared in an endless cycle of trial and failure. And our ancestors will bear the consequences of our neglect.

NOTES

1. For an elaboration of this claim, see Daniel Metlay, "Social Acceptability of Geologic Disposal," in *Encyclopedia of Nuclear Energy*, Ehud Greenspan (ed), (London: Elsevier, 2021), 684–696.
2. Of course, one cannot import international approaches whole cloth. Extra consideration must be given to ensure that prescriptions are well aligned with political culture.

7 Designing an Implementing Organization

Since the dawn of the nuclear era nearly eighty years ago, the federal government has controlled the disposal of high-level radioactive waste (HLW) and spent nuclear fuel (SNF). This arrangement certainly made sense at the time because *the nation as a whole* owned the vast majority of the waste, most of which was located at the weapons-production complex. Fast forward to today, the amount of commercial HLW and SNF, whether measured either by volume or radioactivity, far exceeds the amount of defense nuclear waste. Moreover, *nuclear utilities now* own the growing inventory of the waste. These altered circumstances and conditions present enough new opportunities and tests to warrant a reconsideration of what institution should be in charge.

CHALLENGES FOR DESIGNING IMPLEMENTING ORGANIZATIONS

The very start of this narrative asserted that radioactive-waste management is a messy or wicked problem.[1] According to one well-developed academic framework—contingency theory—the properties of the problem, that is, the degree of value conflict and uncertainty, must align with the decision-making mode necessary to address it.[2] That requisite choice, in turn, has strong implications for organizational arrangements and behavior. So, if problem-solving fundamentally rests on making proper calculations, the appropriate design would be different than if decision-making demands trial-and-error feedback and adaption. Figure 7.1 suggests those relationships.

Although quite valuable in many situations, contingency theory provides only elusive guidance about the appropriate problem-solving mode when choice has to be made under conditions of very high value conflict and very high uncertainty. The scheme typically dodges to vague notions such as charisma, heuristics, or inspiration. As Part I documented, the Department of Energy (DOE) and its predecessors had little choice but to oscillate between problem-solving by, on the one hand, trial-and-error and, on the other, bargaining. One might reasonably ask whether a government agency is the optimal institutional design to do both.

Further, some of the implementors' desired outcomes can only be obtained at the price of others. Political accountability may reduce autonomy. Accessibility to stakeholders and the general public may not be cost effective. A quest for technical excellence may limit financial accountability. Balancing these conflicts is inherently subjective and will depend on the views of the incumbent regime. Therefore, constancy may be at risk.

DOI: 10.1201/9781003528302-9

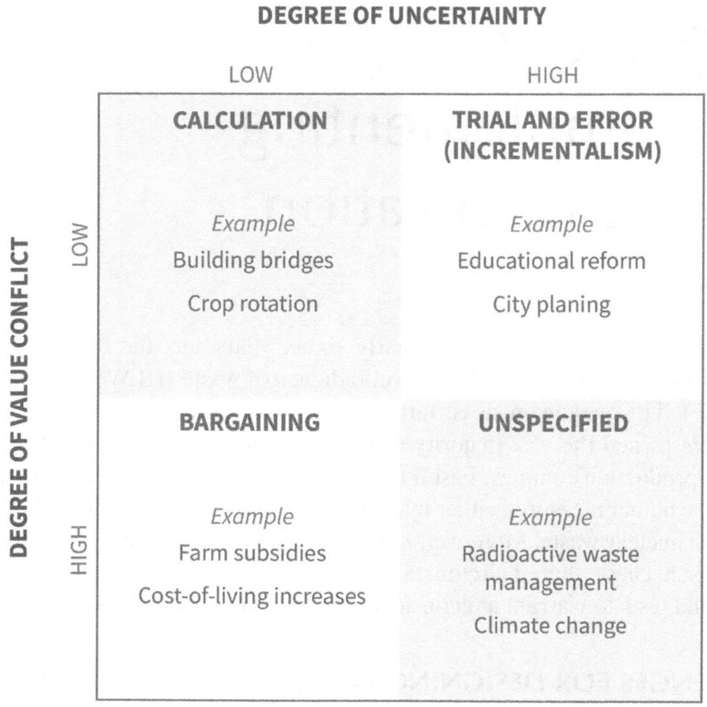

FIGURE 7.1 Character of the problem and decision-making strategy.

The complications brought on by the intrinsic nature of the waste-management endeavors offer at least a plausible explanation for why initial designs have changed so often; it is just tough to find the right alternative at the outset. In the United States, today's federal agency replaced another that itself superseded yet another. In Canada, a company owned by waste producers took the lead from a government-chartered implementor. And that choice was reversed. In Germany, a federal corporation was exchanged for a government agency. In the United Kingdom, a government authority replaced an organization owned by the waste producers. In France, the body developing nuclear weapons was taken over by a public institution reporting to several government ministries.

The remainder of this chapter considers two questions: What was the experience in the United States in selecting an implementing organization? How does that compare with the practices in other nations?

INSTITUTIONAL DESIGN IS NOT A NOVEL ISSUE IN THE UNITED STATES[3]

First Attempts

As described in Chapter 2, the Atomic Energy Commission's Milton Shaw in the late 1960s bureaucratically overcame Ernest Tremmel's advocacy of the private sector by insisting that the federal government become the accountable body. As Chapter 4 relates, a decade later, the presidentially empaneled Interagency Review Group with little internal debate recommended that DOE continue to execute the program.[4]

At about the same time, the logic of that choice began to be called into question. Mason Willrich and Richard Lester undertook one of the first independent examinations of the waste-management issue. They maintained that "[T]he most appropriate organizational scheme for post-fission radioactive waste management appears to be consolidated management of all waste streams under a single authority."[5] Several institutional alternatives, however, might fit that general prescription.

Their analysis started by looking at scenarios in which the federal government's role would be reduced and, subject to strict regulation, handed over to a private industry-run monopoly. Two possibilities presented themselves. First, an independent corporation might be formed with a diverse board of directors. Second, a cooperative venture by nuclear utilities might be founded. In that case, the board of directors would be dominated by the electricity producers, although outside directors might be included. In either case, an independent advisory committee might be created. Any new venture would then charge the federal government for the storage and disposal of the weapons-complex HLW and SNF. Although pointing to some advantages of privatizing waste management, the study concluded, without much analysis, that "it would probably be unacceptable to cut back the scope of the federal government's responsibility…by transferring the task to the private sector."[6]

Rethinking Who Should Be the Waste Management Implementor

Congress ratified the choice of the federal government in the form of DOE's Office of Civilian Radioactive Waste Management (OCRWM) when it passed the 1982 Nuclear Waste Policy Act (NWPA). But already, Willrich and Lester's reservations had risen to the surface. In Section 303 of the legislation, the lawmakers compelled

> The Secretary to undertake a study with respect to alternative approaches to managing the construction and operation of all civilian radioactive waste management facilities, including the feasibility of establishing a private corporation for such purposes.

Alternative Means for Managing Radioactive Waste

Starting in the fall of 1984, an advisory panel representing the nuclear utilities, private sector firms, tribal nations, consumer groups, and academics met regularly and submitted its findings and recommendations to the Energy Secretary Donald Hodel one year later.[7]

The group adopted the classic analytical approach of identifying a range of options and then enumerating their pros and cons. Based on a review of the literature, it started out with ten possibilities.[8] It then collapsed them into four explicitly detailed collectivities: OCRWM, an independent federal agency, a private corporation, and a public-private corporation. The panel next undertook a test-by-test comparison of the options across thirteen critical, but intuitively measured outcomes.[9] Finally, the panel, here denoted as AMFM1, implicitly drew trade-offs among the consequences.

From these estimates emerged its preferred alternative, a federally chartered corporation, called FEDCORP. For some unexplained reason, the advisory committee never explicitly evaluated this option. Nor did the panel even offer an example of this organizational design, although FEDCORP seems to be modeled on the Tennessee Valley Authority. Nonetheless, AMFM1 asserted that it scored highly, especially on measures such as business orientation, involvement with stakeholders, freedom from political influence, complete financial authority, internal flexibility, and responsiveness to regulators.[10] Its major disadvantage was the length of transition.[11]

Two months after Hodel received the report, he appointed a three-person committee of senior DOE political appointees to review it. It rejected each of the panel's findings and evaluations. In the committee's view, OCRWM's credibility was improving; civil servants could provide sufficient continuity; internal flexibility could be gained by seeking exemptions to civil service requirements; political immunity could only be bought at the price of less accountability; and cost-effectiveness could improve as OCRWM proceeded up its learning curve.[12]

The DOE political appointees called attention, without full and explicit evidence, to the states and nuclear industry's strong opposition to change: a transition might take several years to effect. Thus, they advised the new Energy Secretary, John Herrington, that if any institutional adjustment was needed, the modification only should be undertaken once the NRC had granted a construction license for the repository. He agreed with this advice.

Alternative Means for Managing Radioactive Waste Reconsidered

In its report on DOE's 2001 Fiscal Year appropriations bill, the House Subcommittee on Energy and Water directed the Energy Secretary to update the analysis of alternative approaches to managing and financing the nation's nuclear waste disposal program (AMFM2). Because of concerns about "breadth and read-ability [sic], the DOE-contracted study did not attempt to provide the reader with a comprehensive analysis."[13]

The follow-up enquiry rested on a fundamental presumption: OCRWM would remain the implementor for the indefinite future.[14] Consequently, the report takes on a level of abstraction bordering on irrelevance. Nonetheless, faced with a legislative charge, DOE would at least have to go through the motions.

The AMFM2 suggested three long-term options: a performance-based organization within DOE; an internal authority within DOE with broad authority; and an independent federal authority. As with the original study, the new analysis considered a set of performance outcomes.[15] In the end, DOE recommended that it "continue to analyze and refine the concepts of the three longer-term management alternatives."[16]

Blue Ribbon Commission on America's Nuclear Future

The previous chapter reviewed the origins of the President Obama-directed Blue Ribbon Commission on America's Nuclear Future (BRC); it briefly noted one of its major pieces of advice—the importance of establishing a "consent-based" siting process. The group also explored the question of who should be the waste-management implementor.

The BRC report's introductory language to its analysis of institutional designs left no doubt that change should be in the air.[17]

> A new [single-purpose] organization will be in a better position to develop a strong culture of safety, transparency, consultation, and collaboration. And by signaling a clear break with the often troubled history of the U.S. waste management program it can begin repairing the legacy of distrust left by decades of missed deadlines and failed commitments.

Like the 1985 study, the group had its own set of outcomes that the implementor ought to pursue.[18] But the BRC did not assess a number of design alternatives. It simply determined that an independent FEDCORP chartered by Congress "offered the most promising model."[19] It pointed out that striking the right balance between independence and accountability would be a key challenge. In addition, it warned that transferring the task of managing defense HLW and SNF to a new organization outside of DOE had wide-ranging implications.

In response to the Commission's advice, DOE asked the RAND Corporation in 2012 to evaluate several organizational paradigms: a government corporation, such as the now-closed down U.S. Enrichment Corporation; a federally chartered government corporation, such as the Tennessee Valley Authority; and an independent government agency, such as the U.S. Postal Service.[20]

Although the RAND review did not advance a single choice for implementor, it did favor either federally chartered government corporation or an independent government agency. Notably, the study did point to an important observation: how the organization behaved was more important than its form. Put somewhat differently, what counts is the implementor's physiology not its anatomy. One important footnote in the report is worth mentioning. In countries like Canada, Finland, Sweden, and Switzerland, the implementor is a private, non-profit organization owned by the nuclear utilities. For RAND, that model is "unlikely to be a viable option in the United States."[21] No explanation was given.

The following year, DOE released its Strategy for the Management and Disposal of Used Nuclear Fuel and High-Level Radioactive Waste.[22] It agreed that a new single-purpose organization was needed "to provide the stability, focus, and credibility that are essential to get the waste program back on track."[23] The strategy maintained that managing the waste was inherently a governmental responsibility and accepted RAND's finding that either a federal government-charted corporation or an independent government agency were promising "models."

Reset of America's Nuclear Waste Management Strategy and Policy

Over the course of nearly three years beginning in 2015, a small group of specialists from the U.S. and abroad met under the leadership of Rodney Ewing, former chairman of the NWTRB, and Allison MacFarlane, former chairman of NRC.[24] The Steering Committee held five public meetings and heard 75 presentations by internationally recognized experts, government officials, legislators, nongovernmental organizations, tribal representatives, and members of the general public. Although there was a strong consensus (but not unanimous agreement) on most issues, the panel was clearly split on the question of the implementor's institutional form.

It looked favorably on the headway countries such as Finland, France, Sweden, and Switzerland had made. All of them had by then selected a location for developing a deep-mined, geologic repository for high-activity waste. Notably, except for the implementor in France, all of them have adopted a waste-producer-owned, single-purpose, non-profit implementing organization.[25] For that reason, the majority of the Steering Committee observed[26]:

> The track record of nuclear utility-owned implementors abroad, coupled with the intrinsic harmonization of technical and business incentives, makes at least a *prima facie* case for considering [in the United States] the creation of a not-for-profit, nuclear utility-owned implementing corporation (NUCO).

The Steering Committee, however, proposed that a FEDCORP-like organization manage defense SNF and HLW. But it observed that if NUCO demonstrated positive, on-the-ground accomplishments, high-activity waste from the weapons program might at some point be transferred to that organization for disposal upon payment of a fee.[27]

Also of note is a parallel recommendation from the Heritage Foundation, a think tank supporting private sector initiatives.[28]

> When nuclear power companies are responsible for waste management, regulating agencies can then be seen as simply that—regulators with a disinterested goal of protecting health and safety. The government can more transparently play the role of a neutral referee with reliable information. But as both a regulator and repository operator, the government appears to have a bias. Information is easily deemed suspect or distorted due to a conflict of interest, perceived or otherwise.[29]

Figure 7.2 recaps the historical record in the United States: DOE alone endorses its continued involvement as the nuclear-waste implementor.

	ANALYSIS			
INSTITUTIONAL DESIGN	**AMFM (1)**	**AMFM (2)**	**BRC**	**RESET**
OCRWM	OCRWM	***OCRWM***	OCRWM	OCRWM
Autonomous Division with DOE		AUTDIV		
Independent Federal Agency	IFC			
Federal Chartered Corporation	***FEDCORP***	FEDCORP	***FEDCORP***	
Public-Private Corporation	CWM			
Private Corporation	ANCORP			
Corporation Owned by the Waste Producers				***NUCO***

FIGURE 7.2 Alternative institutional designs for waste-management implementor.

Note: Preferred design is ***italicized in bold type***. Blank cells indicate that a design was not analyzed in detail.

INSIGHTS FROM OTHER NATIONS

As of 2025, eighteen nations abroad operated five or more nuclear power plants; of those, eight countries offer reasonable comparisons to the waste-disposition program in the United States.

Figure 7.3 identifies them along with the organizational form of the implementor.

COUNTRY	IMPLEMENTING ORGANIZATION	ORGANIZATIONAL FORM
United States	U.S. Department of Energy	Government agency
Canada	Nuclear Waste Management Organization	Private not-for-profit organization formed by the nuclear electricity producers
Finland	Posiva Ovy	Joint waste management company created by owners of nuclear power plants
France	National Agency for Radioactive Waste Management	Government-funded industrial and commercial institution
Germany	Federal Company for Radioactive Waste Disposal	Government-owned corporation
Japan	Nuclear Waste Management Organization of Japan	Government agency
Sweden	Swedish Nuclear Fuel and Waste Management Company	Private corporation formed by the owners of nuclear power plants
Switzerland	National Cooperative for the Disposal of Radioactive Waste	Public/private consortium of radioactive waste producers, including all the owners of nuclear power plants and the Federal Government
United Kingdom	Nuclear Waste Services	A wholly owned subsidiary of the Nuclear Decommissioning Authority, which is a public sector organization

FIGURE 7.3 Implementor takes on many organizational forms.

Corporations Owned By Waste Producers

A brief assessment of some of the accomplishments and setbacks suggest why many of them have made such significant progress toward developing a deep-mined, geologic repository.

Switzerland

In 1972, the waste producers and the Swiss Confederation created Nagra, the former is chiefly responsible for the long-term management of spent fuel and the latter is accountable for the disposition of low-level material such as medical, industrial, and research waste. The Swiss began searching for a repository site by concentrating on crystalline rock formations. A government review of those investigations concluded that siting feasibility was not fully demonstrated. By 1985, Nagra determined that finding a suitable location in crystalline rock was highly problematic.

Anticipating the infeasibility of identifying a suitable site in granite, the implementor switched its focus to the Opalinus Clay formation in the northern part of the country. Its disposal concept, illustrated in Figure 7.4, relied on an engineered barrier system as well as key properties of the host rock.

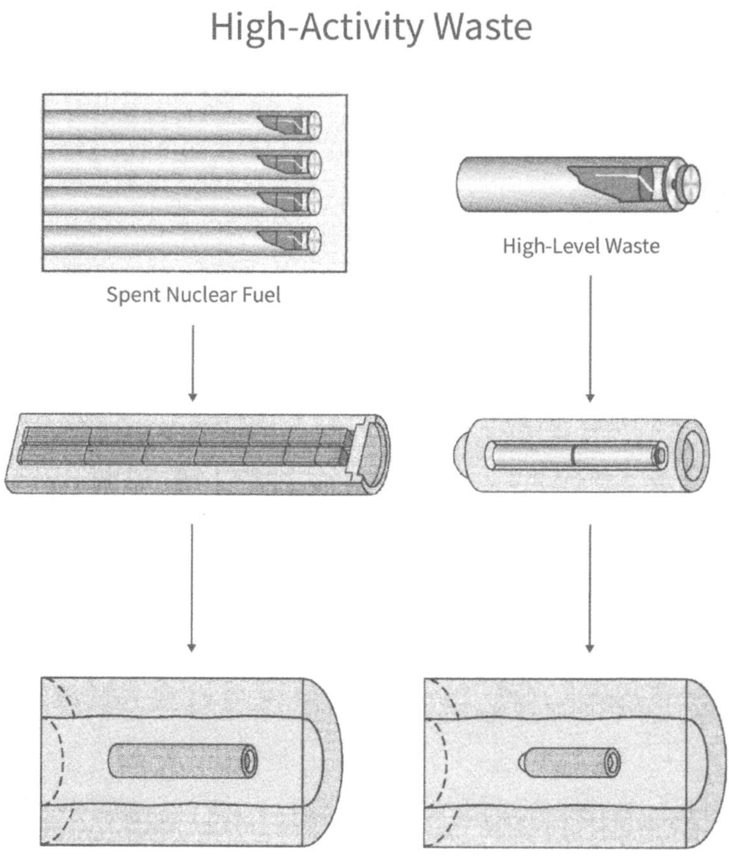

FIGURE 7.4 Swiss disposal concept.

In particular, the location was tectonically stable, lacking in resources, and possessed a predictable structural, hydrological, and geochemical environment. The clay featured low hydraulic conductivity, resulting in slow movement of the waste predominantly by diffusion rather than advection. The hydrogeologic environment was slightly alkaline and reducing, which aided radionuclide absorption and retention. Finally, the Opalinus Clay was self-sealing, minimizing the impact of repository-induced and natural fractures that could provide pathways to the biosphere.

Beginning modestly in 1981, Nagra winnowed down to three locations prospective repository sites for low-level/intermediate-level waste (LILW). The Cantonal Government of Nidwalden, in the pre-alpine area in central Switzerland south of Lucerne, stepped forward to suggest that implementor also consider sites in the Wellenberg municipality.

Intense opposition arose especially in the adjacent municipalities to Wellenberg. One consequence of this resistance was the institution of a cantonal veto right over the selection of a site. The first referendum took place in 1985, when 52 percent voted against selecting the site even as the Wellenberg municipality barely approved. Given the narrowness of the outcome, Nagra and Nidwalden chose to move ahead. In an effort to rescue the situation, Nagra convened a broadly based expert committee, EKRA. It advised that monitored long-term disposal ought to become the baseline for a new repository development process.[30] In a 2002 referendum, 58 percent of the canton rejected the project, although the Wellenberg municipality again approved by a bare majority. Faced with those results, one retrospective case study concluded that a sound balance between technical safety and fair process still needed to be found.

The Nuclear Energy Act took effect in 2005 and was designed to limit the periphery's power. After the Wellenberg experience, the federal government became alarmed that no repository could ever be sited and constructed because of the cantonal veto. In its place, an optional national referendum might be held, but for the same reasons Nevada's objections to Yucca Mountain were overridden by Congress, this formula did not satisfy those still fighting repository development.[31]

When Nagra then recommended a particular location in the Züricher Weinland for developing a repository, the Swiss Government ordered Nagra to slow down: a choice of one is no choice at all. Thus, arose the Sectoral Plan in 2008.[32] It mandated a three-step process that incorporated EKRA's conceptual foundation.

1. Selection of geological siting area for LILW and HLW/SNF (completed in 2011)
2. Selection of at least two sites each for LILW and HLW/SNF (completed in 2018)
3. Site selection and the start of the general licensing procedure (expected to be completed in 2031)

The Federal Council must approve each stage before the process can move to the next one.

Under the first step, Nagra proposed six regions for the HLW/SNF facility: Jura-Süfuss, Wellenberg, Süddranden, Jura Ost, Nőrdlich Lägern, and Zürich Nordost.

It dropped the first three early on. Then the implementor in step two recommended eliminating Nőrdlich Lägern, but the nuclear regulator, Federal Nuclear Safety Inspectorate (ENSI), determined that the third potential site should be retained.[33] In 2022, Nagra announced it would concentrate on that location. The reasons for the choice included the thickness of the Opalinus Clay layer, the extreme stability of the formation, and a high degree of flexibility for the underground layout. Less overtly expressed rationales were the site's closeness to the Paul Scherrer Nuclear Institute and the Zeilag interim storage facility holding SNF and other waste forms.

Participation by interested and affected parties took on a distinctively Swiss flavor. The Sectoral Plan anticipated the formation of Regional Conferences. Six were established; each had approximately 100 members, including advocacy organizations, and, because of proximity of the Opalinus Clay formation to the international border, local German communities. The Conferences' task was to review Nagra reports, hire independent experts, and present ideas for regional development. Importantly, their remit was limited to "having a say" only on decisions related to the siting and design of above-ground facilities. At the urging of one area, however, working groups on "safety" ended up posing critical, but nonetheless informal and non-binding, questions about the repository's postclosure performance.[34] Still left to be accomplished is Step 3 of the Sectoral Plan. Notwithstanding that formality, Nőrdlich Lägern has clearly emerged as the location of Switzerland's repository for HLW, SNF, and intermediate-level radioactive waste.[35]

Sweden

The environmental impacts of producing nuclear electricity and the concomitant issue of managing the waste emerged as a national political controversy in the early 1970s.[36] In 1973, the government appointed the Använt Kärnbränsle och radioaktivt Avfall to assess the subject and make recommendations on a path forward. Compromise legislation passed in 1976; it mandated that reactor owners had to demonstrate "absolute safety" in order to obtain a license to operate new reactors. To do so, they created the Swedish Nuclear Fuel and Waste Management Company (SKB) in the late 1970s.[37] SKB is responsible for research and development, repository construction, and temporary storage of all commercial SNF and a small amount of defense-related high-activity material.[38]

For all practical purposes, the sole geologic formation suitable for constructing a repository is the granitic bedrock. SKB eventually proposed a disposal concept, KBS-3 (Figure 7.5) based on encapsulating SNF in a cast-iron insert and placing that structure in a copper canister, which in the anoxic groundwater conditions prevailing in the host rock, was believed to be highly corrosion resistant. The assemblage is then placed in a borehole drilled out of the crystalline rock and topped off with bentonite clay. Any radionuclides that might escape the canister would be absorbed in the clay; their migration outside the engineered system would be further retarded by the compact granite. Because of that rock's ubiquity, a repository might, therefore, be built virtually anywhere in the country.

By 1980, the implementor began drilling and evaluating boreholes across the nation. Strong local opposition arose in Kynnefjäll and was repeated in three other municipalities. The controversies attracted substantial media attention, prompting a

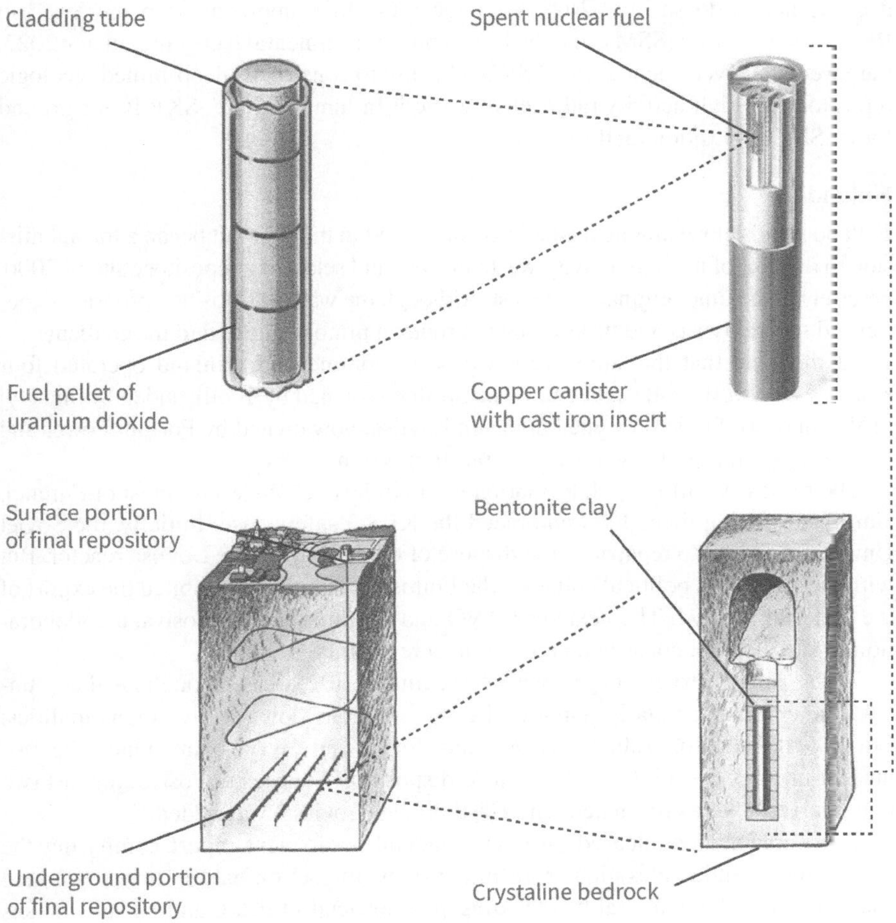

FIGURE 7.5 Swedish disposal concept.

reprimand from the Minister of Energy and Environment for SKB's failure to inform the communities in advance of its plans.

Learning from its earlier missteps, SKB approached all 286 municipalities, asking whether they might be amenable to participating in "feasibility studies." Two localities in northern Sweden, Malå and Storuman, responded positively; referenda in both, however, forced the implementor to graciously withdraw.

SKB's next move in 2001 was to contact the five "nuclear" communities where its owners were deeply entrenched. Ultimately two of them, Oskarshamn and Östhammar, agreed to a full suite of suitability investigations that unfolded over more than a decade. SKB filed an application in 2011 to construct a repository at the latter location, claiming that "the rocks have spoken."

An environmental nongovernmental group (MKG) opposed the application, maintaining that SKB failed to account adequately for the corrosion of the copper canister.[39] Moreover, Östhammar still retained the legal authority to veto the site;

it chose not to do so in 2020.[40] After receiving final approval from the Swedish Radiation Authority (SSM), and the Land and Environmental Court in January 2022, the Swedish Government granted SKB a license to construct a deep-mined, geologic repository for high-activity radioactive waste.[41] In January 2025, SKB broke ground for its SNF disposition facility.

Finland

Finland stands alone among the nations considered in this story. It began a formal mission to dispose of its high-activity waste in 1983 and selected a repository site in 2000, precisely at the time originally forecast. Although the waste-disposition program experienced several twists and turns along the route, it nimbly adjusted to the challenges.

At the time that the site-selection process commenced, Finland operated four reactors: two western-designed units in Olkiluoto owned by Teollisuuden Voima Oyj (TVO) and two Soviet-designed plants in Loviisa, now owned by Fortum Power and Heat.[42] TVO started its own waste disposal program.

The same crystalline rock formation found in Sweden underlays most of Finland. Hardly surprising then, TVO embraced the KBS-3 safety case. Initially, the Soviet Union had agreed to repatriate and dispose of the SNF from the Loviisa reactor. But with that country's political collapse, the Finnish Parliament prohibited the export of the material in 1994. The next year, TVO and Fortum founded Posiva, a collaboration to site, license, construct, and operate a repository.[43]

TVO's siting surveys began with screening of 102 areas throughout the country. The company initially winnowed those locations down to five municipalities: Eurajoki (Olkiluoto), Äänekoski, Kuhmo, Sievi, and Hyrynsalmi. Once the two nuclear utilities consolidated their waste-disposition energies in Posiva, the last two potential sites were eliminated, and Hästholmen (Lovissa) was added.[44]

The implementor released an environmental assessment report comparing the remaining potential sites along four dimensions: impact on nature, impact on land use, effects on human health (including psychosocial effects), and social impacts (including opposition to a repository).[45] Two fundamental conclusions emerged from the analysis. First, although high levels of salinity in adjacent waters might limit a repository's depth at the Eurajoki and Hästholmen sites, "taking into account long-term changes in the prevailing conditions, no considerable differences [in safety] are expected between the alternative sites." Second, "studies show that the population of the nuclear localities of Eurajoki and Hästholmen reacted more positively to nuclear power and final disposal than the population of Kuhmo and Äänekoski."

The second finding took on elevated significance because of a legal provision, similar to the Swedish one, granting a municipality veto power over site-selection. In May 1999, Posiva sought to fulfill the requirement of the 1987 Nuclear Energy Act by submitting an application to the Finnish Government for a "Decision-in-Principle" to construct a repository. One student of the Finnish waste-disposition program detailed "twenty years of arm-wrestling in Eurajoki," which ultimately culminated in the town's acceptance of the site in 2000.[46]

Two years later, the Parliament ratified the application for the final disposal facility for SNF in Eurajoki. Construction of the Onkalo (Finnish for hollow or cave) repository started shortly thereafter. By the end of 2024, the implementor had completed a trial run transferring an unloaded disposal cask from one of the nation's reactors to the nearby encapsulation plant. Posiva projects that initial operations will hopefully begin the following year, when its regulator, the Radiation and Nuclear Safety Authority, completes its review of an operating license.

Canada

As early as 1952, Canada placed responsibility for managing its radioactive waste in the hands of a Federal Crown Corporation, Atomic Energy of Canada Limited (AECL).[47] The company first concentrated its operations on developing and commercializing the CANDU reactor. By the mid-1970s, it had branched out into the waste-disposition arena, projecting selection of a repository site a decade later, construction by the late 1980s, and an operating facility by 2000.[48]

Public opposition to the implementor's preliminary siting studies forced their cancellation. The resistance also triggered a joint Federal/Ontario policy pronouncement that the choice for a repository site would be deferred until a consensus was obtained on a *disposition concept.* As one of the first steps in 1994, AECL prepared a comprehensive environmental impact assessment.[49] Based on the hydrogeologic conditions found in the ubiquitous granitic Canadian Shield, its central finding was:[50]

> [E]nough was known, based upon both modeling work and extrapolations from site specific data, to suggest both the concept itself was adequate and that an assessment methodology was available to ensure safety would be achieved at a chosen repository site.

As is the case of other nations, the Canadian disposal concept relies on a set of engineered barriers: the CANDU fuel pellet, the fuel element and fuel bundle, and the used nuclear fuel steel container with a corrosion-resistant coating, all of which are surrounded by bentonite clay. The entire assemblage is emplaced in the host crystalline rock.

The Federal Government then formed the so-called Seaborn Panel to oversee an extended series of public inquiries on appropriate future stages; these were conducted over the course of two years. The participants in the hearings found them frustrating and irritating. Nongovernmental organizations viewed the entire process as a subterfuge to support the nuclear industry's plans for expansion. AECL considered the proceedings unnecessary and a challenge to its technical competence.[51]

But it was the Seaborn Panel's damning response that proved critical to what later transpired. It maintained[52]:

> From a technical perspective, the safety of the AECL concept has been on balance adequately demonstrated for a conceptual stage of development, but from a social perspective, it has not.
>
> As it stands, the AECL concept for deep geological disposal has not been demonstrated to have broad public support. The concept in its current form does not have the required level of acceptability to be adopted as Canada's approach for managing nuclear fuel wastes.

Ten months after the publication of the Seaborn Panel's report, the Canadian Government accepted the group's advice with respect to social acceptability.

It established in 2002 the Nuclear Waste Management Organization (NWMO), a body mostly funded by the two major generators of high-activity radioactive waste, AECL and (now) Ontario Power Generation.[53]

NWMO crafted what many, including the author, believe is a "gold standard" for a public consultation. The organization made a rigorous commitment to listening carefully first and only afterward proposing a strategy that deeply reflected what was heard. The listening took five years.[54] Along the way, panels of 462 randomly selected citizens met periodically in a dozen locations to discuss long-term disposal of high-activity radioactive waste; face-to-face dialogues took place with a broad range of interested and affected parties; additional conversations were organized via the internet.[55]

Based on these contributions, NWMO proposed to implement a strategy of "adaptive phased management." It argued that its plan had two principal components.[56]

> The *technical method* involves building a deep geological repository in a suitable rock formation to safely contain and isolate used nuclear fuel. The *management approach* involves phased and adaptive decision-making, supported by public engagement and continuous learning.

Importantly, NWMO observed that it was "committed to seeking an informed, willing community to host the long-term management facility ... It is against the backdrop of the community's own vision for its future that we will proceed."[57] The federal government accepted the proposal in June 2007.

Two years later, the organization published a nine-step plan for selecting a repository site.[58] In addition to a set of technical factors favoring choice, such as the characteristics of the host rock, geologic stability, and the low likelihood of human intrusion, NWMO, unique among nations, articulated aspects that "go beyond safety" to consider the *well-being* of the local population. These included social, economic, and cultural effects, long-term sustainability, and the impact on infrastructure. Moreover, NWMO explicitly declared that these latter elements might eliminate an otherwise technically qualified site.

Having erected this elaborate and collaborative structure, NWMO called for communities to express an interest in learning more about the demands and benefits of hosting a deep-mined, geologic repository. Twenty-three of them promptly responded; one was eliminated almost immediately.[59]

The implementor commissioned intensive studies of the technical suitability and community well-being for the remaining 22 locations. Four were dropped in 2013 because of concerns about geologic stability and community well-being. Nine more were excluded either because of unsuitability or for community detriment during the following and the next year. NWMO intensified its evaluation of the remaining nine districts for the subsequent half decade. By 2022, all but two in Ontario Province, Ignace and South Bruce, remained as potential repository sites.

In December, 2023, the implementor published safety assessments for both locations. Both analyses projected a dose level that was approximately three orders of magnitude lower that the regulatory standard.[60] Seven months later, the Township of Ignace confirmed that it was a willing host community. A year later, the Wabigoon

Lake Ojibway Nation also verified its acceptance. The First Nation's leader remarked that the decision is "the most important responsibility of our time."[61] Thereby, Canada became the latest nation to choose the location for its repository of high-activity radioactive waste.

Government Organizations

Four countries abroad with mature programs have chosen a government organization as the waste-management implementor: Germany, France, the United Kingdom, and Japan. It is to their experiences that the chronicle now turns.

Germany

At its peak, Germany operated thirty-six nuclear power plants generating 26.4 MWe.[62] The nuclear industry initially planned to reprocess its SNF, but the government prohibited that activity in 2005.[63]

Prompted by a U.S. National Academy of Sciences' influential 1957 report, West Germany launched its waste-disposition project in the early 1960s (See Chapter 2.).[64] The Federal Institute of Soil Research reinforced the technical suitability of salt formations. For more than a half century, that host rock monopolized every discussion of how to permanently manage the country's HLW and SNF. [65]

By 1976, the search for potential repository sites focused on the *Land* [State] of Lower Saxony. Its government surveyed 140 prospective locations using heavily weighted geotechnical considerations to winnow the search down to six potential settings. Five of them engendered strong public opposition; almost by default, then, the area around the small village of Gorleben, close to the boarder of the former East Germany, became the presumptive choice. Hardly surprisingly, this decision provoked controversy as well. The Lower Saxony Government had to defend what was clearly an *ad hoc* and surreptitious selection process. In addition, unexpected hydrogeologic findings uncovered over the next fifteen years called into question the technical integrity of the site.

The Christian Democratic Union (CDU)/Christian Social Union (CSU) reunified national government collapsed in 1999, ushering in a new coalition composed of the Social Democratic and Green Parties. A pre-election agreement cementing their policy commitments included a pledge to reconsider the country's deep investment in nuclear power and the associated plans for managing the reactors' HLW and SNF.

Upon taking office, the new government formed a siting advisory body, the Working Party on Repository Selection Site, otherwise known as AkEnd. Moreover, the following year the federal government and the nuclear utilities agreed to suspend investigations at Gorleben for at least three and at most ten years. The broadly based committee completed its deliberations in 2002.[66]

As one study observed, AkEnd was established to propose a "siting process designed to identify *the best-possible* repository site in Germany without any spatial preselection and predetermination of the host rock."[67] It offered a set of suitability requirements, including several exclusionary criteria, such as the presence of active fault zones and no evidence of volcanism in the Quaternary Period. It also recommended that intensely participatory mechanisms be put in place. Finally, the panel endorsed the idea of reviving a siting process based on a blank "white map" of Germany; no location, especially Gorleben, would hold a privileged position.

Hopes that the deliberations would lead to consensus among interested and affected groups were dashed when the CDU/CSU withdrew its support. In addition, the industry rejected the recommendations, arguing that the Gorleben site was suitable and, in any evident, investigations of it were already paid for, thereby terminating the owners' further responsibility. Even the sponsoring minister, Jürgen Trittin from the Green Party, recognized that the AkEnd process had reached a dead-end. Instead, he suggested that the nuclear utilities take on the job of managing their own waste as SKB and Posiva did in Scandinavia. This option became moot when a coalition between the CDU/CSU and the Social Democratic parties rose to power in 2005. What followed was six years of not-so-benign waste-disposition neglect.

In the aftermath of the Fukushima Daiichi accident in 2011, the intense national debate over nuclear power was resolved by a plan to shut down the country's reactors.[68] Only then could Germany adopt policies that promised a fresh path toward waste disposal. The Bundestag and Bundesrat authorized the Repository Site Selection Act in 2013. The legislation contained several general provisions, such as a three-stage site-selection procedure starting with a "white map."[69] The process would then look at potential locations in salt, crystalline rock, and clay formations.[70]

The legislation's full implementation awaited recommendations from a newly appointed Commission for the Storage of High-Level Radioactive Waste (EndKo).[71] Given the overlap in membership between the AkEnd and the EndKo panels, it is hardly surprisingly that the two groups reached nearly identical conclusions.[72] But unlike the former, the latter took on real-world importance in framing a path forward. EndKo advanced a transparent multi-level approach, including a National Civil Society Board to oversee the participatory apparatus as well as a re-formed federal implementing authority, the Federal Agency for Final Storage (BGE), as implementor.[73]

EndKo's advice was incorporated in the 2017 amended Repository Site Selection Act. The elaborate new institutional design sought to find "a solution that is supported a broad social consensus and can be tolerated by those who are affected."[74] To safeguard the process, subsequent siting stages must be confirmed by the Bundestag and Bundesrat.

In September 2020, BGE published a "Sub-Areas Interim Report." The document's major conclusion was that the Gorleben site was excluded because it could not meet the law's requirement that it offered the "best possible safety" for a million years. A year later, it put the figurative nail in the Lower Saxony location's coffin by deciding to decommission the site. At the end of 2025, however, half the area of Germany could still be considered suitable for developing a disposition facility.

France

More than any other nation, France depends on nuclear powerplants for roughly 65 percent of its electricity. The country made an early commitment to reprocessing its SNF domestically. It thus must dispose of large quantities of HLW and transuranic (TRU) waste as well as possibly SNF.

France was among the first nations to launch a waste-disposition project in the early 1980s. Its strategy was catholic. The French Atomic Energy Commission

(CEA) by the end of the decade selected four geologic formations for investigation in the following *départements*: Aisne (clay), Deux-Sèvres (granite), Ain (salt), and Maine-et-Loire (schist).[75] The choices were "public but not published." Moreover, the areas affected were not informed about their selection until the CEA drillers appeared.[76] It was no wonder, then, that between 75 and 95 percent of the local populations opposed the investigations.[77] The protests became so heated and disruptive that the Prime Minister Michel Rocard ordered a yearlong halt to the studies at the four sites.

In response, the Parliament passed the Research on Radioactive Waste Management Act in 1991; it profoundly altered the *status quo*. Because of the deep distrust engendered by CEA, the legislation created a new implementor, the, National Agency for Radioactive Waste Management (Andra), a public institution reporting to the Ministries of Research, Industry, and Environment. The Act authorized a three-pronged, 15-year research program, and sought to promote public engagement through the establishment of Local Information and Oversight Committees.

The government appointed Christian Bataille, the Act's chief sponsor, to organize a transparent siting process. He proposed to "verify the geological feasibility of the projects that will be volunteered by interested regions, and not, as was done before, attempt to convince populations [that were] preselected."[78] Bataille conducted public meetings in eight *départements* and ultimately identified four potential sites for an underground research laboratory in Garde (clay), Vienne (granite), Meuse (clay) and Haute-Marne (clay).

Opposition from nearby wine growers eliminated the Garde location. Andra ultimately excluded Vienne because the crystalline rock was unsuitable.[79] The implementor then merged the adjacent Meuse and Haute-Marne locations and decided to center its activities in the border community of Bure in 1998. Although the district around Bure volunteered for a laboratory, it understood that the research facility might be transformed into a repository without its explicit assent. Indeed, that conversion took place in 2006.[80]

Like other nations, France's disposal concept relies on both engineered and natural barriers to prevent the migration of the high-activity radionuclides. The former consists of vitrified high-level reprocessed waste emplaced in welded stainless steel canisters. The packages are buried in the Callovo-Oxfordian clay formation.

In 2006, the National Commission of Public Debate concluded a dialogue from which the idea of a "reversible" repository gained traction. Shortly thereafter, Parliament endorsed the notion when it passed the Radioactive Materials and Waste Planning Law.[81] In 2009, Andra submitted its plan for constructing a disposition facility near Bure called Cigéo. Included in that proposal were ventures to improve the economic wellbeing of the community. In the absence of the legally mandated site-specific information about a repository built in crystalline rock formations, Andra used generic data mostly from Sweden.

A second, but highly controversial and confrontational, Public Debate occurred in 2013. Andra spent the next three years attempting to reassess and reformulate its waste-disposition program in response to the criticisms voiced in the latest Public Debate. A 2016 law formally defined reversibility as a "concept that allows future generations to choose between either continuing the construction and operation of disposal through successive phases or to re-examine the earlier choices and modify the management solutions accordingly."[82] It also provided further details about the Cigéo project.

The implementor submitted its application for a "declaration of public utility" to the Public Debate Commission in mid-2020. The government approved the request two years later. Andra submitted to the Ministry for Energy Transition a license to construct Cigéo in January 2023. The permit is currently under review by the Authority for Nuclear Safety and Radiation Protection (ASNR).[83]

United Kingdom

The waste-disposition experience in the United Kingdom may surpass even that of the United States in terms of conceiving irresponsible ancestors.[84] The nation produced its first HLW and SNF material as part of its defense program in the early 1950s. Shortly thereafter, the initial Magnox civilian reactors came on line.[85] From the start, the country committed to reprocessing its SNF. Although its THORP facility closed for good in 2018, until recently the government has refused to reclassify SNF and separated plutonium as "waste." Moreover, the country still makes a formal distinction between "legacy" and so-called "new-build" waste.[86]

The granitic Mullwharchar Hill south of Loch Doon in Scotland lies in a "wilderness area of natural beauty."[87] It was here in 1976 that the United Kingdom Atomic Energy Agency chose to drill the earliest boreholes to find a site to dispose of its intermediate-level radioactive waste (ILW).[88] Local outrage triggered a Public Inquiry, almost immediately resulting in the cancellation of the drilling.

Contemporaneously with the breakdown of the Scottish venture, the Sixth Report of the United Kingdom's Royal Commission on Environmental Pollution strongly cautioned: "There should be no commitment to a large program of nuclear fission power until it has been demonstrated beyond reasonable doubt that a method exists to ensure the safe containment of long-lived highly radioactive waste for the indefinite future."[89]

It took five years of debate before the private sector Nuclear Industry Radioactive Waste Executive (Nirex) was created to examine more systematically options for radioactive waste disposal. Another five years later, it launched the "The Way Forward" project with a nine-step siting strategy. In the first step, Nirex winnowed nearly 500 potential sites down to two: Dounreay in northern Scotland near the experimental breeder reactor and the Sellafield complex in west Cumbria, home to the country's nuclear-weapons production complex and reprocessing facilities.[90] As one analysis observed[91]:

> The selection of those two sites took place in the face of extensive public consultations that revealed strong opposition in both communities. Based on evaluations of boreholes drilled at both sites in 1989, the implementor concluded that Sellafield was the preferred location.

Nirex then submitted a planning application to the Cumbria County Council to construct a Rock Characterization Facility. The Council opposed the application and, relying on its legal authority, called for another Public Inquiry. That hearing took place over five months, ending in February 1996. The inquiry's moderator determined that the application should be denied, a decision that was ratified by the Secretary of State for the Environment a year later.

These events left the community with a well-deserved lingering sense of distrust. As a confidential Nirex internal report later documented, "A culture of secrecy operated within Nirex which bred public hostility. Incomplete data were released, and the [implementor] admits the case for a deep waste storage at Sellafield 'was not as good as implied by public statements made by Nirex'."[92]

Into the breach stepped the House of Lords, which published a report in 1999 on how to reconstruct the country's waste-management project.[93] The statement's primary advice was that "government needs to build public trust in the concept by starting public consultation on a new national radioactive waste management policy. A central aim of the policy should be to give the public and parliament more say in decision-making than they have had in the past."[94]

The central government formed the independent Committee on Radioactive Waste Management (CoRWM) in November 2003. Its remit was to explore options for the long-term management of HLW and SNF. More importantly, it was to recommend a strategy that could inspire public trust and confidence.

CoRWM presented its recommendations in mid-2006. It emphasized the importance of continued public and stakeholder involvement, a commitment to voluntarism, a partnership approach, the community's right to withdraw up to a predetermined point, and ratification of key decisions by appropriate democratically elected bodies.[95] A year later, the government issued a "White Paper," which largely accepted CoRWM's advice.[96] At the same time, a new organization, the Nuclear Decommissioning Authority (NDA), was selected as implementor.[97]

Three local authorities near and around Sellafield complex—Allerdale Borough Council, Copeland Borough Council, and Cumbria County Council—decided to form a partnership to learn more about hosting a deep-mined geologic repository.[98] The collaboration ultimately featured a total of seven local governments and a dozen nongovernmental organizations. Disinterested observers, including the NWTRB, praised the partnership's seriousness and objectivity. The group released an extensive report, which was forwarded without recommendations to the three founding communities.[99]

The central government responded shortly thereafter. It declared that all three Cumbria authorities had to issue a "green light" before the site-selection process could be completed. Allerdale and Copeland Borough Councils said "yes," but the Cumbria County Council said "no." According to the rule, the partnership screeched to a halt.[100]

The United Kingdom Government tried to revive the siting process in 2013. But its initial labors were mostly in vain, largely because they were tied to failed plans to breathe life into the construction of new reactors. Nonetheless, a new quasi-governmental executor, Nuclear Waste Services, was able to enroll four new partnerships in 2022: Allerdale, Mid-Copeland, South Copeland, and Theddlethorpe, which is situated in Lincolnshire on the east coast.[101] It remains unclear whether any of those efforts will give rise to a technically sound and socially acceptable repository site.

Japan

Lacking indigenous energy sources, Japan embarked on a major program to construct nuclear power plants. At its peak, the country operated 60 reactors with a total capacity of nearly 60,000 MWe.[102] The nuclear utilities opted for a closed fuel cycle with their SNF reprocessed at a large facility located in the Rokkasho Municipality in Amori Prefecture. Construction began in 1993. Thirty years later, the Japan Nuclear Fuel Ltd forecast that the plant's operation would commence in late 2024, yet another prophecy what would go unrealized.[103] In the meantime, most of the 19,000 metric tons of SNF are stored in reactor cooling pools. In addition, significant quantities of transuranic waste (TRU) from the decommissioning of 27 shutdown reactors sit in temporary storage. If reprocessing begins, additional amounts of TRU will also have to acquire a permanent home.

To find a long-term management solution, the Diet passed the 2000 Specified Radioactive Waste Final Disposal Act. It laid out a three-stage process for selecting a repository site: literature surveys, preliminary investigations, and detailed investigations. Shortly thereafter, the government established the Nuclear Waste Management Organization (NUMO), which reports to the Ministry of Economy, Trade, and Industry (METI). NUMO got off to a quick start in 2002, sending an open invitation to more than 3,000 communities. It asked whether they would volunteer for stage-one literature surveys. Nine of them stepped forward; soon, however, all but one, Toyo Town in Kochi Prefecture, lost their early interest.[104]

The municipality's mayor informed METI in 2007 that it was willing to come forward. The backlash was almost immediate. The Prefecture Governor understatedly declared, "I feel a strong sense of doubt. Something is awry with Japan's nuclear energy policy if procedural work is allowed to begin … even though local agreement has not been obtained."[105] The town's citizens recalled the mayor. His replacement was elected with a 2-1 majority, largely based on his anti-repository platform.

METI's first response was to tinker with its voluntarist approach by enhancing its public relations activities. It also suggested that it might "summon" communities to permit literature surveys. When that tack proved fruitless, nuclear-waste policymakers recognized that more fundamental adjustments to the siting process might be essential. The Japanese Atomic Energy Commission called upon the Science Council of Japan (the equivalent to the U.S. National Academy of Sciences) to examine the country's waste-management strategy and suggest improvements. As these events unfolded, the great Tōkoku earthquake and tsunami wreaked havoc on the Fukushima Daiichi nuclear power plant, causing a dramatic decline in public trust in the nuclear authorities and industry.

The Council's "Reply" was delivered in September 2012 and severely disrupted METI's plans.[106] The document made both a procedural argument and a technical one. The former held that "seeking a consensus on an individual issue of selecting the final HLW disposal site before reaching a consensus on broader policies concerning nuclear power generation was procedurally inverted and thus inappropriate." The latter finding maintained that the fundamental question was not social acceptability; rather given Japan's widespread volcanic and seismic hydrogeologic features, "there is a limit to what we can do with currently available scientific knowledge and technological capacities to search out geological formations that will remain stable for tens of millennium."[107]

COUNTRY	INVESTIGATIONS STARTED	SITE SELECTED	YEARS TO SELECT SITE	OPERATIONS COMMENCE
Canada	1975	Yes	49 years	~2040
Finland	1983	Yes	17 years	~2025
France	1983	Yes	18 years	~2030
Germany	1965	No	>65 years	???
Japan	2000	No	>35 years	???
Sweden	1980	Yes	31 years	>2030
Switzerland	1972	Yes	50 years	>2035
United Kingdom	1987	No	>45 years	???

FIGURE 7.6 Progress toward developing a deep-mined, geologic repository.

Note: Countries in the dark-shaded rows have adopted waste producer-owned implementors.

In a partial response to the Science Council's advice, the Japanese Cabinet in 2015 revised the country's strategy. A set of screening criteria were adopted for identifying potential repository locations prior to restarting the three-stage process. In July 2017, METI released a map of regions that are likely to meet the necessary geological requirements for hosting a repository.[108] For all practical purposes, the plot eliminated virtually all of the country's interior.

NUMO also strengthened its public engagement activities. Three years later, Suttu Town and Kamoenai Village on the southern coast of Hokkaido Island volunteered for literature surveys. NUMO opened local offices in 2021 to facilitate exchanges with both communities.[109] That same year, the organization issued a comprehensive report laying out its underlying safety case for a repository situated in three different hydrogeologic formations: plutonic rock, Neogene sedimentary rock, and Pre-Neogene sedimentary rock.[110] NUMO completed the literature surveys in 2024 and forwarded them to Prefecture Governor and the mayors of the two municipalities.

Figure 7.6 summarizes the status of waste-disposition programs in the eight pluralist nations.

The next chapter builds on this descriptive information to gain a firmer grasp of *why* some nations have selected a site for their repositories and others have not.

NOTES

1. See Daniel Metlay and Daniel Sarewitz, "Decision Strategies for Addressing Complex, 'Messy' Problems," *The Bridge*, (42) Fall 2012, 6–16. In a paper drafted for the Secretary of Energy Advisory Board, I first applied this notion to radioactive waste management in 1993. Secretary of Energy Advisory Board, "Earning Public Trust and Confidence: Requisites for Managing Radioactive Waste," November 1993. Achim

Brunnengräber unnecessarily complicates the concept in "The Wicked Problem of Long-Term Radioactive Waste Governance," in Achim Brunnengräber and Maria Rosaria Di Nucci (eds.), *Conflicts, Participation, and Acceptability in Nuclear Waste Governance: An International Comparison* (Wiesbaden: Springer, 2019), 113–140.
2. James Thompson, *Organizations in Action* (New York: McGraw Hill, 1971).
3. Many of these ideas were first presented in Daniel Metlay, "Becoming a Responsible Ancestor," *Bulletin of the Atomic Scientists*, (81) January 2025, 39–35.
4. This outcome is hardly surprising given that DOE's John Deutch was the IRG's chair. As Chapter 3 notes, there were bigger battles to be fought between virtually all of the IRG's members and DOE.
5. Mason Willrich and Richard Lester, *Radioactive Waste: Management and Regulation* (New York: Free Press, 1977), 105. This analysis was paid for by ERDA, the AEC's successor agency.
6. Willrich and Lester, 107. The enquiry also considered making state or international governments the waste-management implementor.
7. Advisory Panel on Alternative Means of Financing and Managing Radioactive Waste Facilities, "Managing Nuclear Waste—A Better Idea [AMFM1]," December 1984.
8. The initial alternatives were: OCRWM, Sub-Cabinet office, Administration within a Cabinet Department, Federal executive agency, Independent Federal Commission, Government-controlled corporation, Government-charted Corporation, Mixed Government-Private Corporation, Utility-type private organization, and Private corporation.
9. AMFM1, Table 8.1, VIII: 14.
10. In a letter from Nunzio Palladino, the NRC Chairman, to John Herrington, Secretary of Energy (Hodel's successor), March 22, 1985, the regulator explicitly declined to comment on whether FEDCORP would be more responsive to NRC than OCRWM.
11. AMFM1, XI: 8-9.
12. D.J. Boggs et al., "Report to the Secretary of Energy on the Conclusions and Recommendations of the Advisory Panel on Alternative Means of Financing and Managing (AMFM) Radioactive Waste Management Facilities," Washington, D.C., 1985.
13. Department of Energy, "Alternative Means for Financing and Managing the Civilian Radioactive Waste Management Program [AMFM2]," DOE/RW-0546, August 2001.
14. AMFM2, 29. The report did state "future decisions will be more informed after the site recommendation process is completed." As Chapter 6 describes, the next year Secretary Spenser Abraham recommended the Yucca Mountain site. Over the following decade, DOE never reconsidered the institutional design for the implementer.
15. AMFM2, 26. As with AM-FM1, many of these outcomes involve trade-offs. There are some important differences between the two analyses. For example, maintaining credibility and responsiveness to regulators are missing in AMFM2.
16. AMFM2, 41.
17. Blue Ribbon Commission on America's Nuclear Future, "Report to the Secretary (BRC)," January 2012, 61.
18. BRC, 62.
19. BRC, 61. "While a corporate structure appears to the Commission to offer particular advantages, previous studies have concluded that a number of different organizational forms could also get the job done." (BRC, 62.)
20. Lynn E. Davis, et al., *Choosing a New Organization for Management and Disposition of Commercial and Defense High-Level Radioactive Materials* (Santa Monica CA: RAND Corporation, 2012).
21. Davis, 63.
22. Department of Energy, "Strategy for the Management and Disposal of Used Nuclear Fuel and High-Level Radioactive Waste," January 2013.
23. Strategy, 9.
24. Steering Committee, "Reset of America's Nuclear Waste Management (RESET)," 2018. The members of the Steering Committee, which included the author, are listed in Appendix A of that study. Not surprisingly, the discussion of the organizational analyses of institutional design draws heavily from that report.

25. The implementer in France is a hybrid organization. Its activities are detailed in the following subsection.
26. Reset, 35.
27. Members of the Steering Committee briefed the one of the nuclear industry’s main lobbying organizations, the Nuclear Energy Institute, shortly after the publication of its report. That group was not enamored at all by the idea of NUCO. Eight years later, it appears that the industry’s attitude may be softening. See the mission statement by the U.S. Nuclear Industry Council Back-End Working Group https://www.usnic.org/working-groups/back-end-fuel-cycle-working-group. Accessed September 24, 2024, 4.
28. Katie Tubb and Jack Spencer, “Real Consent for Nuclear Waste Management Starts with the Free Market,” March 22, 2016. The article can be found at https://www.heritage.org/environment/report/real-consent-nuclear-waste-management-starts-free-market. Accessed on September 17, 2024.
29. As the reader will recall from the discussion in the previous chapter, this scenario is not an abstract or theoretical one.
30. Expert Group on Disposal Concepts for Radioactive Waste (EKRA), “Disposal Concepts for Radioactive Waste: Final Report,” On behalf of the Federal Department for the Environment, Transport, Energy and Communication, 2000.
31. See Krütli et al as well as Sophie Kuppler et al., “Who Decides What is Safe? Experiences from Radioactive Waste Governance in Switzerland,” in Maarten Arentsen and Rinie van Est (eds), *The Future of Radioactive Waste Governance: Lessons from Europe* (Wiesbaden: Springer, 2023), 169–198.
32. Swiss Federal Office of Energy (SFOE), “Sectoral Plan for Deep Geological Repositories: Conceptual Part,” 2008.
33. See https://nagra.ch/en/knowledge-centre/site-selection-process/#:~:text=The%20search%20for%20the%20most,deep%20geological%20repository%20in%202022. Accessed November 4, 2024.
34. See Peter Hocke and Sophie Kuppler, “Participation under Tricky Conditions: The Swiss Nuclear Waste Strategy Based on the Sectoral Plan,” in Achim Brunnengräber et al. *Nuclear Waste Governance: An International Comparison* (Wiesbaden: Springer, 2015), 157–176.
35. Nagra filed a general license application with the Swiss Federal Office of Energy to construct the repository in November 2024.
36. Thomas Kaiserfeld and Arne Kaijers, “Changing the System Culture: Mobilizing the Social Sciences in the Swedish Nuclear Waste System,” *Nuclear Technology*, (207) September, 2021, 1,456–1,468; Gäran Sundqvist, *The Bedrock of Opinion: Science, Technology, and Society in the Siting of High-Level Nuclear Waste* (Dordrecht: Kluwer Academic Publishers, 2002).
37. Vattenfall AB (36%), Forsmarks Kraftgrupp AB (30%), OKG Aktiebolag (22%), and Sydkraft Nuclear Power AB (12%) are the current owners. See https://skb.com/about-skb/organisation. Accessed October 10, 2024.
38. Under the Nuclear Activities Act passed in 1984, the requirement for absolute safety was replaced with a risk limit on the radioactive material released into the environment. That limit must be complied with for at least 100,000 years. In 1999, the Swedish Parliament passed the Environmental Code, which required the approval of a specialized court for major infrastructure projects including a repository. See Chapter 10 for additional information.
39. See https://www.mkg.se/en/the-swedish-government-disregards-the-opinion-of-the-environmental-court-and-approves-the. Accessed November 4, 2024.
40. Oskarshamn agreed in 2018 to forego its veto power over the construction of a spent fuel encapsulation facility.
41. A full discussion of the licensing process in Sweden appears in Swedish National Council for Nuclear Waste, “Licensing under the Environmental Code and the Nuclear Activities Act of a Final Repository for Spent, Nuclear Fuel,” 2013. See https://www.karnavfallsradet.mkg.se/sites/default/files/documents/report_2011_2.pdf. Accessed November 4, 2024.

42. A fifth reactor at Olkiluoto began generating electricity in 2022.
43. The Finnish-led consortium, Fennovoima, contracted with Russia in 2013 to construct a reactor in northern Finland. Intermittent negotiations between the firm and Posiva ensued for nearly a decade to determine what terms, if any, Fennovoima would have to accept to dispose of its SNF in Onkalo; the discussions were unsuccessful. In 2022, Fennovoima cancelled the transaction.
44. McEwen and Äikäs, 35–43. See also Matti Kojo, "The Strategy of Site Selection for the Spent Nuclear Fuel Repository in Finland," in Matti Kojo (ed), *The Renewal of Nuclear Power in Finland* (London: Palgrave Macmillan, 2009), 161–191.
45. TVO considered but dismissed a range of technical alternatives along with KBS-3: hydraulic cage, deep boreholes, reprocessing and final disposal, and nuclide separation and transmutation. See Posiva, "The Final Disposal of Spent Nuclear Fuel: Environmental Impact Assessment Report," 1999, 24–25. See also Matti Kojo and Anna Oksa, "Adaption of the Swedish KBS Disposal Concept to Finland: A Technology Transfer Case Study," 2014. https://trepo.tuni.fi/handle/10024/95667. Accessed November 11, 2024.
46. Tapio Litmanen et al., "The Rationality of Acceptance in a Nuclear Community: Analyzing residents' Opinions on the Expansion of the SNF Repository in the Municipality of Eurajoki, Finland," *International Journal of Nuclear Governance, Economy, and Ecology*, (3) 42–58, 2010.
47. The most comprehensive discussion of the Canadian nuclear waste-management program is William Leiss, *Deep Disposal: A Documentary Account of Burying Nuclear Waste in Canda* (Montreal: McGill-Queen's University Press), 2024.
48. Darrin Durant, "Responsible Action and Nuclear Waste Disposal," *Technology in Society*, (31) 2009, 150–157.
49. Atomic Energy Canada Limited. "Environmental Impact Statement on the Concept for Disposal of Canada's Nuclear Fuel Waste," AECL-10711, 1994.
50. Durant, 152. Although designed for the essentially the same host rock, the Canadian safety case differs somewhat from KBS-3. AECL proposed at the start to deploy titanium alloy canisters. The current implementor has decided to use a carbon-steel container coated with copper to replace the Swedish copper canister. Unlike SKB's plan to place the canister in a borehole drilled in the bedrock and topped with bentonite, NWMO intends to place a loaded bentonite buffer box directly on the host rock.
51. Durant offers the most comprehensive discussion of the hearings. See especially 151–154.
52. Seaborn Panel, "Report of the Nuclear Fuel Waste Management and Disposal Concept Environmental Assessment Panel," 1998. https://iaac-aeic.gc.ca/archives/evaluations/431C8844-1/default_lang=En_n=0B83BD43-1_printfullpage=true.html#ws6B85477A. Accessed November 12, 2024.
53. New Brunswick Power Corporation and Hydro-Québec are small minority participants.
54. See Thomas Webler et al., "Technical Appendix: Case Studies of Public and Stakeholder Engagement: A Technical Report Prepared for the Blue Ribbon Commission of America's Nuclear Future," Social and Environmental Research Institute, Greenfield, Massachusetts, 2011. https://cybercemetery.unt.edu/brc/20120620223108/http:/brc.gov/sites/default/files/documents/webler_etal_pse_report16june11.pdf. Accessed November 13, 2024.
55. To frame the discussions, NWMO released a variety of documents. Three of the most important are Nuclear Waste Management Organization, "Asking the right questions? The Future Management of Canada's Used Nuclear Fuel," 2003; Nuclear Waste Management Organization, "Understanding the Choices: The Future Management of Canada's Used Nuclear Fuel," 2004; and Nuclear Waste Management Organization, "Choosing a Way Forward: The Future Management of Canada's Used Nuclear Fuel (Final Study)," 2005.
56. See https://www.nwmo.ca/canadas-plan. Accessed November 13, 2014. It is worth noting that NWMO's safety case only departed narrowly from AECL's.

57. NWMO (2005), 40.
58. Nuclear Waste Management Organization, "Moving Forward Together: Designing a Process for Selecting a Site," 2009. See also Nuclear Waste Technical Review Board, "Designing a Process for Selecting a Site for a Deep-Mined, Geologic Repository for High-Level Radioactive Waste and Spent Nuclear Fuel: Detailed Analysis," Washington D.C., 2015.
59. NWMO eliminated Red Rock, Ontario, almost immediately.
60. See Nuclear Waste Management Organization, "Confidence in Safety—Revell Site—2023 Update," NWMO-TR-2023-07, December 2023 and Nuclear Waste Management Organization, "Confidence in Safety—South Bruce Site—2023 Update," NWMO-TR-2023-06, December 2023.
61. Quoted in *The Economist*, December 7, 2024, 24.
62. World Nuclear Association, "Nuclear Power in Germany." See https://world-nuclear.org/information-library/country-profiles/countries-g-n/germany. Accessed February 12, 2025.
63. International Atomic Energy Administration, "Germany: National Framework for Management of Radioactive Waste and Spent Fuel." See https://sris.iaea.org/country-overview/introduction/DE/Germany. Accessed February 12, 2025.
64. See Chapter 2.
65. NWTRB (2015) provides a detailed discussion of the early evolution of Germany's waste-management program. See especially 98–101. Also, Brunnengräber and Di Nucci (eds.), *Conflict, Participation and Acceptability in Nuclear Waste Governance: An International Comparison* (Wiesbaden: Springer, 2019). assembled a useful collection of articles related to the German waste-disposition program: Daniel Häfner, "The Future is Still Unwritten—History Too: Overcoming the Conflicts of the Past in Germany;" Anselm Tiggemann, "The Elephant in the Room—The Role of Gorleben and its Site Selection in the German Nuclear Waste Debate;" Jan-Hendrix Kamlage et al., "Chances, Challenges, and Choices of Participation in Siting a Nuclear Waste Repository: The German Case;" Ana Maria Isidoro Losada et al., Experts and Politics in the German Nuclear Waste Governance: Advisory Bodies between Ambition and Reality;" Julia Olliges, "A 'Deliberative Turn' in German Nuclear Waste Governance? The Participation Process of the Commission on the Storage of Radioactive Waste;" Miranda Schreurs and Jordina Suckow, "Bringing Transparency and Voice into the Search for a Deep Geological Repository;" and Peter Hocke and Achim Brunnengräber, Multi-Level Governance of Nuclear Waste Disposal: Conflicts and Contradictions in the German Decisionmaking System."
66. Committee on a Selection Procedure for Repository Sites (AkEnd), "Selection Procedure for Repository Sites: Final Report for the Federal Ministry for the Environment, Nature Conservation, and Nuclear Safety," 2002.
67. European Commission, Cooperative Research on the Governance of Radioactive Waste (COWAM 2), Final Report, Brussels, Belgium, 2006. See https://cordis.europa.eu/docs/projects/files/508/508856/99009291-6_en.pdf. Accessed November 20, 2024.
68. Dorothy Nelkin and Michael Pollack, *The Atom Besieged: Antinuclear Movements in France and Germany* (Cambridge, Massachusetts: Massachusetts Institute of Technology Press, 1981). Germany closed its last operating reactor in April 2023.
69. The law did not exclude the Gorleben location. To do so, its authors believed, would undermine the technically driven "white map" siting process.
70. The geology of Germany features significant clay formations in the north, a narrow band of salt mid-country, and crystalline formations in the southeast, centered in Bavaria. See Maria Rosaria Di Nucci and Achim Brunnengräber, "The Long Road Towards the Soft Nuclear Repository State: Nuclear Waste Governance in Germany," in Arentsen and Van Est (eds), 125.
71. The commission's report is not available in English. For the German text, see https://www.bundestag.de/resource/blob/434430/bb37b21b8e1e7e049ace5db6b2f949b2/drs_268-data.pdf. English version produced by Google Translate. Accessed November 22, 2024.

72. Although the membership of the committee was broad, antinuclear groups largely boycotted the panel. See Klamage et al., 103.
73. See Olliges for a nice discussion about whether the EndKo exercises fulfills Habermas' conceptualization of a "deliberative democracy." She concludes that it mostly did not.
74. Translation of §5(2) provided by Kamlage et al., 102.
75. Ray Kemp, *The Politics of Radioactive Waste Disposal* (Manchester University Press: Manchester, England, 1992).
76. Clair Mays, "Where Does It Go: Siting Methods and Social Representations of Radioactive Waste Management in France." in A. Boholm and R. Löfstedt, eds., *Facility Siting: Risk, Power, and Identity in Land-Use Planning*. 21–43. (London: Earthscan, 2004).
77. Swedish National Board for Spent Nuclear Fuel (SKN) 'Survey of Siting Practices for Selected Management Projects in Seven Countries," Report 60, Stockholm, 1992.
78. Quoted in Mays (2004), 26.
79. The 1991 legislation called for the comparison of potential sites in two hydrogeologic formations. Despite Christian Bataille's best efforts, no community overlaying crystalline rock formations volunteered to host an underground research facility.
80. At the same time, Andra released its "Argillite Dossier" to justify its selection of clay over granite as the preferred hydrogeologic formation. Given that it was prevented from conducting studies in crystalline rock formation, Andra used representative data from Sweden. See https://international.andra.fr/sites/international/files/2019-03/6-%20Dossier%202005%20Argile%20Tome%20-%20Safety%20evaluation%20of%20a%20geological%20repository_0.pdf. Accessed December 5, 2024.
81. At the time, Andra noted that reversibility "is not the direct result of a technical or scientific need, but of social and political choice, which engineers and scientists are duty-bound to accept." Quoted in footnote 8 in Markku Lehtonen, "Megaproject Underway: Governance of Nuclear Waste Management in France," in Achim Brunnengräber et al., *Nuclear Waste Governance: An International Comparison* (Wiesbaden: Springer, 2015), 117–138.
82. Lehtonen, 239–240.
83. Andra specifically emphasizes its intergenerational obligations. "Over and above the political, economic, scientific, technical and technological aspects, a failure to deal with this waste here and now would pass on the responsibility for this task to future generations when all the technical and technological expertise as well as the financial possibilities are already available setting up Cigeo." See https://international.andra.fr/projects/cigeo/protection-most-hazardous-radioactive-waste/geological-disposal-protection#:~:text=The%20Cigeo%20deep%20geological%20disposal,stable%20for%20160%20million%20years. Accessed January 8, 2024.
84. Further complicating matters is Scotland's refusal to accept disposal of high-activity waste in a deep-mined, geologic repository. The country's position is to promote long-term surface storage of that material.
85. The Magnox design was named after the magnesium–aluminum alloy used to clad the fuel. The reactors used unenriched uranium and were cooled using carbon dioxide and moderated using graphite. The last unit closed in 2015. In the early 1960s, the United Kingdom switched reactor design to the Advanced Gas-Cooled Reactor, eight of which are still operating. A more traditional Pressurized Water Reactor also is producing electricity.
86. Despite many attempts since 2006, only one reactor is under construction. It is plagued with substantial delays and cost overruns.
87. Scottish Campaign to Resist the Atomic Menace, "Poison in Our Hills," 1980, 7.
88. The reasoning behind this focus on ILW was the belief that reprocessed HLW waste would decay to ILW along before there was a need to dispose of it.
89. Royal Commission on Environmental Pollution, "Nuclear Power and the Environment," 1976, 131.
90. Boreholes were also drilled at ten other sites, but their location was kept secret until 2005. https://webarchive.nationalarchives.gov.uk/ukgwa/20150806091433/http://www.nda.gov.uk/publication/geological-disposal-historic-list-of-possible-locations-for-a-radioactive-waste-repository. Accessed November 27, 2024.

91. Daniel Metlay, "Social Acceptability of Geologic Disposal," in *Encyclopedia of Nuclear Energy*, Ehud Greenspan (ed.) (Amsterdam: Elsevier, 2021), 4.
92. See https://www.theguardian.com/uk/2001/jul/23/nuclear.world. Accessed November 27, 2024.
93. See https://publications.parliament.uk/pa/ld199899/ldselect/ldsctech/41/4101.htm. Accessed November 27, 2024.
94. The report itself is not available online. But see a summary https://www.nature.com/articles/18499. Accessed November 27, 2024.
95. Committee on Radioactive Waste Management (CoRWM), "Managing Our Radioactive Waste Safely: CoRWM's Recommendations to Government," 2006.
96. Department for Environment, Food, and Rural Affairs, "Managing Radioactive Waste Safely: A Framework for Implementing Geological Disposal." 2008.
97. Subsequently, Radioactive Waste Management, a public organization within the NDA, took over the responsibility for planning and delivering geological disposal. In 2022, NDA merged that agency with the larger Nuclear Waste Services.
98. The collaboration was called the West Cumbria MRWS Partnership.
99. West Cumbria MRWS Partnership, "Final Report of the West Cumbria Managing Radioactive Waste Safely Partnership," 2012.
100. See NWTRB (2015) for a much more detailed description of these events.
101. Stephen Thomas, "UK Nuclear Waste Policy: 50 Wasted Years," in Maarten Arentsen and Rinie van Est (eds), 199–228. In 2023, Nuclear Waste Services withdrew from Allerdale because the hydrogeology was unsuitable.
102. See https://world-nuclear.org/information-library/country-profiles/countries-g-n/japan-nuclear-power. Accessed November 15, 2024. After the Fukushima Daiichi nuclear accident, twenty-seven of them, generating slightly of 17,000 MWe, were shutdown.
103. The construction cost of the reprocessing plant has ballooned from $5 billion to nearly $21 billion. See https://www.asahi.com/ajw/articles/15183716. Accessed November 17, 2024.
104. For a fuller discussion of this saga, see NWTRB (2015), especially 132–133.
105. Reported in Japanese Atomic Industrial Forum, "NUMO Accepts Toyo Town's Application to Become Final Radwaste Disposal Site," *Atoms in Japan*, January 26, 2007.
106. No English translation of the *Reply from the Science Council of Japan to the Japanese Atomic Energy Commission* (JAEC) is available. It is summarized in *Japan for Sustainability Newsletter 124*, December 2012. A JAEC Statement, "Renewing Approaches to Geological Disposal of High-Level Waste," December 18, 2012, also describes the *Reply's* main recommendations. See https://www.aec.go.jp/jicst/NC/about/kettei/121218_e.pdf. Accessed November 18, 2024.
107. NWTRB (2015), 132.
108. See https://www.numo.or.jp/en/jigyou/Nationwide_Map_of_Scientific_Features.pdf. Accessed November 18, 2024.
109. In June 2024, NUMO began a literature survey in Genkai on Kyushu Island. See https://japannews.yomiuri.co.jp/society/general-news/20240802-202369. Accessed November 18, 2024.
110. Nuclear Waste Organization of Japan (NUMO), "The NUMO Pre-Siting SDM-Based Safety Case," 2021; Nuclear Energy Agency, "The Nuclear Waste Management Organizations of Japan's Pre-Siting Safety Case Based on the Site Descriptive Model: An International Peer Review of the NUMO Safety Case," 2023; and Tetsuo Fujiyama and Kenichi Kaku, "Current Status of the Geological Disposal Programme and an Overview of the Safety Case at the Pre-Siting Stage in Japan," *Rock Mechanics Bulletin*, (2) 2023, 1–15.

8 Selecting a Repository Site

A Talmudic legend claims that "it took God twenty-seven tries to create the world. Twenty-six attempts failed, which is encouraging to all creators: nothing great was ever done on the first try."[1] With the exception of Finland, the global development of a disposition facility substantiates this insight of the sages.

Choosing a location for a deep-mined, geologic repository in Canada, Finland, France, Sweden, and Switzerland demanded the efforts of at least one generation and sometimes two. For those countries still searching, even more time has elapsed.[2] What was once optimistically envisioned as a straightforward, even trivial, scientific and engineering task has mostly been transformed into a controversial, fraught, and complicated sociotechnical undertaking.[3] This chapter offers some explanations for that shift.

MODELS AND METAPHORS

In Figure 8.1, a repeat of Figure 3.3, the disconnect between the solid-colored discs, which delineate a set of potential sites, signifies that a disposition facility must demonstrate both technical suitability and social acceptability; otherwise, its development will almost certainly encounter opposition on either or both accounts.[4]

Figure 8.2, a repeat of Figure 3.4, suggests that site-selection is an iterative process. Implementors begin with multiple possible settings and then winnow them down. Importantly, the order in which settings "pass" through the technical and social "filters" may vary from country to country. In some nations, efforts begin by seeking suitable locations and then determining whether any of them fall within the boundaries of jurisdictions prepared to accept them. In other countries, that order is reversed.

STRATEGIC CHOICES

National laws, regulations, and traditions dictate how two foundational decisions are reached. First, should the hunt for sites be carried out in a *serial* or *parallel* fashion? Second, should the exploration *focus initially* on finding technically suitable ones or socially acceptable areas?

Process

Figure 8.3 illustrates the difference between a serial and parallel process. If the former process is embraced, then the search terminates as soon as a single qualified (i.e., a technically suitable and socially acceptable) location is found. If the latter

DOI: 10.1201/9781003528302-10

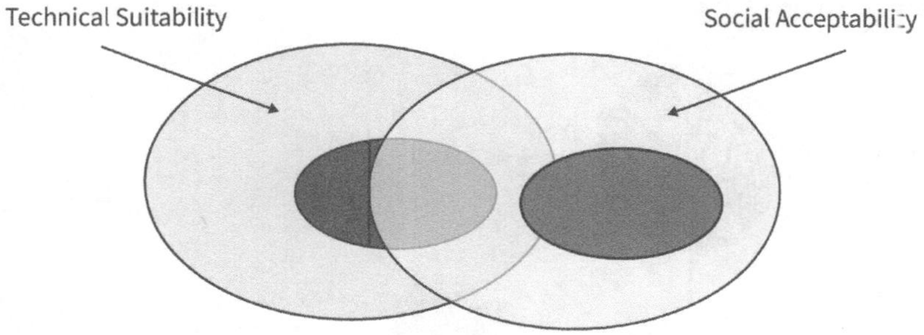

FIGURE 8.1 Prerequisites for choosing a repository site.

approach is adopted, then implementors identify and compare a suite of qualified sites before choosing a final setting.

Early attempts to identify a disposition location pursued a serial process in nations such as the United States, Germany, Switzerland and the United Kingdom. As Chapter 3 documented, the Atomic Energy Commission (AEC) concentrated on an area around Lyons, Kansas, until technical and political obstacles arose. Even afterward, the so-called Deutch report during the Carter Administration proposed developing a single site before considering alternatives.[5] As Chapter 7 detailed, the Lower Saxony Government and the German Federal Republic were prepared to select a location near Gorleben without extensively studying other settings. The Swiss implementor, Nagra, planned to construct a repository in the Züricher Weinland until the central government proclaimed that a choice of "one is no choice at all." In the United Kingdom, a single site in Cumbria might have been advanced had the County Council agreed.

Following the recommendations of President Carter's Interagency Review Group and consistent with the provisions of the NRC's procedural rule, Congress enacted the 1982 Nuclear Waste Policy Act; it was the first legislation worldwide to mandate a parallel process for choosing repository sites. Since then, every nation that has adopted a site-selection process has implemented a parallel one. (See Figure 8.4.)

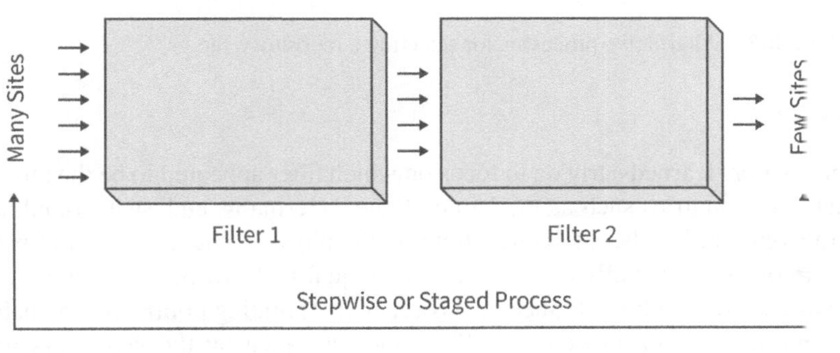

FIGURE 8.2 Winnowing many potential repository sites.

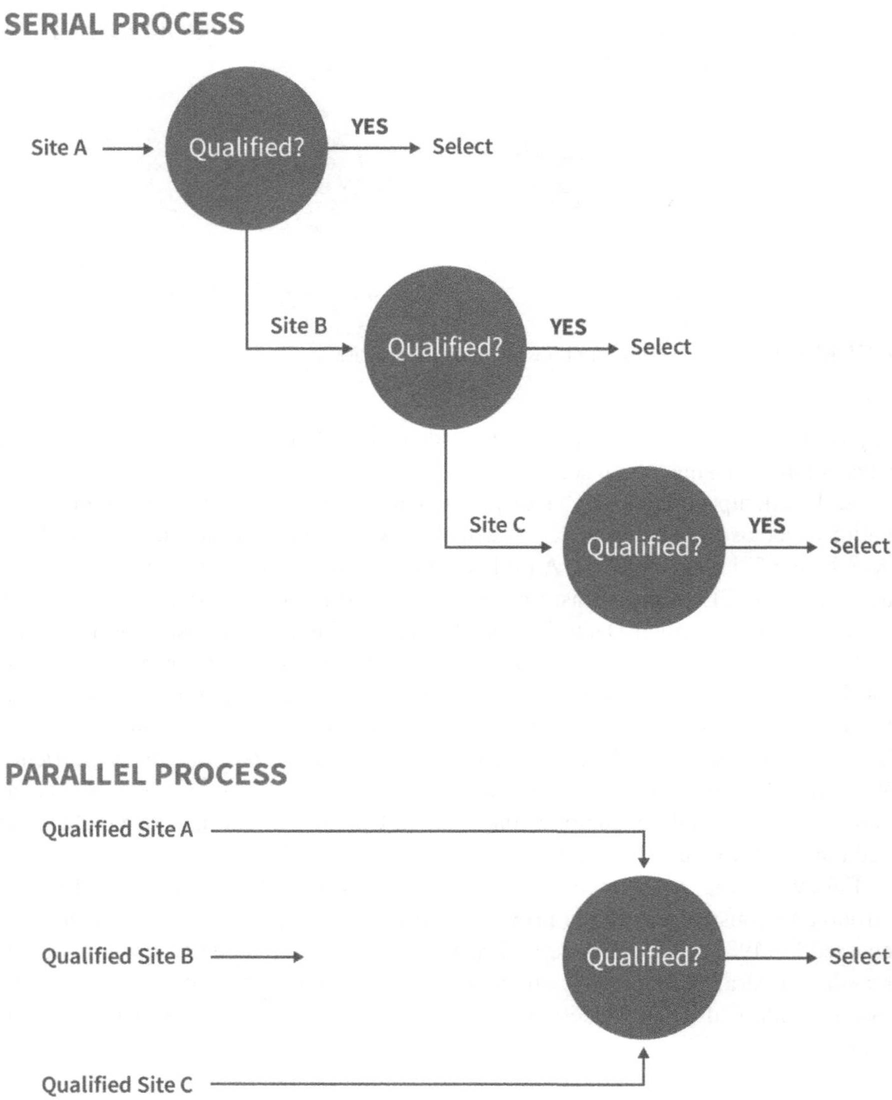

FIGURE 8.3 Alternative processes for selecting a repository site.

Order

Implementors learned early on to focus on which filter appeared to be the most problematic. In countries such as the United States, Germany, and Switzerland, a technocratic culture led them to concentrate on the physical characteristics of potential sites. In others, especially those with a single qualified hydrogeologic environment, such as Canada, Finland, France, and Sweden, pinpointing politically amenable settings ultimately took precedence. The approach taken by the waste-management programs in Japan and the United Kingdom so far has been to solicit volunteers.

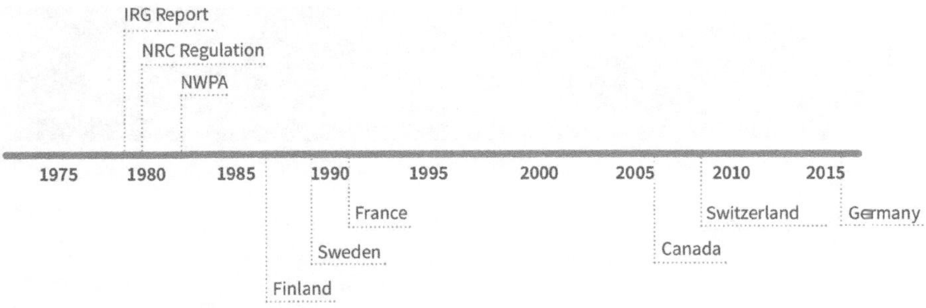

FIGURE 8.4 Over time, national waste-disposition programs have elected to employ a parallel process.

Note: The process finally adopted in the United Kingdom and Japan depends on how many qualified sites volunteer.

Initial Focus on Technical Suitability

As Chapter 2 explained, the 1957 report from a National Academy of Sciences-convened panel set the tone for early attempts in all countries to find a potential setting for a deep-mined, geologic repository.[6] Concentrating almost exclusively on proposing a "technical fix," the document defined the waste-management challenge as simply finding good rock. But what exactly constitutes such a formation?

As in the choice of process, explicit rules governed the answer. Implementors often commenced the search by applying exclusion criteria to eliminate sites whose hydrogeologic characteristics almost automatically precluded the development of a repository. (See Figure 8.5.)

For countries having a single qualified hydrogeologic environment, such as Sweden, Finland, and France, repository developers also used qualitative or quantitative *host rock-specific criteria* to identify potential locations. Examples of this type of condition included the following:

- The repository should be located less than 800 m below the surface.
- The specific hydraulic gradient in the isolating rock zone should be less than 10^{-2}.
- The overburden should have low permeability.
- The choice of sites should minimize groundwater flow.
- The effective diffusion coefficient in the isolating rock zone should be as low as possible (less than $10^{-11} m^2/s$).

When the implementors have the option of adopting more than one disposal concept and must compare sites in different hydrogeologic environments, they employed *generic criteria*. Because these criteria must be applied to more than one type of host rock, they were usually qualitative in nature. Consequently, it was extremely difficult to specify in advance the values for various rock properties.[7] Chapter 5 interprets the years-long battle to ratify generic Site-Suitability Guidelines in the United States.[8]

COUNTRY	CANADA	FRANCE	JAPAN	SWEDEN	SWITZ.	US*	UK
Fast and/or significant ground-water flow	X	X		X		X	
Unfavorable ground-water chemistry				X			
Tectonic activity	X	X	X	X	X	X	X
Inadequate depth and/or extent of the host-rock formation					X	X	
Significant faulting in the host rock	X	X	X				X
Presence of natural resources	X		X			X	X
Volcanic activity			X				

FIGURE 8.5 Exclusion criteria that disqualify a site for repository development.

Note: X = Exlcusion criteria. *Prior to 2001.

Initial Focus on Social Acceptability

Unlike the benchmarks that stipulated technical suitability, no single route or activity offers an appropriate measure of social acceptability. Over the course of time in many nations, these events had a determinant impact on whether the siting process continued.

- Public protests and demonstrations
- Partnership missions
- Local referenda
- Community opinion surveys
- National legislation

Indeed, the typical absence of social acceptability, a circumstance that will be elaborated upon below, plagued preliminary struggles to select a qualified repository site. As Chapter 7 described, the implementor's defeats at the hands of Swedish protestors in Kynnefjäll as well as setbacks in plebiscites in the Malå and Storuman municipalities ground SKB's efforts of the that nation's implementor to a halt. In

France, the disruptions in the Aisne, Deux-Sèvres, Ain, and Maine-et-Loire *départements* compelled the Prime Minister to terminate the investigations of the implementor, Andra. In Japan, the Governor of the Kochi Prefecture forced the Nuclear Waste Management Organization (NUMO) to discontinue the siting process even before it started. The Seaborn panel's conclusion that the Atomic Energy of Canada's disposal concept "lacked broad public support" drove the central government to terminate the process. In the United Kingdom, a special public inquiry into the first implementor's (Nirex) planning application in West Cumbria forced the country to start from scratch.

All these nations subsequently adopted processes that involved some form of voluntarism. The Swedish executor and the Finnish implementor "rediscovered" the potency of the municipalities' veto power. The 1991 legislation in France provided for a community to volunteer to host an underground research laboratory but denied it the power to reject that facility's conversion into a repository. In Japan, the executor reinvigorated efforts to find provinces that might agree to literature surveys of their hydrogeologic environments. Supported by an amended law, a new organization launched an elaborate campaign to ask Canadian communities whether they would agree to become a "informed and willing host" to a disposition facility. In the United Kingdom, a reformed establishment found several localities that might bid on hosting a deep-mined, underground repository. Notably, they were mostly still in West Cumbria.

What is particularly intriguing is the situation in the technocratic nations of the United States, Germany, and Switzerland. Despite the acceptability crisis and the Blue Ribbon Commission's (BRC) forceful advice, the Department of Energy has largely deferred to Congress' lack of appetite for opening a consent-base siting initiative for a disposition facility. The new Swiss legislation not only missed an opportunity to empower cantons but explicitly denied them veto power. In Germany, the siting process is still evolving, but it is unclear how much actual power communities will be able to exercise in the choice of disposal sites.

CONSIDERATIONS INFLUENCING THE OUTCOMES OF SITE-SELECTION ENGAGEMENTS

The nine nations featured in our story undertook at least twenty-three discrete site-selection engagements. Their outcomes revealed several striking patterns. (See Figure 8.6.) Only once (four percent of the time) did a site so quickly fail to pass through the technical suitability filter that its social acceptability was never put to the test. In three cases (13 percent), sites were technically suitable but socially unacceptable. In another three situations (13 percent), the search for sites is still ongoing. In four engagements (17 percent), settings failed both the suitability and the acceptability criteria. In five instances (22 percent), locations successfully passed through both the technical suitability and social acceptability filters. Seven times (30 percent), a location failed to pass through the social acceptability filter, which meant that its technical suitability was not formally evaluated.

If the calculations eliminate the three siting efforts that are now in progress (Germany, Japan, and the United Kingdom), then *fully three-quarters* of the engagements resulted in a site's failed passage through the social acceptability filter. No wonder so many national programs shifted their initial focus from technical suitability to social acceptability. As other analysts have observed, the 1957 National

ENGAGEMENT	FILTER PASSAGE	
	TECHNICAL	SOCIAL
United States		
Lyons, Kansas	Failed	Failed
First repository	Passed*	Failed
Second repository	Undetermined	Failed
Canada		
AECL concept	Passed	Failed
Adaptive Phased Management	Passed	Passed
Finland		
Posiva search	Passed	Passed
France		
CEA search	Undetermined	Failed
Andra search in clay	Passed	Passed
Andra search in granite	Failed	Failed
Germany		
Gorleben	Failed	Failed
Search in multiple hygdogeologic environments	Undetermined	Undetermined
Japan		
Tokyo Town	Undetermined	Failed
Volunteers in literature survey	Undetermined	Undetermined
Sweden		
Covert drilling	Undetermined	Failed
Måla and Storuman	Undetermined	Failed
Nuclear communities	Passed	Passed
Switzerland		
Search in granite	Failed	Undetermined
Wellenberg	Undetermined	Failed
Züricher Weinland	Passed	Failed
Sectoral Plan	Passed	Passed
United Kingdom		
NIREX in West Cumbria	Failed	Failed
West Cumbria Partnership	Undetermined	Failed
Radioactive Waste Management Services	Undetermined	Undetermined

FIGURE 8.6 Outcomes of siting engagements.

Note: *Based on NRC staff's evaluation of the Yucca Mountain license application.

Academy of Sciences' study now appears to have been turned on its head; the waste management challenge is not finding good rock but developing persuasive and credible arguments why a polity should host a repository.

FAILURE OF SITES TO PASS THROUGH THE TECHNICAL SUITABILITY FILTER

Five times the outcome from siting engagements resulted in a location that failed the test of technical suitability. The French implementor, Andra, misjudged the appropriateness of the Vienne location. Its oversight authority, the National Review Board, in 1997 issued an unvarnished opinion on the Vienne site; it maintained that the risks of fluids circulating between the granite and the aquifers in the sedimentary overlying formations were too high. A consultation commission headed by Christian Bataille reported in July 2000 that none of other communities controlling crystalline rock were willing to volunteer to host an underground research laboratory.

In many respects, the implementor in Switzerland encountered a similar restrictions. Motivated by the progress achieved in Sweden and Finland, the Swiss implementor, Nagra unveiled *Projekt Gewähr* (Guarantee) in 1978.[9]

> Despite all the differences, many similarities can be identified between KBS-3 and Project Guarantee; both consider crystalline as host rock, both place great emphasis on the material bentonite, with which the storage tunnels are to be filled, and the safety analyses are similarly structured in both.

Nagra advanced two postclosure and one preclosure host-rock specific criteria: tectonic stability, adequate depth, and ease of construction. Granitic rock formations meeting these requirements could only be found in a relatively small areas in four cantons. During a period starting in October 1982 and ending in February 1985, the implementor drilled a half-dozen boreholes. These investigations discovered that a large sedimentary (clay) trough intersected the granite. The regulator concluded that this hydrogeologic formation unnecessarily complicated the safety analysis of the settings, "It would be wrong to commit to crystalline as the host rock from the outset."[10] Nonetheless, despite this caution, Nagra continued to pursue these investigations for another decade before finally abandoning granite formations.

Chapter 3 briefly told the saga of the AEC's investigation of a possible repository site in Kansas. Already under criticism from the state's Geological Survey and its Director, William Hambleton, two new technical issues came to the fore by midsummer 1970.[11] The first was the question of the number and location of the boreholes that were sunk in search for oil and gas deposits. If any of them went unplugged, surface water could flow down and dissolve the salt at the repository horizon, thereby compromising the system's ability to isolate and contain the waste.

The second was the uncertainty caused by the endemic secrecy surrounding salt mining operations. The owner of the American Salt mine, adjacent to the AEC's proposed location, reported that it had introduced more than 175,000 gallons of water to suspend and then recover the mineral. The company, however, had no idea where the liquid went. The unexplained location of the water intensified concerns among geologists that the solution mining and hydrofracturing "could create '*jugs*' (lakes) that could cause surface subsidence," which in turn might degrade the disposition facility.[12] Moreover, it was unclear where the boundary lay between the American Salt mine and AEC's proposed repository.

These newly revealed quandaries gave weight to critics of the Lyons site. They also forced Congress to have second thoughts about moving forward in Kansas. As the Executive Director of the powerful Joint Committee on Atomic Energy informed his members, "Reconsideration of the Lyons site and looking for alternative sites are not only reasonable [options] but realistic ones."[13] Within six months, the Lyons location had disappeared from the AEC's agenda.

The previous chapter succinctly reported how the United Kingdom's first implementor, the Nuclear Industry Radioactive Waste Executive (Nirex), proposed to build an exploratory chamber as a precursor for developing a deep-mined, geologic repository for intermediate-level reprocessing waste. Its site-selection process began in 1987 by examining 537 locations stretched broadly across the country. Using a formal multiattribute utility decision analysis, Nirex sifted through the locations.[14] In 1988, its Board considered 11 settings, which were eventually winnowed down to two—Dounreay in northern Scotland and Sellafield in West Cumbria. Two years later, Nirex restricted further investigations to the latter location.

In June 1994, the implementor applied to the Cumbria County Council for "planning permission" to bore an investigatory tunnel, dubbed the Rock Characterization Facility. The Council denied the application for a variety of technical, social, and legal reasons.[15] The company appealed, and British Government opened a Public Inquiry that lasted five months. The Inspector overseeing the review prepared a nearly 300-page report that evaluated the positions of the appellant and interested and affected parties voicing opposition.

Among his conclusions, two appear especially compelling. First, he criticized Nirex for ignoring a European Union directive, its enactment into British law, and other regulatory and legislative mandates, all of which obliged a developer to consider alternative sites. (Put somewhat differently, the rules of the game demanded a parallel not a serial process.)[16]

> Crucially, it is obvious that the appeal site is not uniquely suitable for a repository, and that a variety of potential locations could be chosen depending on the importance attached to different factors. Cumulatively, these propositions make an overwhelming case for examining the merits of alternative sites... The law requires these alternatives to be examined sooner rather than later, so that they must be looked at now instead of waiting for the inquiry into the construction of the repository.

Second, the Inspector took Nirex strongly to task for its overly optimistic analyses of the site's technical suitability. It is worth citing his language at length not only for the insights it provides in this discussion but for its relevance as our tale's narrative continues in later chapters.[17]

> A type of geological environment, which eventually scored consistently well in Nirex's site selection exercise, is termed "basement rock under sedimentary cover." This seemingly could offer a range of inland locations. Nirex appears to misunderstand the concept, by claiming that the appeal site is within such an environment, whereas the area has never been so designated by its geological consultants...

> In the later stages of the [siting] process, the crucial point is that safety was not treated as the most important discriminative factor. This role was given instead to the costs to be borne by the generations which would utilize the repository, and yet that is contrary to the principle of sustainability…
>
> All the work and cooperation have not led Nirex to a sufficient understanding yet of the groundwater conditions in and around the appeal site. Another fundamental point is that the area is extensively faulted but there is a considerable amount of speculation as to the influence which any local fault has on groundwater flow. Nirex does not know whether the faults in general or particular increase or reduce or deflect flow, nor does it know where all the faults are…
>
> Nirex does not seem to have fully appreciated these limitations on its understanding of the appeal site, nor the significance of the limitations. Partly in consequence of its undue optimism, it is not planning to take all the steps necessary to put these deficiencies right.

It is hardly surprising then that the Inspector concluded, "The indications are overwhelmingly that this site is not suitable for the proposed repository and that investigations should now be moved to one of the more promising sites elsewhere."[18]

The fifth set of engagements where a site failed to pass through the technical suitability filter was the assessments of salt formations in northwest Germany. (Chapter 7 also summarizes these.)

After more than ten years of research ending in early 1977, the Prime Minister of Lower Saxony, Ernst Albrecht, declared the salt dome at Gorleben, a tiny speck of a village on the Elbe River bordering on German Democratic Republic, as the most suitable potential location for a reprocessing plant and for disposing of its HLW. Almost immediately, the Federal Republic of Germany eagerly accepted the choice.[19] As in the United States and the United Kingdom, the determination involved a serial not a parallel process.

As opposition mounted, Lower Saxony organized in March and April 1979 a public examination of its proposal. As two analysts observed, "The hearing demonstrated that even the proponents of nuclear energy had some doubts about the feasibility of the integrated design."[20] Consequently, shortly after the hearing concluded, Albrecht decided to abandon the reprocessing component but to move forward with a repository.

Between 1979 and 1983, surface-based investigations were conducted at Gorleben.[21] Then a report from the Federal Institute of Physical Technology held that the site was *Eignungshöffigkeit* (high likelihood of suitability or suitable enough).[22] But a 2007 study highlighted the importance of the so-called "Gorleben channel," an erosional structure that runs through the salt dome.[23] The channel likely allowed groundwater to pass through the formation. Its potential instability could thereby threaten the repository's capacity to isolate and contain any emplaced radioactive waste.

The adoption of the 2017 Repository Site Selection Act established a process that, in stark bureaucratic language, announced Gorleben's fate[24]:

> The Gorleben salt dome has not been included as a sub-area based on the geoscientific weighing criteria according to Section 24 of the [the Act]. The provision set out in Section 36 para. 1s.5 no. 1 of [the Act] shall therefore apply, and the Gorleben salt dome is excluded from the procedure.

A year later, the decommissioning of Gorleben began. As the cognizant Federal State Secretary reflected with greater insight[25]:

> For three decades, Gorleben represented a major social conflict in Germany. With regard to the search for a final repository site, politicians have learned from this conflict: in the end, the reasons why a certain site was chosen need to be clearly understandable.

It is to that critical question that our tale now turns.

Beliefs About Technical Suitability Shape Attitudes About Social Acceptability

Opposition to the siting of a deep-mined, geologic repository has called forth an embarrassment of acronyms: NIMBY (Not In My Backyard); LULU (Locally Unwanted Land Usage); BANANA (Build Absolutely Nothing Anywhere Near Anyone); and NOTE (Not Over There Either). All of these slogans connote an unreasonable and capricious set of beliefs. But in fact, upon closer examination, public reactions possess a basis in well-defined, even rational, attitudes.[26] (Box 8.1 reminds the reader that attitudes about social acceptability can also influence beliefs about technical suitability.)

Quantitative Evaluations

With few exceptions, waste-management implementors commission surveys to gauge how citizens respond to their siting engagements. Regrettably, these studies are mostly proprietary. Disinterested social scientists rarely have the resources to carry

BOX 8.1 THE INTERDEPENDENCE OF ATTITUDES TOWARD ACCEPTABILITY AND BELIEFS ABOUT SUITABILITY

Footnote #4 of this chapter alerted the reader to the prospect that the two filters might not independent.

Both the quantitative and qualitative analyses that will be reviewed below deliberately chose to frame the relationship between the two filters as a recursive one. That is, suitability influences acceptability. But is not hard to imagine that the relationship might just as well be reciprocal. In such a circumstance, attitudes about acceptability color beliefs regarding suitability. For example, opponents to siting a repository in the Garde Dępartement came to believe that radionuclides might be released and thereby stigmatize the region's highly regarded wine.

Although the evidence here is a bit more speculative, a case might be made that opponents in other nations might be similarly driven not only with respect to sigma but also with respect to attitudes toward nuclear power in general.

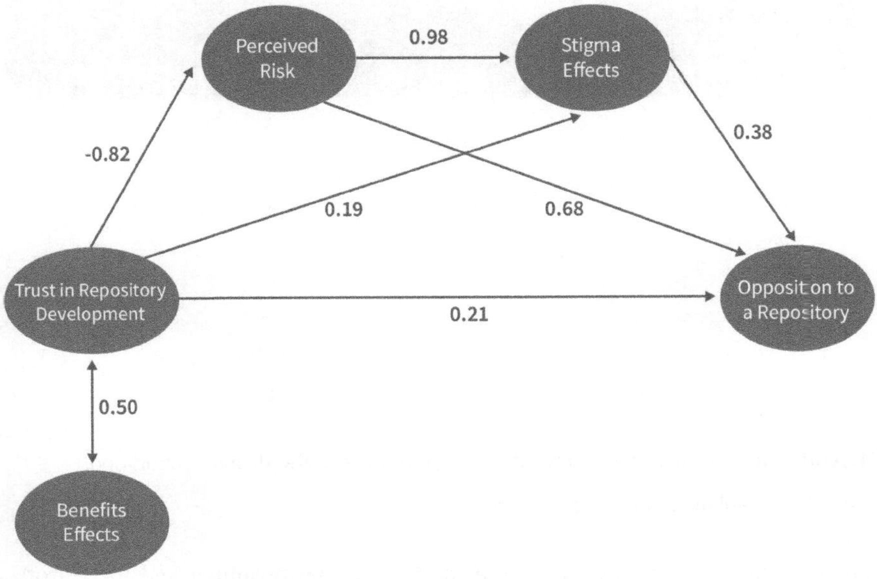

FIGURE 8.7 Causal analysis of opposition to the Yucca Mountain repository.

Note: Standardized regression coefficients. Sample of Nevada population.

out similar investigations. However, three of the best available analyses explore support/opposition to the proposed Yucca Mountain repository, a generic repository in Sweden, and the siting process in Germany.

In Figure 8.7, the dominant paths begin with trust and then either affect attitudes toward Yucca Mountain directly or indirectly influence them through technical suitability (perceived risk) and stigma. Importantly, the provision of benefits had virtually no effect on attitudes toward the planned repository.[27]

The Swedish inquiry asked whether local municipalities should go ahead with site exploration.[28] However, as noted in Chapter 7, in some nations, policies toward radioactive waste management were entangled with policies related to the production of nuclear electricity; therefore, this independent variable was included as well in the Swedish analysis.

The resulting analysis produced equivalent results as the Nevada study for perceived risk and trust. But it also revealed the independent impact of more generalized attitudes toward nuclear power production. Once again, however, benefits from the Swedish government and the European Union had virtually no impact on support/opposition to surface investigations. (See Figure 8.8.)

The most recent survey of public attitudes toward nuclear waste-management was a major study in Germany.[29] The investigators' objective was to revisit the exploration undertaken by Flynn and his colleagues. Their findings differ somewhat but are generally consistent with the earlier research. Trust and risk perceptions still strongly influence opposition to a repository. However, unlike the earlier inquiries, Seidl and his collaborators found no direct connection between the two

INDEPENDENT VARIABLE	STANDARDIZED REGRESSION COEFFICIENT
Risk Perceptions	
Dread	-0.212
Interference with Nature	-0.095
Stigma	-0.119
Trust	
Trust in Management	0.138
Trust in Operations	0.087
Attitudes toward Nuclear Power	0.140

FIGURE 8.8 Local attitudes toward a repository in your Swedish neighborhood.

Note: Total Explained Variance = 0.485.

independent variables. Benefits from the German Government and the European Union had a trivial influence either on whether the siting decision was acceptable or on support for a repository in an individual's municipality. Somewhat surprisingly given the history of German waste-management activities, however, Seidl and his collaborators chose not to include attitudes toward nuclear power in their inquiry. (See Figure 8.9.)

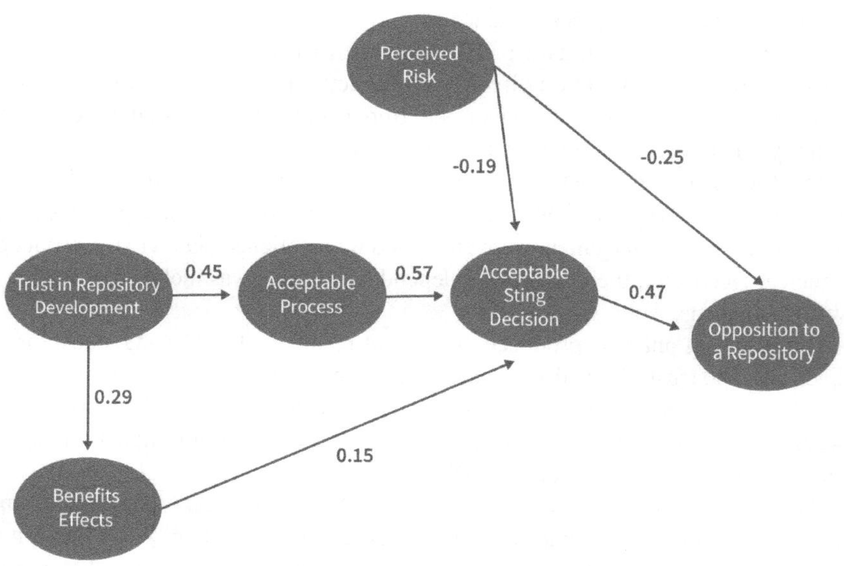

FIGURE 8.9 Attitudes of the German public toward a local repository.

Note: Standardized regression coefficients.

Qualitative Evaluations

The narratives in Chapter 7 reinforced the quantitative conclusions: risk perceptions and stigma, trust, and attitudes toward nuclear power production strongly influence beliefs about social acceptability.

The historian Spencer Weart convincingly argued that the general public looks at things nuclear through a different—strongly negative—lens than it perceives other technological hazards.[30] (See Figure 2.7.) *Risk perceptions and alarms about stigmatization* could prompt local and state/provincial leaders to reject efforts to site a repository. In Nevada, they voiced concerns that a route transporting waste through downtown Las Vegas might negatively brand the area and thus drive tourists away from gambling casinos. In France, plans in the late 1980s to study the Bresse location aroused opposition because the French especially valued "Bresse chickens.' Even changing the area's name to Ain did not muffle the protests. In Germany, the Lower Saxony Ministry of Agriculture objected to studying salt formations because radionuclide releases might stain the reputation of locally produced food.

Implementors must merit *public trust and confidence*. The BRC made it clear that[31]:

> Trust and confidence in the federal government's basic commitment and competence to deliver on its waste management obligations have all but completely eroded since 1987. Restoring that trust and confidence must be the government's first priority and is essential for getting all aspects of the nation's nuclear waste program back on track.

So did the United Kingdom's Committee on Radioactive Waste Management. "There should be continuing public and stakeholder engagement, which will be essential to build trust and confidence in the proposed long-term management approach, including siting of facilities."[32]

Canada's Nuclear Waste Management Organization also emphasized that it had to establish the trust of any community that might be willing to host a repository.[33] "We expect that only through successful engagement and collaborative decision-making in the early stages of implementation can we begin to build trust and confidence in the process. Ultimately, it will be essential in maintaining the momentum for action."

The connection between *nuclear energy production and attitudes toward the management of the generated waste* varies from nation to nation and from time to time. The link appeared to be strongest in Sweden in the early 1970s, in Germany up until 2016, in the United Kingdom until the early 2020s, and in Japan up until to this very time.

As detailed in Chapter 7, Sweden forged an unambiguous link between future nuclear power generation and efforts to dispose of its HLW and SNF. It was "established in direct response to uncompromising government legislation introduced in 1977 demanding that prior to the commissioning of further nuclear capacity, the reactor owners present technical designs demonstrating how and where radioactive waste can be finally stored with absolute safety."[34] Although the connections were slightly loosened subsequently, the culture of the country's waste managers continues to recognize that the production of waste remains contingent upon the vigorous development of a disposition facility.

Two German observers noted how nuclear accidents abroad altered attitudes.[35]

> The Chernobyl accident in 1986 strengthened the wide spread nuclear skepticism in the German society and the political controversy. The Fukushima Daiichi accident in March 2011 marked a further turning point. Impacted by this event, political majorities were formed for new phase-out agreements and for restructuring nuclear waste management policy.

The United Kingdom has determined that its existing HLW and SNF will eventually be disposed of in a deep-mined, geologic repository. However, the fate of the waste from any "new build" reactors still awaits decisions on whether those power plants will ever be erected. As CoRWM also emphasized[36]:

> New build wastes would extend the timescales for implementation, possibly for very long, but essentially unknowable, future periods. Further, the political and ethical issues raised by the creation of more wastes are quite different from those relating to committed–and, therefore, unavoidable–wastes.

Japan's nuclear waste-management program is just restarting the search for a site for its repository. As one analyst concluded, "As it currently functions, the system is mitigating the liability of the electric power companies, while increasing the authority of the state ... If mandatory generator liability is relaxed and electric power companies are relieved of their responsibilities, the only outcome will be a continuation of irresponsible power generation."[37]

Passage of a Site Through the Technical Suitability and Social Acceptability Filters

Precisely how a repository was finally selected varied from nation to nation. Finland became the forerunner when Parliament passed the Decision-in-Principle in 2002. By that time, the Eurajoki municipality, home to the Olkiluoto nuclear power plant, had given its formal consent, thereby passing of the site through the social acceptability filter.[38]

The executor, Posiva, sponsored detailed technical safety analyses comparing four potential locations: Eurajoki (Olkiluoto), Loviisa, Äänekoski, and Kulmo. The primary conclusion of those evaluations was that, from the standpoint of postclosure safety, "all candidate sites are suitable to host a repository for spent fuel. The safety assessments do not give reasons to reject any of them, neither do the release and transport analyses of radionuclides provide firm grounds to rank one site above the others."[39] A typical safety analysis is illustrated in Figure 8.10. It indicates that releases from a repository are more than four magnitudes below the regulatory limit of 10^{-4} Sieverts/year for a million years.

Posiva also prepared an environmental impact assessment comparing the potential sites along other dimensions, such as social impacts and psychosocial effects.[40] Public opinion surveys exposed strong differences among residents of the locations. In the two nuclear communities (Eurajoki and Loviisa) approval ran from 59 to 62 percent, whereas in the two non-nuclear districts (Kuhmo and Äänekoski) support was roughly

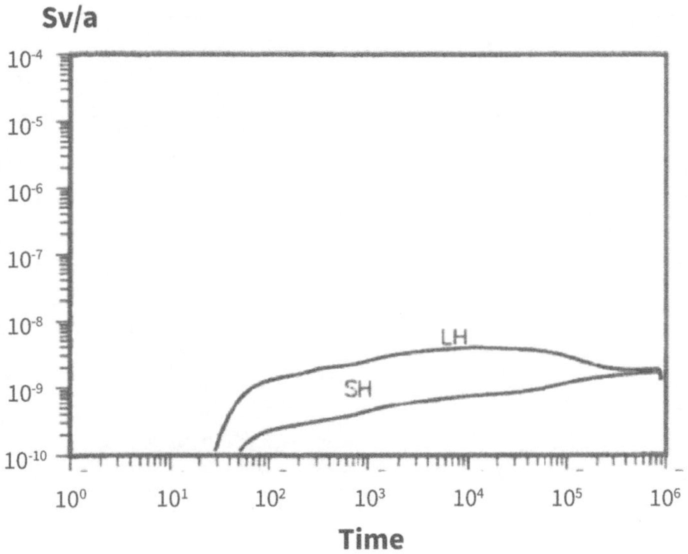

FIGURE 8.10 Dose after one million years under two scenarios.

Note: SH = Small hole in the waste package. LH = Large hole in the waste package.

27 percent lower. In addition, the evaluation weighed positive economic benefits against negative safety worries and adverse effects on the image and character of the area. It determined that siting a repository in Eurajoki would be the least disruptive choice.[41] Based on this information, the Olkiluoto setting passed through the technical suitability filter.

In sum, through this parallel siting process, Finland picked where it would construct its deep-mined, geologic repository for SNF.

As Chapter 7 detailed, Andra initially explored four potential repository settings: Vienne (crystalline rock) and Meuse, Haute-Marne, and Garde, all in clay formations. The implementor could not overcome the opposition from wine producers near Garde. Christian Bataille's so-called "Granite Mission" also failed to another discover a community willing to host a repository.

Almost by default then, Andra had little choice under its controlling legislation but to submit to the Council of State in early 1999 a license application to approve the construction of an underground research laboratory in the Callovo-Oxfordian clay formation overlaying Bure. Although Haute-Marne reached agreement with its *département* to support the research facility, Meuse itself never voted.

Conseil d'Etat ratified the application later that year and sent it to the responsible ministers for signature. The Minister of the Environment, Dominique Voynet, a Green Party member, withheld her assent until the government made "reversibility" an integral part of future repository planning. Once that condition was met, the Bure site passed through the social acceptability filter.

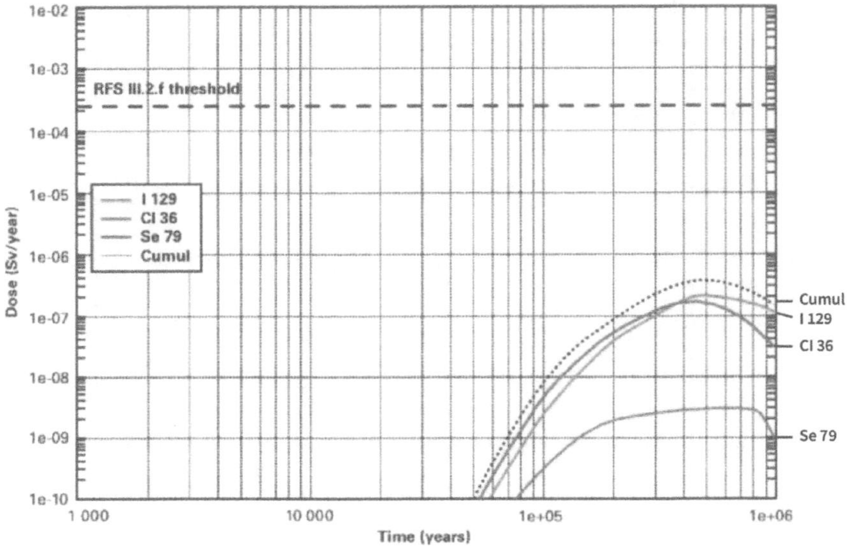

FIGURE 8.11 Dose after one million years.

For the next six years, Andra conducted safety analyses of the Bure setting and a generic location in a granite formation.[42] For a disposition facility set in clay, the calculations evaluated the thermal, hydraulic, chemical, and mechanical evolution of the system over several millennia. For the nonspecific crystalline rock alternative, similar estimates were reckoned using data from the Swedish, Finnish, and Swiss programs.

Figure 8.11 clearly indicates that a repository constructed in the vicinity of Bure would likely release a dose to the environment over a million years that was roughly three orders of magnitude below the limit set by the regulator. For the French, that was the only condition for a setting to satisfy the technical suitability filter. Thereby, using a modified serial siting process, France selected the location of its disposition facility for HLW, SNF, and transuranic waste.

Unlike those in Finland and France, SKB did not have to seek the *imprimatur* of the state to select the locations for its plant to encapsulate SNF and the potential repository. Although for more than a decade it had interacted with the two finalist municipalities, Oskarshamn and Östhammar, SKB claims that it only tacitly comprehended their views about in-the-ground progress.

Indeed, without the implementor's full knowledge, the mayors of the two municipalities worked to create the "Added Value Programme." They announced that they would "reject the final repository unless some form of compensation would be offered also to the community that would lose the 'race' for the repository." SKB had to set aside between EUR 160–216 million, of which the "also-ran" would get 75 percent for its good-faith participation in the siting process. Once SKB agreed, the setting in the Östhammar municipality passed through the social acceptability filter.

SKB advanced four factors that would influence the selection of a repository site: safety-related site characteristics, social resources, technology for execution, and health and environment.[43] It then maintained that safety would be determinative. Only if the sites displayed "no decisive difference" in safety, would the other factors come into play.

The implementor acknowledged that many of the parameters affecting safety were similar in the Östhammar and Oskarshamn settings.[44]

> There are, however, differences in groundwater conditions that have great consequences for the assessment of long-term safety in scenarios where canister damage could occur in the long term. The frequency of water-conducting fractures at repository depth is significantly lower in Östhammar than in Oskarshamn. This affects the environment for the buffer, which can in turn affect how long it takes for canister damage to occur and the number of canisters that may be damaged.

As SKB's Managing Director put it: "The rocks have spoken." (See Figure 8.12.) Thus, using a parallel siting process, Sweden selected the location of its disposition facility for SNF.

By 2018, the Swiss Federal Council, modifying a recommendation from Nagra, winnowed five potential disposition settings down to three: Jura Ost, Zürich Nordost and Nőrdlich Lägern. None of the areas had any formal say in that decision. Four years later, Nagra submitted its siting proposal recommending the last location for development of a repository. In 2024, the implementor tendered its application to the Federal Office of Energy for a general license to develop a deep-mined, geological repository in the last location.[45] Once other federal agencies certify to completeness

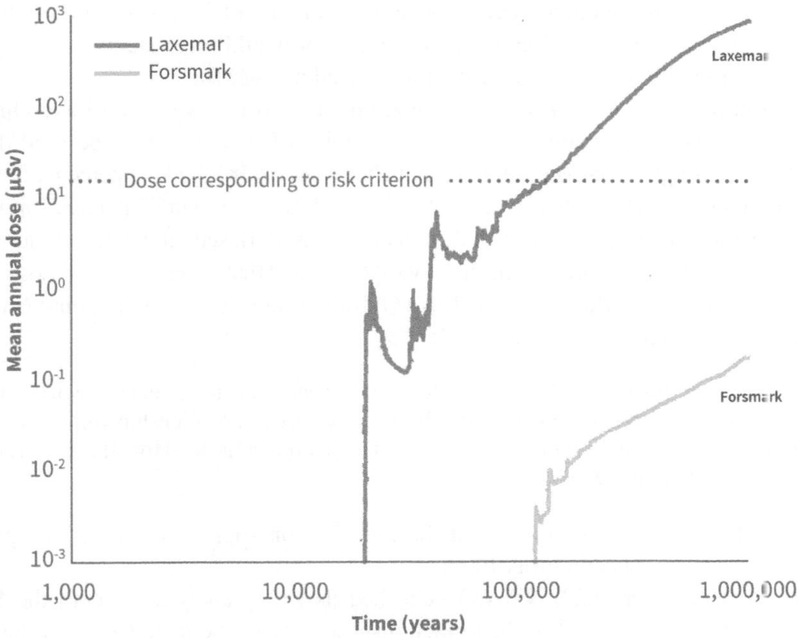

FIGURE 8.12 Performance assessment comparing the Laxemar (Oskarshamn) and the Forsmark (Östhammar) sites.

of the submission and the regulatory authority approves, that setting will have passed through the social acceptability filter.

Nagra based its proposal on three technical considerations. First, the Opalinus Clay formation at Nőrdlich Lägern was significantly deeper than that at either Jura Ost or Zürich Nordost. Further, the so-called confining geological units located above and below the Opalinus Clay were superior in the proposed location. Second, the long-term stability of the clay was greater in the proposed location than in the other two potential settings. That strength provided additional protection from erosion. Third, Nőrdlich Lägern offered greater engineering flexibility than either Jura Ost or Zürich Nordost. The Nördlich Lägern siting region had the largest connected area with no indications of major faults. For this reason, adaptability with regard to the repository layout is greatest in that region. In Nagra's view at least, its recommended area passed through the technical suitability filter.

Thereby, using a parallel siting process, Switzerland contingently selected the location of its disposition facility for SNF.

As Chapter 7 explained, NWMO by the end of 2019 eliminated all but two of the original 23 potential repository settings, leaving just two locations in Ontario Province, South Bruce and Ignace (Revell setting).

From the very start of the siting activity, NWMO insisted that the local community and any First Nation held veto rights over the choice of site. Although the South Bruce community seemed to support a disposition facility for used nuclear fuel, the neighboring Saugreen Ojibway First Nation's stance was more ambiguous. In January 2020, the indigenous group rejected Ontario Hydro's plans for constructing a low- and intermediate-level repository near one of its nuclear power plants. Never tested was the question whether that action would foreshadow the Saugreen Ojibway's reaction to a system for disposing used nuclear fuel.

In contrast, the views of the Ignace community and the Wabigoon Lake Ojibway First Nation began to manifest themselves within a few years after the final down-selection. Through a contractor, Ignace launched a survey in which 62 percent of the eligible individuals participated (roughly 1,000). Slightly more than 77 percent supported the repository, and slightly less than 21 percent opposed. Based on that information, the Ignace Council in July 2024 unanimously told NWMO that it was a willing host.

Four months later, the Wabigoon Lake Ojibway followed suit. As the First Nation's Chief Clayton Wetelainen remarked[46]:

> We have been involved in this process for over 12 years and our members have decided to continue on this path. As Anishinaabe we have a sacred relationship with the land. We belong to the land, and we have a sacred duty to protect the land for all our relations and for generations yet to come.

Once the Ignace community and the First Nation spoke, the Revell site passed through the social acceptability filter.

In late December 2023, NWMO published its safety analyses of both the South Bruce and Revell settings.[47] At the former, disposal would be effected at the Ordovician sedimentary rock (clay) layer. At the latter, disposition would be executed within the Revell batholith, a rock with favorable characteristics, including a near-homogeneous granitoid composition.

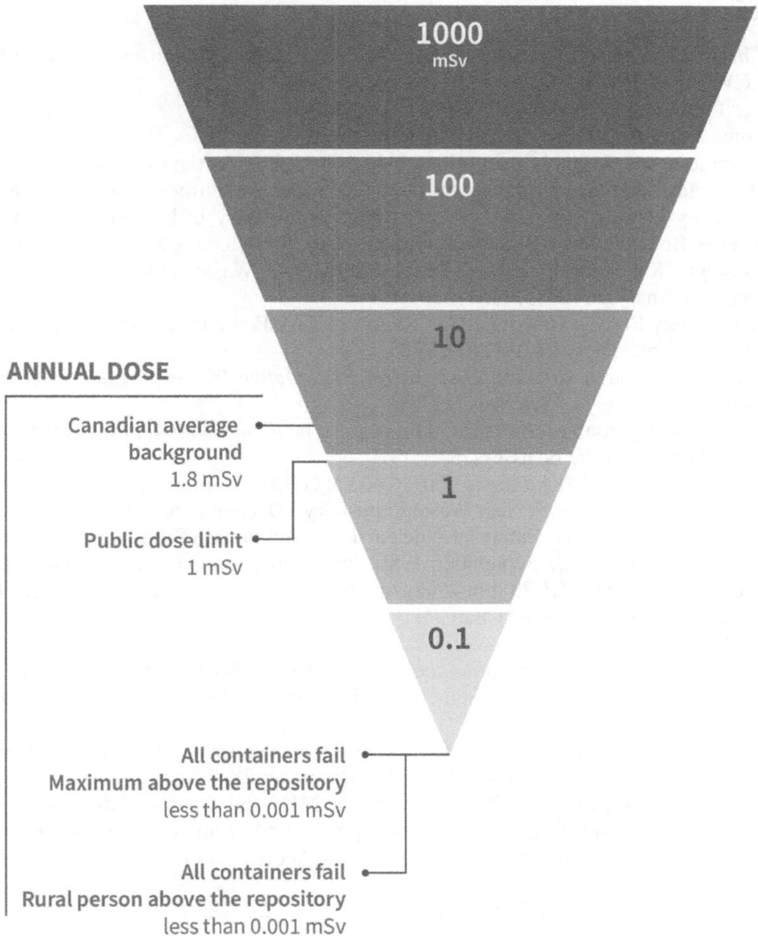

FIGURE 8.13 Safety analysis of the Revell site.

In a very real sense, comparing the dose calculations was a moot point; the Saugreen Ojibway's unspoken resistance meant that South Bruce could never be a contender. But NWMO also had to demonstrate that the Revell site was more than technically suitable. Indeed, it fell four orders of magnitude below the regulatory requirement at one million year had. These results are shown in Figure 8.13.[48]

It only took six days after the Wabigoon Lake Ojibway gave its consent for NWMO to announce that, through a parallel siting process, Canada selected its setting for a repository for used nuclear fuel. NWMO's leader reflected: "Canadians and Indigenous Peoples have been clear that it is essential to take responsibility now, in this generation, to safely manage Canada's used nuclear fuel for the long term."[49]

Once a site has been chosen, the next phase is to seek a license from the implementor's regulatory authority. It is to that pursuit that our tale now turns.

NOTES

1. Christophe LeBold, *Leonard Cohen: The Man Who Saw the Angels Fall*, (Toronto: ECW Press, 2024).
2. See Figure 7.5.
3. Some analysts use the term "deliberative turn."
4. In some respects, this is a simplified model/metaphor. As the remainder of the chapter elaborates, the two filters may not be independent; sometimes views of one are influenced by views of the other. See also the commentary in Nuclear Waste Technical Review Board (NWTRB), "Designing a Process for Selecting a Site for a Deep-mined, Geologic Repository for High-Level Radioactive Waste and Spent Nuclear Fuel: Detailed Analysis," 2015, 145–156.
5. Interagency Review Group (IRG), "Report of Task Force for Review of Nuclear Waste Management," DOE/ER-0004/D, 1978.
6. National Research Council, *Disposal of Radioactive Waste on Land*, (Washington: National Academy of Sciences, 1957).
7. A particularly troublesome issue in this case is how the isolation properties of the engineered barrier are to be assessed.
8. 10 *Code of Federal Regulations* 960, "General Guidelines for the Preliminary Screening of Potential Sites for a Nuclear Waste Repository," December 6, 1984.
9. Main Division for the Safety of Nuclear Installations (HST), *Expert Report on Project Guarantee 1985*, September 1986, https://ensi.admin.ch/fr/wp-content/uploads/sites/4/2012/05/hsk_23-28_gutachtenzumprojektgewaehr1985.pdf. Accessed January 22, 2025. Translated using ChatGPT.
10. HST, 85.
11. A. A. Albert de la Bruhèze, *Political Construction of Technology: Nuclear Waste Disposal in the United States*, (Twente, Netherlands: University of Twente Press, 1992), 139–172.
12. Bruhèze, 161.
13. Memorandum from Edward Bauser, Executive Director of the Joint Committee on Atomic Energy, to Committee Members, September 28, 1971.
14. This methodology is virtually identical to the MUA that Chapter 5 discusses.
15. See the Inspector's Report on Nirex's appeal found at https://www.davidsmythe.org/nuclear/inspector's_report_complete.pdf, 1–2. Accessed January 23, 2025.
16. Inspector's Report, 267.
17. Inspector's Report, 274–275 *passim*.
18. Inspector's Report, 276.
19. Even today, it is unclear why Lower Saxony was so enthusiastic about the Gorleben setting. Some observers suggest that Albrecht deliberately tried to annoy the leaders of the former German Democratic Republic.
20. Peter Hocke and Ortwin Renn, "Concerned Public and Paralysis of Decision-Making: Nuclear Waste Management Policy in Germany," *Journal of Risk Research*, (12) 2009, 921–940. Citation is at 925.
21. The best English-language discussion of the domestic and international politics surrounding Gorleben's selection can be found in Astrid M. Eckert, *West Germany and the Iron Curtain*, (New York: Oxford University Press, 2019), 201–243. She maintains that, according to the nuclear utilities' own calculation, the industry paid roughly 1.8 billion euros to conduct the site investigations at Gorleben. Cited on 331.
22. Hocke and Renn, 925.
23. Federal Company for Radioactive Waste Disposal (BGE), "Hydrogeology of the Overburden of the Gorleben Salt Dome," Hannover, Germany, 2007.
24. BGE, "Sub-areas Interim Report Pursuant to Section 13 StandAG," Hannover, Germany, September 28, 2020.
25. Federal Office for the Safety of Nuclear Waste Management, "Press Release," September 17, 2021. Https://www.base.bund.de/shareddocs/pressemitteilungen/en/2021/0917_gorleben.html. Accessed January 27, 2025.

26. Much of this discussion draws from Daniel Metlay, "Social Acceptability of Geologic Disposal," in Ehud Greenspan (ed), *Encyclopedia of Nuclear Energy*, Volume 2, (Amsterdam: Elsevier, 2021), 684–697.
27. James Flynn et al., "Trust as a Determinant of Opposition to a High-Level Radioactive Waste Repository: Analysis of a Structural Model." *Risk Analysis* (12) 1992, 417–429.
28. Lennart Sjőberg, "Local Acceptance of a High-Level Nuclear Waste Repository." *Risk Analysis*, (24) 2004, 737–749.
29. Roman Seidl et al., "The Role of Trust and Risk Perception in Current German Nuclear Waste Management," *Risk Analysis*, (42) 2022, 2704–2719.
30. Spenser Weart, *Nuclear Fears: A History of Images* (Cambridge MA: Harvard University Press, 1988).
31. Blue Ribbon Commission on America's Nuclear Future (BRC), *Report to the Secretary of Energy*, 2012, 40.
32. Committee on Radioactive Waste Management (CoRWM), "Managing Our Radioactive Waste Safely: CoRWM's Recommendations to Government," 2006, 113.
33. Nuclear Waste Management Organization (NWMO), "Choosing a Way Forward: The Future Management of Canada's Used Nuclear Fuel," 2005, 42.
34. Yannick Barthe et al., "Technological Fix or Divisible Object of Collective Concern? Histories of Conflict over the Geological Disposal of Nuclear Waste in Sweden and France," *Science as Culture*, (29) 2020, 205.
35. Peter Hocke and Beate Kallenbach-Herbert, "Always the Same Old Story? Nuclear Waste Governance in Germany," in Achim Brunnergräber et al. (eds), *Nuclear Waste Governance*, (Wiesbaden: Springer VS, 2015), 197.
36. CoRWM, 209.
37. Lila Okamura, "False Premise, False Promise: Governance and Management of Nuclear Waste in Japan," in Achim Brunnengräber et al. (eds), *Nuclear Waste Governance: An International Comparison*, (Wiesbaden: Springer VS, 2018), 114.
38. Matti Kojo's dissertation provides the best analysis of local decisionmaking in Finland. See "The Site Selected: The Local Decision-Making Regarding the Siting of the Spent Nuclear Fuel Repository in Olkiluoto." Https://www.osti.gov/etdeweb/servlets/purl/20789738. Accessed February 19, 2025.
39. Timo Viena and Henrik Nordman, "Safety Assessment of Spent Fuel Disposal in Hastholmen, Kivetty, Olkiluoto and Romuvaara," TILA-99, 1999, 218.
40. Posiva, "The Final Disposal of Spent Nuclear Fuel: Environment Impact Assessment Report," 1999. This evaluation has an extremely valuable discussion of calculated versus perceived risked. See 165–170.
41. Posiva, 172.
42. Andra (2005); Andra, "Dossier 2005 Argile Synthesis: Evaluation of the Feasibility of a Geologic Repository in an Argillaceous Formation," Paris, 2005b."
43. Swedish Nuclear Fuel and Waste Management Company (SKB), "Site Selection—Siting of the Final Repository for Spent Nuclear Fuel," TR-11-07, 2011.
44. SKB, 4–5.
45. Nagra, "The Site For the Deep Geological Repository: Nagra's Proposal," 2002, 1–35.
46. World Nuclear News, "Wabigoon Lake Ojibway Nation Says Yes to Next Phase of Repository Process," November 22, 2024. Https://www.world-nuclear-news.org/articles/wabigoon-lake-ojibway-nation-says-yes-to-next-phase-of-repository-process. Accessed February 21, 2025.
47. Nuclear Waste Management Organization (NWMO), "Confidence in Safety—Revelle Site: 2023 Update," NWMO-TR-2023-07, 2023a, 1–120; Nuclear Waste Management Organization (NWMO), "Confidence in Safety—South Bruce Site: 2023 Update," NWMO-TR-2023-08, 2023, 1–118.
48. The Canadians did not report the traditional dose-likelihood curve. Instead the diagram points out the conclusion that the expected dose is three orders of magnitude below the regulatory standard.
49. See https://www.nwmo.ca/news/the-nuclear-waste-management-organization-selects-site-for-canadas-deep-geological-repository. Accessed February 21, 2025.

9 Phased Repository Development

The foundational *raison d'être* for developing a deep-mined, geologic repository is to ensure that the current generation and perhaps the next does not export the entailed risk further into the future. At the same time, civil society's angst about simply "throwing the keys away" prompts demands that avenues be orchestrated for addressing flawed repository-performance expectations. The waste-management choices taken in the nine nations studied here reflect a wide range of responses to balancing those two imperatives.

LINEAR VERSUS PHASED REPOSITORY DEVELOPMENT

Consider a thought experiment in which a legislature authorizes the development of a repository. The implementor then prepares a multi-thousand-page document that prescribes a set of investigations to establish the technical suitability of a predetermined site. That work garners such complete scientific and engineering accord that virtually all interested and affected parties endorse the project. Contemplate as well a second thought experiment in which explicit, broadly participatory decisions are taken, perhaps annually, to ascertain how or whether the enterprise should continue.

Figure 9.1 incorporates both conceptual possibilities. The first—linear repository development—eliminates for all practical purposes the "reevaluate," "go back," and "continue on a modified paths". The second—phased repository development—asserts that those latter paths doom the "following the reference" choice, Of course, in application, implementors adopt approaches that fall somewhere in between the two extremes.

The diagram as a whole illustrates the classic incremental, trial-and-error model of decision-making; it posits continuous surveillance of an organization's physical and political environments to detect deviations from anticipated repository performance as well as actions to remediate those departures.[1]

SHIFTING THE PARADIGM

The notion of phased repository development is a venerable one, by now nearly a half-century old. In one of its earliest initiatives, the Nuclear Regulatory Commission (NRC) formed a task force in 1976 to propose goals and objectives for managing the country's high-level radioactive waste and spent nuclear fuel. Among the group's prominent observations was the following: "It is prudent to build technological systems so that one can reverse the effects they cause, or at least so the direction or impact of those effects can be changed or modified. This is

DOI: 10.1201/9781003528302-11

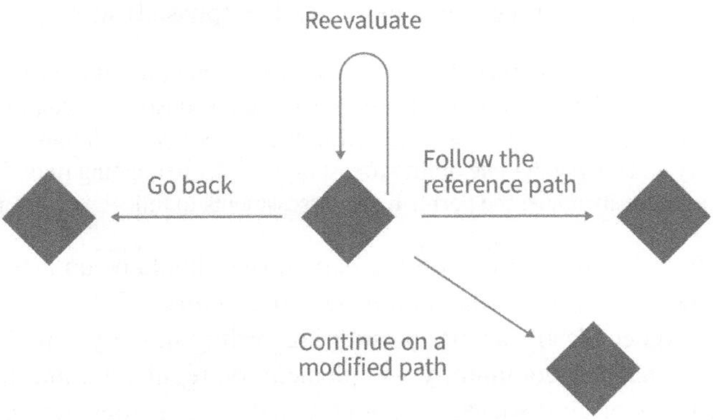

FIGURE 9.1 An incremental approach for developing a repository.

why retrievability is attractive."[2] NRC eventually incorporated retrievability into its licensing procedures.[3]

The Interagency Review Group on Nuclear Waste Management, established during President Jimmy Carter's administration, presumed that "conservatism would be applied in repository design and development." It further recommended that "the development of a full-scale repository should proceed in a stepwise manner...This procedure would provide greater assurance than would a single step from site characterization to operations that all factors were identified and adequately considered."[4]

One of the earliest external advocates of this approach was a 1990 report issued by the U.S. National Academy of Sciences (NAS).[5] It opened with a note of caution. "There is no scientific or technical reason to think that a satisfactory geological repository cannot be built. Nevertheless, the U.S. program [Yucca Mountain Project], as conceived and implemented over the past decade, is unlikely to succeed." The document sharply criticized the program's strategy, which was uncomfortably close to the first thought experiment, on several grounds. The implementor produced computer models and geophysical calculations that "have been inappropriately applied." In the report's view, such efforts lacked scientific credibility and would lead to unsound engineering practice. Moreover, unlike all other nations, the nuclear regulator [NRC] promulgated rules and specifications "before all the data were in." Consequently, the program was compelled to pursue requirements that "may be impossible to meet."

NAS proposed abandoning the "perfect knowledge" approach simply because it was unrealistic. Proof of safety could not be obtained in the usual sense; projections of performance could only be judged many millennia into the future. Improved hydrogeologic understanding unequivocally demonstrated the importance of variability and unpredictability, each of which can produce major uncertainties. Most often, this ambiguity only permitted both the implementor and regulator to set bounds on the behavior of a complex engineered and natural system. Failure to recognize these technical limitations inexorably leads to failures, delays, and thus to public dissent and mistrust.

In its place, NAS recommended a more flexible, (phased) strategy.[6]

> In a program governed by this alternative approach, change would not be seen as an admission of error; the system would be receptive and responsive to a continuing stream of information from site characterization...But the necessary conditions of the system are flexibility and resiliency—flexibility to respond rapidly to ongoing findings and resiliency to continuously adjust the performance assessments to reflect new information.

NAS, in short, claimed that the programmatic thrust should be on *resolving* complications as opposed to *anticipating and forestalling* setbacks.[7]

Over the next decade and a half, questions regarding strategy percolated throughout the broader nuclear community—implementors, regulators, and interested and affected segments of civil society. What emerged were lessons drawn mostly from failures in waste-management programs across the globe.

Monitored Retrievable Repository

Following the 1995 vote in a public referendum in the Nidwalden canton that rejected a low- and intermediate-level repository in Wellenberg, the Swiss Federal Department of the Environment, Transport, Energy, and Communications chartered an expert group, EKRA, to compare three alternative concepts for managing radioactive waste: geological disposal, monitored and retrievable long-term disposal, and interim storage.[8] The primary and general considerations in organizing the comparison were active and passive safety, monitoring and control, and retrievability of the waste.

The expert group recommended that the second option—monitored and retrievable long-term disposal—be selected as the Swiss strategy for developing a repository. See Figure 9.2. Under that approach, test and pilot facilities would be constructed in

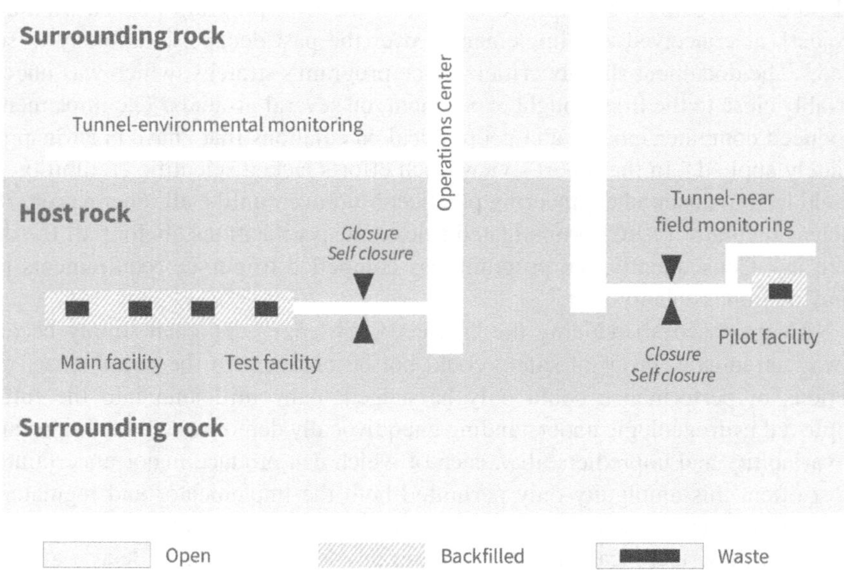

FIGURE 9.2 Swiss concept of a monitored retrieval repository.

the Opalinus Clay. The former would serve as a rock laboratory for site-specific studies that would provide input for the safety demonstration. The latter would monitor the behavior of representative volumes of wastes with the aim of "validating' long-term predictions. Both structures would continue to provide safety-related evidence until the repository was finally closed.

As an official from the Swiss Federal Office of Energy told an international conference is 2012, "For society's decision-making process, it is clear that reversals are part of the "game" and must be addressed knowing that a reversal is not the "end of the game" but a step towards a final solution."[9]

French Parliament Mandates Reversibility

The National Commission for Public Debate held a series of hearings between September 2005 and January 2006. Their aim was to obtain the views of interested and affected parties on possible revisions to the 1991 legislation that had restarted the nation's waste-disposition program. Among those were provisions to mandate, not just waste retrievability, but full reversibility of the entire endeavor.

The deliberations were colored by the refusal of some parties to participate. Groups opposed to the continued generation of nuclear electricity unsuccessfully sought to transform the forum into an argument over energy policy. When that tactic failed, they asserted that their continued participation was a sham. Nonetheless, more than 3,000 individuals in a dozen locations put forth their observations. Hardly surprisingly, the Commission observed that[10]

> As reassuring as it may seem, this concept of reversibility is not advocated by everyone. On the contrary, some voices are being raised to say that the very spirit of geological storage must be strictly applied: to eliminate all human intervention as quickly as possible in order to trust a geological layer whose stability and compactness have been tested. Therefore, society must close the storage as quickly as possible. Seeking reversibility, or even closing it in stages over too long a period, would only increase the risks unnecessarily.

But advocates of reversibility prevailed. Following an even more heated Public Debate in 2013 on the repository, the implementor, Andra, advanced its definition of reversibility. "Reversibility is the ability to leave future generations options regarding long-term management of radioactive waste, including sealing off disposal structures or retrieving waste packages."[11] In practical terms, the implementor would continuously improve its knowledge, support incremental development and gradual construction, promote operational flexibility, and maintain retrievability until society makes a decision to finally seal the disposition facility.

Choosing A Way Forward

In November 2005, the Canadian Nuclear Waste Management Organization (NWMO) sent a letter to the Minister of Natural Resources Canada forwarding the final study on how to revitalize the nation's waste-disposition program.

Core to NWMO's mission was the belief that the current generation had a moral obligation to start a process for managing fission's waste. In particular, it should not make irrevocable choices that deprived future generations of the opportunity to make decisions that were best suited to their circumstances. For that reason, NWMO proposed to establish a new development strategy that combined a technical approach with organizational transformation. The two elements were termed "adaptive phased management."

In the implementor's view, it had fashioned[12]

> A responsible path forward that intends to meet rigorous standards of safety and security for people and the environment. It embraces the precautionary principle. It is grounded in concepts of continuous learning and adaptive management. We believe this is the strongest possible foundation for managing the risks and uncertainties that are inherent in the very long time-frames over which used nuclear fuel must be managed with care.

The reader might recall the 1998 Seaborn Report, which concluded that the AECL's proposal for disposing of Canada's used fuel waste in the nation's granitic shield was a technically suitable plan to start. (See Chapter 7.) Not unexpectedly then, NWMO allotted only eleven percent of its commentary to assessing this idea. For the remainder of its study, NWMO had to work against a backdrop of the scathing language of the earlier inquiry regarding social unacceptability.

The new management strategy would combine the now familiar attributes.

- Phased and adaptive decision-making
- Continuous monitoring
- Provision for retrievability
- Citizen engagement

Importantly, NWMO maintained strongly that the new organizational arrangements were the "most responsive to citizens and the times."

NWMO anticipated that the plan would unfold in three stages. The first, lasting about 30 years, would involve preparations for centralized fuel management. The second, continued for the next 30 years, would demonstrate centralized storage and disposal. The third, perhaps persisting for 60 years, would entail long-term containment, isolation, and monitoring.

Some Nations Continue To Adhere To The Linear Strategy

United States

In the fall of 2000, the director of DOE's Office of Civilian Radioactive Waste Management (OCRWM) sought advice from NAS on strategies that the Office might pursue for "staging the design, construction, operation, and closure of a repository in a safe, secure, cost-effective, and societally acceptable fashion." Slightly over two years later, a panel from the National Research Council released its report.[13]

In preparing this report, the members of the committee were unquestionably motivated to find a remedy for what they believed were serious, perhaps fatal, institutional flaws affecting the Yucca Mountain Project. They were concerned that the project was too schedule-driven, had difficulties incorporating new data, and had manifestly failed to create either public trust or public acceptance, especially in

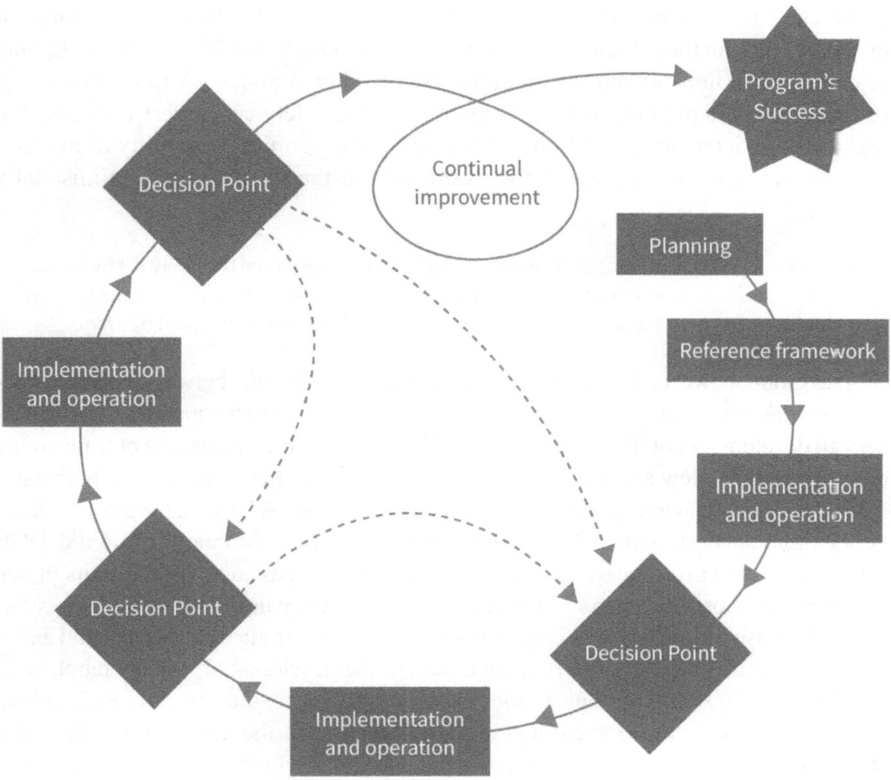

FIGURE 9.3 Phased development of a repository in the United States.

Nevada. As a remedy for those weaknesses, the panel proposed a phased strategy. Figure 9.3 illustrates the approach that the panel had in mind.

The group dubbed the reformed approach "adaptive staging." It was the simultaneous presence of an organizational commitment to learning, flexibility, reversibility, auditability, transparency, integrity, and responsiveness. In sum, adaptive staging appeared to require a fundamental transformation of the Department of Energy's operating culture, something that the agency might not be capable of doing. (This last point, it should be noted, was never made explicit in the report.)

This organizational reinvention emerged as a prerequisite for what would ensue. Explicit "decision points" separated steps in the developmental process. At those periods of time, the implementor structured its choice using a set of activities that provided information, considered a range of options, and engaged interested and affected parties. Noteworthy was the panel's definition of programmatic success: "Achieving the required degree of technical and social consensus to begin waste emplacement [emphasis added] and the incremental improvement of waste emplacement operations."[14]

Essential to staging is the iteration of the safety case. Here, the group envisioned more than a quasi-quantitative probabilistic risk analysis; it called for a more comprehensive assessment that marshaled reinforcing and independent lines of evidence; this methodological approach is widely adopted in Europe but is largely rejected by the implementor and regulator in the United States. (A fuller discussion of this point is found in Chapter 10.) Iteration of the safety case offers the hope that additional understanding might be gained about the technical suitability of the chosen site and repository design. Under those circumstances, the panel optimistically asserted[15]

> It may be possible to develop and maintain a technical consensus on the likely performance of the repository system, so that the adaptive process would, over time, build increasing public confidence in the behavior of the repository geology and engineered barriers.

The group, however, failed to create convincing causal links between valued objectives and the use of adaptive staging. For example, it noted in an appendix that the successful development of the Waste Isolation Pilot Plant for the disposition of transuranic defense waste in New Mexico took place using a strategy that was essentially linear.[16] That conclusion, however, raised unanswered questions. Why exactly did the panel believe that such an approach would *not* work at Yucca Mountain? Why did DOE often appear so unresponsive to new information in that case and why were its public involvement efforts so anemic? Was it because the agency used a linear staging strategy or because the agency was subject to extreme and contradictory internal and external pressures that would be exerted regardless of the development strategy employed?

Moreover, in their comments, both DOE and NRC pressed virtually an identical argument: phased development was already the letter of the law.[17] DOE followed a staged process that winnowed nine potential sites for the first repository down to five and then to three. Following the will of Congress, it then recommended the Yucca Mountain location. The State of Nevada (unsuccessfully) exercised its legal right to object to that determination. DOE then submitted a construction license application to NRC. Although the permitting process eventually ground to a halt, the regulator at some point must determine whether DOE can load and emplace waste in some future repository and whether it can finally seal the facility. At each of those stages, updated performance assessments would provide the opportunity to review projections of repository behavior. Moreover, in virtually all of those actions, extensive comments from interested and affected parties would be received and evaluated.

Sweden

A certain ambivalence toward phased repository development cloaked the Swedish waste-management undertaking.[18] As a senior official in the Ministry of the Environment put it: "There is no formal requirement in the Swedish legislation saying that a repository should, or should not, be designed to make retrievability possible." The program's non-regulatory oversight body, the National Council for Nuclear Waste, however, set forth a more nuanced position:

> A repository should be designed and constructed so that monitoring and remedial actions are not necessary in the future. However, future generations, probably with better knowledge and other values, must still have the freedom to make their own decisions; we should therefore not make monitoring and remedial actions unnecessarily difficult.

Notwithstanding this point of view, the Environment Ministry's representative maintained that retrievability, and thus reversibility, was historically a marginal issue. A very large part of the reason for this benign neglect was the widespread confidence in the KBS-3 safety case.[19] Even before closely evaluating the potential for the copper canister to corrode under anoxic conditions, the implementor, SKB, mined a section of the relevant bedrock. It exposed untouched copper nodules that had been present in the formation for millennia. SKB displayed this sample in exhibits that traveled throughout the country.

In presentations to the international waste-management community, SKB felt obliged to argue that its strategy was not an extreme outlier. It depicted a set of scenarios in which it might have to retrieve the waste canisters before and after repository closure.[20] It mapped its strategy to the Nuclear Energy Agency's so-called retrievability scale. By doing so, SKB was able to conclude that its strategy enjoyed similar features to those contained in the Swiss, French, and Canadian staged approaches. Waste could be recovered if needed but, as with those other nations, with increasing cost and technical difficulty as time advanced.

Finland

Given their parallel courses, it is hardly remarkable that Finland has taken a similar stance as Sweden when it came to phased repository development.[21] As part of its deliberations over the Decision-in-Principle authorizing development of a disposition facility in Eurajoki, the government insisted that the implementor, Posiva, make plans for the reopening of a repository. Significantly, the Finnish motivation for recoverability was grounded in the possibility that more advanced technologies could be deployed in the future. Thus, the option of retrievability needed to be maintained even after the structure was finally sealed. Once the engineered barriers no longer ensured essentially complete containment of the spent nuclear fuel, however, retrievability ceased to be a practicable option. In any event, planning for removal of the waste could never compromise the long-term safety of the repository. Like Sweden, the concept of reversibility is not adopted in Finland.

IS STAGING PROMISING?

Our appraisal and critique of the strategy begins with an exploration of practical considerations and then transitions into an assessment of the scheme's workability and applicability. In both respects, serious questions about the practice's appropriateness still demand serious answers.

Lessons From the Practical World

Students of policymaking in pluralistic societies have documented convincing cases of how budgeting, scientific research, and transportation planning proceed in a step-by-step manner. In the context of creating a disposition facility, however, the lessons are less clear *even* presuming that the differing priorities of implementors, regulators, and civil society can somehow be reconciled.

Decision-making Overload?

The discussion at the beginning of Chapter 7 called attention to the prospect that no commonly understood decision-making mode appropriately fits a disposition project when the degree of uncertainty and value conflict are both high. Under those conditions, an implementor may have no choice but to oscillate between incrementalism and bargaining. Staging adopts incrementalism; bargaining, however, is inescapable from power.

Two highly influential scholarly contributions have crucial implications for phased repository development. The Nobel Prize-winning administrative theorist Herbert Simon elaborated the concept of "bounded rationality" in 1957. He argued that *homo economicus*, the all-knowing utility maximizer, was an illusory and distorted figure. Instead, real-world choice-makers can do no better than "satisficing," making decisions that are "good enough" considering the incomplete knowledge base, uncertain costs, and cognitive constraints that abound.

Reinforcing Simon's analysis, a young economist/political scientist Charles Lindblom two years later authored a paper advancing incrementalism or the "science of muddling through." He too rejected *homo economicus*. Instead, he advocated a policymaking strategy in which important alternative potential policies were neglected, important possible outcomes were neglected, and important affected values were neglected. These decision-making aspects were disregarded, not because they were unimportant but because it was impossible and unworkable to accommodate them. By accepting small changes in plans and programs, major unpleasant surprises could be forestalled.[22]

It is instructive to compare these realistic expectations about policymaking with one of the decision-making flowsheets included in the NAS's 2003 report. (See Figure 9.4.) The model clarifies and makes plain what a phased strategy might actually entail. It calls out fifteen discrete decision points associated with just the determination whether to seal a repository. There can be no doubt that faithful implementation of staging can be quite taxing and time-consuming. Some shortcuts will have to be taken. Are some decisions more important than others? What should be insisted upon at each moment of choice? Who dictates the answers to these questions?

Preparation Is Essential

The German experience in reversing the disposal of several radioactive waste forms in the Asse II mine can enlighten those who might propose staging in the future.

The Asse II mine is located in Wolfenbüttel in southeastern Lower Saxony; its original purpose was to extract potash and rock salt. The Federal Republic of Germany purchased the pit in 1965 as a site for research into the disposition of low- and intermediate-level radioactive waste. Although officially termed an exploratory facility, the mine became a *de facto* repository after 126,000 waste-containing carbon-steel drums had been emplaced between 1967 and 1978. That inventory included an estimated 11.6 kilograms (about 25 pounds) of plutonium.

Even before the sale of the mine, several reports asserted that salt-saturated radioactive groundwater flowed from the underground cavities. More definitive accounts of leaks began to emerge in 1988. The mine's operator finally admitted to additional

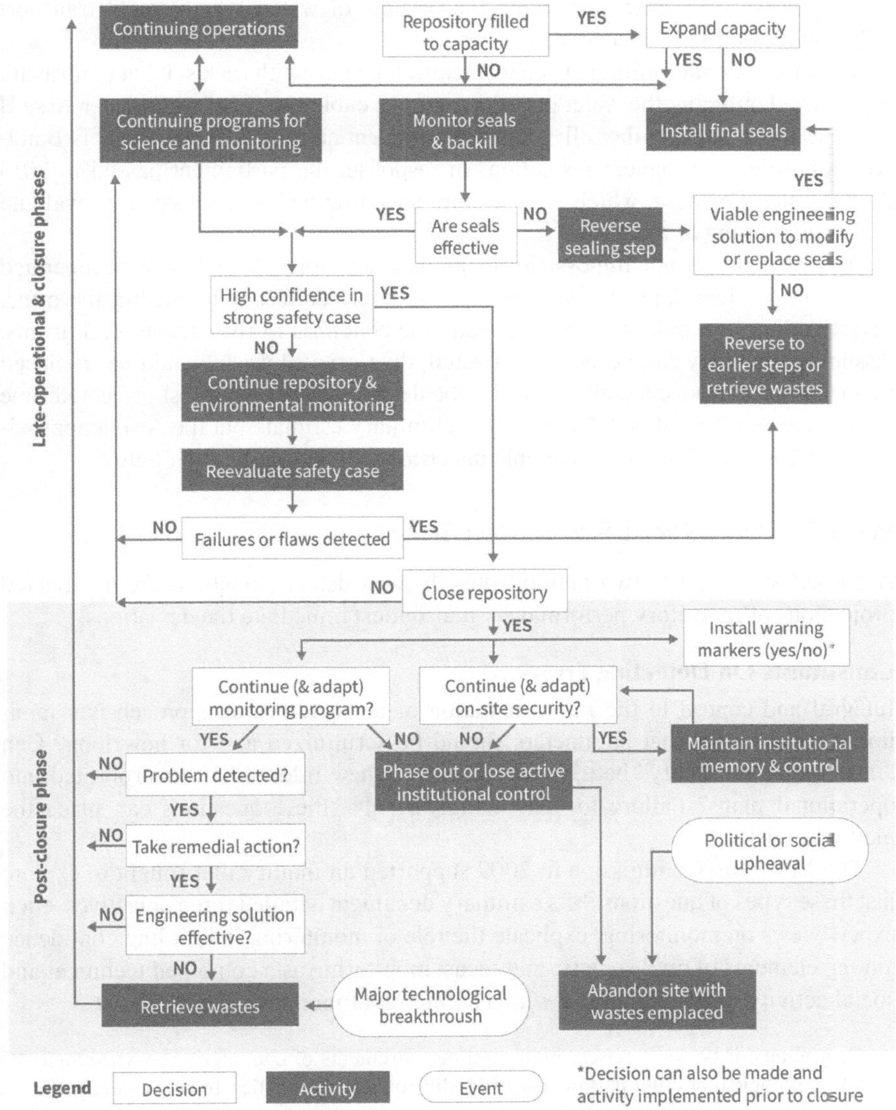

FIGURE 9.4 Phased development of a repository in the United States during post-closure.

discharges in 2008. Its spokesman, under pressure from the press, conceded that before then, "We did not consider that the leaks were worthy of a declaration to the press. We did not have the feeling that the public would be interested in knowing that radioactive brine is leaking in Asse II."[23]

As earlier chapters described, salt rock is plastic, that is, any cavities will eventually fill up with the mineral over time. The plasticity creates significant mechanical stresses that build up between the "surrounding diapir and the mine construction."[24]

The pressures form cracks allowing the outflow of water, which might transport radionuclides into the environment.

The mine operator initially tried to backfill the voids with rock salt but to no avail. It then tried pumping the water out. Currently 12 cubic meters of water enter Asse II per day, about 50 bathtubs full; roughly 95 percent are pumped out. But this Band-Aid is hardly a permanent resolution. In response, the parliament passed in 2013 the so-called *Lex Asse*, which calls for the "speeding up" of the waste retrieval and decommissioning of the mine.[25]

It is difficult, if not impossible, to predict how long Asse II can be operated safely and for how long the operators can securely collect water within the mine. Investigations are underway to understand the condition of the carbon-steel drums. Assuming that they can be safely recovered, the retrieved waste would be emplaced in a new mine, whose location is still to be determined. The earliest projected time to commence retrieval is 2033. A very preliminary estimate put the cost at approximately EUR 5.0 billion. Considerable uncertainty is attached to that figure.

What Else Does Social Science Have To Say

A phased strategy has two prerequisites. It must detect deviations from expected projections of repository performance. It also must remediate the deviations.

Constraints On Detecting Errors

Integral and central to the implementation of an incremental approach is a monitoring blueprint. What parameters should be scrutinized and for how long? Can "interpretive flexibility" be minimized? How is new information incorporated into operational plans? Failure to convincingly resolve these questions can place the method at risk.

The European Commission in 2009 supported an inquiry that sought to explore just these types of questions.[26] Its summary document detailed three objectives: elicit expert views on monitoring; explicate the role of monitoring in building confidence among elements of civil society; and recast monitoring as a combined technical and social activity. The study began with a definition of monitoring[27]:

> Continuous or periodic observations and measurements of engineering, environmental, radiological or other parameters and indicators/characteristics, to help evaluate the behavior of components of the repository system, or the impacts of the repository and its operation on the environment, and to help in making decisions on the implementation of successive phases of the disposal concept.

Based on a review of international technical material and interviews with eighteen specialists, the study team reached two significant, interrelated conclusions. First, monitoring after the repository has been closed was "a major challenge," "unrealistic," and "potentially counterproductive.[28]" Second, during the short period following the sealing of the disposition facility, the general position taken by the experts interviewed was that it will be possible to identify measurable parameters that would enable them to validate (and, if need be, adjust) the models on which they built their safety cases, but only a very few.

Given the narrow value of *postclosure* surveillance, what might be the prospects of *preclosure* error detection? Implementors, regulators, and members of civil society can certainly monitor some parameters, such as temperature and radioactivity, with a high degree of precision. Other phenomena, especially those that derive from experiments and simulations, may have sizable uncertainties surrounding their "correct" value. That ambiguity may formidably resist objective reconciliation.

More than a century ago, the economist Thorstein Veblen aptly observed that specialists can fall victim to a "trained incapacity to think."[29] One well-known example of this tendency involved the destruction of Challenger 11 shuttle flight. As the sociologist Diane Vaughn persuasively described, the scientists and engineers managing the launch had all the information needed to decide if the rocket should take off under low-temperature conditions.[30] After prolonged debate, they gave the go-ahead. The outsider physicist Richard Feynman, a member of the Presidential Commission examining the accident, later on took but a few seconds to immerse a critical O-ring in a glass of ice water. He conclusively demonstrated that the part would lose much of its required flexibility at the launch temperature of 36° F. Absent that elasticity, gaps would open between the rocket's junctures allowing flammable gases to escape.

Vaughn's research drew heavily on a *corpus* of studies by cognitive psychologists and sociologists of science and technology.[31] (See Figure 9.5.) The former group fostered the notion of heuristics and their consequent biases. Augmenting Simon's research, it identified mental shortcuts or rules of thumb that streamline an implementor's choice. Adopting heuristics often led to biases, resulting in systematic errors of judgment. In the Challenger case, cognitive dissonance likely explained the managers' behavior.

The latter group explored how a wide range of external factors can influence the way that knowledge is produced. These social constructivists argued that gender, political ideology, and training, among many other factors, swayed conclusions about the validity of technical hypotheses. The scientific community's rejection of Barbara McClintock's "jumping genes" theory in the 1950s was, in their view, most likely prompted by her gender.[32] (She won the Nobel Prize in Physiology or Medicine in 1983.)

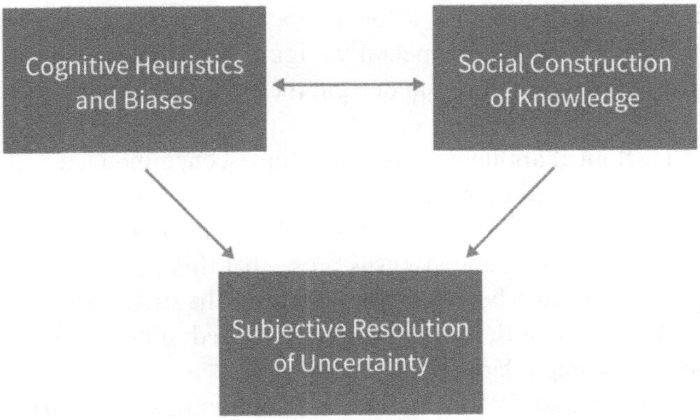

FIGURE 9.5 Mechanisms for subjectively resolving technical uncertainty.

Routine quality control audits, of course, are designed to catch at least some of the errors arising from the subjective resolution of uncertainty. But these internal checks can never elude the organizational milieu; they are dependent on a broader culture that may be systematically error prone. Advice from external peer reviewers, such as NAS, the UN-connected International Atomic Energy Agency, and the Organization for Economic Cooperation and Development's Nuclear Energy Agency, may be valuable in identifying miscalculations and technical misconceptions. But whether their recommendations are adopted typically depends solely on the goodwill of the implementor.

Constraints On Remediating Errors

The 2003 NAS report presented two forceful comments about the organizational limitations on correcting errors that might arise.[33] The first emphasized the importance of meriting and sustaining the trust of wide swaths of civil society. Without that confidence, the political costs of change may become unacceptably high. The second stressed the need for institutional constancy. After all, by its very nature, staging is a multigenerational strategy. Should institutional memory be lost, compromised, or distorted, whatever virtues that phased development might enjoy would certainly evaporate.

The experience in the United States provides rich lessons on how resistant remediating errors can be. One of them involves the exchange between DOE and the NWTRB over whether the drifts at the proposed Yucca Mountain repository should be maintained at an above- or below-boiling temperature. Another encompasses the regulatory transformation following NAS's publication of the congressionally mandated study, *Technical Bases for Yucca Mountain Standards*.[34] Information about both of these case studies can be found in Chapter 6. For the reader's convenience, the key events are recapitulated here.

NWTRB raised two related reservations about DOE's high-temperature repository design.[35] First, the proposal would introduce greater uncertainties into the probabilistic performance assessment. Initially, the heat would drive the humid air within the repository drifts into the rock that lay above. As the repository cooled, the water would re-enter the excavated openings. Modeling such phenomenon would be more difficult than for a low-temperature design, and thus the TSPA would be less definitive. (See especially Figure 6.5.) Although DOE did make modifications to its original thermal design, the Board's concerns were never fully addressed.

The second technical argument was that deliquescence-induced corrosion caused by calcium chloride would corrode the surface of the Alloy-22 waste package. The Board maintained that the corrosion would only occur in an above-boiling repository design. A broad scientific consensus arose that this corrosion mechanism was largely insignificant even when the temperature in the drifts was elevated. In this instance, the Board's error-detecting function misfired, although it never explicitly acknowledged its oversight lapse.

Congress required that EPA promulgate an environmental dose/risk standard for the proposed Yucca Mountain repository. The mandated NAS report, albeit a bit imprecisely, advised EPA to set the compliance period at the time of peak dose or roughly 1,000,000 years. The agency, however, asserted, that[36]

> *Despite NAS's recommendation* [emphasis added], we conclude that there is still considerable uncertainty as to whether current modeling capability allows development of computer models that will provide sufficiently meaningful and reliable projections over a time from up to tens-of-thousands to hundreds-of-thousands of years

Thus, EPA issued an individual-protection standard that was capped at 10,000 years.

The State of Nevada and the Natural Resources Defense Council filed suit in the Court of Appeals for the D.C. Circuit asking, among other things, that EPA's final determination of the compliance period for Yucca Mountain be vacated. On July 9, 2004, the court rejected most of Nevada's claims but agreed that the compliance period was unlawful.[37]

> It would have been one thing had EPA taken the Academy's recommendations into account and then tailored a standard that accommodated the agency's policy concerns. But that is not what EPA did. Instead, it unabashedly rejected NAS's findings and then went on to promulgate a dramatically different standard, one that the Academy had expressly rejected. Although [the 1992 Energy Policy Act's] instruction that the standard should be "based upon and consistent with" standard does not require EPA to walk in lock-step with the Academy, we think it entirely unreasonable for EPA to have acted *inconsistently* with NAS findings and recommendations.

Although strictly speaking the agency committed a legal error, it had enormous technical implications. DOE, EPA, and NRC took advantage of interagency consultations through the presidential Office of Management and Budget to minimize the technical turbulence that further regulatory change might trigger. (Those negotiations remain unavailable to the public, but a senior DOE waste-management leader alluded to them at an NWTRB meeting.) Perhaps ironically, the reissued standard that emerged still drew its own legal critique.[38]

From these events, two lessons can be drawn. First, error-detectors may themselves be fallible. NWTRB members and staff never closely probed whether calcium chloride particles were, in fact, carried in the dust blowing in the vicinity of Yucca Mountain. If absent, they could hardly migrate within the upper layer of the formation, land on the Alloy-22 waste packages, deliquesce, and then initiate corrosion mechanisms. Significantly, the Board's chairman reaffirmed NWTRB's position in a hearing before the Subcommittee on Energy and Air Quality of the House Committee on Energy and Commerce in March 25, 2004.[39] At his side was the Acting Chairman of the NRC, who took strong exception to his colleague's remarks.

Second, if detecting errors likely demands the existence of internal and/or external *independent* bodies, correcting mistakes requires *interdependence* between the operational arm and the monitors to enforce change. Juggling both prerequisites turns out to be an intrinsically challenging act.

In the thermal design debate, NWTRB's strong public criticism of the DOE's performance assessments, especially the one undertaken in conjunction with the

Secretary's site recommendation, forced the department to try to find some common ground. But reputational power may not be determinative; that NRC never mentioned similar issues meant that the implementor could rest assured that its ultimate objective—an approved construction license application—was not in danger if the NWTRB's advice was not fully accepted.

Much the same dynamic played out in the second example, whether the Appeals Court could force EPA (and by extension NRC and DOE) to alter its environmental standard to make it consistent for NAS's *Technical Basis Report*. Here the three agencies negotiated—colluded might be too strong a verb—to minimize the most substantial downsides of regulatory change, thereby undermining at the very least the *Report's* spirit.

In the next chapter, our narrative explores in greater detail to the next step in the repository-development process: licensing the construction of a disposition facility.

NOTES

1. John Steinbruner's landmark study demonstrates how social and psychological forces influence both behaviors. See John Steinbruner, *The Cybernetic Theory of Decision: New Dimensions of Political Analysis,*" (Princeton: Princeton University Press, 1974). Steinbruner inspired my own early work on error correction in bureaucracy.
2. William P. Bishop et al., "Proposed Goals and Objectives for Radioactive Waste Management: A Report to the U.S. Nuclear Regulatory Commission," NUREG-0300, 1978. The quotations is found at 72. See also a set of commentaries by the same group, "Essays on Issues Relevant to the Regulation of Radioactive Waste Management," NUREG-0412, 1978. The author was a member of this interdisciplinary group. See https://www.nrc.gov/docs/ML2130/ML21308A053.pdf. Accessed March 6, 2025.
3. Nuclear Regulatory Commission, "Disposal of High-Level Radioactive Wastes in Geologic Repositories: Licensing Procedures," *Federal Register*, (48) February 25, 1981, 13,971–13,987.
4. Interagency Review Group on Nuclear Waste Management, "Subgroup Report on Alternative Technology Strategies for the Isolation of Nuclear Waste, TID-28818 (Draft), October 1978, 33, 35–36.
5. Board on Radioactive Waste Management, "Rethinking High-Level Radioactive Waste Disposal: A Position Statement," 1990.
6. Board on Radioactive Waste Management, 29.
7. DOE's Office of Civilian Radioactive Waste established a Task Force on an Alternative Strategy to argue for program modifications that were better aligned to the NAS's recommendations. It met an unreceptive fate. See Thomas Isaacs and Maxwell Blanchard, "A Proposed Alternative Strategy for the Department of Energy's Civilian Radioactive Waste Management Program," March 31, 1993. See https://adamspublic.nrc.gov/navigator. Accessed March 4, 2025.
8. Expert Group on Disposal Concepts for Radioactive Waste (EKRA), "Disposal Concepts for Radioactive Waste: Final Report," On behalf of the Federal Department for the Environment, Transport, Energy and Communication, 2000. (See Chapter 7.)
9. Nuclear Energy Agency. "Reversibility and Retrievability in Planning for Geological Disposal of Radioactive Waste: Proceedings of the "R&R" International Conference and Dialogue December 14–17, 2010, Reims, France" NEA-6993, 2012, 31.
10. Commission on Public Debate, "Report on the Public Debate on General Options for Radioactive Waste Management," Paris, 2006, 38. See https://cpdp.debatpublic.fr/cpdp-dechets-radioactifs/docs/pdf/compte-rendu.pdf. Accessed March 11, 2025. English version from Google Translate.

11. Andra, "Position Paper on Reversibility," January 2016. See https://international.andra.fr/sites/international/files/2019-03/reversibility_paper.pdf. Accessed March 11, 2025.
12. Nuclear Waste Management Organization (NWMO), "Choosing a Way Forward: The Future of Canada's Used Nuclear Fuel: Final Study," 2005, 4.
13. National Academy of Sciences, "One Step at a Time: The Staged Development of Geologic Repositories for High-Level Radioactive Waste," 2003.
14. National Academy of Sciences (2003), 3.
15. National Academy of Sciences (2003), 61.
16. National Academy of Sciences (2003), 164.
17. DOE's comments can be found on the Licensing Support Network, See https://adams-public.nrc.gov/navigator. Accessed March 14, 2025.
18. Ansi Gerhardsson, "Reversibility and Retrievability—The Swedish View," in Nuclear Energy Agency (2010). Quotations are found on 45 and 46 respectively.
19. The environmental advocacy organization, MKG, was a notable exception to the consensus. The issue of copper corrosion will be described in the Chapter 11. See also Swedish National Council for Nuclear Waste, "Mechanisms of Copper Corrosion in Aqueous Environments," Report 2009:4, 2010.
20. See Erik Setzman, "Swedish Retrievability Scenarios," in NEA (2010), 113–116.
21. See Jaana Avolahti, "Policy Perspective on R&R in the Finnish Context," in NEA (2010), 27–28 and Risto Paltemaa, in NEA (2010), 211–212.
22. Switzerland and France are still awaiting final approval of their respective construction license applications. Understanding just how those nations will implement a staged strategy must therefore await future programmatic decisions. But a substantial set of challenges will unavoidably arise.
23. Bellona, "20-Year-Long German Leak Scandal Engulfs Country and Disturbs Europe," News Release, September 8, 2008. See https://bellona.org/news/nuclear-issues/2008-09-20-year-long-german-nuclear-leak-scandal-engulfs-country-and-disturbs-europe. Accessed February 24, 2025.
24. A diapir is a type of intrusion in which a more mobile and ductilely deformable material is forced into brittle overlying rocks.
25. Three options for dealing with the waste were advanced after the passage of the *Lexe Asse*: complete backfilling, relocation of the material within Asse II, and retrieval of the waste. An economic analysis of these options can be found in Patrick Ilg et al., "Nuclear Waste Management under Approaching Disaster: A Comparison of Decommissioning Strategies for the German Repository Asse II," *Risk Analysis*, (37) 1213–1232, 2016.
26. Anne Bermans et al., "Monitoring the Safe Disposal of Radioactive Waste: A Combined Technical and Socio-Political Activity," 2012. See https://cordis.europa.eu/docs/projects/files/232/232598/07-modern-d1-3-1-basis-for-stakeholder-communication_en.pdf. Accessed March 26, 2025. See also M.J. White (ed.) "Monitoring During the Staged Implementation of Geological Disposal: The MoDeRn Project Synthesis" 2014. See https://igdtp.eu/wp-content/uploads/2018/09/MoDeRn-D.6.1-MoDeRn-Project-Synthesis-v1.1.pdf. Accessed March 26, 2025.
27. Bermans et al., 8.
28. Bermans et al., 22.
29. Thorstein Veblen, "Why is Economics Not An Evolutionary Science?" *Quarterly Journal of Economics*, (12) 1898, 373–397.
30. Diane Vaughn, *The Challenger Launch Decision: Risky Technology, Culture, and Deviance at NASA*, (Chicago: Chicago University Press, 1997).
31. An important introduction to the question of interpretative flexibility can be found in Imre Lakatos and Alan Musgrave (eds), *Criticism and the Growth of Knowledge*, (London: Cambridge University Press, 1970). See also the groundbreaking research Amos Tversky and Daniel Kahneman, "Judgment Under Uncertainty: Heuristics and Biases," *Science*, (185) September 27, 1974, 1124–1131. A useful catalogue of several dozen heuristics and biases is https://thedecisionlab.com/biases. Accessed April 2, 2025. A

valuable collection of research on the social construction of science and technology is Sheila Jasanoff et al., (eds) Handbook of Science and Technology Studies, (Thousand Oaks, CA: Sage, 1995).

32. Evelyn Fox Keller, *A Feeling for the Organism: Life and Work of Barbara McClintock*, (New York: W.H. Freeman, 1983).
33. See especially National Academy of Sciences (2003), 38–41.
34. National Research Council, *Technical Bases for Yucca Mountain Standards*, (Washington D.C.: National Academy Press, 1995).
35. Letter from Michael Corradini to Margaret Chu and the attachment "An Evaluation of Key Elements in the U.S. Department of Energy's Proposed System for Isolating and Containing Radioactive Waste," November 25, 2003.
36. Environmental Protection Agency, "Public Health and Environmental Radiation Protection Standards for Yucca Mountain, NV: Final Rule," *Federal Register* (66) June 13, 2001, 32,074–32,135.
37. *Nuclear Energy Institute, Inc. v. Environmental Protection Agency*, 373 F.3rd 1251, 1270.
38. See *State of Nevada v. Nuclear Regulatory Commission*, Court of Appeals for the District of Columbia Circuit, USCA Case #09-1133, (2016). One major point of controversy was EPA's setting a much larger environmental standard for the post-10,000 year compliance period.
39. See https://www.nwtrb.gov/docs/default-source/testimony/con209.pdf?sfvrsn=c61cf605_8. The hearing and response from Nils Diaz, Acting NRC Chairman, can be found at https://www.govinfo.gov/content/pkg/CHRG-108hhrg93300/pdf/CHRG-108hhrg93300.pdf. Both sites accessed April 24, 2025. See also the presentation of Board member Thure Cerling at the NWTRB meeting on May 18, 2004.

10 Repository Licensing

Regardless of whether a nation opts for a linear or a phased strategy to develop a deep-mined, geologic repository, each country's implementor has or will have to submit to an independent regulator an application to construct that facility.[1] Although the UN-affiliated International Atomic Energy Agency (IAEA) has provided guidance for protecting people and the environment from radionuclide releases, each nation must codify its own rules.[2]

As a consequence, national regulatory approaches vary along four dimensions. (1) What level of dose/risk is considered "acceptable" over what period of time?[3] (2) What methodology must the implementor employ to demonstrate compliance with an individual-protection dose limit? (3) What procedure must be followed as part of the licensing process? (4) How prescriptive should the rules be? Our story considers each question in turn.

FOR HOW LONG MUST A REPOSITORY LIMIT RELEASES TO AN ACCEPTABLE DOSE?

Both the so-called compliance period and the level of acceptable dose are fundamentally social judgments. As a group of prominent analysts put it: "Even in the same situation and at a single time, different people with different values, beliefs, objectives, or decision methods might disagree on which option is best. In short, the search for absolute acceptability is misguided."[4]

Licensing Conditions in the United States

The 1982 Nuclear Waste Policy Act (NWPA) continued EPA's authority to determine both the compliance period and an individual-protection dose limit. The law also instructed the Nuclear Regulatory Commission (NRC) to align its repository-licensing rules with EPA's decrees. (See Chapter 4.)

It should not come as a surprise that some technical specialists tried to fashion a "scientific" rationale for determining the first parameter. Arguing that it was not the responsibility of generators of nuclear electricity to "purify" the earth,[5]

> They promoted the position that a repository need not sequester waste beyond the time when it became less hazardous than the uranium ore from which it was derived. A rough measure of hazard was constructed, the water dilution volume. It was the amount of liquid it would take to reduce the concentration of waste to acceptable drinking water standards.

Drawing graphs such as Figure 10.1, those advocates compared the hazard associated with the ore with the hazard associated with the waste as both decayed over

DOI: 10.1201/9781003528302-12

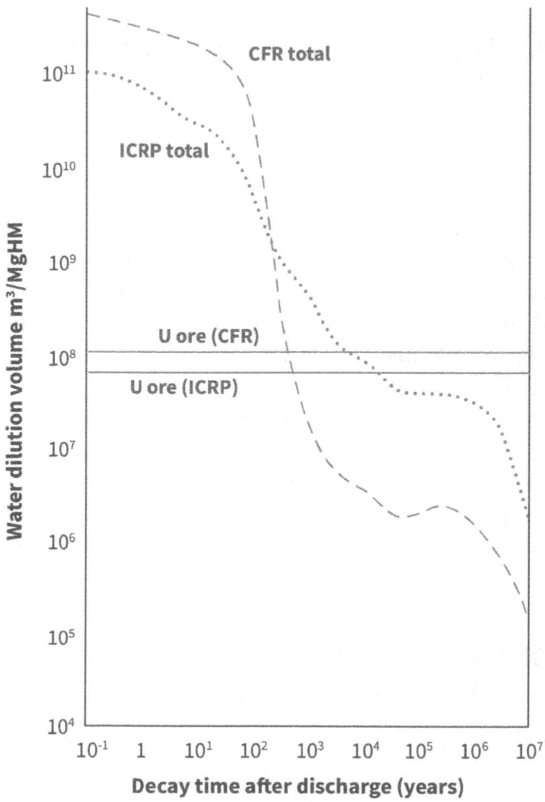

FIGURE 10.1 Decay time after discharge (years).

time. Depending on what regulatory assumptions were used, the hazards appeared roughly comparable sometime no later than 10,000 years after the fuel had been discharged from the reactor.

Likely influenced by such arguments, EPA in 1985 issued its first regulation. (See Chapters 4 and 6.) It set the compliance period at 10,000 years. It also promulgated limits on the release of selected radionuclides. Further, it instituted an individual-protection dose limit of 0.25 millisieverts (mSv)/year.[6]

In 2003, following congressional direction and the release of the National Academy of Science's report, *Technical Bases for the Yucca Mountain Standards* (TYMS), EPA established a revised Yucca Mountain-only dose limit of 0.15 millisieverts/year but only for the first 10,000 years after the repository was sealed.[7] Beyond that time, EPA decided that no formal dose constraint needed to be set.

As Chapter 9 explained, that regulation was overturned the following year in *Nuclear Energy Institute, Inc. v. E.P.A.* EPA would have to prescribe an individual-protection dose limit up to the time of peak dose, roughly 1,000,000 years for Yucca Mountain.

In 2005, EPA attempted to modify the rule once again. This time the agency proposed establishing the individual-protection dose limit at 0.15 millisieverts/year for the first 10,000 years. It also suggested a 3.5 millisieverts/year individual-protection dose limit after 10,000 years but within the "period of geologic stability."[8] The final rule, issued three years later, lowered the post-10,000-year dose limit to 1.0 millisieverts/year.[9] NRC subsequently conformed its licensing rules to make them consistent with EPA's determinations.[10] With the collapse of the Yucca Mountain Project, however, the issue became moot.[11] Any effort to restart a repository-development program at a site other than Yucca Mountain will have to include, among many other things, the promulgation of a new compliance period and individual-protection dose limit.[12]

Licensing Conditions Abroad

The regulators in the other eight nations featured in this narrative did not experience nearly as much turbulence as they decreed compliance periods and individual-protection dose constraints. Heavily influenced by the 2007 recommendations of the International Commission on Radiological Protection, most of those countries have promulgated their own licensing rules.[13] (See Figure 10.2.)

COUNTRY	DOSE CONSTRAINT	COMPLIANCE PERIOD
Canada	An upper close limit of 1.0 mSv/year established; implementer is required to provide a rationale for the dose constraint, which is a fraction of the close limit	Not specified
Finland	Less than 0.1 mSv/year for normal events Release limits for various radionuclides established Impacts should be comparable to those arising from natural radioactive materials but should remain insignificantly low	Not specified
France	0.25 mSv/year for normal scenarios	10,000 years
Germany	0.01 mSv/year for probable developments; 0.1 mSv/year for less probable developments	1,000,000 years
Japan	No decision has been made	No decision has been made
Sweden	Less than 0.02 mSv/year	Minimum of 100,000 years and can exceed up to a million
Switzerland	0.1 mSv/year for expected scenarios	1 million years
United Kingdom	0.5 mSv/year for the operational period	Not specified

FIGURE 10.2 Health requirements for disposal.

Thus, the highest acceptable individual-protection dose limit is less than a factor of four of the lowest. The range of compliance periods, however, is much greater, going from 10,000 to 1,000,000 years. Of some note, the U.S. standard is generally more restrictive than the regulations adopted by France or the United Kingdom but less restrictive than the rules established in Germany or Switzerland.[14]

WHAT METHODOLOGY DEMONSTRATES COMPLIANCE?

Demonstrating conformity with the individual-protection dose limit throughout the compliance period demands a multidisciplinary analysis that evaluates the disposition facility's behavior both spatially and temporally. At the core of all the appraisals is a total system performance assessment, which might be supplemented by independent lines of evidence.

Methodology Adopted in the United States

As discussed in Chapter 4, EPA initially cast its individual-protection standard in terms of limiting the release of selected radionuclides. A set of algorithms instructed how acceptable discharges should be calculated as a function of the amount of waste disposed and the level of the spent nuclear fuel's burn-up within a reactor. In a noteworthy comment, EPA observed that "substantial uncertainties are likely to be encountered in making these predictions. In fact, sole reliance on these numerical predictions to determine compliance may not be appropriate."[15] For that reason, the implementor would also have to carry out a performance assessment to establish that its proposed disposition facility met the then-regulatory constraint of 0.25 millisieverts/year.

At the time, the implementor and regulators directed most of their attention to understanding and calculating radionuclide release levels. The ramifications of conducting performance assessments, in contrast, drew relatively less consideration. That circumstance fundamentally changed with the release of the TYMS report.[16]

> To judge compliance against a risk-based standard of the type proposed, a risk analysis including treatment of all scenarios that might lead to releases from the repository and to radiation exposures is, in principle, required. To include them in a standard risk analysis, all these scenarios need to be quantified with respect to the probabilities of scenario occurrence and the probability distribution of their consequences to humans, such as the health effects of radiation doses.

By mandating performance assessment as the methodology for evaluating compliance, the TYMS panel, not incidentally, found a reason to jettison the subsystem performance requirements that NRC had imposed in its 1983 regulations: "[B]ecause it is the performance of the total system in light of the risk-based standard that is crucial, imposing subsystem performance requirements might result in suboptimal repository design."[17] (See Chapter 4.)

In the wake of *NEI v. EPA*, EPA also had to wrestle with writing methodological rules for two different compliance periods. For the 10,000-year compliance period, it retained its first Yucca Mountain-only procedure.[18] For the post-10,000-year timeframe, EPA proposed to exercise its discretion to direct how the performance assessment should be accomplished. It laid out how the effects of climate change must be calculated; it narrowed the considerations that structured the analyses of seismic and igneous activity; and it detailed how the effects of generalized corrosion should be represented. EPA adopted these provisions in 2008.[19]

Similarly, NRC initially maintained its first Yucca Mountain-only regulations for the first 10,000 year compliance period.[20] For example, it adopted EPA's characteristics of the reference biosphere, adding that biosphere must be arid or semi-arid. It embraced EPA's definition of the "reasonably maximally exposed individual" as the "receptor" of the doses. On its own volition, NRC determined that those individuals lived in a community consisting of 15 to 25 farms. In 2009, NRC issued its conforming implementing regulations for the post-10,000-year period.[21]

Methodology Adopted Abroad

What distinguishes the United States from many other countries with advanced nuclear waste-management programs is the latter's prominence given to the "safety case."[22] That weight derives from a desire to communicate *confidence* in the performance assessment's conclusions in a manner that goes beyond a statistical summary of Monte Carlo iterations.[23]

As the Nuclear Energy Agency of the Organization for Economic Cooperation and Development put it, the safety case is the "*synthesis* [emphasis added] of evidence, analyses, and arguments that quantify and substantiate a claim that the repository will be safe after closure and beyond the time when active control of the facility can be relied on."[24] One commentary likened the safety case to a "legal brief, advancing a compelling argument for safety. Through a safety case, an implementor reveals its understanding of the repository site and how it expects the radionuclides to behave in the repository over long periods of time."[25] Figure 10.3 illustrates the relationship between performance assessments and the safety case.

To be sure, performance assessments, sometimes called safety assessments, continue to be an integral element of the safety case. But the safety case incorporates plain-language explanations of why the site remains as technically suitable and socially acceptable as it was when it was first selected. It makes clear why the individual barriers isolate and contain the radionuclides. Especially critical is the safety case's explication of whether the multiple natural and engineered barriers are independent of each other, that is, redundant, and thereby whether they provide a measure of defense-in-depth.[26]

The dialectics stimulated by the safety case become even more valuable if potential critics of the disposition facility have access to independent specialists. One way to accomplish this end is be to make available money and other resources. In Sweden, for example, two environmental groups obtained funds to participate in the site selection and licensing activities. In Canada, local communities also received support from the implementor to contribute to the Nuclear Waste Management Organization's adaptive phased management process. (In the United States, the State of Nevada received resources through the NWPA.)

Constructing the safety case, then, provides a means through which an implementor engages not only the regulator but also interested and affected members of the

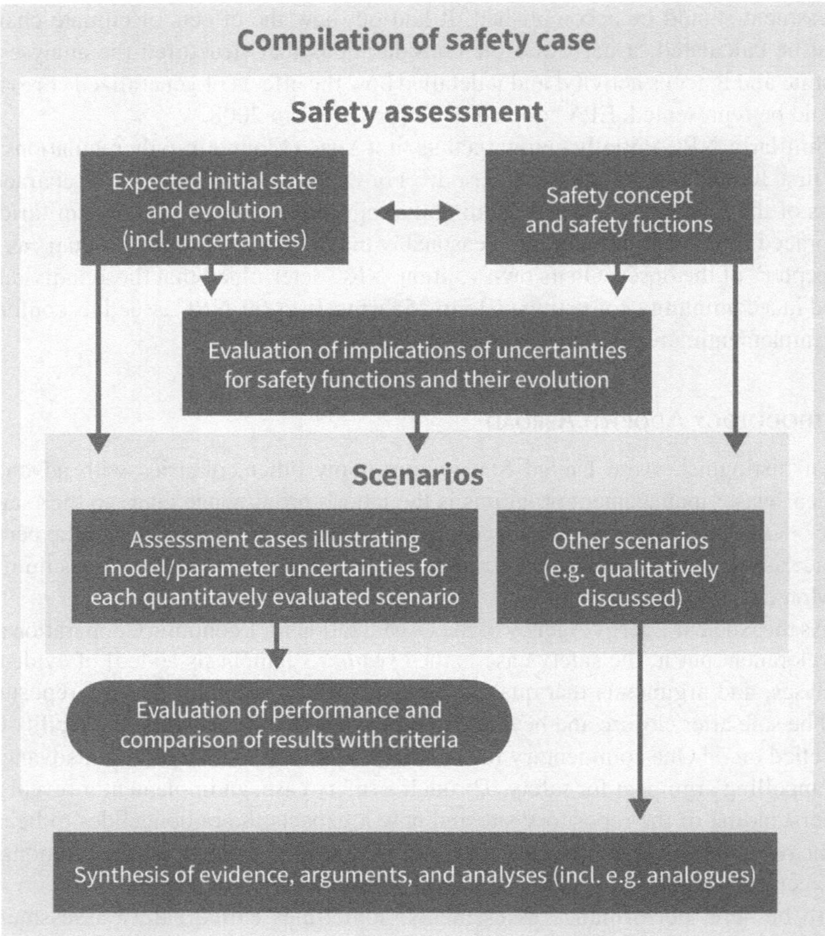

FIGURE 10.3 Constructing a safety case.

public. Among the dialogues' most notable benefits are opportunities for the repository proponents to build trust and confidence, which, as Chapter 8 makes abundantly clear, remains a critical element contributing to a repository program's ultimate fate.

In the sections below, our narrative explores how implementors in two nations have advanced claims that their proposed repository satisfied the regulatory benchmarks for acceptable dose across the compliance period.

The first, Finland, sought to build confidence by using analogues. The second, France, relied wholly on the safety assessment.

Finland

In 2012, the Finnish implementor, Posiva, submitted a construction-license application to the country's Council of State. Included in the tender was a postclosure safety case, which was based on the KBS-3 disposal concept.[27] The request consisted, among other material, of information on the design basis, performance assessments and formulation of scenarios, analyses of releases, and models and data. For the purposes of this narrative, a document focusing on "complementary considerations" is most relevant.[28]

That report's objective was stated unambiguously at the start.

> Complementary considerations address diverse and less quantifiable types of evidence and arguments that can be made to enhance confidence in the outcome of the safety assessment. In particular, demonstration of repository safety at times greater than a few thousands of years is not possible using conventional measures such as dose to an exposed population, so other arguments are required that rely on evidence gained from systems that have operated over geological timescales.

These matters were not only essential to the implementor's interactions with the Radiation and Nuclear Safety Authority (STUK) but they also were indispensable to convince "less technical audiences who need to trust in the safety of the repository."

Natural and archaeological analogues, in particular, presented opportunities to test whether existing tools and models could accurately forecast the copper waste canisters' evolution over time. To do so, Posiva studied equivalents that might provide insights into the suitability of the repository design and materials on the one hand and the limited rates of radionuclide migration on the other.

A fundamental safety dispute was whether the native copper canister corrodes in pure anoxic—oxygen-free—groundwater.[29] Considerable data from laboratory studies as well as conclusions derived from thermodynamic theories seemed to strongly support copper's durability. (But see Chapter 11.)

But for Finland at least, analogues put the final nail on the technical debate's coffin.[30] The Swedish warship *Kronan* sank in 1676 during the Battle of Öland. She carried more than 120 bronze (96 percent copper and 4 percent tin) cannons. Even in an oxidizing environment, the cannon exhibited extremely low corrosion. In a reducing environment, corrosion almost certainly would be less. This study, as well as others involving archeological analogues "increases confidence in the copper's stability as a canister material."[31]

Archeological analogues can provide information about century-long processes. They are less persuasive in extrapolating to millennia-long developments. Here natural analogues can often fill the gaps. Elemental copper has persisted in a range of geological environments, ranging from Michigan to England to Finland. Posiva conducted investigations in the English mud (clay) flats around Devon and the Palmottu site in southern Finland.

The Devon analogue not only offered evidence about copper corrosion but also about the interaction between the metal and the clay, which is similar to, but much less effective, than the bentonite employed in the KBS-3 concept. The copper remained relatively inert and isolated for at least 170,000,000 years, much longer than demanded by the Finnish repository design. Likewise, the Palmottu analogue showed signs of corrosion in reducing conditions due to the presence of sulfides, a well-understood scenario. Again, Posiva concluded that "corrosion has been shown to proceed at a very slow rate, possible due to passivation mechanisms."[32]

In addition to the integrity of the engineered barrier, the Finnish implementor had to address another crucial issue: how are radionuclides transported outside the repository system through the natural geologic barriers and then potentially expose biosphere inhabitants to toxic doses. The key variable here is the groundwater travel time of each component of the waste. This depends on whether each radionuclide moves by matrix diffusion, by advection in porous media, or by fracture flow. Transport times in each of these processes also can be reduced if the material is absorbed, adsorbed, or chemically bonded to the surrounding geology. Over time, each radionuclide will decay into other species, a phenomenon that also must be accommodated in the calculations.

Posiva examined 26 different potential natural analogues across five continents, including the natural reactor at Oklo in Gabon and Cigar Lake in Canada. Although the investigations overall suggested groundwater travel through the granite found at the Olkiluoto repository location was relatively slow and that colloid-associated transport time was similarly protracted, Posiva seemed reluctant to draw firm or detailed conclusions on this matter.

France

As Chapter 7 briefly detailed, the French nuclear waste-management implementor, Andra, focused the country's search for a disposition a facility in the Callovo-Oxfordian clay formation near the small town of Bure. In 2005, it submitted to the government an evaluation of the feasibility of a geological repository in that hydrogeologic environment.[33] Andra tendered an application to construct the repository called Cigéo, to its regulator in January 2023. As of the end of 2025, it is still under review.[34]

The implementor maintains that compliance with the individual-protection dose limit depends on four safety functions[35]:

1. Isolate waste from surface phenomena and human actions
2. Limit the circulation of water
3. Limit the release of radionuclides and toxic chemical and immobilize them
4. Delay and mitigate the migration of radionuclides

According to Andra, the first safety function is satisfied because Cigéo is installed approximately 500 meters below the surface away from erosion zones. It is also located far from extractable underground and geothermal resources. The combination of the two factors minimizes the likelihood of inadvertent human intrusion. That the implementor has committed to maintaining memory of the site also contributed to the low probability of disturbance.

The second safety function is fulfilled by the properties of the Callovo-Oxfordian formation itself. As in all clay structures, it has a very low permeability coupled with a low hydraulic head gradient. Cigéo is located away from potentially hydraulically active features. By sealing the surface-to-bottom connections, groundwater flow is minimized. The disposal cells are isolated from one another, thereby further separating the waste packages.

Limiting the release of radionuclides, the third safety function, is achieved through several means. The materials used to construct the disposal cells protect the waste packages from physiochemical alteration. As a result, the obstacles to discharge are increased. Low groundwater flow, diffusive transport, and reducing groundwater chemistry all contribute to low release and mobility.

The delivery of the waste material is delayed for extended periods, the last safety function. Radionuclides must travel through at least 130 meters of clay before they might reach the biosphere. During that passage, the material is diluted so that exposures to humans, flora, and fauna are reduced. Indeed, placing the underground facility in the middle of the retentive Callovo-Oxfordian formation limits migration and lessens the concentration of the waste.

Unlike Posiva's explicit discussion of natural and archeological analogues, Andra's main report filed as part of its license application is silent about complementary considerations. However, the implementor's publication of an earlier analysis notes in the Glossary potential independent lines of evidence regarding radionuclide migration (Oklo), bituminous marl formations (Jordan), silica dissolution (volcanic phenomena and stained glass from the Middle Ages), and iron corrosion (metalwork in Europe between 500 and 300 B.C.E.).[36]

WHAT IS THE STRUCTURE OF THE LICENSING PROCESS?

Although every implementor must present a construction application to its independent regulator, the procedures subsequently followed vary from nation to nation. In some countries, the process wards off interested and affected parties. In others, outsiders with a stake are closely involved.

The Process in the United States

NRC has interpreted the Atomic Energy Act to require trial-like hearings for contested applications to construct nuclear power plants. These adjudicatory inquiries allow testimony, discovery of information, and cross-examination of witnesses. Neither that law nor the NWPA, however, speaks to whether that format is also mandated for hearings involving the licensing of a disposition facility for high-activity radioactive waste. Anticipating the submission of a construction-license application for the proposed Yucca Mountain repository, NRC Commissioners chose to order a formal review.

In that case, a specially appointed hearing board oversaw the proceeding. That process entailed the filing of a petition to intervene, which set forth with "particularity" the contentions the intervenor sought to raise. For instance, the States of Nevada and California advanced more than 300 contentions, all of which both DOE and the NRC staff attempted to dismiss.[37] Ultimately, the vast majority were admitted. The parties to the hearing spent the next two years battling back and forth over a wide range of technical, environmental, and legal issues, including whether DOE had the right to withdraw its application. In the end, the case was suspended because of budgetary constraints on DOE.

The Processes Adopted Abroad

None of the nations outside the United States have embraced nearly as formal a process. Nonetheless, important distinctions can be drawn as our tale examines the procedures adopted in the three countries whose implementors have filed construction-license applications.

Finland

In late 2012, Posiva presented its request to the Finnish Government to construct the ONKALO repository near Olkiluoto on the Baltic Sea.[38] Although the dossier was incomplete, STUK agreed to begin its review eighteen months later. At that time, it requested additional information on the facility's design, construction plans, operational safety assessment, and the long-term safety case.

In late 2015, STUK released its evaluation report. The regulator's central conclusion was generally positive[39]:

> Based on STUK's review of the safety case documentation, the postclosure safety of the facility has been analyzed in a sufficient manner for the purposes of the construction license stage. The methods and analyses used in the safety case are, in general, representative of the current international level and the results demonstrate that, after the closure, the facility is safe to people and other living nature in the surroundings as required by the Government Decree.

Notwithstanding this overall conclusion, the section of the report addressing the performance assessment identified several lacunae that still needed to be filled in prior to the operation of the repository. For instance, Posiva had to reconsider its safety function and performance target approach to clarify its safety argumentation and to "remove internal inconsistencies."[40] Moreover, "the connection between the performance targets and design specifications remains unclear. Consequently, it is difficult for Posiva to construct the facility and demonstrate its acceptability."[41]

Nine months would pass before the Finnish Government granted the construction license in November 2015. It did so, acknowledging that the regulations were still in flux and that STUK's review had identified a raft of technical issues that were still unresolved. (One might speculate whether that outcome would have emerged had the licensing process not barred external interested and affected parties from participating.)

Just before the end of 2021, Posiva submitted an application for an operating license for the ONKALO repository. Currently, STUK expects to complete its review by the end of 2025.

Sweden

The implementor, SKB, tendered in 2011 an application to both the Radiation Safety Authority (SSM) and a Land and Environmental Court appointed to supervise this particular controversy. The regulator reviewed the submission in light of the requirements of the Act on Nuclear Activities and the Radiation Protection Act, laws ratified in the mid-1980s. The workings of SSM were closed to the public.

In parallel, the Land and Environmental Court reviewed the application under the requirements of the Environmental Code, which passed in 1998. Interested and affected parties were permitted to make statements in support or opposition to SKB's petition to the court. Hearings were conducted in 2017 and 2018. The court initially declined to certify that the provisions of the Environmental Code had been fulfilled. It asked the implementor to provide additional information about the anoxic corrosion of the copper canister in pure groundwater. SKB did so in 2019; the following year the court confirmed that SKB had satisfied its environmental obligations. (See also the discussion in Chapter 11.)

At this point, Oskarshamn, the site of the Encapsulation Plant, and Östhammar, the location of the disposition facility, decided not to exercise their functionally unconditional veto power. In 2022, the Swedish Government approved the construction of the deep-mined, geologic repository. The conditions under which the repository would be built are still under negotiation. Nonetheless, excavations began in 2025. See Figure 10.4.

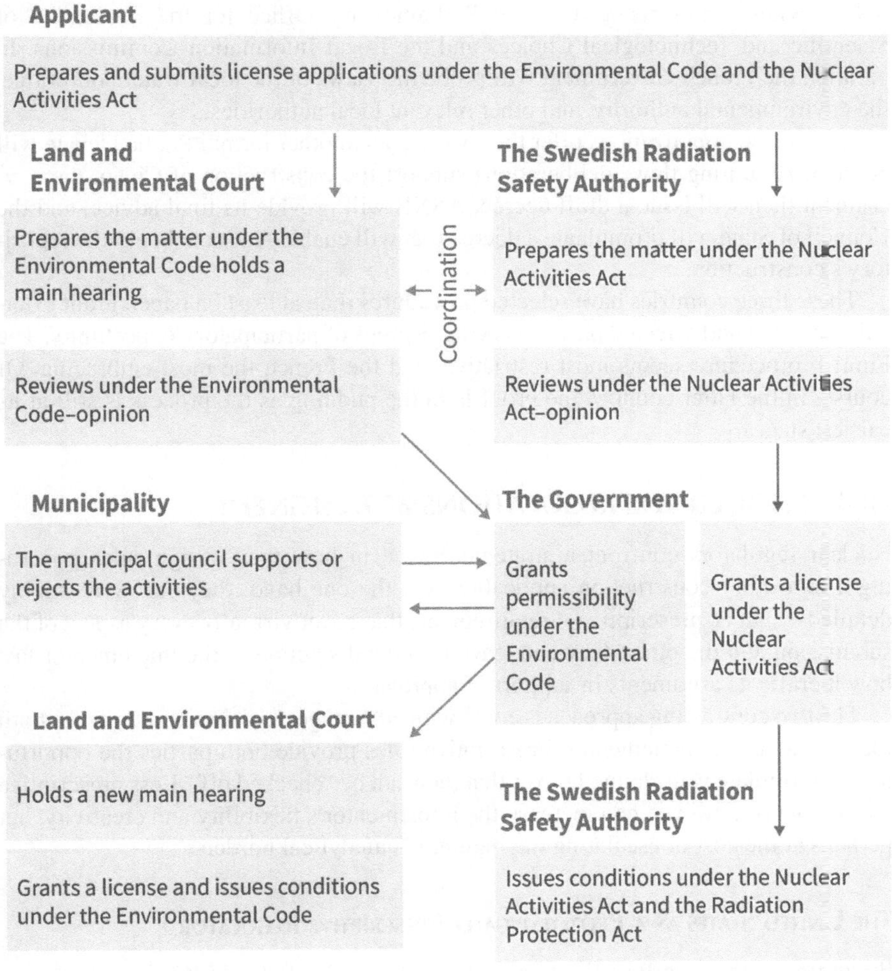

FIGURE 10.4 Flow diagram of the licensing process in Sweden.

France

Andra forwarded its construction-license application to the Minister for Energy Transition in early January 2023. The report included an environmental assessment, a safety analysis, and a risk management study of preclosure operations. The ministry initially assigned the application's review to the Nuclear Safety Authority (ASN), which subsequently was reconstituted as the Nuclear Safety and Radiation Protection Authority (ASNR).[42]

The dossier is currently working its way through a complicated multistage evaluation process.[43] The regulator has held meetings with Andra. It has prepared a set of recommendations to which Andra largely concurs. The application is now under review by ASNR's Advisory Committee, the Permanent Group of Experts. The scope of that assessment was established through public consultations.[44]

Once the regulator has provided its evaluation, several other bodies weigh in. The Committee for National Evaluation, an advisory body to the French Government, will provide comments as will the Parliamentary Office for the Evaluation of Scientific and Technological Choices and the Local Information Commissions. In parallel, the French Government will probe the views of the local water committee, the environmental authority, and other relevant local authorities.

Once those organizations offer their views, yet another formal Public Debate will be held. Assuming those deliberations support the construction of Cigéo, the government then will issue a draft decree, ASNR will provide its final advice, and the Council of State will promulgate a decree that will enable Andra to begin the repository's construction.

These three countries have selected procedures that, at least on paper, grant external interested and affected parties varying degrees of participatory opportunity. The Finnish procedure seems most restrictive, and the French the most embracing. Of course, in the latter country, the proof is in the pudding as the process is still at its earliest stages.

HOW SHOULD THE REGULATIONS BE DESIGNED?

Nuclear regulators confront a strategic choice in how they design rules governing a repository-construction application. On the one hand, they can write highly detailed—that is, prescriptive—requirements that cover virtually every aspect of the submission. On the other, they can leave it to the discretion of the implementor just how it crafts its arguments in support of approval.

The two contrasting approaches come with advantages and disadvantages for both the regulator and implementor. Prescriptive rules provide both parties the opportunity to formulate well-defined boxes that then can be "checked off." Less prescriptive requirements, however, can increase the implementor's flexibility and creativity, but perhaps at the cost of escalating the regulator's analytical burden.

The United States as a Prototypically Prescriptive Regulator

Beginning with its earliest licensing rules in the early 1980s, NRC elected to write "general and detailed prescriptive requirements, derived from the Commission's experience and practice in licensing other facilities."[45] In adopting its final permitting rules two years later, it backed off from some of those constraints, such as those dealing with subsurface ventilation and shaft seals. Nonetheless, the Commission's regulations remained highly prescriptive. Two provisions illustrate the broader point.

Section 60.113 of the regulation, for instance, contains the three notorious subsystem performance goals, substantially complete containment, annual release limit from the waste package, and groundwater travel time. (See Chapter 4.)

Section 60.134 lays out criteria for the waste package and its components. Included are these considerations.

> The design shall include but not be limited to consideration of the following factors: solubility, oxidation/reduction reactions, corrosion, hydriding, gas generation, thermal effects, mechanical strength, mechanical stress, radiolysis, radiation damage, radionuclide retardation, leaching, fire and explosion hazards, thermal loads, and synergistic interactions.

The regulator's preference for prescriptive rules was moderated with its promulgation of the November 2001 and March 2009 Yucca Mountain-only final licensing rules. But the former still takes 24 triple-columned single-spaced, small-font pages to enumerate.

In a number of instances, NRC had no choice but to conform to EPA's final Yucca Mountain-only rule, 40 CFR 197.[46] For example, NRC adopted EPA's specification of the arithmetic mean as the basis for determining compliance with the individual-protection dose limit. NRC accepted EPA's weighting factors for estimating potential radiation for members of the general public. It also had to concur with EPA's requirement to assess the effects of climate change by evaluating increased water flow through the repository.

Nonetheless, NRC laid out multitudinous other obligations that DOE would have to fulfill. For example, Section 63.21 contains two dozen listings of what material had to be included in the license application.[47] Among them, the implementor has to design "controls that DOE will apply to restrict access and to regulate land use at the Yucca Mountain site and adjacent areas, including a conceptual design of the monuments that would be used to pinpoint the site after permanent closure."[48]

Section 63.114 delineates the set of features, events, and processes to be incorporated in a performance assessment. The applicant must "provide the technical basis for either inclusion or exclusion of degradation, deterioration, or alteration processes of engineered barriers in the performance assessment, including those processes that would adversely affect the performance of natural barriers.[49]"

How Prescriptive Are the Waste-Management Regulations in Other Countries?

By almost any metric one might pick—page length is adopted here—the rules in the United States are remarkably prescriptive. But one should only compare apples to apples not to oranges. So, for the purposes of our narrative, that measure considers only the regulations themselves, not their derivative guidance, explanations, or their informal precedents. Using that very imperfect calculus, most of the requirements abroad do not even come close to this country's level of prescription.

The Swedish requirements, for instance, cover just over three pages of text.[50] They specify that the repository's design must consist of a system of passive barriers. "The function of each barrier shall be to, in one or several ways, contribute to the containment and prevention or retention of dispersion of radioactive substances, either directly or indirectly by protecting other barriers in the barrier system." The system must be structured so that the necessary safety is maintained despite a single deficiency in [any] one barrier. A safety analysis must be conducted; it must examine the features, events, and processes "that can lead to the dispersion of radioactive substances after closure." An Appendix to the rules describes the methodology used to carry out the safety and environmental analyses.

It is worth quoting in some length how one analyst from a nuclear regulator described the differences between the approach taken in the United States and that taken in Sweden.[51]

> The difference between the Swedish system and the U.S. system is that we take a less prescriptive role. We make recommendations to SKB, but these are rather general and cover the overall direction of the program, progressing in sort of the right direction. We don't always comment on the technical details. But we have our own experts, with whom we discuss these matters, and we write review statements. The intention of being less prescriptive is that we should not go so far that we start to design the repository from the regulatory side. Then we would have difficulty in reviewing it objectively when we get the license application.

The licensing rules in Finland are a bit more prescriptive; they take up nine double-columned pages. The regulation not only dictates an individual-protection dose constraint but also limits on the release of radionuclides. It instructs the implementor to consider expected and unlikely events. Further, the safety analysis must utilize computational methods that the results with "high degree of certainty and overestimate the radiation exposure or radioactive release likely to occur." The rule also demands the design of multiple, redundant barriers, so that "a deficiency in one does not jeopardize the long-term safety" of the disposition facility. Noteworthy is the implementor's obligation to use independent lines of evidence such as "natural analogues or observations of the geological history of the disposal site."[52]

The rules in France are somewhat more prescriptive. They are laid out in 27 pages of single-columned text, covering the following topics: reversibility conditions; radiation protection criteria; safety principles and functions; and demonstration of safety following closure. For instance, "with regard to the design of the engineered components, the corresponding requirements shall be taken into account: removal of the heat and gases given off by the waste packages; reduction of the intensity of the mechanical stresses created; and maintaining the physical-chemical conditions favorable to limiting the corrosion of the containers and the migration of the radionuclides." In addition, two appendices stipulate actions the waste producers must take in preparing their waste packages for disposal, investigations to be carried out on the site, and the scenarios that must be evaluated in the safety analysis.[53]

In Switzerland, the Federal Nuclear Safety Inspectorate, ENSI, promulgated rules that take up roughly 13 pages.[54] They cover matters such as protection criteria in the postclosure phase, the design of the disposition facility, retrievability, assumptions about climate development, and preclosure operations. Of special interest is the following: "The safety case must include an evaluation of the safety analysis methods and the data used. Where appropriate, supporting arguments for the principles and results of the safety analysis must be provided. Where possible, the conclusions of the safety analyses must be supported by observations in nature (natural analogues)."

Finally, the Canadian Nuclear Safety Commission released its 21-page requirements for preparing an application to dispose radioactive waste in January 2021.[55] It defines the safety case as

> An integrated collection of arguments and evidence to demonstrate the safety of a facility and the meeting of all applicable regulatory requirements. A safety case normally includes a safety assessment, but could also typically include information (such as supporting evidence and reasoning) on the robustness and reliability of the safety assessment and the assumptions made therein.

The regulation covers the general requirements and the components of the safety case, the postclosure safety assessment, and the importance of defense-in-depth. Perhaps the rule's most prescriptive requirement is found in Section 7.10, which outlines how the safety arguments are integrated.

As the licensing process evolved in these five nations, technical debates surrounding the long-term performance occupy center stage. In the next chapter, our narrative focuses on two of them, whether native copper corrodes under anoxic condition in Finland and Sweden and the measurement of percolation flux at the proposed Yucca Mountain repository.

NOTES

1. In the United States, the Atomic Energy Commission was split in 1975 into an implementor, the Energy Research and Development Administration, and an independent regulator, the Nuclear Regulatory Commission. Among the nations considered in this narrative, Japan and Germany were the last to create an independent nuclear regulator, the former in 2011 and the latter in 2014.
2. See, for example, International Atomic Energy Agency, *IAEA Safety Standard for Protecting People and the Environment: Geological Disposal Facilities for Radioactive Waste*, Specific Safety Guide-14, 2011.
3. Some countries establish a dose limit and others a risk limit. The two parameters differ by a constant factor. To convert the former to the latter, see https://www.wise-uranium.org/rdcri.html. Accessed April 22, 2025. Except as context requires, the remainder of this discussion uses the term "dose" exclusively.
4. Baruch Fischhoff et al., *Acceptable Risk* (Cambridge, England: Cambridge University Press, 1981), 3.
5. Daniel Metlay, "Obligation to Future Generations: The Interplay of Science and Policy," Presentation to the 3rd International Conference on Politics and Information Systems, Orlando, Florida, 2005, 3. Available from the author.
6. Environmental Protection Agency, "Environmental Standards for the Management and Disposal of Spent Nuclear Fuel, High-Level and Transuranic Radioactive Wastes," *Federal Register*, (50) September 19, 1985, 38,066–38,089. See especially 38.087-38,089. EPA's thinking regarding the compliance period is found at 38,066.

7. Environmental Protection Agency, "Public Health the Environmental Radiation Protection Standards for Yucca Mountain, NV," *Federal Register*, (66) June 13, 2001, 32,074–32,135.
8. Environmental Protection Agency, "Public Health and Environmental Radiation Protection Standards for Yucca Mountain, NV: Proposed Rule," *Federal Register*, (70) August 22, 2005, 49,014–49,065.
9. Environmental Protection Agency, "Public Health and Environmental Radiation Protection Standards for Yucca Mountain, Nevada: Final Rule," *Federal Register*, (73) October 15, 2008, 61,256–61,289.
10. Nuclear Regulatory Commission, "Implementation of a Dose Standard After 10,000 Years," *Federal Register*, (74) March 13, 2009, 10,811–10,830.
11. The downfall of the project also shunted NRC's licensing regulation into oblivion at least for the moment. Guarding against the resurrection of the Yucca Mountain Project, the State of Nevada filed a lawsuit to abrogate that final rule. See *State of Nevada v. U.S. Nuclear Regulatory Commission*, Case No. 09-1113, U.S. Court of Appeals for the District of Columbia Circuit, May 12, 2009. That case is currently suspended.
12. A study group formed by the American Nuclear Society recommended that a generic repository standard have an acceptable dose limit between 0.15 and 1.0 mSv per year and a compliance period of 10,000 years. See American Nuclear Society Special Committee on Generic Standards for the Disposal of High-Level Radioactive Waste, "Recommendations on Postclosure Aspects of Generic Standards for the Permanent Disposal of Spent Nuclear Fuel and High-Level and Transuranic Radioactive Wastes in the United States," August 2023. See https://www.ans.org/file/12339/1/ANS%20Disposal%20Standards%20Report%20final%20booklet-online.pdf. Accessed April 24, 2025.
13. International Commission on Radiological Protection, "The 2007 Recommendations of the International Commission on Radiological Protection," Publication 103, March 2007. See https://journals.sagepub.com/doi/pdf/10.1177/ANIB_37_2-4. Accessed April 23, 2025. See as well, International Commission on Radiological Protection, Radiological Protection in Geological Disposal of Long-Lived Solid Radioactive Waste," Publication 122, January 2010. See https://journals.sagepub.com/doi/pdf/10.1177/ANIB_42_3. Accessed May 12, 2025.
14. Swedish regulations impose a risk constraint such that the annual risk of harmful effects may not exceed 10^{-6} (one in a million) for a representative individual in the group exposed to the greatest risk. This limit roughly translates to 0.02 mSv/year.
15. Environmental Protection Agency (1985), 38,088.
16. National Research Council, *Technical Bases for Yucca Mountain Standards* (Washington D.C.: National Academy Press, 1995).
17. National Research Council, 13.
18. Environmental Protection Agency (2001).
19. Environmental Protection Agency (2008).
20. Nuclear Regulatory Commission, "Disposal of High-Level Radioactive Wastes in a Proposed Geologic Repository at Yucca Mountain," *Federal Register*, (66) November 2, 2001, 55,732–55,816.
21. For the post-10,000-year compliance period, see Nuclear Regulatory Commission (2009). For the pre-10,000-year-compliance period, see Nuclear Regulatory Commission, "Disposal of High-Level Radioactive Wastes in a Geologic Repository at Yucca Mountain, Nevada," *Federal Register*, (66) November 2, 2001, 55,792–55,816.
22. Intermittently DOE couched its analyses as components of a safety case. It did acknowledge that independent lines of evidence could be developed outside of performance assessments. See, for example, presentation by Steve Hanauer, "Multiple Lines of Evidence," at the April 13, 2001 meeting of the Nuclear Waste Technical Review Board. The more typical view was that, if any "independent" lines of evidence existed, they would be incorporated into the performance assessment. Therefore, for example, conclusions about defense-in-depth and barrier capability drew on that probabilistic technique. See the presentations by Russell Dyer, Peter Swift, and Patrick Brady made at the September 27, 2006 meeting of the Nuclear Waste Technical Review Board in Amargosa Valley, Nevada.

23. In its appraisal of an early version of NRC's Yucca Mountain Review Plan, the Office of Civilian Radioactive Waste Management objected to the use of the term "safety case," arguing that the concept was alien to the promulgated regulations. In subsequent revisions, NRC eliminated any mention of "safety case" as part of its evaluation of construction-license application. See Nuclear Regulatory Commission, "Yucca Mountain Review Plan: Final Report," NUREG-1804, Revision 2, July 2003.
24. Nuclear Energy Agency, *Postclosure Safety Case for Geological Repositories*, NEA-3679, 2004, 7.
25. Steering Committee, *Reset of America's Nuclear Waste Management: Strategy and Policy*, October 2018, 81.
26. The aim of the multibarrier concept is to confine the radionuclides so that the failure of one component does not jeopardize the safety of the containment system as a whole. Note that a repository system-design could contain multiple barriers but not provide defense-in-depth.
27. Posiva, "Safety Case for the Disposal of Spent Nuclear Fuel at Olkiluoto Radionuclide Migration Parameters for the Geosphere," February 2014.
28. Posiva, "Safety Case for the Disposal of Spent Nuclear Fuel at Olkiluoto: Complementary Considerations 2012," Posiva Oy 2012-11, December 2012. See https://inis.iaea.org/records/14ea2-vf044. Accessed May 12, 2025. Quotations are at 14 and 17.
29. Chapter 11 details why this issue was uncontroversial in Finland and extremely contentious in Sweden.
30. This discussion draws from Posiva (2012), 89–95. For a fuller picture of how Posiva evaluated analogues, see 83–158.
31. Posiva (2012), 91.
32. Posiva (2012), 95.
33. Andra, "Dossier 2005 Argile," December 2005.
34. On January 1, 2025, the regulator (ASN) and its technical support organization (IRSN) merged to form the Authority for Nuclear Safety and Radiation Protection (ASNR).
35. Andra, "Security Options File : After-Closing Part, 2023." See https://www.andra.fr/sites/default/files/2018-04/dossier-options-surete-apres-fermeture_0.pdf. Accessed May 20, 2025. See especially 47–58. English version from Google Translate.
36. Andra (2005), 234.
37. "State of Nevada's Petition to Intervene as a Full Party," NRC Docket 63-001, December 19, 2008. See https://anp.nv.gov/uploadedFiles/anpnvgov/content/Legal/NRC_Licensing/Contentions_NV.pdf#:~:text=Nevada%20requests%20a%20formal%20adjudicatory%20hearing%20on%20each%20of%20its%20contentions&text=Nevada's%20case%20against%20Yucca%20Mountain%20from%20individual%20contentions%2C%20we. Accessed May 212, 2025.
38. This section draws heavily on Juhani Vira, "Geological Repository for High-Level Nuclear Waste Becoming a Reality in Finland," in *Geological Repository Systems for Safe Disposal of Spent Nuclear Fuels and Radioactive Waste* (Sawston, U.K.: Woodhead Publishing 2017), 645–666.
39. Radiation and Nuclear Safety Authority, "STUK's Review on the Construction License Stage Postclosure Safety Case of the Spent Fuel Disposal in Olkiluoto," STUK-B 197, November 2015, 3.
40. Radiation and Nuclear Safety Authority (2015), 101.
41. Radiation and Nuclear Safety Authority (2015), 126.
42. One close observer questioned whether the independence of the merged technical support organization would now be lost.
43. See https://www.asn.fr/espace-professionnels/installations-nucleaires/projet-de-centre-de-stockage-en-couche-geologique-profonde-cigeo#:~:text=Pour%20obtenir%20cette%20autorisation%2C%20l,premiers%20colis%20de%20d%C3%A9chets%20radioactifs. Accessed May 23, 2025.
44. The consultations took place with a range of twenty nongovernmental organizations, including the Local Information Commissions near Bure and environmental protection associations.

45. Nuclear Regulatory Commission, "Disposal of High-Level Radioactive Waste in Geologic Repositories: Technical Criteria," *Federal Register*, (48) June 21, 1983, 28,194–28,229. Quotation is found at 28,198. NRC wrote these when the EPA standards were still in draft form. See Chapter 4.
46. Environmental Protection Agency (2008).
47. Nuclear Regulatory Commission, "Disposal of High-Level Radioactive Wastes in a Proposed Geologic Repository at Yucca Mountain, NV," *Federal Register*, (66) November 2, 2001, 55,732–55,816.
48. Nuclear Regulatory Commission (2001), 55,798.
49. Nuclear Regulatory Commission (2001), 55,807.
50. Swedish Radiation Safety Authority, "Regulations Concerning Safety in Connection with the Disposal of Nuclear Material and Nuclear Waste," SSMFS 2008:21, January 30, 2009.
51. Statement by Bo Strőmberg in Swedish National Council for Nuclear Waste, "Mechanisms of Copper Corrosion in Aqueous Environments," Report 2009:4e, November 2009, 109.
52. Radiation and Nuclear Safety Authority, "Long-term Safety of Disposal of Spent Nuclear Fuel," YVL 8.4, May 23, 2001. See https://ohjeisto.stuk.fi/YVLold/YVL8.4en_2001-05-23.pdf. Accessed May 25, 2025.
53. Nuclear Safety Authority, "Safety Guide Concerning the Disposal of Radioactive Waste in a Deep Geological Formation," February 12, 2008. See https://www.google.com/search?q=asn+safety+guide+nuclear+waste+repository+2008&rlz=1C1ONGR_enUS1101US1101&oq=&gs_lcrp=EgZjaHJvbWUqCQgCECMYJxjqAjIJCAAQIxgn-GOoCMgkIARAjGCcY6gIyCQgCECMYJxjqAjIJCAMQIxgnGOoCMgkIBBAjGC-cY6gIyCQgFECMYJxjqAjIJCAYQIxgnGOoCMgkIBxAjGCcY6gLSAQkyMTM2a-jBqMTWoAgiwAgHxBXF_tuaZw8qW&sourceid=chrome&ie=UTF-8. Accessed May 25, 2025. English version from Google Translate.
54. Federal Nuclear Safety Inspectorate, "Specific Design Principles for Deep Geological Repositories and Requirement for the Safety Case," ENSI-G03, April 2009. See https://ensi.admin.ch/de/wp-content/uploads/sites/2/2011/08/g03_d.pdf. Accessed May 25, 2025. English version from Google Translate.
55. Canadian Nuclear Safety Commission, "Waste Management, Volume III: Safety Case for the Disposal of Radioactive Waste," REGDOC-2.11.1, Version 2, January 2021. See https://www.cnsc-ccsn.gc.ca/eng/acts-and-regulations/regulatory-documents/published/html/regdoc1-2-3. Accessed May 5, 2025.

11 Addressing Technical And Scientific Controversies

The opening pages of our story termed the long-term management of high-activity radioactive waste a "wicked problem." As such, it is characterized by substantial technical uncertainties, long time-constants of feedback, and significant conflicts over values. The licensing of a disposition facility brings to the fore some of the implications of trying to tackle that class of challenges.

This chapter explores two pivotal permitting issues: one in Finland and Sweden and the other in the United States. The first, whether the elemental copper canisters corrode in pure anoxic groundwater, presents an opportunity to assess how the licensing process itself influences the manner in which a technical controversy is shaped. The second considers the magnitude and movement of the percolation flux within the unsaturated zone at Yucca Mountain. As Chapter 6 maintained, the inability to fully understand that phenomenon contributed to a fundamental shift in DOE's concept of waste isolation and containment.

Both case studies involved drawing conclusions from experiments that were at the cutting edge of science and engineering. Confounding interpretations were enmeshed in what one philosopher of science called auxiliary hypotheses.[1] By examining the "thinking" of the protagonists, the sociology of science can be illuminated.

ELEMENTAL COPPER'S FATE IN PURE ANOXIC GROUNDWATER

In the early 1980s, several peer-reviewed papers appeared in Sweden suggesting that elemental copper could corrode in anoxic pure groundwater. Most telling was a paper by Gunnar Hultquist.[2] Interest in that experimental result faded over the next two decades until the finding became central to the credibility of the KBS-3 disposal concept.

It is not often in the social sciences that one encounters a natural "experiment." Understanding how this technical controversy unfolded in Finland and Sweden is just this type of black swan. Both countries opted (with slight differences) for the KBS-3 concept. The hydrogeological environments of their proposed disposition facilities were virtually identical. Moreover, the early steps of the repository-development process were comparable. In the two Nordic efforts, each implementor was obligated to prepare a document outlining its research strategy over the following three years. They circulated the plan to cognizant government agencies. In principle, interested and affected parties could also comment. But as the previous chapter detailed, what differed were the final stages of the licensing process. In Sweden, the procedure was

DOI: 10.1201/9781003528302-13

open to the influence of external nongovernmental parties while in Finland it was not.[3]

The next section of our narrative explains how the Finnish implementor and regulator responded to the technical controversy. The subsequent section portrays the Swedish reaction.

The Finnish Implementor and Regulator Weigh In

The Finnish implementer, Posiva, produced its first blueprint for researching the suitability of a repository in 2003.[4] The document conveyed an optimistic outlook, expecting favorable results from future experiments and improved insights into the corrosion mechanisms. Posiva went so far as asserting that, "available evidence supports the claim that the [copper] canister can hold the waste for more than 100,000 years."[5]

At the same time, the implementor acknowledged that the corrosion issue needed further consideration. It committed to investigating, in collaboration with the SKB, whether copper might degrade under oxygen-free and saline conditions.[6] It conceded that existing experiments had produced "contradictory results."[7]

The Radiation and Nuclear Safety Authority, STUK, responded the following year. In contrast to Posiva, it saw the report as merely providing a general overview of the issue rather than claims that corrosion would not render the canisters vulnerable to radionuclide releases. The regulator sought assurances that Posiva would also examine the potential for localized deterioration such as microbial and stress corrosion cracking. STUK made special mention that the service-time of the canister was a much longer span "compared to the operation time of any other industrial product."[8]

Posiva released the next research plan in 2006.[9] That report suggested that Posiva paid increased awareness to how copper behaved over the long-term under shifting repository conditions. Stress corrosion cracking seemed unlikely. Although more data had been gathered about copper's corrosion in anoxic and saline groundwaters, Posiva observed that substantial uncertainties persisted.

STUK's review the following year acknowledged that the new information was valuable but maintained that additional data had to be acquired about numerous copper-corrosion-related issues. Even the seemingly settled question of corrosion due to sulfide-containing bentonite demanded further study. Moreover, extensive mathematical modeling still had to be carried out. In short, the regulator told Posiva that it had not completed its homework; whether the copper canister barrier would function as declared remained an open question.

Three years later, Posiva published the latest iteration of its research agenda.[10] Unlike earlier accounts, this one was more responsive to STUK's comments. It provided additional information about how the bentonite buffer would protect the canister. It expanded investigations to include the likelihood of degradation caused by chloride ions and other corrosive agents. It analyzed whether oxygen in glacial meltwater could initiate corrosion. The statement addressed the various types of corrosion (general, localized, intergranular, and microbial) as they might arise at various phases of the repository-development process.

The regulator simultaneously responded to both the investigations mounted in the previous three years and to Posiva's submission of a document outlining the thrust of its preliminary draft of a construction-license application. Comments on the work undertaken since 2006 were generally positive, although supplementary studies still needed to be performed on corrosion in pure oxygen-free groundwater and on the implications for the canister of evolving environmental conditions.

STUK's assessment of Posiva's construction-licensing intentions was more critical. The regulator noted that the report on canister design and corrosion was delayed. The implementor failed to close several safety issues regarding copper corrosion. STUK complained that the safety case was not well documented and was difficult to trace. It also criticized Posiva for not sufficiently demonstrating the technical validity of its models of safety-affecting conditions.

In 2012 and 2013, a crisis between Posiva and STUK surfaced.[11] The latest risk dialogue took on a testier tone. From its point of view, the implementor argued that it had carried out extensive research on copper oxidization, including sulfide, pure water, and stress corrosion cracking. Indeed, as Chapter 10 described, the implementor attempted to bolster its safety case by providing evidence of complementary considerations, particularly the value of natural and artificial analogues. As Posiva maintained,

> The proposed repository design provides a safe solution for the disposal of spent nuclear fuel, and the performance assessments are fully consistent with all legal and regulatory requirements... Moreover, the level of confidence in the demonstration of safety is appropriate and sufficient to submit the construction license application to the authorities.[12]

STUK rejected Posiva's initial tender. In its opinion, the implementer underestimated the time necessary to perform crucial studies. It also failed to clearly link its research to its safety case. In response, Posiva delivered additional information to its regulator, who cautioned that the entire licensing dossier would require a stringent review. But for the moment, Posiva could submit its construction-license request, which it did in December 2012. By then, it had become evident that the two organizations' positions were converging; both placed considerable weight and importance on the development of the repository.[13]

STUK published its statement and safety assessment of the proposed Olkiluoto disposition facility in November 2015.[14] Given the intense decade-long exchanges between implementer and regulator regarding copper corrosion, STUK's cursory discussion of the issue is somewhat puzzling. Distilling its comments to the core, the regulator determined that Posiva's safety case was sufficiently acceptable for the construction-license milestone, but that it would have to be substantially augmented at the operating-license stage.[15]

The Technical Community in Sweden Confronts the Question of Copper Corrosion

Three technical specialists from the Swedish Royal Institute of Technology (KTH), one of whom had initially raised the issue of copper corrosion in the 1980s, published in 2007 a second peer-reviewed study reiterating the earlier claim.[16] They grounded their conclusion on an increase in pressure of hydrogen gas within their experimental apparatus. SKB strongly rebutted the finding, asserting, among other things, that it violated the law of thermodynamics.[17]

The advisory body to government, the Swedish National Council for Nuclear Waste, formerly known as KASAM, sponsored a symposium to ventilate the conflicting technical stances.[18] It invited four international specialists to present papers. It also organized a panel that included additional experts to discuss informally what was the state-of-the-science regarding copper corrosion in pure anoxic groundwater. The group also replied to comments and questions from scientists at KTH and personnel and consultants from SKB. Included in the Council's report was an edited transcript of the conversation.

One of the authors of the 2007 paper, Gunnar Hultquist, kicked off the panel's discussion. He set forth his central hypothesis[19]:

$$\textbf{Copper} + \mathbf{H_2O} \Rrightarrow \textbf{Copper hydroxides} + \mathbf{H_2} + \textbf{Hydrogen in Copper Metal}$$

Hultquist then argued that the water molecules dissociate into H^+ and OH^- ions. The former either end up as hydrogen molecules, whose presence is measured by increased pressure within the experimental apparatus or are "absorbed into the copper metal." In that case, a solid reaction product is created, which can also be measured. Further, Hultquist maintained that "first-principles simulation results in the strong bonding of the OH^- to the copper surface, leading to the formation of a three-dimensional growth of monovalent copper hydroxide."

Hultquist's colleague and the first author of the 2007 paper, Peter Szakálos, picked up where his collaborator left off. He asserted that the KTH findings hardly represented a novel outcome. "It was already known 30 years ago that copper was not thermodynamically immune in pure O2-free water, and that this fact is undisputed among thermodynamic experts."[20] In addition, Szakálos criticized the experimental designs of two studies that SKB cited as contradicting the KTH research. Finally, he maintained that the natural and archeological analogues, such as those advanced by Posiva, were inapt.[21]

The reactions of the other panelists to the Hultquist-Szakálos talks ranged from skeptical to intensely critical. Of course, scientific "truth" is not determined by a plebiscite, certainly not one in which many of the "voters," either by design or coincidence, were linked to the Scandinavian implementors. Nonetheless, in this non-specialist's view, the KTH investigators never persuasively rebutted the detractors' points.

For instance, one scientist, Gaik Khuan Chuah, who was a coauthor on some of the KTH papers from the late 1980s, argued that "Hultquist and Szakálos have proposed the formation of an amorphous copper hydroxide H_xCuO_y. However, such a hydroxide has yet to be directly detected."[22] Another corrosion specialist, Ronald Latanision, reinforced Chuah's position, "The hydrogenated copper oxide reaction product is not known in nature and, thus, this reaction is unknown in the thermodynamics of copper corrosion. I have seen no compelling data to suggest that this reaction product has been produced and identified in any of the KTH team's publications."[23]

A second line of critique questioned whether the laboratory findings were applicable to what might transpire in a repository. The moderator of the panel discussion, David Duquette, observed[24]:

> The real question is whether the results that have been generated both by the SKB and by the KTH teams are relevant to the vault. In other words, if you're primarily looking at whether it's going to be a safe disposal of copper containers or not. Are these the experiments that should be being performed? Are these the appropriate experiments to justify that?

Another specialist, David Shoesmith, advanced an incisive response that is worth citing at length.[25]

> It isn't simply a corrosion issue. The multibarrier system alone is meant to make sure that we don't rely on one barrier. However, that doesn't mean to say that you shouldn't make a really good attempt to determine the specific mechanisms and appropriate experiments for individual barriers. When you make the extrapolation from whether we have the right issue on the table here to whether there is a solution to long-term containment of radioactive waste, you're asking a much bigger question. Such as: Do we accept the interpretations of these corrosion processes? I don't think the experiments are off-base, no. I think they are fine, but it will always be difficult to bridge that gap between something that you want to get, where you have a practical number from the large system, and then the experimental system which will generate the data. It will always end up in some computational model. It will always be obscure.

The Licensing Process Commences

Because of its remit, SKB had an incentive to try to establish common ground among all the central scientific and engineering actors. In an effort to do so, it sponsored a "Reference Group" in the spring of 2010.[26] Local, regional, and national authorities and two nongovernmental organizations as well as the SKB and KTH investigators participated.[27] Neither the National Council nor the regulator, the Radiation Protection Authority (SSM), chose to join. The Reference Group met until December 2012.

The Reference Group's agenda focused initially on two experiments, one conducted by the SP Technical Research Institute of Sweden and the other undertaken by scientists from Uppsala University. Regrettably, the former study was plagued by methodological and experimental flaws. The latter one was still underway when the Reference Group disbanded. So, it was no wonder that reaching any sort of agreement on the technical controversy proved to be a "bridge too far."

Even before the collapse of the Reference Group, SKB moved forward and tendered its construction-license application to the Swedish government in March 2011. That achievement set in motion the bifurcated process described in Chapter 10 and illustrated in Figure 10.4. For the purposes of our tale, the discussion that follows is limited to events surrounding the arbitration before the Land and Environmental Court.[28]

The information supporting SKB's application for a permit under the Environment Code exceeded 10,000 pages.[29] It included background material, a discussion of the

KBS-3 system, alternative strategies for the final disposal of spent nuclear fuel, and non-radiological environmental impacts and consequences. The submissions covered the expansion of the interim storage facility (CLAB), the construction of a canister encapsulation plant (CLINK), and the final repository itself.

The Swedish National Council for Nuclear Waste restricted its comments to procedural matters. It maintained that the license application, in its present state, was severely lacking in information. Therefore, it concluded that the permit could not be assessed in a purposeful or conclusive way.

Two nongovernmental organizations—MKG and the Swedish Society for Nature Conservation—intervened to oppose the implementor's license request and introduced a joint rebuttal.[30] Hardly surprisingly, their arguments concentrated on the vulnerability of the copper canister but with a bit of a twist.[31]

> Until just the last few years, the copper capsule has been considered the most important protection in the barrier system. Even if the clay buffer would not work perfectly, it was believed that 5 cm of copper would hold tight for long periods of time. With the increasing knowledge in recent times that there may be more and interacting corrosion processes than the applicant has assumed in its safety analysis, the importance of the [bentonite] clay being completely tight for very long periods of time has increased. From the time when the copper canister was "king," now the clay has taken over the throne.[32]

Yet, according to the objectors,

> The knowledge of how the clay will behave is extremely lacking, especially during the first 1,000 years. The assumptions made in the safety assessment that the clay buffer in all disposition holes will reach the ideal state are highly theoretical with very limited experimental support. The [intervenors] believe that the knowledge of how the copper canister will be affected in the repository environment and especially during the first 1,000 years is also very lacking. Even here it is primarily theoretical and largely lacking experimental support.[33]

The Initial Verdict

The Land and Environmental Court issued its opinion on January 23, 2018.[34] The bottom-line verdicts were:

> The investigation shows that there are uncertainties, or risks, regarding how much certain forms of corrosion and other processes can impair the ability of the canister to contain the nuclear waste in the long term. Overall, these uncertainties about the canister are significant and have not been fully taken into account in the conclusions of SKB's safety analysis....
>
> The Land and Environmental Court considers that there is some leeway for accepting further uncertainties. The uncertainties surrounding certain forms of corrosion and other processes are, however, of such gravity that the Court cannot, based on SKB's safety analysis, conclude that the risk criterion in the Radiation Safety Authority's regulations has been met. In the context of the comprehensive risk assessment required by the Environmental Code, the documentation presented to date does not provide sufficient support for concluding that the final repository will be safe in the long term.[35]

In particular, the Court identified the following key uncertainties that SKB had to address:

- General corrosion due to reactions in oxygen-free water
- Localized pit corrosion due to reactions with sulfide
- Stress corrosion cracking due to reactions with sulfide
- Hydrogen embrittlement
- Effect of ionizing radiation on pit corrosion, stress corrosion, and hydrogen embrittlement

The Implementor Augments Its Case

The Ministry of Environment and Energy in June asked SKB to supplement its application consistent with the Court's opinion. It set a deadline of no later than the end of April 2019. SKB finalized its response several weeks ahead of time.[36]

The implementor dealt first with its evaluation of the potential for copper corrosion in pure anoxic groundwater.[37]

> There is no reason to assume that this corrosion mode occurs to an extent that exceeds that predicted by established thermo dynamic data. The conclusion is based on a thorough evaluation of available scientific evidence on the matter. It was also concluded that data, used by a few researchers to claim corrosion, is greater by many orders of magnitude than established scientific views. [That claim] would cause corrosion of only around one millimeter [out of five centimeters] of copper during the one million year period covered by the safety assessment. Based primarily on the former of these two conclusions, corrosion of copper in pure, oxygen-free water is not further considered in the report, since established thermodynamic data predicts a corrosion depth under repository conditions that is completely negligible during the assessment period.

SKB then addressed sulfide-induced copper corrosion.[38]

> Since localized corrosion in the form of micro-galvanic sulfide corrosion, could not be conclusively ruled out for the highest sulfide fluxes that could be expected in the repository environment, an updated dose consequence calculation based on a pessimistic interpretation of available data on micro-galvanic pit formation was carried out. The results of the calculation demonstrate that the calculated peak dose is only marginally affected compared to the value in the safety assessment SR-Site, meaning that the corresponding risk is still below SSM's risk criterion by about one order of magnitude.

SSM's response to the Court's opinion contained no new revelations.[39] The regulator had never regarded copper corrosion as a significant or disqualifying phenomenon. But its reaction was still unambiguous.

> SSM considers that SKB has satisfactorily investigated and answered the court's specific questions about the integrity of the canister against possible degradation and corrosion processes…
>
> According to SSM's assessment, the corrosion process of copper in pure oxygen-free water is of negligible importance with regard to the radiation safety of a final repository with spent nuclear fuel enclosed in copper canisters…the extent of the reaction according to established knowledge is so small that it can hardly even be measured, let alone affect canisters with a five-centimeter-thick copper casing placed in a final repository at a depth of 500 meters.

The National Council for Nuclear Waste observed that SKB had produced "important calculations."[40] Nonetheless, the advisory body still maintained that unanswered questions about copper corrosion remained. Although, the government might decide to proceed, "there is currently no scientific evidence that outstanding questions about the protective capacity of the copper capsule will be answered during the step-by-step review process that awaits in that case."

MKG chose not to relitigate the results of both SKB and KTH's earlier experiments. Instead, it argued that the implementor had failed to disclose the findings of the so-called LOT investigations. In the environmental organization's view, those outcomes were inconsistent with SKB's copper corrosion-model. Also implicit in MKG's comments was the charge that the implementor, consciously or not, was concealing critical information.[41]

Almost four years after SSM and the Land and Environmental Court forwarded their opinions to the Swedish Government, it approved the construction of the final repository in Östhammar, the extension of CLAB, and the construction of CLINK (the last two in Oskarshamn).[42] The Land and Environment Court then established conditions for the facilities. The Swedish Radiation Safety Authority decided on permit conditions under the Nuclear Activities Act. On October 26, 2024, SKB received final approval, and three months later, the implementor broke ground for the disposition facility. A recent public opinion survey showed that 86 percent of the Östhammar population supported the development.[43]

MEASURING PERCOLATION FLUX AT YUCCA MOUNTAIN

Earlier, our narrative examined the rise of Yucca Mountain as the presumptive location for the first deep-mined, geologic repository for high-activity waste in the United States.[44] DOE initially planned to characterize a location in the saturated zone within the water table. Based on the advice of the U.S. Geological Survey, however, it decided to explore the possibility of locating the repository horizon within the thick unsaturated zone (UZ) above the water table.[45] "In the UZ, water and gases are present in the pore spaces in varying proportions, but always at less than saturation, and the water either does not move at all through the pore spaces, or it moves extremely slowly."[46]

Early Estimates of Percolation Flux

The validity of DOE's waste-disposal concept rested on the argument that the amount of water reaching the repository level would be very small and that a repository could be designed so that any such water would pass into the permeable rocks within the UZ, with minimal contact with the canisters of waste. If site studies bore that conjecture out, then the UZ would form a natural barrier that could promote the isolation of the waste. In doing so, it would contribute to the repository's defense-in-depth.

The flow of water, percolation flux, strongly affects the performance of a repository at Yucca Mountain. The relationship between percolation flux and radionuclide release is not linear. The more water contacts the waste packages, the more they will corrode and the more likely waste material will be mobilized and released from the repository. As a practical matter, however, percolation flux cannot be measured directly because of the inaccessibility of the rock units in the UZ and the slow movement of waste through those formations. Although laboratory investigations could evaluate the parameter with some uncertainty, mathematical modeling had to be relied on to obtain firmer values.

Over the course of a decade and a half starting in 1983, DOE and its contracting community attempted to compute what level of percolation flux would violate EPA's then-dose constraint of 0.25 millisieverts/year over the 10,000-year compliance period. Using expert judgment, scientists from the Geological Survey put the number at 4 mm/year in 1983. A year later, two of their colleagues argued that the value was closer to 1 mm/year.[47]

By the early 1980s, DOE had carried out a series of performance calculations to identify the magnitude of percolation flux that might breach the standard. In 1984, DOE released its draft Environmental Assessment for Yucca Mountain. In its view, "Despite the uncertainty about the exact conditions and processes of the geohydrologic system at Yucca Mountain, especially in the unsaturated zone, the conservatism of the assumptions and the analyses allows confidence in the general conclusions [i.e., low percolation flux] about the hydrogeologic system."[48] In fact, the implementor held that there would be *zero release* to the environment.[49]

By the time DOE published the final Environmental Assessment in 1986, DOE held sufficient confidence in its performance calculations that it would claim that "the EPA's limits for the release rates from the engineered barrier system can be met *without any engineered barriers other than the waste form* [emphasis added] because the amount of water likely to be in contact with the waste is insufficient to cause higher rates of waste dissolution."[50]

The importance of the UZ as a barrier took on greater significance when DOE began to highlight the possibility of lateral diversion of water by the tin-roof Paintbrush unit, which lay just above the repository horizon in the Topopah Spring layer.[51] In fact, that phenomenon coupled with lower estimates of rainfall allowed DOE to assert that "the percolation flux may very well be about or much less" than initially projected.[52] Within the space of four years, then, the implementor's estimates of percolation flux fell by a factor of 40 from 4 mm/year to 0.1 mm/year.

Sensitivity of Radionuclide Releases to Percolation Flux

Over the next seven years, DOE believed that the percolation flux varied from 0.02 to 1.0 mm/year. It incorporated those values into a set of performance assessments. In 1991, it conducted two calculations. The first used 0.01, 0.05, 0.1, and 0.5 mm/year. Each suggested that the repository complied with the regulatory standard.[53] The second, however, was less optimistic. DOE adopted a distribution of values for percolation flux, ranging from 0.0 to 39 mm/year with a mean of 1.0 mm/year. The distribution reflected the so-called "weeps" model that permitted the water to travel not only through the pores of the UZ but through random and discrete fractures. Under that scenario, the individual-protection dose limit might not be heeded.[54]

Two years later, DOE contractors exercised two new performance assessments. The first adopted a value for the percolation flux of 0.0 to 7 mm/year.[55] In the second one, percolation flux ranged from 0.0 to 3.0 mm/year.[56] In both, the mean value of the distributions was 0.5 mm/year, which was still eight times lower than the Geological Survey's 4 mm/year. The contractors anticipated that the regulators might now fix the compliance period to be more than 10,000 years. Under that circumstance, the surprising conclusion from both evaluations was that if percolation flux exceeded 0.1 mm/year, the Yucca Mountain repository would not comply with new regulatory requirements.

Similar findings emerged from a third set of performance assessments carried out in 1995.[57] Those projections led the Nuclear Waste Technical Review Board (NWTRB) to conclude, "If it can be shown definitively that the percolation flux at the repository horizon is primarily matrix [pore] flow of less than 0.1 mm/year, then it probably will be difficult not to deem the site 'suitable' on hydrogeologic grounds."[58] That inference proved to be short-lived.

New Experimental Evidence Clouds the Estimates of Percolation Flux

As DOE continued its site investigations at Yucca Mountain, it prioritized the excavation of the Exploratory Studies Facility (ESF), a nine-mile tunnel that paralleled the eastern boundary of the proposed repository emplacement area. The passageway allowed investigators to gain access deep within the Topopah Spring unit. Scientists from Los Alamos National Laboratory (LANL) began collecting samples every 200 meters, aiming to ascertain the presence of the radioactive isotope chlorine-36.

That species could serve as a marker for how fast water moved from the surface to the repository horizon because it was almost exclusively produced as a byproduct of the atmospheric testing of nuclear weapons. Thus, if it were found, it must have travelled for less than a half-century, calling into question whether water moved through the UZ only through the matrix and whether it was diverted by the Paintbrush's tin roof.

At an NWTRB meeting in October 1996, a DOE manager and the principal Lawrence Berkeley National Laboratory (LBNL) scientist showed new data, Figure 11.1, collected by Los Alamos National Laboratory (LANL).[59] In their view, $^{36}Cl/Cl$ ratios of greater than 1250×10^{-15} were consistent with the presence of bomb-pulse of ^{36}Cl. The spikes appeared associated with fractures and other transmissive geological features. Taken at face value, the experiments overturned DOE's prevailing presumptions about percolation flux. As the DOE manager explained[60]:

> We've had a lot of discussion about whether this is a new model, whether it's an evolving model, whether it's an alternative model, and, very frankly, I would prefer not to get too involved in those semantic discussions. We've got a lot of information coming in. I use the simple term evolving to mean it's changing with time. That's basically the way I view it.

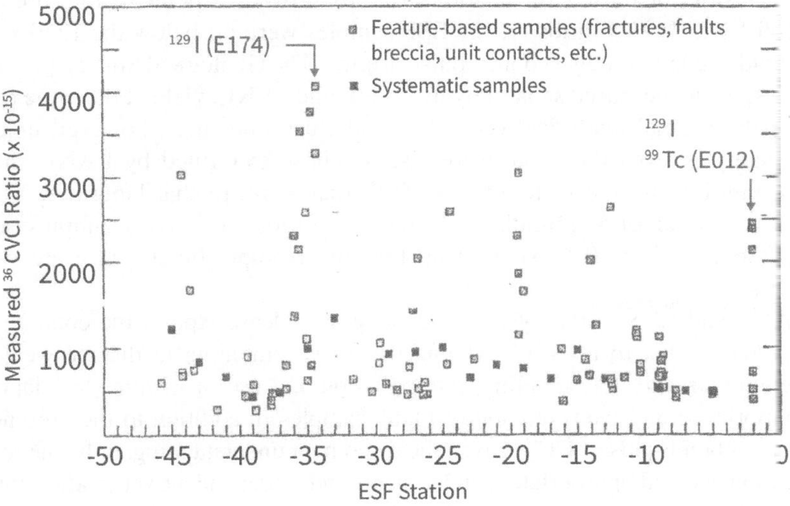

FIGURE 11.1 Samples of Clorine-36 and other tracers.

For the Yucca Mountain Project, the latest findings forced reconsideration of the dominant disposition hypothesis. But, in and of itself, they did not appear to render the site technically unsuitable; the implementor could fall back to other strategies. One of them would be to downgrade the importance of the UZ natural barrier and to enhance the role the engineered components might play in isolating and containing the radioactive waste. But first DOE needed to confirm, at least in general terms, the LANL results.

A new team from LANL conducted what it maintained was an "independent validation" inquiry in 2003.[61] This study varied the leaching method (active vs. passive), the leaching time (0.5 hours to 165 hours), and the particle size (dust to ½ inch). Among the conclusions was that the $^{36}Cl/Cl$ ratio was most likely to exceed 1250 × 10^{-5} level if the specimen was leached passively for a minimal amount of time. The most recent LANL results, however, still did not provide a definitive answer to whether bomb-pulse chlorine migrated to the repository horizon in approximately 50 years. Additional analyses would have to be undertaken.

To do so, DOE commissioned a new set of investigations, this time undertaken jointly by the Lawrence Livermore National Laboratory (LLNL), the Geological Survey, and Atomic Energy of Canada Ltd.[62] They drilled 50 boreholes within a section of the ESF that, based on the earlier experiments, ostensibly contain elevated $^{36}Cl/Cl$ ratios. Of critical consequence, the researchers applied a methodology that, among other things, passively leached the samples for 1 hour as opposed to the 48-hour leaching technique employed by LANL. In addition, the two national laboratories received a split sample from a niche in the ESF that were processed by the Geological Survey.

The results of these "verification" studies proved difficult to interpret. The USGS/LLNL $^{36}Cl/Cl$ ratios for the ESF samples were far below the 1250×10^{-15} threshold needed to suggest bomb-pulse origins. The Geological Survey-processed samples, when measured separately by LANL and LLNL, yielded the same results within the normal analytical error. Contradictory outcomes, however, emerged when samples from the niche were tested. Those examined by LANL showed ratios, ranging from 1016 to 8558×10^{-15}, that were, in that laboratory's view, a clear indication of bomb-pulse chlorine. The ratios for LLNL's samples ranged from 226 to 717×10^{-15}, suggesting that the isotope did not migrate within 50 years.

LANL and LLNL attempted to reconcile or at least explain the contradictory interpretations but to no avail. Among the areas remaining in dispute were "the conceptual strategy for sampling, interpretation and use of tritium (^{3}H) data, and the importance and interpretation of blank samples in addition to the presence or absence of bomb-pulse ^{36}Cl."[63] Given the discrepant understanding, only one recommendation seemed appropriate, conduct a second, independent verification investigation using new samples.[64]

Heeding that advice, DOE sought out the Harry Reid Center at the University of Nevada Las Vegas to bring order out of the interpretive chaos. The study's objective was "to investigate the source or sources of the conflicting results from the previous validation study, and to obtain additional data to determine whether or not there are bomb-pulse isotopes at the repository horizon."[65] To optimize the chances of finding bomb-pulse ^{36}Cl, the new effort modified LANL's methodology and collected rock samples on or adjacent to geologic features.

As work proceeded, several obstacles surfaced. The ESF had to shut down for extended periods, limiting the researchers' access. Further, some samples appeared to be contaminated by the presence of ^{36}Cl in the Reid Center's laboratory. Attempts to pinpoint the cause were unsuccessful. The studies resumed after moving the workshop to a new building that was not tainted.

In the end, the Nevada scientists could analyze only seven samples from the ESF. Of them, only one was barely above the 1250×10^{-15} threshold. However, a nearby specimen was below the divide. Rocks taken from around other geological structures failed to even approach the limit. In conclusion, the investigators could only advance an on-the-one hand and on-the-other conclusion.[66]

> Detection of a bomb-pulse $^{36}Cl/Cl$ ratio in the UNLV study supports the LANL assessment that water traveled from the surface to specific locations at repository horizon depths within the last 50 years. However, it should be emphasized that because of time we were not able to replicate the results, and these few data points are insufficient to conclude whether either interpretation is legitimate.

At a meeting of the Nuclear Waste Technical Review Board on January 24, 2007, the Yucca Mountain Project's Chief Scientist tried to summarize the bomb-pulse chlorine state of play.[67] He described LANL's 1996 original discovery, an external peer review of those findings, the USGS/LLNL's "enigmatic" judgments, and

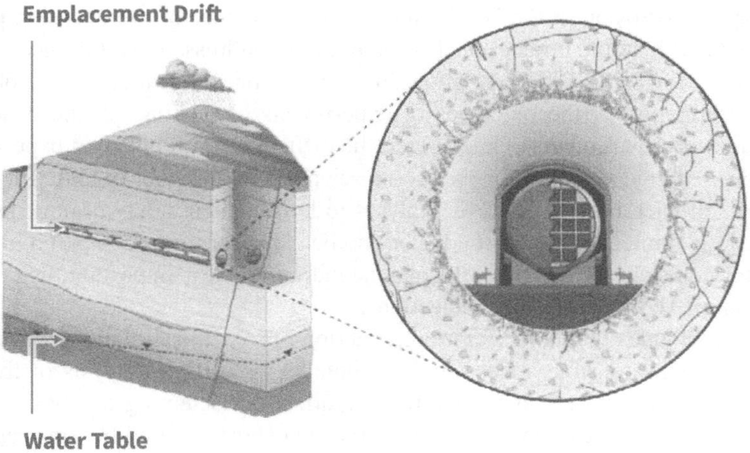

FIGURE 11.2 Revised conceptual model for waste isolation and containment under anticipated scenarios.

ambiguous appraisals from the Harry Reid Center. In a tone that suggested resignation, he told the Board that[68]:

> Our take is that we have got bomb pulse Chlorine-36 detections at a limited number of locations in the ESF, which indicates the presence of a few fast flow paths within the repository host rock. So, how do we accommodate that in the current models? Well, the current unsaturated zone flow and transport models reflect this chlorine-36 data in what we think is a reasonable and conservative manner. We have about 1 percent of the fast and transient flow paths–well, fast and transient flow paths carry about 1 percent of the water, primarily through faults and fractures, in our UZ flow and transport model, but they don't significantly affect the overall flow paths in the unsaturated zone. And, where we stand right now is we're not pursuing the chlorine-36 issue further at the moment. We think we've adequately addressed it in the existing state of models that we have.

Although many contributing factors influenced DOE's reconceptualization of its disposal concept, the ambiguity surrounding the chlorine-36 experiments undoubtedly played a role. Figure 11.2 portrays the modified waste-disposition system. By the turn of the century, DOE added a corrosion-resistant waste package fashioned out of Alloy-22 and envisioned a titanium drip shield to further guard against the packages' deterioration. Collectively, the engineered barriers dominate projections of repository performance.

OBSERVATIONS

It is always dangerous to draw "lessons" from just two case studies. But these technical controversies are so fertile that the rewards gained just might justify the risk borne. (Of course, the insights cannot be universal, only suggestive.) In the debate over copper corrosion in the Nordic nations, the structure of the licensing process itself dictated how the broader nuclear community addressed pivotal uncertainties. The copper-corrosion and percolation-flux controversies demanded the application and interpretation of the outcomes of experiments employing advanced or novel techniques. Both arguments underscored the difficulties of replicating experiments, the professed gold standard for producing scientific knowledge. In both, the implementors fell back on conservative analyses to bolster their safety arguments. Both revealed the limited influence of advisory bodies to government compared to formal regulatory agencies. Put differently, understanding was less important than compliance. Figure 11.3 captures the connections.

Thus concludes the historical reconstruction of why this country devolved into a nation of irresponsible ancestors. It completes as well an analysis of five core activities—designing an implementing organization, selecting a repository site, phasing repository development, licensing a repository, and addressing technical controversies—that any new radioactive waste-management program will have to undertake. The next chapter combines the two discussions and offers prescriptions about how this nation might transcend its morally uncomfortable circumstances. It also allows a deeper exposition of a topic so far touched only tangentially: social acceptability in the United States.

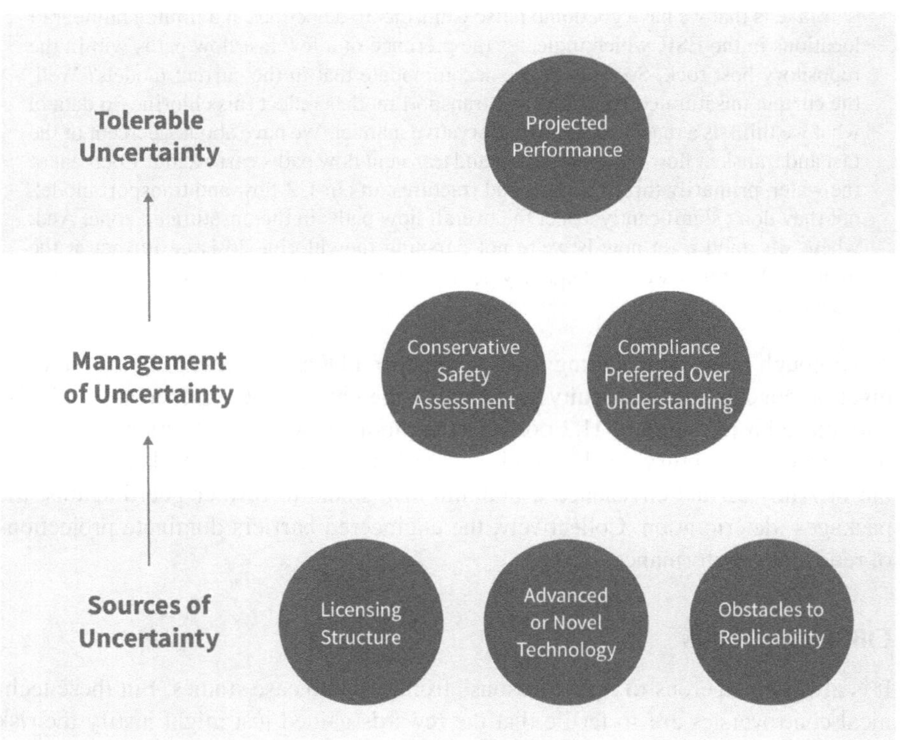

FIGURE 11.3 Dealing with uncertainty in technical controversies.

NOTES

1. See Imre Lakatos. "Falsification and the Methodology of Scientific Research Programmes," in Imre Lakatos and Alan Musgrave (eds.), *Criticism and the Growth of Knowledge*, Cambridge University Press, 1970, 59–90.
2. Gunnar Hultquist, "Hydrogen Evolution in Corrosion of Copper in Pure Water," *Corrosion Science*, (26) 1986, 173–177. See the references contained in the Swedish National Council for Nuclear Waste, "Mechanisms of Copper Corrosion in Aqueous Environments," Report 2009:4e, November 2009, 47–50.
3. The Swedish and Finnish nuclear-waste management programs differed in one essential respect. As is the case in a half-dozen nations, the former country chartered an advisory body to government, the National Council for Nuclear Waste, earlier known as KASAM. The organization prepared reports and organized seminars until it was disbanded in 2023 after the regulator, SSM, approved SKB's license application.
4. This discussion draws heavily on Tapio Litmanen et al., "Socio-technical Risk Governance Through Dyadic Risk Dialogue: Copper Corrosion as a Safety Challenge in the Geological Disposal of Spent Nuclear Fuel," Working Paper for the InSOTEC Project, June 19, 2014. See https://www.researchgate.net/publication/263408541_Socio-technical_risk_governance_through_dyadic_risk_dialogue_Copper_corrosion_as_a_safety_challenge_in_the_geological_disposal_of_spent_nuclear_fuel. Accessed May 30, 2025.
5. Posiva, "Nuclear Waste Management of the Olkiluoto and Loviisa Power Plants: Programme for Research, Development and Technical Design for 2004–2006," TKS-2003, 2003, 267.
6. There was little controversy about whether the copper would corrode when the copper canister came in contact with sulfides contained in the bentonite buffer. It would not. However, the impact of localized mechanisms such as microbial and stress corrosion cracking was uncertain.
7. Posiva (2003), 120–121.
8. Litmanen et al., 17.
9. Posiva, "Nuclear Waste Management of the Olkiluoto and Loviisa Power Plants: Programme for Research, Development and Technical Design for 2007–2009," TKS-2006, 2006.
10. Posiva, "Nuclear Waste Management at Olkiluoto and Loviisa Power Plants Review of Current Status and Future Plans for 2010–2012," TKS-2009, 2009.
11. Litmanen et al., 23–25.
12. Quoted in Litmanen et al., 25.
13. Litmanen et al., 28–31.
14. Radiation and Nuclear Safety Authority, "STUK's Statement and Safety Assessment on the Construction of the Olkiluoto Encapsulation Plant and Disposal Facility for Spent Nuclear Fuel." STUK-B 196. November 2015.
15. Radiation and Nuclear Safety Authority. Posiva submitted its operating-license application just as 2021 ended. STUK asked its overseeing Ministry of Economic Affairs and Employment to extend its review deadline to the end of 2025.
16. Peter Szakálos et al., "Corrosion of Copper by Water," *Electrochemical and Solid-State Letters*, (10) 2007, C63–C67.
17. Fraser King, "Critical Review of the Literature on the Corrosion of Copper by Water," SKB TR-10-69, December 2010.
18. Swedish National Council for Nuclear Waste (2009) provides a succinct description of the technical issues involved in assessing whether copper corrodes in anoxic groundwater. See 7–17.
19. The quotations below are in Swedish National Council for Nuclear Waste (2009), 53–54.
20. For a fuller exposition of the KTH position, see Swedish National Council for Nuclear Waste (2009), 57–64.
21. See the discussion in Chapter 10.

22. Swedish National Council for Nuclear Waste (2009), 21.
23. Swedish National Council for Nuclear Waste (2009), 26. The head of the anti-repository nongovernmental organization visited the United States as these events unfolded. The U.S. Nuclear Waste Technical Review Board supported Latanision's visit to Washington to discuss face-to-face the copper corrosion issue. The MKG representative was almost completely dismissive of Latanision's position. That reaction was to be expected as the NGO hung its opposition squarely on that technical claim.
24. Swedish National Council for Nuclear Waste (2009), 94.
25. Swedish National Council for Nuclear Waste (2009), 94–95.
26. Kjell Andersson, "Copper Corrosion in Nuclear Waste Disposal: A Swedish Case Study on Stakeholder Insight." *Bulletin of Science, Technology, and Society*, (33) 2013, 85–95. Andersson was the moderator of the Reference Group.
27. MKG left the Reference Group in October 2012 because of concerns that SKB was not making all of its experiments open to the public. See Andersson, 93–94.
28. The first round of hearings before the Land and Environmental Court took place between September 5 and October 27, 2017.
29. The application can be found at https://www.skb.com/wp-content/uploads/2017/01/1282835-MB-ansőkan-på-engelska.pdf. Accessed June 14, 2025.
30. MKG and Swedish Society for Nature Conservation, "Statement to the Land and Environmental Court, June 6, 2012. See https://www.mkg.se/uploads/Yttrande_MMD_(M_1333-11)_SSM_i_sak_Naturskyddsforeningen_och_MKG_160531.pdf. Accessed June 14, 2025. English version from Google Translate. Page numbers cited refer to the translation.
31. Bentonite is a clay that absorbs water and swells when it is in a wet environment—cat litter is often bentonite clay—and the idea is that after disposition in the final repository, the clay buffer will swell and then become dense and prevent harmful substances from reaching the copper canister.
32. MKG and Swedish Society for Nature Conservation, 11.
33. MKG and Swedish Society for Nature Conservation, 11.
34. See the opinion in Swedish at https://www.mkg.se/uploads/Aktbilagor/842_Nacka_TR_M_1333-11_MMDs_yttrande_till_regeringen_180123.pdf. Accessed June 14, 2025. MKG's translation of a summary can be found at https://www.mkg.se/uploads/Summary_opinion_Swedish_Environmental_Court_regarding_proposed_final_repository_spent_nuclear_fuel_Forsmark_Jan_23_2018_(unofficial_translation_MKG).pdf. Accessed June 14, 2025. Based on secondary sources, MKG's translation appears largely faithful to the original.
35. MKG translation, 4.
36. Swedish Nuclear Fuel and Waste Management Company, "Supplementary Information on Canister Integrity Issues," TR-19-15, March 2019. See https://www.skb.com/wp-content/uploads/2021/02/TR-19-15.pdf. Accessed June 15, 2025.
37. Swedish Nuclear Fuel and Waste Management Company, 3.
38. Swedish Nuclear Fuel and Waste Management Company, 4.
39. Swedish Radiation Safety Authority, "Response to the Land and Environmental Court' Opinion," September 30, 2019. Quotations are at 7. See https://www.mkg.se/uploads/M2018_00217_Me/M2018_00217_075_Yttrande_SSM_190930.pdf. Accessed June 15, 2025. English version from Google Translate.
40. National Council for Nuclear Waste, "Nuclear Waste Council's Response," September 13, 2019. See https://www.mkg.se/uploads/M2018_00217_Me/M2018_00217_066_Remissvar_Karnavfallsradet_190913.pdf. Accessed June 15, 2025. English version from Google Translate.
41. MKG, "Responses Regarding Quality Assurance and Reporting of Results from the LOT experiment," December 12, 2020. See https://www.mkg.se/uploads/M2018_00217_Me/M2018_00217_098_Bilaga1.pdf. Accessed June 15, 2025. English version from Google Translate.

42. See https://www.mkg.se/m201800217me-regeringsprovning-enligt-miljobalken. Accessed June 17, 2025.
43. See https://www.nucnet.org/news/support-for-repository-facilities-higher-than-ever-poll-says-5-2-2023. Accessed June 17, 2025.
44. Chapter 6 and especially Figure 6.3 describe the DOE's initial waste-isolation concept. This section draws heavily from Daniel Metlay, "From Tin roof to Torn Wet Blanket: Predicting and Observing Groundwater Movement at a Proposed Nuclear Waste Site," in Daniel Sarewitz et al., (eds), *Prediction: Science, Decision Making, and the Future of Nature*," (Covelo, CA: Island Press, 2000.)
45. Eugene Roseboom, Jr., "Disposal of High-Level Nuclear Waste Above the Water Table in Arid Regions," *Geological Survey Circular 903*, 1983.
46. Metlay (2000), 281–282.
47. Parvis Montazer and William Wilson, "Conceptual Hydrogeologic Model of Flow in the Unsaturated Zone, Yucca Mountain, Nevada," USGS-WRI-84-4345, 1984.
48. Department of Energy, "Draft Environmental Assessment, Yucca Mountain Site, Nevada Research and Development Area," DOE/RW-0012, 1984, 6–120.
49. Department of Energy (1984), 6–247.
50. Department of Energy, "Environmental Assessment, Yucca Mountain Site, Nevada Research and Development Area," DOE/RW-0073, 1986, 6–295.
51. Department of Energy (1986), 6–137.
52. Department of Energy (1986), 3–208.
53. Pacific Northwest Laboratory, "Example of Post-Closure Risk Assessment Using the Potential Yucca Mountain Site," Richland, WA., 1992.
54. Sandia National Laboratory, "An Initial Total-System Performance Assessment for Yucca Mountain," SAND-91-2795, 1992.
55. Sandia National Laboratory, "Total-system Performance Assessment for Yucca Mountain—SNL Second Iteration," SAND-93-2675, 1994.
56. Thompson, Ramo, and Woolridge (TRW), "Total Systems Performance Assessment–1993: An Evaluation of the Potential Yucca Mountain Repository," 1993.
57. Thompson, Ramo, and Woolridge (TRW), "Total Systems Performance Assessment–1995: An Evaluation of the Potential Yucca Mountain Repository," 1995.
58. Nuclear Waste Technical Review Board, *Report to the U.S. Congress and the U.S. Secretary of Energy*, 1996, 28.
59. June Fabryka-Martin et al., "Summary Report of Chlorine-36 Studies: Systematic Sampling for Chlorine-36 in the Exploratory Studies Facility," March 29, 1996. See https://www.nrc.gov/docs/ML0314/ML031410299.pdf. Accessed June 23, 2025.
60. Transcript, Nuclear Waste Technical Review Board Meeting, October 9, 1996, 97.
61. See the presentation by Robert Roback, Nuclear Waste Technical Review Board meeting, September 16, 2003. See https://www.nwtrb.gov/docs/default-source/meetings/2003/september/roback.pdf?sfvrsn=3ba4f505_5. Accessed June 24, 2025.
62. Department of Energy, and U.S. Geological Survey, "Chlorine-36 Validation Study at Yucca Mountain, Nevada," TDR-NBS-HS-000017 REV00, August 2006. See https://www.nrc.gov/docs/ML0314/ML031410299.pdf. Accessed June 23, 2025.
63. Nevada System of Higher Education, "Bomb-Pulse Chlorine-36 at the Proposed Yucca Mountain Repository Horizon: An Investigation of Previous Conflicting Results and Collection of New Data," TR-06-002, Rev 0, July 28, 2006. See https://www.nrc.gov/docs/ML0915/ML091590561.pdf. Accessed June 23, 2025.
64. Department of Energy and U.S. Geological Survey, 60.
65. Nevada System of Higher Education, 2.
66. Nevada System of Higher Education, 3.
67. Transcript, Nuclear Waste Management Review Board Meeting, January 24, 2007, 64–69. See https://www.nwtrb.gov/docs/default-source/meetings/2007/january/07jan24049e9148a-bad67069638ff00003d5b4b.pdf?sfvrsn=7a30f605_6. Accessed June 24, 2025.
68. Transcript (2007), 68–69.

Part III

What Can We Learn?

12 To Become a Responsible Ancestor

The overarching theme of this narrative is a moral one, something that goes beyond calculations of "intergenerational equity." Whatever the computed risk that the *status quo* of extended storage of high-activity waste poses (and that is subject to some dispute), it also places substantial and uncertain burdens on our descendants. Indeed, beneath all the quibbling about identifying the whereabouts for a centralized interim facility or the necessity to preserve spent nuclear fuel because of its potential energy content, this generation of Americans, *alone* among citizens of those pluralistic democracies that have adopted nuclear energy, has abandoned its obligations to their offspring by turning a blind eye to the long-term management of its radioactive waste.

To be sure, nascent messages from the Nuclear Waste Technical Review Board, the Trump Administration, and the U.S. Nuclear Industry Council have called attention to the disposition-gap in the nuclear fuel cycle.[1] But symbolic expressions of support for a repository are not a substitute for on-the-ground actions. It is here that our story might make a contribution.

At the start, candor requires an acknowledgement that any reform initiative will be highly controversial. Among current interested and affected parties are DOE, the national laboratories, nuclear utilities, and citizenry of storage and potential repository locations throughout the country.[2] Their material and attitudinal interests will prompt them to highlight all the grounds for opposing change, many of which, it must be conceded, have some merit. But against those criticisms, two incontrovertible facts stand clear. First, the federal government has failed for more than a half-century to establish a robust and credible repository-development program. Second, other nations have succeeded, at least thus far.

A STRATEGY FOR CHANGE

Part I of our tale surveyed how and why the United States has fallen into this negligent situation. Part II of this story has clearly demonstrated that nations can do better, albeit understanding that the experiences in other countries cannot be exported whole cloth to ours. So, what is to be done?

The 1982 Nuclear Waste Policy Act (NWPA) contained many sensible provisions, but its four strategic bargains have long since been corrupted or ignored. (See Chapters 5 and 6.) The breach of one of those bargains by the passage of the 1987 Nuclear Waste Policy Amendments Act further exacerbated problems. So, any attempt to resurrect the U.S. disposition-program must begin with a fundamental revision of those laws in a manner that captures contemporary trade-offs and understanding.

DOI: 10.1201/9781003528302-15

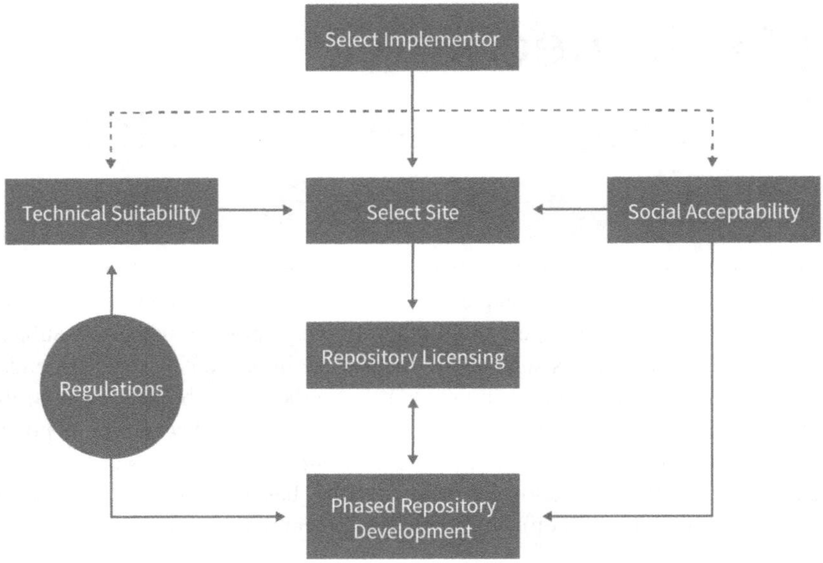

FIGURE 12.1 Interactions among the elements of a waste-disposal program.

Note: Solid lines represent causal links; dashed lines represent reciprocal links.

The task here is not to craft statutory language or to specify the myriad details that will ultimately have to be negotiated; instead, it is to paint broad strokes that underscore critical choices. If implemented with fidelity, those judgments can establish an honest, robust, credible, and technically sound path leading toward becoming a responsible ancestor. Figure 12.1 provides both a map detailing the interactions of the essential elements of a rejuvenated waste-disposal program as well a framework for advancing prescriptions suggested by our narrative.

Replace the Department of Energy as the Waste-Disposition Implementor

One of the central conclusions of Chapter 7 was that only the Department of Energy had endorsed itself as the implementor of the waste-disposition program going forward. (See Figure 7.2.) By the early 2010s, however, Energy Secretary Steven Chu acknowledged the widespread dissatisfaction with DOE by favoring either an independent federal agency (IFA) or a federally chartered corporation (FEDCORP) as appropriate substitutes. The movement away from DOE has since gained momentum with virtually all interested and affected parties advocating some form of implementor other than the current one.

Standing Up a New Implementor: The Frontrunners

At least 35 independent federal agencies operate today, including such familiar ones as the CIA, EPA, the Federal Reserve System, NASA, the Postal Service, and USAID.[3] Importantly, two DOE predecessors, the Atomic Energy Commission and the Energy Research and Development Administration, were independent federal agencies; neither sport a particularly promising history when it comes to launching a waste-disposition system. Nearly a dozen public/private corporations offer a range of financial services. Another 17 government-owned corporations, including the Tennessee Valley Authority, provide a range of activities, again most of them concentrate on financial matters.[4]

Yet among the roughly 65 IFAs and FEDCORPs, none seem to provide a persuasive model for an organization that must site a repository and conduct state-of-the-art scientific and engineering research, all the while simultaneously meriting the trust and confidence of parties whose interests may often be in conflict.[5] Adopting either of the frontrunner institutional forms would likely unleash a significant design experiment whose outcome may not be preferable to the *status quo*. Indeed, it is noteworthy that only one of the studies promoting a change in implementor point to a *specific existing organization* as a promising model.

Standing Up a New Implementor: A Dark-Horse Alternative

The discussion of waste-management implementors in Chapter 7 took special note that utility-owned, non-profit organizations—NUCOs—execute many programs abroad. Those in Finland, Sweden, Switzerland, and Canada represent four out of the five endeavors that have resulted in the selection of sites for a repository. The implementor in the first country, Posiva, holds a construction license and has submitted an application for an operating permit. The implementor in the second, SKB, has been granted authority to construct a disposition facility. Given this promising track record, can such an organizational form be adapted to the U.S. context?[6]

Critical choices will have to be made about NUCO's structure. Reactor owners would manage the undertaking. But, for example, should other interested and affected parties be on NUCO's Board of Directors or should they be included in a strong advisory body? How should voting power be distributed among the diverse set of owners? How should mismanagement be cured? How and when should NUCO report its activities to Congress?

Although those questions are important, it must be emphasized that NUCOs abroad bring to the repository-development process a set of highly relevant attributes and competencies. They do not change either form or policy with each new political administration. They draw on the experiences of their owners to transport SNF, to design and construct nuclear-quality facilities, to submit and defend a license application before their regulatory authorities, and to load and handle casks containing high-activity radioactive waste. These NUCOs also interact with other segments of the industry, whose involvement is critical for a successful repository program. In addition, the utilities engage nearby communities, albeit with mixed outcomes.

Moreover, utilities' experiences in siting a reactor presents some of the same geological challenges as selecting a location for a repository, such as identifying seismic and volcanic risks. To the extent that this form of implementor lacks the internal capacity to investigate hydrogeologic features deep underground, it can bring that expertise in-house or contract not only with national laboratories and geologic surveys, but also with a wider range of private-sector vendors around the world.

In short, there is no reason to believe that a NUCO in the United States could not operate in much the same manner as its sister organizations abroad.

Nonetheless, any decision to stand up a new implementor is likely to impose transition costs. For NUCO in particular, those "frictions" are reflected in three questions. Is it financially sustainable? Can it manage both defense and commercial waste? Why should nuclear powerplant owners take on this controversial task?

Financial Sustainability: The NWPA established a structure through which reactor owners contribute into the Nuclear Waste Fund (NWF) $0.001 per kilowatt-hour of energy generated.[7] That account pays the life-cycle outlays for disposing of the utilities' high-activity material. General appropriations cover the expenditures for isolating the nation's defense-origin waste. In 1985, President Ronald Reagan determined that defense and commercial waste should be disposed of in the same repository.[8]

Over the years, Congress has used the NWF to camouflage the level of the federal budget deficit and has been reluctant to appropriate adequate amounts to support repository development or to cover the costs of disposing defense-origin waste.[9] In 2013, after the demise of the Yucca Mountain Project, a federal court ordered DOE to stop collecting the fee.[10] Nonetheless, as of September 2024, the NWF had a balance of slightly more than $52 billion, including nearly $3 billion still owed by nuclear utilities. The account's growth from the previous year was approximately $2 billion.[11] For context, the House Appropriations bill for FY2026 allocates $12 million from the NWF.

One option for ensuring NUCO's financial sustainability would be for Congress to enact legislation requiring the transfer to the new implementor of no less than the *growth* of the NWF every year.[12] If this approach were adopted, NUCO's budget would be roughly four to five times the typical size of DOE's appropriations from the NWF. By only removing a small fraction of the NWF, Congress' appetite for using the account to obscure the federal deficit may be whetted. The federal government, of course, would retain some auditing authority but would leave programmatic control to NUCO's management. (One cannot prevent subsequent Congresses from altering this arrangement, but should it do so without a credible Plan B, its abdication of responsibility would be plain to all. In any event, this proposal seems preferable to the annual budgetary chaos under the *status quo*.)

Management of Defense-Origin High-Activity Radioactive Waste: President Reagan based his 1985 decision to commingle defense and commercial waste in a single disposition facility on the six factors specified in the NWPA.[13] In 2014, however, DOE used those same factors to recommended a "strategy for disposal of some DOE-managed HLW and SNF in a separate deep-mined, geologic repository."[14] The following year, President Barack Obama signed a Presidential Memorandum determining that "the development of a repository only for the disposal of high-level radioactive waste resulting from atomic energy defense activities is required."[15] That judgment persists as national policy, although when, how, or if it will be implemented remains unclear.

The creation of NUCO delivers clarity to the present circumstances. Standing up a new implementor will demand extraordinary energy and hard work. To address the technical and political intricacies of disposing of the varieties of defense waste undoubtedly will undermine the focus of the new implementor.[16] It is a distraction that best should be avoided at least in the short term.

Thus, DOE should retain its authority to establish a defense-waste-only repository. The cost of that endeavor will continue to be borne by taxpayers through the appropriations process. Although the two disposition programs could learn from each other and stimulate headway, a strong possibility may arise, perhaps within a short time, that the federal effort will come to be viewed as an unnecessary and costly redundancy. If that proves to be the case, the new legislation should stipulate that DOE has the authority to contract with NUCO to dispose of the government-owned waste.

Why Would NUCO Take Over the Disposal of Commercial Waste? If truth be told, the owners of nuclear power plants currently find themselves in an extremely enviable position. Early reactor adopters continue to defer payments into the NWF at relatively modest expense. All of them no longer have to pay the $0.001 per kilowatt/hour fee for generating electricity. Moreover, a series of court cases has imposed liability on the federal government for failure to accept spent nuclear fuel for disposal by the end of January, 1998.[17] Those damages—now estimated at $65 billion—are paid not through the NWF or DOE appropriations but through the so-called Judgment Fund, which, after court review, automatically dispenses compensation to entities harmed by federal actions.[18] Given all the benefits of the existing state of affairs, NUCO's taking over the commercial-waste disposal responsibility seems irrational, counterproductive, and senseless.

Appealing to the industry's better angels, as is the case of the most forward-looking national programs, may not outweigh the material costs. Thus, policymakers may need to provide the industry with offsetting material benefits. What they might be would be subject to intense negotiations. Among the possibilities, none of which are guaranteed to be effectual, might be some or all of the following[19]:

- Additional support for nuclear power innovations, such as small modular reactors
- Transferring the remaining NWF *corpus* to NUCO or its owners after a predetermined period of successful repository operation[20]
- Subsidies for enrichment and fuel acquisition costs

Ultimately, however, the nuclear renaissance that is seemingly underway will almost certainly be cut short if the waste-management issue is not addressed. One need only note that community interest in hosting a federal centralized interim storage facility seems to be contingent at the very least on having a viable repository program. The federal government's track record in this arena is too well-known. And small modular reactors are still nuclear reactors. Their attendant risk perceptions, as charted in Figure 2.7, are unlikely to wane; these mindsets seem to be indelibly imprinted in large and more significantly, intense segments of the general public. Already, an environmental activist organization in Texas has sued to halt the construction of four pebble-bed reactors, partly because the proponent has not committed to removing the waste from the site.[21]

Transform the Process for Siting a Deep-Mined, Geologic Repository

Earlier our tale established the prerequisites for selecting a sustainable location for a disposition facility: demonstrating technical suitability and social acceptability.

(See Figure 8.1.) Implementors have encountered stumbling blocks as they sought to accomplish both tests. (See Figure 8.6.) In the two sections immediately below, each requirement is examined. The discussion further below on addressing technical controversies explores their reciprocal interactions.

Facilitating a Finding of Technical Suitability

It is abundantly clear that implementors can site repositories in a range of hydrogeologic environments. For some of those settings, however, evaluating technical suitability is less persuasive than others. For example, the NRC staff was convinced that the elaborate natural/engineered system in the Yucca Mountain's volcanic tuff would satisfactorily isolate and contain the waste for as much as one million years. Its conclusions presumed that the Alloy-22 waste packages would perform as expected and that the titanium drip shield would be effectively installed many decades after the facility began operations. Both premises were open to challenge, making it more difficult to obtain a broad consensus supporting technical suitability.

The influential 1957 report from the National Academy of Sciences panel promoted salt formations to a preferred position among potential repository environments. Until the mid-2010s, it appeared that Germany would select the now-eliminated Gorleben site as the location for its repository. No rigorous safety case, however, was ever prepared to confirm the technical suitability of that choice. Nor has any other country so far investigated siting a disposition facility in a similar formation. Thus, there is a shortage of experience applicable to demonstrating the technical suitability of a salt site.[22]

In sharp contrast to either volcanic tuff or salt, the global nuclear waste-management community has acquired considerable understanding about how to evaluate the technical suitability of a repository in crystalline rock and clay. Finland, Sweden, and Canada have validated arguments for a repository in granite-like formations. France and Switzerland have established that at least two types of clay can contain high-activity waste for many millennia.

Thus, when this country restarts it disposition efforts, it should concentrate on both hydrogeologies, which are ubiquitous in the United States. Narrowing the search in this way minimizes the likelihood that any one state can claim that it is being targeted. (In Nevada's words, screwed.) It would also reinforce perceptions that technical considerations, not political ones, are driving the siting process. Moreover, contentions about suitability rest on a very well-understood propositions that do not depend on advanced or novel techniques.

All of the nations that have chosen repository sites have compared multiple areas prior to deciding on the final spot. Indeed, a parallel rather than a serial process seems now to be the international standard. (See Figure 8.3.) A reinvigorated disposal effort in the United States should adopt this approach as well.

This country is the only one whose regulations require underground investigations prior to picking a site.[23] That policy seems prudent as geoengineers hold that surprises will almost certainly arise if only surface-based tests are conducted. As the serial process gradually winnows potential locations, preferred sites in crystalline rock and clay will emerge. Each environment should be studied at depth. Ordinarily, this rule might impose such great costs that it might be deemed impractical and too expensive. But given international experiences, both surface and subsurface studies can be focused on the parameters known to dominate performance projections. As a consequence, the costs for in-depth studies are likely to be substantially lower than those incurred by the Yucca Mountain Project. Moreover, NUCO will not have to acquiesce in fulfilling the sometimes pork-barrel investigational demands from the national laboratories.

Securing Social Acceptability

Figure 8.6 makes clear that the overwhelming majority of engagements to site a repository have failed to gain social acceptability or their outcomes were never determined. Why should one expect any different result if NUCO takes on the responsibility for siting?

In the United States, three jurisdictions have a stake in where a repository is located: the local community or tribal nation, the state, and the federal government. The siting process should better reflect all of those interests by redistributing power from the center to the periphery. The experience in Canada, Finland, France, Sweden, and Switzerland suggests one approach.[24]

Communities and tribal nations would be invited to express an interest in hosting a repository. If none do, NUCO would still have the authority to propose potentially suitable sites based on literature reviews. It would also have permission to carry out surface-based tests to better understand the hydrogeology of those locations. Notwithstanding whether the community or the tribal nation volunteers, NUCO would offer financial and other resources to create formal partnerships with the local interested and affected parties. To the extent possible, the partnership would explore with NUCO issues such as technical design, monitoring, and local authority over operational decisions.

After some minimal time, perhaps two or three years, the partnership would recommend to the local authorities whether NUCO should be allowed to continue its siting investigations, especially those involving studies at the repository horizon. At this point, the community or tribal nation could exercise an *absolute veto* over NUCO's activities. The rationale for allowing NUCO unfettered investigatory access for a reasonable period is twofold. First, the national interest requires that a potential repository site not be eliminated at the start. Second, the two to three-year period provides the locality with crucial information about NUCO's trustworthiness. Failure to merit it sends an unmistakable signal that the long-term relationship is problematic. If efforts continue, the local jurisdictions would continually review the implementor's investigations. The community or tribal nation's power to exercise an absolute veto would expire once NUCO submits a license application to the Nuclear Regulatory Commission.

The regulatory authority would review the permit using an adjudicatory process. Participants would include NUCO, the regulatory staff, the locality, members of the partnership, the host state, and other interested parties that present accepted contentions. If a license is granted, the host state would have a fixed period to disapprove the site. That disapproval could only be overridden by super-majorities in both the House of Representatives and the Senate.

The siting process herein recommended protects local, state, and national interests. It is also time-consuming. But because developing a repository is a multigenerational effort, the consent of multiple jurisdictions is a precondition for creating a promising environment for positive results.

Modify the Regulatory Regime

Three federal bureaucracies—the Environmental Protection Agency, the Nuclear Regulatory Commission, and the Department of Energy have promulgated interlocking rules germane to the disposition of high-activity radioactive waste.[25] They all narrowly focus on performance assessment as the sole indicator of compliance.

That approach places the United States at odds with many other pluralistic democracies. (See Chapter 10.) These prepare safety cases, which typically include independent lines of evidence that strengthen *confidence* that the repository system can meet social judgments about the individual protection dose and compliance periods. The Finnish implementor, for example, published a separate "complementary considerations" document when it submitted its construction license to its regulator.[26] When it tendered its operating license, it updated that document.[27]

The United States should require NUCO to prepare a more comprehensive safety case. This need not be especially onerous. For both crystalline and clay hydrogeologies, that path has already been blazed by other nations.

A number of provisions in the current regulatory regime do not need to be altered. These include[28]:

- Individual health consequences
- Reasonable expectation and risk-informed decision-making
- Characteristics of a reasonably maximally exposed individual
- Screening of relevant features, events, and processes
- Human intrusion scenario
- The one-million-year or time-to-peak-dose compliance period

The following changes, however, will need to be made. They undoubtedly increase the conservatism of the new rules. Given the longstanding failure of the waste-disposition program in the United States, especially with respect to social acceptability, that tack may be unavoidable.

- Retain the 0.15 millisievert/year individual protection dose limit for the entire compliance period of one million years
- Require NUCO to calculate the dose beyond 100,000 years by using conservative bounding parameters
- Set forth an obligation to conduct adjudicatory hearings
- Improve the requirements for demonstrating a repository's defense-in-depth

Since NUCO will become the implementor, these tightened rules will also be indispensable to establish trust in the regulator.

But this increased stringency is unlikely to increase NUCO's compliance burden. Figures 8.10, 8.11, 8.12, and 8.13 make clear that a disposition facility sited in either crystalline rock or clay hydrogeologies can be designed to meet even these tougher requirements.

Increase the Efficacy of Phased Repository Development

For more than 50 years, phased repository development seemed to be the holy grail that would unlock progress toward a functioning disposition facility. (See the discussion in Chapter 9.) Although the waste-management programs in Switzerland and Canada claim to have adopted this strategy, only the one in France has taken concrete measures so far to illustrate how a stepwise approach might be executed. That nation's implementor, Andra, has designed critical technologies to accomplish this task. It also has "demonstrated" their operation under controlled conditions. By undertaking these actions, it anticipated and softened criticism that its acceptance of the outcome of the 2013 Public Debate was largely symbolic.

But the work of some important social scientists has called into question many of the premises upon which stepwise development rests. First, contingency theory maintains that incrementalism or trial-and-error decision-making should be applied only in cases where the feedback from prior actions arises quickly and visibly. Further, it should also be employed only when a consensus more or less obtains about how conflicting values are traded. (See Figure 7.1.) Second, decision-makers lack the cognitive capacity to formulate comprehensive or synoptic choices. Thus, some options and their accompanying consequences inevitably vanish from their calculations.

Third, to take the stepwise approach seriously demands preparation and allocation of substantial resources ahead of time just in case retrieval of waste or programmatic reversal becomes imperative. Implementors' optimism that their plans will succeed almost certainly will discount importance of preparation and resource allocation. Fourth, phased development poses problems for organizational architects. Detecting mistakes demands *independence* from other internal elements to forestall "groupthink," while correcting errors requires *interdependence* with other units to enforce change. Combining both attributes in a single administrative body is no mean feat.[29]

One ancillary benefit of creating NUCO from scratch in the United States is that an opportunity arises to mold it into a learning organization. It overcomes the conceptual weakness of incrementalism by engaging interested and affected parties about short-term tradeoffs that might be broadly acceptable. It takes to heart Charles Lindblom's notion of the "intelligence of democracy" to overcome cognitive limitations.[30]

At first glance, it might appear that NUCO might be more inclined to discount preparation and resource allocation than the Department of Energy. But a "guaranteed" income stream might offset that temptation. Finally, the nuclear utility-owned Institute for Nuclear Power Operations might provide a useful model for how the possibility of "off-normal" events can be reduced.[31]

On balance, then, the advantages of adopting an overt step-by-step repository-development process likely outweigh the costs, even if it turns out to be flawed. Self-conscious action is always preferable to allowing institutional momentum to push matters forward without deliberation. But NUCO's organizational culture will always have to be safeguarded to make phased development promising.

Minimize Technical and Scientific Controversies

Restricting the siting of a repository to crystalline rock and clay hydrogeologies lessens generic technical controversies; the foundational science and engineering for those disposal concepts have been well constructed in Canada, Finland, France, Sweden, and Switzerland. The United States can easily piggyback on that knowledge and experience as it reinaugurates its own long-term waste-management program.

This tactic, of course, does not imply that repository development will avoid all technical disputes. Whether a specific location matches the *desiderata* of either disposition models will always be an open issue at least until detailed and rigorous investigations can be completed above and below ground. But, with increased focus, NUCO will not get bogged down by extraneous studies and thus will be able to substantially reduce the time (and expense) it takes to nominate technically suitable areas.[32]

Here the so-called "advisory bodies to government" can play a constructive role. These organizations—the Committee on National Evaluation in France, the National Council on Nuclear Waste in Sweden, and the Nuclear Waste Technical Review Board in the United States in addition to several others—are independent of their corresponding implementor and therefore are well positioned to identify errors.[33] They typically lack, however, the interdependence to compel the implementor to alter its course. Indeed, the effectiveness of these peer-reviewers tends to be strongest at the start of the national programs when implementors are most likely to defer to outside advice; it generally diminishes as time goes on. This shortcoming became clear in Chapter 11's discussion of the copper corrosion and percolation flux controversies.

In contrast, regulatory bodies hold the ultimate power over implementors. They determine what investigations have to be carried out, how the data gathered are analyzed and interpreted, and whether conclusions reached are supported and justified. But, as Chapter 7 noted, the central question is whether regulators are, in actuality, independent of the implementor. Creating NUCO changes the optics but fails to conclusively resolve the question. It is, nonetheless, at least an improvement over the *status quo*.[34]

CAN THIS GENERATION OF AMERICANS BECOME RESPONSIBLE ANCESTORS?

It has become something of a truism to assert that the United States needs to "find the will" to effect a coherent enterprise for managing its high-activity radioactive waste over the long term. This story provides one systematic basis for translating the cliché into actionable reform. Yet it is crucially important to understand the implications of the five major recommendations contained in this final chapter. These are *not items* on a menu from which to choose the courses of a dinner. Rather they are a *recipe* where the absence of any one ingredient spoils the dish.[35]

Kicking the can down the road, of course, is one moral choice that this generation of Americans might make: it can remain imprudent forebearers. For this narrator, that option seems discordant with our nation's tradition of *trying* to maintain accountability for its actions. Perhaps others will see the virtue of this point of view and join together to do what it takes to elevate this generation to a position of responsible ancestors. Let the dialogue begin.

NOTES

1. Letter from Peter Swift, Chairman, Nuclear Waste Technical Review Board, to Mike Johnson, Speaker of the House, Chuck Grassley, Senate President Pro Tempore, and Christopher Wright, Secretary of the Department of Energy, March 18, 2025. See https://www.nwtrb.gov/docs/default-source/correspondence/pns001vf-nwtrb-mar2025-letter-report.pdf. Accessed July 5, 2025. Donald Trump, "Deploying Advanced Nuclear Reactor Technologies for National Security," May 23, 2025. See https://www.whitehouse.gov/presidential-actions/2025/05/deploying-advanced-nuclear-reactor-technologies-for-national-security. Accessed July 5, 2025. United States Nuclear Industry Council, "New Policy: Transforming the Back End of the U.S. Nuclear Fuel cycle," June 13, 2025. See https://cdn.prod.website-files.com/5ff2c071d62f1697a64e8c45/68519bbb561b400c9262f281_USNIC%20Letterhead_USNIC%20Backend%20Plan%20for%20Backend%202025%200613.pdf. Accessed July 5, 2025.
2. Neighbors of shutdown powerplants might very well welcome reform as it may hold a more realistic prospect for removing spent nuclear fuel.
3. See https://en.wikipedia.org/wiki/Independent_agencies_of_the_United_States_federal_government#cite_note-18. Accessed July 11, 2025.
4. Lynn E. Davis, et al., *Choosing a New Organization for Management and Disposition of Commercial and Defense High-Level Radioactive Materials*, (Santa Monica CA: RAND Corporation, 2012), 91–92. This analysis properly acknowledges that the listed IFAs and FEDCORPs are subject to change.
5. The obstacles EPA encountered in regulating the air and water quality of plants owned by TVA is documented in Robert F. Durant, *When Government Regulates Itself :EPA, TVA, and Pollution Control in the 1970s*," (Knoxville: University of Tennessee Press, 1985).
6. Davis et al. present a well-structured analytical framework for evaluating organizational designs. As appropriate, that outline guides the assessment of the dark horse alternative. A preliminary analysis of this issue can be found in Daniel Metlay, "Becoming a Responsible Ancestor," *Bulletin of the Atomic Scientist*, (81) 2025, 29–35.
7. Nuclear Waste Policy Act, Section 302. Among its provisions, the law allows utilities to defer payment to dispose of the waste produced prior to the law's passage.
8. The Nuclear Waste Policy Act mandated this decision in Section 8(b).
9. Blue Ribbon Commission on America's Nuclear Future, "Report to the Secretary of Energy," 2012, 70–80.
10. *National Association of Regulatory Utility Commissioners v. U.S. Department of Energy*, No. 11-1066, Court of Appeals for the D.C. Circuit, 2013.
11. "Audit Report: The Department of Energy Nuclear Waste Fund's Fiscal Year 2024 Financial Audit," DOE-OIG-25-03, November 2024. See https://www.energy.gov/ig/articles/financial-statement-audit-doe-oig-25-03. Accessed July 18, 2025.
12. I have been informed that several bills have advanced this proposal over the years. But it would be scored as a mandatory appropriation that requires a mandatory offset, either taxes or mandatory spending cuts. It has been an insurmountable obstacle so far. Also, since the waste fee is no longer collected, the interest alone doesn't affect the deficit, because it is an intra-governmental transfer. Of course, any revision of waste-management legislation would need to address this issue. Indeed, many proposals for ensuring a steady budget for a new implementor face the same obstacles. My thanks to Mark Holt for this information.
13. NWPA Section 8(b)(1).
14. Department of Energy, "Assessment of Disposal Options for DOE-Managed High-Level Radioactive Waste and Spent Nuclear Fuel," October 2014. See also Department of Energy, "Report on Separate Disposal of Defense High-Level Radioactive Waste," 2015.
15. Barack Obama, "Presidential Memorandum—Disposal of Defense High-Level Radioactive Waste in a Separate Repository," March 24, 2015. See https://obamawhitehouse.archives.gov/the-press-office/2015/03/24/presidential-memorandum-disposal-defense-high-level-radioactive-waste-se. Accessed July 19, 2025.

16. Nuclear Waste Technical Review Board, "Management and Disposal of U.S. Department of Energy Spent Nuclear Fuel: A Report to the United States Congress and the Secretary of Energy," December 2017.
17. See, for example, *Indiana Michigan Power Co. v. Department of Energy*, 88 F.3d 1272 (D.C. Circuit 1996).
18. See https://fiscal.treasury.gov/judgment-fund. Accessed July 19, 2025. The Government Accountability Office estimates that the federal government's liability will reach approximately $65 billion by 2030. See Government Accountability Office, "Commercial Spent Nuclear Fuel: Congressional Action Needed to Break Impasse and Develop a Permanent Solution," GAO-21-603, September 2021. Under the financial terms of the instant proposal, the damages would still be paid to reactor owners from the Judgment Fund unless that arrangement is changed through negotiations between NUCO and the federal government.
19. Nuclear opponents will argue that these possibilities promote a controversial and risky technology. Because of growing concerns about climate change, that attitude appears to be changing, perhaps dramatically. Further, if the nation effectively addresses the waste-management question, one rationale for opposing nuclear power could be moderated.
20. The same problem that was described in Footnote 10 would arise for this option under current legislation.
21. See https://sanantoniobaywaterkeeper.org/local-group-files-first-ever-legal-challenge-to-smr. Accessed August 25, 2025.
22. The Waste Isolation Pilot Plant in New Mexico is located in a salt formation. The material disposed there generates significantly lower heat than high-activity waste. Extrapolating from the former waste type to the latter may raise its own set of problematic and technical issues.
23. See Nuclear Regulatory Commission, "Disposal of High-Level Radioactive Wastes in Geologic Repositories: Licensing Procedures," *Federal Register*, (46) February 25, 1981, 13,972–13,973.
24. See Steering Committee, "Reset of America's Nuclear Waste Management: Strategy and Policy," 2018, 68–71.
25. The latest versions are in turn 40 CFR 197, 10 CFR 63, and 10 CFR 963.
26. Posiva, "Safety Case for the Disposal of Spent Nuclear Fuel at Olkiluoto. Complementary Considerations 2012," POSIVA 2012-11, 2012. See https://inis.iaea.org/records/14ea2-vf044. Accessed May 12, 2025.
27. Posiva, "Safety Case for the Operating License Application—Complementary Considerations," POSIVA 2021-02, 2021. See https://cms.posiva.fi/safety-case/complementary-considerations-cc. Accessed August 8, 2025.
28. The first five points are consistent with the recommendations of American Nuclear Society Special Committee on Generic Standards for the disposal of High-Level Radioactive Waste, "Recommendation of Postclosure Aspects of Generic Standards of the Permanent Standards for the Permanent Disposal of Spent Nuclear Fuel and High-Level and Transuranic Radioactive Wastes in the United States," 2023, x-xi.
29. See Daniel Metlay," Error-Correction in Bureaucracy," Ph.D. dissertation to the Department of Political Science, University of California, Berkeley, 1978.
30. Charles Lindblom, *The Intelligence of Democracy*," (New York: Free Press, 1965).
31. See Joseph V. Rees, *Hostages to Each Other: The Transformation of Nuclear Safety Since Three Mile Island*, (Chicago: University of Chicago Press, 1994). A less disinterested account is Philip L. Cantelon, *Core of Excellence: A History of the Institute of Nuclear Power Operations*, (Atlanta: INPO, 2012).
32. The multi-volume Yucca Mountain Site Characterization Plan, DOE/RW-0199, became an infamous example of an unconstrained research agenda.

33. A former leader of the Organization for Economic Cooperation and Development's Nuclear Energy Agency, the sponsor of these advisory bodies, threatened in the late 2010s to invite implementors to the group's meetings. Led by the United States and Sweden, the advisors were determined to meet independently of the Agency if the threat were carried out. It was not.
34. A prime example of the appearance—and possibly the reality—of collusion was the Nuclear Regulatory Commission staff's objection to virtually all of the more than 300 contentions filed by the States of Nevada and California in the Yucca Mountain adjudication.
35. A nod to the Secretary of Energy Advisory Board, "Earning Public Trust and Confidence: Requisites for Managing Radioactive Waste," 1993, for this formulation.

Index

Note: Page numbers in *italics* refer to figures.

A

B

C

D

E

S

T

U

W

Y

Latin America and the International Court of Justice

This book aims to evaluate the contribution of Latin America to the development of international law at the International Court of Justice (ICJ). This contemporary approach to international adjudication includes the historical contribution of the region to the development of international law through the emergence of international jurisdictions, as well as the procedural and material contribution of the cases submitted by or against Latin American states to the ICJ to the development of international law. The project then conceives international jurisdictions from a multifunctional perspective, which encompasses the Court as both an instrument of the parties and an organ of a value-based international community. This shows how Latin American states have become increasingly committed to the peaceful settlement of disputes and to the promotion of international law through adjudication. It culminates with an expansion of the traditional understanding of the function of the ICJ by Latin American states, including an analysis of existing challenges in the region.

The book will be of interest to all those interested in international dispute resolution, including academic libraries, the judiciary, practitioners in international law, government institutions, academics, and students alike.

Paula Wojcikiewicz Almeida is Professor of International Law and Jean Monnet Chair, sponsored by the European Commission at Fundação Getulio Vargas Law School in Rio de Janeiro, Brazil. She is an Associate Researcher at the Institut de Recherche en Droit International et Européen de la Sorbonne (IREDIES) and received her doctorate *summa cum laude* from the Sorbonne Law School, Université Paris 1, France.

Jean-Marc Sorel is Professor of Public Law at Sorbonne Law School, Université Paris 1, France. He is a former Director of the Institut de Recherche en Droit International et Européen de la Sorbonne (IREDIES), Counsel and Advocate of States before the International Court of Justice, and author or editor of 28 books and 150 articles in different fields of International Law.

Routledge Research in International Law
Series Editor: Professor Surya P. Subedi, O.B.E.

A full list of titles in this series is available at: www.routledge.com/Routledge-Research-in-International-Law/book-series/INTNLLAW

Power and Law in International Society
International Relations as the Sociology of International Law
Mark Klamberg

The International Criminal Court and Global Social Control in Late Modernity
International criminal justice in late modernity
Nerida Chazal

International Law and Boundary Disputes in Africa
Gbenga Oduntan

Means of Transportation and Registration of Nationality
Transportation registered by international organizations
Vincent P. Cogliati-Bantz

Reciprocity in International Law
Its impact and function
Shahrad Nasrolahi Fard

The Responsibility to Protect in International Law
An emerging paradigm shift
Susan Breau

Seeking Justice in International Law
The significance and implications of the UN Declaration on the Rights of Indigenous Peoples
Mauro Barelli

Legal Accountability and Britain's Wars 2000–2015
Peter Rowe

Fragmentation vs the Constitutionalisation of International Law
A practical inquiry
Edited by Andrzej Jakubowski and Karolina Wierczyńska

Prosecuting Juvenile Piracy Suspects
The International legal framework
Milena Sterio

Latin America and the International Court of Justice

Contributions to international law

Edited by
Paula Wojcikiewicz Almeida
and Jean-Marc Sorel

Routledge
Taylor & Francis Group
LONDON AND NEW YORK

First published 2017
by Routledge

2 Park Square, Milton Park, Abingdon, Oxfordshire OX14 4RN
711 Third Avenue, New York, NY 10017

Routledge is an imprint of the Taylor & Francis Group, an informa business

First issued in paperback 2018

British Library Cataloguing in Publication Data
A catalogue record for this book is available from the British Library

Library of Congress Cataloging in Publication Data
Names: Almeida, Paula Wojcikiewicz, 1980-editor. | Sorel, Jean-Marc, editor.
Title: *Latin America and the International Court of Justice: Contributions to International Law*/Edited by Paula Wojcikiewicz Almeida and Jean-Marc Sorel.
Description: New York: Routledge, 2016. | Series: Routledge research in international law | Includes bibliographical references and index.
Identifiers: LCCN 2016026067 | ISBN 978-1-138-85857-2 (hbk) | ISBN 978-1-315-71789-0 (ebk)
Subjects: LCSH: Latin America—Foreign relations. | International Court of Justice. | International law—Latin America.
Classification: LCC KZ4116 .L38 2016 | DDC 341.5/52098—dc23
LC record available at https://lccn.loc.gov/2016026067

ISBN: 978-1-138-85857-2 (hbk)
ISBN: 978-1-138-61374-4 (pbk)

Typeset in Galliard
by FiSH Books Ltd, Enfield

Contents

Notes on contributors

Contributors' biographies by order of appearance in the book:

Liliana Obregón is an Associate Professor at the Universidad de los Andes in Bogotá, Colombia and was a Postdoctoral Researcher at the University of Helsinki, Finland. She received an SJD at Harvard University; an MA, SAIS at The Johns Hopkins University (Bologna-Washington, DC); and a Law degree from the Universidad de los Andes.

Ricardo Abello-Galvis is Principal Professor of Public International Law of the Universidad del Rosario Law School in Bogotá, Colombia. He is a Member of the Permanent Court of Arbitration (PCA) (2014–2019); Director/Editor of the Anuario Colombiano de Derecho Internacional (ACDI); and Former President of the Colombian Academy of International Law.

Walter Arévalo-Ramírez is a Professor of Public International Law – Comparative Law and a Member of Anuario Colombiano de Derecho Internacional (ACDI). He received LLM Master of Laws in International Law (Summa Cum Laude) and LLB Political Scientist, Postgraduate in Constitutional Law at Stetson College of Law, Gulfport, Florida and obtained his PhD at the Universidad del Rosario Law School in Bogotá.

Antônio Augusto Cançado Trindade is a Judge of the International Court of Justice, a Former President of the Inter-American Court of Human Rights, Emeritus Professor of International Law of the University of Brasilia and of Brazil's Diplomatic Institute Rio Branco, a Member of the Curatorium of the Hague Academy of International Law and of the Institut de Droit International.

Rosa Riquelme is a Professor of International Law at the University of Murcia (Spain) and a Visiting Fellow at the Human Rights Implementation Centre of the University of Bristol, UK (2013).

Miriam Cohen is an Assistant Professor at Bora Laskin Faculty of Law at Lakehead University in Thunder Bay, Ontario, Canada and formerly served as Associate Legal Officer at the International Court of Justice.

María Teresa Infante Caffi is a Professor of International Law at the University of Chile, a former Director of Frontiers and Limits Ministry of Foreign Affairs and Former Director of the Institute of International Studies University of Chile.

Karel Wellens is an Emeritus Professor of Public International Law at the University of Nijmegen in the Netherlands. His most recent publication is *Negotiations in the Case Law of the International Court of Justice, A Functional Analysis.*

Leonardo Nemer Caldeira Brant is Professor of International Law at UFMG and PUC Minas, Belo Horizonte, Brazil. He is President of the Brazilian International Law Center (CEDIN), Chief Editor of the *Brazilian Yearbook of International Law* and served as legal advisor to the International Court of Justice.

Beatrice I. Bonafé (PhD, EUI) is an Associate Professor of Public International Law at Sapienza University of Rome, Italy, where she teaches International Dispute Settlement and International Organizations Law.

Luis González García is a member of Matrix Chambers, UK. He specialises in international dispute resolution and advises and acts for governments, international organisations and corporations in international law matters. Prior to joining Matrix, he acted as counsel and lead counsel for Mexico in NAFTA arbitrations.

Karin Oellers-Frahm is a Senior Research Fellow at the Max-Planck Institute for Comparative Public Law and International Law in Heidelberg, Germany. She has written widely on issues of international law, focusing in particular on aspects of public international law and the International Court of Justice.

Cristina Hoss is a Legal Adviser to Judge Bruno Simma at the Iran–US Claims Tribunal. Prior to joining the Tribunal, she was a Legal Officer at the International Court of Justice and a Research Fellow at the Max-Planck Institute in Heidelberg, Germany.

Pablo Sandonato de León is an Associate with a Swiss law firm and was formerly Associate Professor of Public International Law (Uruguay). He is a Member of the Bar of Uruguay and of the Geneva Bar Association ('Ordre des Avocats de Genève').

Guy S. Goodwin-Gill is Emeritus Fellow of All Souls College, Oxford, Emeritus Professor of International Refugee Law and Honorary Associate of the Refugee Studies Centre, University of Oxford. He is also a barrister at Blackstone Chambers, London, where he practises in public international law, human rights, citizenship, refugee and asylum law.

Cesare P.R. Romano is a Professor of Law and W. Joseph Ford Fellow at Loyola Law School, Los Angeles, US, and a Senior Fellow, iCourts, University of Copenhagen, Denmark, and Pluricourts, University of Oslo, Norway.

Alfred M. Boll, SJD, is a United States Foreign Service Officer currently serving as Deputy Director of the Department of State's Office of International Migration. He is the author of *Multiple Nationality and International Law* (2007). He was previously a legal adviser to the International Committee of the Red Cross.

Efthymios Papastavridis is a Postdoctoral Research Assistant in International Law of the Sea at the Faculty of Law of Oxford University, a part-time Lecturer in Public International Law at the Democritus University of Thrace and a Research Fellow at the Academy of Athens.

Virginie J.M. Tassin is a lawyer admitted to the Paris Bar, an award-winning specialist in Law of the Sea (INDEMER Prize 2011), and a Researcher at DIRNAT (Brazil). She served in the International Tribunal for the Law of the Sea's Legal Office as an Associate Legal Officer.

Tullio Scovazzi is a Professor of International Law at the University of Milano-Bicocca, Milan, Italy.

Chie Kojima, PhD (Chuo), LLM (Yale) and JSD (Yale), is an Associate Professor in the Faculty of Law at Musashino University in Tokyo, Japan.

Yoshifumi Tanaka, PhD and DES (Graduate Institute, Geneva), LLM (Hitotsubashi University, Tokyo) is a Professor in the Faculty of Law at the University of Copenhagen in Denmark.

Antonio Remiro Brotóns is an Emeritus Professor of Public International Law in the Universidad Autonoma of Madrid, Spain, and a member of the Institut de Droit International.

Ximena Fuentes is a Professor of Public International Law at Universidad de Chile.

Mariana Durney, LLM int., (Heidelberg University/University of Chile) is a Professor at the Diplomatic Academy of Chile.

Martin Dawidowicz, PhD (Cantab), M.Jur (Oxon), Jur.Kand (Stockholm) is a Departmental Lecturer in Public International Law at the University of Oxford, UK.

Sandrine Maljean-Dubois is a Researcher at the Centre National de la Recherche Scientifique (CNRS) and teaches International and European Environmental Law at the Faculty of Law and Political Sciences of Aix-Marseille University, France.

Vanessa Richard is a Researcher at the Centre National de la Recherche Scientifique (CNRS) specialising in international water law and international environmental law. She has been awarded an ERC Grant for the International Grievance Mechanisms and International Law & Governance (IGMs) project (ERC GA no. 312514.)

List of abbreviations

Institutions and bodies

AFDI	American Freedom Defense Initiative
AG	Spanish abbreviation for 'Asamblea General'
Aladi	Spanish abbreviation for 'Asociación latino-americana de integración'
APRA	Spanish abbreviation for 'Allianza Popular Revolucionaria Americana'
ASIL	American Society of International Law
CACJ	Central American Court of Justice
CAJP	Spanish abbreviation for 'Comisión de Asuntos Jurídicos y Políticos'
CAN	Spanish abbreviation for 'Comunidad Andina'
CARU	Joint Commission of the River Uruguay
CEESI	Spanish abbreviation for 'Comisión Especial para Estudiar el Sistema Interamericano y Proponer Medidas para su Reestructuración'
CELAC	Community of Latin American and Caribbean States
CIJ	French abbreviation for 'Cour internationale de justice'/ Spanish abbreviation for 'Corte Internacional de Justicia'
CLCS	Commission on the Limits of the Continental Shelf
CMB	Celulosas de M'Bopicuá S.A.
CP	Consejo Permanente
DINAMA	Uruguayan National Directorate for the Environment
ECHR	European Court of Human Rights
EIA	Environmental Impact Assessment
ELSI	Elletronica Sicula S.p.A.
ENCE	Spanish abbreviation for 'Empresa Nacional de Celulosas de España'
EWCA Civ	Court of Appeal of England and Wales, Civil Division
GA	General Assembly
GTAN	Spanish abbreviation for 'Grupo Técnico de Alto Nivel'
ILA	International Law Association

ICJ	International Court of Justice
IDI	French abbreviation for 'Institut de Droit International'
ILC	International Law Commission
IMF	International Monetary Fund
ISIS/ISIL	Islamic State of Iraq and Syria/Islamic State of Iraq and the Levant
ITLOS	International Tribunal for the Law of the Sea
I.U.S.C.T.	Iran–US Claims Tribunal
Mercosur	Spanish abbreviation for 'Mercado Común del Sur'
MRE	Spanish abbreviation for 'Ministerio de Relaciones Exteriores'/Portuguese abbreviation for 'Ministério de Relações Exteriores'
OAS	Organization of American States
ODECA	Organización de Estados Centroamericanos (Organization of Central American States)
OEA	Spanish abbreviation for 'Organización de los Estados Americanos'
PCA	Permanent Court of Arbitration
PCIJ	Permanent Court of International Justice
SC	Security Council
SDN	French abbreviation for 'Société des nations'
SICA	French abbreviation for 'Système d'intégration centraméricain'/Spanish abbreviation for 'Sistema de la Integración Centro Americana'
SRE	Spanish abbreviation for 'Secretaría de Relaciones Exteriores'
SRFC	Sub-Regional Fisheries Commission
TAHM	Spanish abbreviation for 'Tribunal arbitral ad hoc del Mercosur'
UN	United Nations
UNC	United Nations Charter
UNCLOS	United Nations Convention on the Law of the Sea
UNGA	United Nations General Assembly
UNITAR	United Nations Institute for Training and Research
VCCR	Vienna Convention on Consular Relations
WHO	World Health Organization

Journals, publishers and conventions

ACDI/CYIL	Anuario Colombiano de Derecho Internacional/ Colombian Yearbook of International Law
AFDI	Annuaire français de droit international
AJIL	American Journal of International Law
ARSIWA	Articles on the Responsibility of States for Internationally Wrongful Acts

BYIL	British Yearbook of International Law
CEDIN	Centro de Direito Internacional
CUP	Cambridge University Press
EJIL	European Journal of International Law
Eur J Int Law	European Journal of International Law
FILJ	Fordham International Law Journal
HHRJ	Harvard Human Rights Journal
ICLQ	International and Comparative Law Quarterly
IFC	Institución Fernando El Católico
I.U.S.C.T.R.	Iran-US Claims Tribunal Reports
LJIL	Leiden Journal of International Law
LOSC	Law of the Sea Convention
MPEPIL	Max Planck Encyclopedia of Public International Law
OUP	Oxford University Press
PUF	Presses Universitaires de France
RCADI	Recueil des cours de l'Académie de droit international
Rec.	Recueil
RGDIP	Revue générale de droit international public
RIAA	Reports of International Arbitral Awards
RTNU	Recueil des traités des Nations Unies
UNJYB	United Nations Juridical Yearbook
UNTS	United Nations Treaty Series
VCLT	Vienna Convention on the Law of Treaties
YbILC	Yearbook of the International Law Commission
ZaöRV	Zeitschrift für ausländisches öffentliches Recht und Völkerrecht

List of maps

Editors' introduction

Latin America[1] prides itself in being a peaceful region. Historically resistant to any kind of external interference, Latin American states have tended toward the principles of national sovereignty, non-intervention, and peaceful settlement of disputes, which are deeply embedded in their political and juridical cultures. These principles were recognised in the agreements that established the Organization of American States (OAS) in 1948 and have also been codified in the OAS Charter. Moreover, they are strongly rooted in the region's diplomatic and legal cultures, as enshrined in many Latin American constitutions.

While Latin American states have always opted for the peaceful settlement of disputes, they have more generally contributed to the development of international law from this perspective. Their universal presence at the 1907 Second Hague Peace Conference and contribution to the work and outcome of the conference are well known: they encouraged the recourse to arbitration and non-use of force, the principle of juridical equality of states, the strengthening of international jurisdiction, and the direct access of individuals to international justice.[2] During the 1907 Hague Conference, these states advanced the idea that international courts should not operate on the exclusive basis of the consent of disputing states.[3]

1 Latin America is here understood as to cover all American states resulting from Spanish, Portuguese, and French colonization. According to Liliana Obregón, 'the concept of "*Latin* America" emerged in the second half of the nineteenth century as a creole adaptation of French "Latinism" in response to the USA's expansionist policy' (Obregon, Liliana (2016) 'Identity formation, theorization and decline of a Latin American international law'. In: Wojcikiewicz Almeida, Paula and Sorel, Jean-Marc (eds) *Latin America and the International Court of Justice: Contributions to International Law*, Oxford: Routledge, pp. 3–14).

2 Trindade, Antônio A. Cançado (2008) 'The presence and participation of Latin America at the Second Hague Peace Conference of 1907'. In: Daudet, Yves, *Topicality of the 1907 Hague Conference, The Second Hague Peace Conference*, Leiden/London: Martinus Nijhoff Publishers.

3 Société des Nations/CPJI, Documents au sujet de mesures prises par le Conseil de la Société des Nations aux termes de l'article 14 du Pacte et de l'adoption par l'Assemblée du Statut de la Cour permanente, Geneva, SDN/CPJI, 1920, p. 142. *Dissenting opinion of*

The old debate regarding Latin American international law also demonstrates the region's willingness to influence the development of norms in the field of international law.[4] Concretely, Latin Americans have advanced recognition of the compulsory jurisdiction of the Permanent Court of International Justice (PCIJ) and the future International Court of Justice (ICJ). The ingenuous formula, known as 'declarations recognizing the jurisdiction of the Court as compulsory,' was proposed by the Brazilian jurist Raul Fernandes in order to overcome a deadlock within the Advisory Committee of Jurists responsible for drafting the Statute of the PCIJ. The referred formula contributed to attracting the acceptance of compulsory jurisdiction of the PCIJ by a total of 45 states and was firmly supported by Latin American states. The same Latin American formulation of 1920 was maintained in the present Statute of the ICJ due to the intransigent position of the more powerful states.

It followed that after the Second World War, several international organizations were created in Latin America, such as the OAS, the Central American Integration System (SICA), the Andean Community (CAN), the Common Market of South America (Mercosur), and, more recently, the Union of South American Nations (UNASUR). Most of these organizations developed their own dispute settlement mechanisms, enhancing the region's tradition of peaceful settlement of disputes. The same rule applies to the UN, under the auspices of which Latin American states have been strongly active. Indeed, these states are involved in 29 cases since the creation of the ICJ, most of which concern maritime and territorial disputes.

In this context, the present book is focused on the contribution of Latin America to the ICJ since these states tend to submit their disputes to the World Court, even though other regional or sub-regional dispute settlement systems are eminently capable of solving matters. The fact that Latin American states are currently the most active litigants before the Court shows their belief in it as a legitimate public authority that not only resolves disputes among states but also, more generally, contributes to the development of international law.[5] The project then conceives of the Court as a

Judge Cançado Trindade, Application of the International Convention on the Elimination of All Forms of Racial Discrimination (Georgia v. Russian Federation), Preliminary Objections, Judgment, I.C.J. Reports 2011, p. 70.

4 Lorca, A. B. (2006) 'International Law in Latin America or Latin American International Law? Rise, Fall, and Retrieval of a Tradition of Legal Thinking and Political Imagination', *Harvard International Law Journal*, 47.1: 284–305.

5 For an assessment of the idea of international public authority, see Von Bogdandy, Armin and Venzke, Ingo (2012) 'In Whose Name? An Investigation of International Courts' Public Authority and Its Democratic Justification', *The European Journal of International Law*, 23.1: 7–41; Von Bogdandy, Armin and Venzke, Ingo (2012) 'On the Functions of International Courts: An Appraisal of Their Burgeoning Public Authority', *ACIL Research Paper* No 2012-10: 1–29; Von Bogdandy, Armin and Venzke, Ingo (eds) (2011) 'Beyond Dispute: International Judicial Institutions as Lawmakers', Special Issue, *German Law Journal*, 979: 12; Von Bogdandy, Armin, Dann, Philipp, and Goldmann, Matthias (2008)

multifunctional actor responsible for the protection and development of the international community and its values.[6]

This contemporary approach to international judicial institutions goes beyond the traditional understanding that depicts international courts only as adjudication bodies limited to the bilateral dispute settlement between states. It includes their contribution to social interaction with the intention of taking into account other important functions that international courts currently perform in order to achieve international adjudication's full potential, which is the realization of justice.[7] Dispute settlement remains an important contribution, as the Court is responsible for the pacific settlement of international disputes (Chapter VI of the UN Charter). However, this traditional function cannot solely capture the full relevance of ICJ decisions since there are disputes that may continue even after a decision has been rendered.[8]

The *Nicaragua case* represents a famous example of how the function of international adjudication at the ICJ is not limited to resolving disputes between states. Although unable to reach a settlement between Nicaragua and the US, this judgment exhorted the principle of prohibition of the use of force and restricted the possibility of states in acting on the right to self-defence.[9] This interpretation reaches beyond the individual case submitted to the ICJ, serving as a precedent to future case law. It also illustrates the Court's influence on the development of international law and its capability to act as an 'agent' of legal development.[10]

The *Avena case* also demonstrates that non-compliance with ICJ judgments does not compromise its contribution to international law: the case has indeed contributed to clarify the meaning of 'without delay' under the Vienna Convention on Consular Relations and has been considered as a turning point in the discussions of how international law and domestic law interface, and what place international law occupies within the US legal

'Developing the Publicness of Public International Law: Towards a Legal Framework for Global Governance Activities', *German Law Journal*, 9: 1375. See also: Lauterpacht, Sir Hersch (1958) *The Development of International Law by the International Court*, London: Stevens & Sons Limited; Trindade, Antônio A. Cançado (2010) *International Law for Humankind, Towards a New* Jus Gentium, Leiden/Boston: Martinus Nijhoff Publishers.

6 Von Bogdandy, Armin and Venzke, Ingo (2012) 'In Whose Name? …', op. cit., p. 38.

7 Functions can be understood as contributions of the particular 'institution or its characteristic activity to a larger whole' (Von Bogdandy, Armin and Venzke, Ingo (2014) 'In Whose Name? A Public Law Theory of International Adjudication', Oxford: Oxford University Press, p. 11). See also Von Bogdandy, Armin and Venzke, Ingo (2012) 'On the Functions of International Courts…', op cit., pp. 4–5.

8 Ibid. p. 15.

9 Papastavridis, Efthymios (2016) '*Military and Paramilitary Activities in and against Nicaragua (Nicaragua v. United States of America)*'. In: Wojcikiewicz Almeida, Paula and Sorel, Jean-Marc (eds), op. cit., pp. 211–222.

10 See Tams, Christian, and Tzanakopoulos, Antonious (2010) 'Barcelona Traction at 40: The ICJ as an Agent of Legal Development', *Leiden Journal of International Law*, 23: 782. See also Lauterpacht, Sir Hersch (1958) *The Development of International Law by the International Court*, London: Stevens & Sons Limited.

system.[11] In the case concerning *Territorial and Maritime Dispute (Nicaragua v. Colombia)*, although Nicaragua introduced a new claim against Colombia shortly before the effective denunciation of the Pact of Bogota by Colombia, this case allowed the ICJ to address and analyze, for the first time in the history of the international settlement of disputes, the applicable law to the delimitation of the extended continental shelf between an UNCLOS State Party (Nicaragua) and a non-State Party (Colombia).[12] These are only examples among others in which Latin American cases clearly prompted the material development of international law.

It is worth emphasizing that the Latin American contribution is not limited to material international law. Indeed, the Court's procedural rules on third party intervention (discretional intervention and the right to intervene) and on demands for judicial interpretation must be interpreted more broadly. These rules are the product of the Court's own making, being an important source of its independence.[13] By promoting wider participation in judicial proceedings before the Court, Latin American states recognized that the effects created by judgments go far beyond the parties immediately involved in a particular dispute. They finally contribute to reasserting law's validity and to expanding the Court's traditional function.

Ultimately, the Court's decisions have significantly affected the region, allowing the effects of international adjudication on domestic governance to become visible.[14] This is particularly applicable to Latin American cases before the ICJ as the Court's decisions have significantly affected the course of regional relations and the construction of a regional legal framework. The assessment of the impact of ICJ's decisions on the region includes the evaluation of how governments, parliaments, national courts, and other domestic actors and institutions interact with the authority of ICJ decisions.

The book makes accessible for the first time an up-to-date systematic and comprehensive analysis of the Latin American contribution to international law and to the peaceful settlement of disputes before the ICJ from a multifunctional perspective. It includes the historical contribution of the region to the development of international law through the emergence of international jurisdictions (Part I), as well as both the procedural (Part II) and material contribution (Part III) of the cases submitted by or against Latin American states to the ICJ to the development of international law.

11 Romano, Cesare (2016) '*Avena and Other Mexican Nationals (Mexico v. United States of America), 2003*'. In: Wojcikiewicz Almeida, Paula and Sorel, Jean-Marc (eds), op. cit., pp. 183–194.

12 Tassin, Virginie (2016) '*Territorial and Maritime Dispute (Nicaragua v. Colombia)*'. In: Wojcikiewicz Almeida, Paula and Sorel, Jean-Marc (eds), op. cit., pp. 225–236.

13 Sorel, Jean-Marc (2009) 'International Courts and Tribunals, Procedure'. In: R. Wolfrum (ed.), *Max Planck Encyclopaedia Public International Law*; Oxford University Press, online (www.mpepil.com), 2008, 11p. Von Bogdandy, Armin and Venzke, Ingo (2012) 'In Whose Name? …', op. cit., p. 25.

14 Von Bogdandy, Armin and Venzke, Ingo (2014) 'In Whose Name? …', op. cit., pp. 21–22.

This regional approach was explored by contributors coming from diverse continents and backgrounds who were invited to participate in this project. The analysis undertaken by the invited contributors, be they academics or practitioners, took into account two complementary aspects: the procedural or material contribution of the cases to international law and their impact on the region. It intends to show how Latin American states have become increasingly committed to the peaceful settlement of disputes and to the promotion of international law through adjudication. It culminates with an expansion of the traditional understanding of the function of the ICJ by Latin American states, including an analysis of existing challenges in the region.

Paula Wojcikiewicz Almeida
Getulio Vargas Foundation Law School, Rio de Janeiro, Brazil

Jean-Marc Sorel
École de droit de la Sorbonne, Paris, France

Acknowledgements

I would like to first thank Dr. Antônio Augusto Cançado Trindade, judge at the International Court of Justice, for his constant guidance, personal attention and constructive suggestions during the planning and development of this book.

I am also grateful to Dr. Karin Oellers-Frahm, Senior Research fellow at the Max Planck Institute for Comparative Public Law and International Law for her constant encouragement and full support for the conception of this Project. Her willingness to give her time so generously has been very much appreciated. I would also like to express my very great appreciation to Dr. Armin von Bogdandy, Director at the Institute, and Mariela Morales Antoniazzi, Senior Research fellow of the referred institute for making it possible for me to work as a post-doctoral visiting fellow at the Institute where I devoted most of my time researching for this project. My grateful thanks are also extended to Dr. Anne Peters, Managing Director of the referred Max Planck Institute, for her advice and enthusiastic encouragement of this research.

I would also like to express my gratitude to Dr. Eduardo Valencia-Ospina, Member of the International Law Commission and former president of the Latin American Society of International Law (LASIL), for moderating a pre-launch event devoted to this book during the LASIL Biennial Conference, in August 2016, in Santiago del Chile.

Finally, I wish to warmly thank all contributors for their full commitment and genuine interest in this project.

Paula Wojcikiewicz Almeida

Part I

The historical contribution of Latin America to the development of international law through adjudication

1 Identity formation, theorization and decline of a Latin American international law

Liliana Obregón[1]

I Latin American proposals for international justice

Many Latin American heads of state, diplomats, and lawyers actively participated in the development of a permanent system of international justice. During the 1907 Hague Conference, half of the forty-four states were from the region and eager to promote an international court of justice.[2] However, Ruy Barbosa, Brazil's Minister of Foreign Affairs and head of his delegation, criticized the 'Great Power representation' proposal for the court because by institutionalizing world hierarchies, it violated the equality principle and would lead to decision deadlock and an arms race.[3] Barbosa's position brought him regional acclaim and the 'Hague Eagle' nickname.[4] After that, most Latin Americans states ratified the 1907 Convention for the Pacific Settlement of International Disputes whereas no other non-European states ratified it until after World War II.[5]

Soon thereafter, the Central American Peace Conference proposed to establish a Central American Court of Justice as a permanent tribunal with compulsory jurisdiction, continuous functions, and equal representation of justices from each state.[6] Eleven Latin American states were part of the thirty-two League of Nations' founders, and six (Argentina, Chile, Colombia, Paraguay, El Salvador, and Venezuela) were invited to accede the

1 The author thanks Nicolás Forero-Villarreal for research and editing assistance.

2 W. Hull, *The Two Hague Conferences and Their Contributions to International Law*, Boston: Ginn and Cpy, 1908, p. 14.

3 R. Barbosa, 'The Equality of Sovereign States,' *The Independent* 64, 1907, 75; A. Becker-Lorca, *Mestizo International Law*, Cambridge: CUP, 2015, pp. 158–164.

4 S. Dennison, *Joaquim Nabuco: Monarchism, Panamericanism and Nation-Building in the Brazilian Belle Epoque*, Bern: Peter Lang, 2006, p. 180.

5 Convention for the Pacific Settlement of Disputes, 18 October 1907, 36 US Stat. 2259, UKTS 1910 No. 6.

6 J. Esquirol, 'Latin America', in B. Fassbender, A. Peters (ed.) *The Oxford Handbook of the History of International Law*, Oxford: OUP, 2012, pp. 553–572; M. Hudson, 'The Central American Court of Justice,' *AJIL* 26, 1932, p. 761.

covenant.[7] Latin America had the highest regional representation (20/51) of the UN founding members in which they continued to promote a permanent court of justice.[8]

Regional solidarity has been a constant since colonial times, as Latin Americans often used their ideas of exceptionalism in response to external domination.[9] The continuous reconstruction of identities allowed for a variety of projects, including a regional perspective of international law, in which Latin Americans searched for inclusion despite the discipline's Eurocentrism.[10] This chapter reviews the stages of identity (nineteenth century) and theorization and decline (twentieth century) of a Latin American international law.

II Early nineteenth century: Spanish Americans and the law of nations

Immanuel Kant's 'On Perpetual Peace' (1795) argued for a confederation of states as a solution to the problem of war, an idea taken by independence leader Simón Bolívar who in 1815 wrote about Spanish American unification.[11] In 1824, Bolívar sent out invitations for the 1826 Panama Congress to bring the former Spanish colonies together as an 'union, league and confederation,' an 'eternal' government that would celebrate Columbus' world, make diplomatic alliances and international treaties, and provide advice and mediation for regional conflicts in a 'society of sister nations.'[12]

7 Annex The Versailles Treaty June 28, 1919: Part I. Online. Available at: http://avalon.law.yale.edu/imt/parti.asp

8 International courts, nonetheless, are often viewed as politically motivated despite the non-national basis selection criteria for judges. See, G. Hernández, *The International Court of Justice and the Judicial Function*, Oxford: OUP, 2014, p. 6.

9 E. Dussel, 'Philosophy in Latin America in the Twentieth Century: Problems and Currents,' in E. Mendieta (ed.) *Latin American Philosophy Currents, Issues, Debates*, Bloomington: Indiana University Press, 2003, p. 15. In a recent work, Dussel extended the time and geography of Latin American philosophy from 1300 to include the Caribbean and U.S. Latinos. E. Dussel, E. Mendieta and C. Bohórquez, *El pensamiento filosófico latinoamericano, del Caribe y "latino" (1300–2000)*, México: Siglo XXI, 2011.

10 L. Lund, 'Barbarian Theorizing and the Limits of Latin American Exceptionalism,' *Cultural Critique* 47, 2001, 54–90, 76.

11 I. González Niño, *Simón Bolívar, precursor del derecho internacional americano*, Bogotá: Instituto Colombiano de Estudios Latinoamericanos y del Caribe, 1985, pp. 1–131; M. Mackenzie, *Los ideales de Bolívar en el derecho internacional americano*, Bogotá: Biblioteca del Ministerio de Gobierno: Colección Bolivariana, 1955, pp. 1–548; J.M. Torres-Caicedo, *Unión latino-americana, pensamiento de Bolívar para formar una liga americana. Su origen y sus desarrollos (...)*, Paris: Rosa y Bouret, 1865, pp. 1–385; J.M. Yepes, *Del Congreso de Panamá a la Conferencia de Caracas, 1826–1954: el genio de Bolívar a través de la historia de las relaciones interamericanas*, Caracas: Cromotipo, 1955.

12 A. De la Peña y Reyes, *El Congreso de Panamá y algunos otros proyectos de unión hispanoamericana*, México: Publicaciones de la Secretaría de Relaciones Exteriores, 1926, pp. 1–262; G.A. De la Reza, *El Congreso de Panamá de 1826 y otros ensayos de integración latinoamericana en el siglo XIX: estudio y fuentes documentales anotadas*, México:

For Bolívar, the Panama Isthmus was an ideal venue because it was situated 'midway between Mexico City and Buenos Aires;' had symbolic value as the 'centre of the world,' and, was a geographic and political 'parallel to the Isthmus of Corinth.'[13] Bolívar's union protected a fragile independence in face of still-threatening Spain and uncertain Europe as the congress was 'destined to form the widest, most extraordinary ... league ever appeared to this day on earth ... The relations of political societies will receive a code of public law that will rule universal behaviour ... the New World will be constituted by independent nations, bound together by a common law defined by their external relations.'[14]

Although the Panama Congress never achieved Bolívar's expectations, such a regional utopia may shed light into what I have described elsewhere as a 'creole legal consciousness' that allowed for the building of a Latin American international law. By the time of independence, close to 17 million people, inheritors of three hundred years of a complex caste system of political, legal, and social stratification, lived in the former Spanish colonies.[15] The legal and political elite was composed of Spanish and creole (*criollo*) functionaries. In the sixteenth century, *criollo* was the colonizers' pejorative term for a person born in America of Spanish parents, considered as an impure or defective European due to suspect mixture with the indigenous or slave population. The interactive and dialogic character of the *criollo* makes it misleading to define it as a racial or biological category, monolithic subject, or homogenous elite. Therefore, I prefer to ascribe creoles a social and legal standing because – like other colonial subjects – they were defined in legal terms through separate jurisdictions, privileges, and restrictions. As Spanish Americans, creoles were a minority of the Spanish Republic that had access to education and land ownership. Should other social conditions be met (such as honour, purity of blood, legitimacy of birth, family and business connections), *criollos* were considered worthier than *castas* (mixed peoples),

UNAM-Azcapotzalco y Ediciones y Gráficos Eón, 2006; E.Vieira Posada, *La integración de América Latina: del Congreso Anfictiónico de Panamá en 1826 a una comunidad Latinoamericana o Sudamericana de Naciones en el año 2010*, Bogotá: Editorial Univ. Javeriana, 2004, p. 43.

13 S. Bolívar, *Coleccion de documentos relativos a la vida pública del libertador de Colombia y del Perú para servir a la historia de la independencia del suramérica: Tomo Tercero*, Carácas: Impresa de Devisme Hermanos, 1826, p. 178.

14 G.A. De la Reza, 'Simon Bolívar: Un pensamiento sobre el Congreso Anfictiónico de Panamá', in G.A. De la Reza (ed.) *Documentos sobre el Congreso Anfictiónico de Panamá*, Carácas: Fundación Biblioteca Ayacucho/Banco Central de Venezuela, 2010, pp. 84–5.

15 Though numbers should be viewed with caution considering the limits of census and resistance to racial categories 7.5 million Indians, five million castes (or *mestizos*, 'mixed' peoples), three million creoles, and 800,000 slaves of African descent were said to live in the region. See, M. Lucena Giraldo, *A los cuatro vientos: las ciudades de la América hispánica*, Madrid: Marcial Pons Historia, 2006, p. 245; D.J. Robinson (ed.), 'Social Fabric and Spatial Structure in Colonial Latin America,' in *Dellplain Latin American Studies* 1, Department of Geography, Syracuse University, 1979, pp. 17–22.

Indians (as part of the 'Republic of Indians'), and slaves or free blacks. Creoles viewed themselves as persons of Spanish descent and of 'white' culture and physical presence, but also as American patriots.[16]

By the independence period, creole patriotism continued to permeate the elite's regional identity. Bolívar best explained the creole's dilemma in a famous phrase:

> We ... are not Indians nor Europeans, but a middle species between the legitimate owners of the country, and the Spanish usurpers, in short, we being Americans by birth, and our rights derived from Europe, we have to dispute this land against the natives of this country and against the invaders ... so we are in the most extraordinary and complicated case.[17]

Bolívar viewed the creole minority as natural leaders, indigenous peoples as litigious opponents of land rights, Europeans as potential invaders, and he did not acknowledge *mestizos* or blacks as citizens, despite constituting the majority population.[18] Thus, the idea of a union based on a common origin, customs, and language was an extraordinary utopia but Bolívar proceeded to send invitations to Bolivia, Central America, Chile, Colombia, Empire of Brazil, Mexico, Peru, and the United Provinces of Río de la Plata.[19] The US,[20] Great Britain, and the Netherlands were invited as observers.[21]

16 On the construction of the category of the creole; see E. Kuznesof, 'Ethnic and Gender Influences on Spanish Creole Society in Colonial Spanish America,' *Colonial Latin American Review* 4, 1995, pp. 153–68; B. Lavalle, *Las promesas ambiguas: ensayos sobre el criollismo colonial en los Andes*, Lima: Pontificia Universidad Católica, 1993; J.A. Mazzotti (ed.), *Agencias criollas: la ambigüedad 'colonial' en las letras hispanoamericanas*, Pittsburgh: Instituto Internacional de Literatura Iberoamericana, 2000. See also L. Obregón, 'Between Civilisation and Barbarism: Creole interventions in international Law,' *Third World Quarterly* 27, 2006, 818–819.

17 S. Bolívar, 'Contestación de un americano meridional a un caballero de esta isla (Jamaica)', in *Coleccion de documentos relativos à la vida pública del libertador de Colombia y del Perú para servir a la historia de la independencia del suramérica: Tomo Vigésimo Primero*, Carácas: Impresa de GF Devisme, 1830, p. 21.

18 *Mestizo* was a racial and legal category used to describe people of mixed (white, indigenous, African) ancestry.

19 D. Bushnell, *Simón Bolívar: hombre de Caracas, proyecto de América: una biografía*, Buenos Aires: Editorial Biblos, 2002, p. 130.

20 Letter of Henry Clay, Secretary of State, to Richard C. Anderson, US Minister to Colombia, in W.R. Manning (ed.), *Diplomatic Correspondence of the US Concerning the Independence of the Latin-American Nations*, New York: OUP, 1925, Document 145, p. 253. For a discussion on US participation in Panama, see J.J. Malanson, 'The Congressional Debate over US Participation in the Congress of Panama, 1825–1826: Washington's Farewell Address, Monroe's Doctrine, and the Fundamental Principles of U.S. Foreign Policy,' *Diplomatic History* 30, 2006, 813–838.

21 J. Schoonhoven and C. Tymen De Jong, 'The Dutch Observer at the Congress of Panama in 1826', *The Hispanic American Historical Review* 36, 28–37. British citizens were invited to support the union through their 'character and habits' and received commercial incentives and equal citizenship. G. De la Reza, op. cit., p. 84.

Haiti's exclusion from the Congress of Panama gives further insight into creole identity. In 1816 three decades into the revolution, and ten years after its declaration of independence, Haiti's first president, Alexandre Pétion, supplied Bolívar with men, ships, money, guns, and a printing press in return for his vow to free all slaves.[22] Though Bolívar recruited slaves in exchange for freedom[23] and honoured Pétion as 'the author of our independence,'[24] he never complied with his promise.[25] The Haitian government expected an invitation to Panama, but Bolívar and other creoles feared that an implied recognition would promote slave uprisings across the continent.

Six years after the Panama Congress, the Venezuelan, Andrés Bello, wrote the first international law treatise published in the Americas, under the title *Principios del Derecho de Gentes.* The book was later titled *Principios del Derecho Internacional* in its second (1844) and third (1864) editions. In this book, Bello appropriated European and US texts and rewrote, edited, and organized it with a Spanish American perspective as part of his region-building project.[26] Bello's treatise was the most reprinted, distributed, plagiarized, and taught in the Spanish-speaking world and anticipated similar courses in Europe or the US.[27]

Bolívar, Bello, and other creole's efforts to promote regional perspectives of the law of nations or international law could be understood as belonging to a 'creole legal consciousness,' or a set of ideas on the law shared by creole

22 S. Bolívar, 'Latin-American Correspondence: Simon Bolívar and Alexandre Sabes Pétion,' *Phylon* 7 (2), 1946, 196.

23 S. Bolívar, 'Decreto sobre libertad de los esclavos – Carúpano, 2 de junio de 1816,' in V. Lecuna (ed.), *Bolívar, Simón: Proclamas y discursos del Libertador, 1811–1830*, Los Teques: Biblioteca De Autores y Temas Mirandinos, 1983, pp. 148–149.

24 S. Bolívar, Carta al General Alejandro Pétion, Feb 8, 1816, *Escritos del Libertador*, Vol 9, Sociedad Bolivariana de Venezuela: Caracas, p. 13.

25 None of the states represented in the Panama Congress abolished slavery. Slavery and the slave trade lasted the nineteenth century. Mexico abolished slavery in 1829, Argentina in 1853, Peru in 1855, and Brazil in 1888. Bolívar's territories (Colombia, Panama, Ecuador, and Venezuela) abolished slavery between 1851 and 1854, but paid compensation to slave owners.

26 I. Jaksic, *Andrés Bello: la pasión por el orden*, Santiago de Chile: Editorial Universitaria/Imagen De Chile, 2001, p. 323; L. Obregón, 'Construyendo la región americana: Andrés Bello y el derecho internacional,' in S.B. González and J. Poblete (eds), *Andrés Bello y los estudios latinoamericanos*, Pittsburgh: Instituto Internacional de Literatura Iberoamericana, 2009, pp. 189–18.

27 In France, Germany, and England the first international law courses were taught in the second half of the nineteenth century, while in the US they began with Yale in 1846, Harvard in 1863, and Columbia University in 1865. By 1907, ten of the 81 law schools in the US taught international law, a course which was considered a 'luxury' and continues to be an elective course in most US law schools. M. Koskenniemi, *The Gentle Civilizer of Nations: The Rise and Fall of International Law 1870–1960*, Cambridge: CUP 2001; J.M. Raymond and B.J. Frischholz, 'Lawyers Who Established International Law in the US, 1776–1914,' *AJIL* 76, 1982, 802–29.

lawyers in the post-independence period.[28] A 'creole legal consciousness' is characterized by the creoles' belief of superiority; a legacy of Indigenous and Spanish law; an understanding of Roman law as their historical and 'civilized' roots; and a practice of the law as in the use of foreign sources to solve local problems.[29]

Fundamental to the development of a creole legal consciousness in the nineteenth century was the appropriation of the European discourse of civilization, which expressed the idea of progress and the perfectibility of man as a universal ideal achieved through law and institutions in opposition to barbarism.[30] In order to avoid exclusion from the rights assigned (by Europe) to members of the 'community of civilized nations', creoles demonstrated a desire to 'improve civilization' through legal, economic, religious, educational, and social policies.[31]

III Late nineteenth century: Latin American professionals of international law

The global scenario changed mid-century. Until then, the creole elite considered the US as a successful model of statehood that inspired new constitutions, forms of government, and adopted the Monroe doctrine against European imperialism.[32] However, by mid-nineteenth century, the Monroe doctrine was used by the US in order to intervene in Mexico (1848), resulting in the loss of California and Texas. Another major intervention occurred in Nicaragua in 1855 with the purpose of reinstating slavery and annexing it to the US.

It is no surprise then that the concept of '*Latin* America' emerged in the second half of the nineteenth century as a response to US expansionist policies. French economist Michel Chevalier (1806–79) and Ernest Renan (1823–92) promoted the idea of *Pan-Latinism* as a union among 'Latin'

28 Legal Consciousness is a 'particular form of consciousness characteristic of advocates of a social group at a given time'; D. Kennedy, 'Toward a Historical Understanding of Legal Consciousness: The Case of Classical Legal Thought in America 1850–1940,' *Research in Law and Sociology* 3, 1980, 24–23.

29 L. Obregón, 'Completing Civilization: Creole Consciousness and International Law in Nineteenth-Century Latin America,' in A. Orford (ed.), *International Law and Its Others*, Cambridge: Cambridge University Press, 2006, pp. 247–64.

30 For an extensive conceptual history in international law, see L. Obregón, 'The Civilized and the Uncivilized in the History of International Law,' in B. Fassbender and A. Peters (eds), op. cit., pp. 917–39.

31 C. Rojas, *Civilization and Violence: Regimes of Representation in Nineteenth-Century Colombia*, Minneapolis: University of Minnesota Press, 2001.

32 US constitutionalism spread through A. de Tocqueville, *De la démocratie en Amérique*, Paris, 1836, Spanish translation by A.S. de Bustamante, *De la democracia en América del Norte*, Paris: Rosa, 1837, which had a broad circulation among creoles, see H.F. Zamudio (ed.), 'Función del poder judicial en los sistemas constitucionales Latinoaméricanos,' in *Latino América: constitución, proceso y derechos humanos*, México: Porra, 1988, p. 223.

states based on the ethnic and cultural categories of European historicism. Pan-Latinism projected France's opposition to the continental domination of Great Britain and the US. At the same time, Pan-Latinism was a geo-ideological model, used to legitimize France's expansionist economic interests and cultural patronage.[33] Chevalier argued that France needed to counter the advance of the 'Anglo-Saxon race' because of the imbalance of power.[34] Although Chevalier's Pan-Latinism defended France's imperial role, his description of the former Spanish colonies as part of the 'Latin family' found resonance among the creole elite living in Paris.

Chilean diplomat and lawyer, Francisco Bilbao, was one of the first to use the term '*Latin* America' and '*Latin* Americans' in a lecture given before the diplomatic community in Paris in 1853. Bilbao used '*Latin* America' as opposed to '*Anglo* America' in a manner similar to Chevalier's description, but he did not allocate France or any other European nation a role against US expansionism. Bilbao's solution was to revive the proposal for a union of America's '*Latin*' peoples against the US threat.[35]

Bilbao's conference, like Chevalier's books, had a major impact on other creoles. Colombian diplomat José María Torres-Caicedo separated the term from its connotation with French imperialism and used it in opposition to US imperialism and in support of a new regional project of unification.[36] In the 1856 poem 'Las dos Américas,' Torres-Caicedo glorified Spanish American unification based on common language, religion, law, and traditions as a response to US imperialism.[37]

In a manifesto circulated in Paris, Torres-Caicedo promoted a union of Central and South America as one nationality that included: a) the same civil and political rights, regardless of place of birth; b) the adoption of *uti possidetis* rule to define territorial limits; c) the unification of codes, dimensions, weight, and money; d) the establishment of a single supreme court of justice; e) a uniform system of postal conventions; f) the binding nature of all public and private contracts celebrated in America; g) a liberal system of trade; h) a compulsory education system; i) freedom of conscience and religious tolerance; j) common rules for extradition; k) the common defence system; l) uniform principles applied to commercial treaties, conventions, and nationality; m) no responsibility or equal compensation by legitimate governments

33 J.L. Phelan, *El origen de la idea de América*, Mexico: Universidad Nacional Autónoma de México, 1979.

34 For Chevalier, race was a cultural and ethnic concept, rather than a matter of skin colour. M. Chevalier, 'Sobre el progreso y porvenir de la civilización', *Revista Española de Ambos Mundos* 1, 1853.

35 F. Bilbao, *Obras completas de Francisco Bilbao vol 1*, Buenos Aires: Imprenta de Buenos Aires, 1866, p. 293.

36 J.L. Phelan, op. cit.

37 A.J. Rivadeneira Vargas, *El bogotano JM Torres Caicedo (1830–1889): La multipatria latinoamericana* (*Colección Lecturas De Bogotá*), Bogotá: Academia Colombiana de Historia, Alcaldia Mayor de Bogotá, 1989, pp. 61–73.

for damage caused to foreigners during civil wars or riots; n) abolition of the (black, yellow, or white) slave trade; and o) the establishment of a French newspaper to defend American interests and enforce the laws, resources, institutions, geography, and topography of the states of Latin America.[38]

Another early promoter of '*Latin* America' was the Argentine publicist, Carlos Calvo, who used the term in the title of his 1862 Spanish American treaty compilation in defence of the region's sovereignty and independence. After Calvo's publication, the Colombian José María Samper published a review of Calvo's book where he argued the creation of a regional international law was of the utmost importance as the highest aspiration of civilization.[39] Bilbao, Torres-Caicedo, Calvo, Samper, and many other's writings illustrate the late-nineteenth century elite's appropriation of international legal discourse. Like many Europeans of their time, Latin American lawyers were convinced international reform could result from their society, history, human nature, and institutional modernity. Nonetheless, though critical of European and US interventions in the region, they justified the appropriation of indigenous lands as a form of civilizing their own countries and supported the colonization of Africa.

IV Early twentieth century: rise of a 'Latin American International Law' (LAIL)

From 1905 until his death in 1960, Alejandro Álvarez promoted a LAIL through publications,[40] at international organizations, lectures, and scientific meetings. In a representative text, he argued that Latin American legal unity was based on different points of view, *sui generis* problems, distinct interpretations of freedom and sovereignty, a regional homeland with civilizing origins, a mutual desire to achieve independence, and the same moral, geographical, material influences, and needs. For Álvarez, Latin Americans were as civilized as Europeans and possessed a 'soul' that was 'more idealistic, imaginative and cosmopolitan than the Anglo-American.'[41]

After World War II, Álvarez continued to place Latin America at the forefront of international law in his dissenting votes as a judge of the

38 Transcribed in E.R. Pérez Calvo and L.R. Pérez Calvo, *Vida y trabajos de Carlos Calvo: Los Calvo en el Río de la Plata*, Buenos Aires: Ediciones Dunken, 1996, p. 152. Calvo argued that Torres-Caicedo borrowed many of his ideas from his 1862 publication.

39 J.M. Samper (ed.), 'Derecho público latino-americano,' in *Miscelánea o colección de artículos escogidos de costumbres, bibliografía, variedades y necrología*, Paris: Eugene Vanderhaeghen, 1869, pp. 350–59, published in 1862 in *El Comercio de Lima*.

40 A. Álvarez, 'Origen y desarrollo del derecho internacional americano,' in *Tercer Congreso Científico Latino Americano*, Rio de Janeiro, 1905; A. Álvarez, *Le droit international américain: son fondement, sa nature: d'après l'histoire diplomatique des états du nouveau monde et leur vie politique et économique*, Paris: A Pedone, 1910, pp. 1–386.

41 A. Álvarez, *International Law and Related Subjects from the Point of View of the American Continent: a Report on Lectures Delivered in the Universities of the US, 1916–1918 (...) vii*, Washington: Carnegie Endowment for International Peace, 1922, pp. 14, 28, 79.

International Court of Justice (ICJ). Álvarez, like others in the creole tradition, recognized the inequalities of power but did not present the Spanish American states as backward but rather viewed them as more advanced because the civilization–barbarism dichotomy had been reversed as shown through European participation in two world wars. However, there was both adherence and opposition to the Álvarez project. Lawyers such as the Brazilian diplomat, Manóel de Souza Sá Vianna, opposed LAIL arguing that common problems or historical experiences did not constitute a basis for a regional approach; while others, such as the Colombian Jesús María Yepes, followed and promoted his school of thought in legal arguments, teachings and publications.

V Late twentieth century: fall of LAIL

By the mid-twentieth century, the belief in a LAIL declined due to the emergence of the UN and other international organizations, the fragmentation of international law, and the death of Álvarez, its main promoter. Following the emergence of the Organization of American States (OAS) in 1948, the region's unifying approach was more economic with the establishment of organizations such as the Latin American Free Trade Association (1960), Central American Common Market (1960), Andean Community of Nations (1969), Cartagena Agreement (1980), and Mercado Común del Sur (MERCOSUR) (1986). During the 1980s and 1990s, Latin American lawyers participated in international organizations and in new fields, such as international human rights and environmental law, but their efforts were no longer unified.

VI Conclusion

As this book shows, Latin Americans participated in the development of the ICJ due to their long-standing belief in the international rule of law. This chapter highlighted the origins of a regional consciousness (that led to principles such as sovereignty, non-intervention, solidarity, and peaceful settlement of disputes, among others). Fifteen Latin American states participated in the League of Nations while five judges served on the first Permanent Court of International Justice (PCIJ). Of the fifty-one founding members of the UN, twenty were from Latin America and all promoted the creation of the ICJ. These facts do not mean that Latin American states have attained utopic unification, because history shows how regional solidarity ends when economic or political interests are at stake. The region's enormous disparities in the distribution of wealth and continuous problems of discrimination on the basis of gender, race, or ethnicity, continue. Nonetheless, regional consciousness has also been the source of many international law initiatives, which look to integration and solidarity in support of a permanent system of justice.

Bibliography

A. Álvarez, 'Origen y desarrollo del derecho internacional americano,' in *Tercer Congreso Científico Latino Americano*, Rio de Janeiro, 1905.

A. Álvarez, *International Law and Related Subjects from the Point of View of the American Continent: a Report on Lectures Delivered in the Universities of the US, 1916–1918*, Washington: Carnegie Endowment for International Peace, 1922.

R. Barbosa, 'The Equality of Sovereign States,' *The Independent* 64, 1907.

A. Becker-Lorca, *Mestizo International Law*, Cambridge: CUP, 2015.

F. Bilbao, *Obras completas de Francisco Bilbao vol 1*, Buenos Aires: Imprenta de Buenos Aires, 1866, p. 293.

S. Bolívar, *Coleccion de documentos relativos a la vida pública del libertador... Tomo Tercero*, Carácas: Impresa de Devisme Hermanos, 1826.

S. Bolívar, 'Decreto sobre libertad de los esclavos – Carúpano, 2 de junio de 1816,' in V. Lecuna (ed.), *Bolívar, Simón: Proclamas y discursos del Libertador, 1811–1830*, Los Teques: Biblioteca De Autores y Temas Mirandinos, 1983.

S. Bolívar, 'Latin-American Correspondence: Simon Bolívar and Alexandre Sabes Pétion,' *Phylon* 7 (2), 1946.

J. Brown Scott, 'The Central American Peace Conference of 1907,' *AJIL* 2(1), 1908.

D. Bushnell, *Simón Bolívar: hombre de Caracas, proyecto de América: una biografía*, Buenos Aires: Editorial Biblos, 2002.

A.S. de Bustamante (trans), *De la democracia en América del Norte*, Paris: Rosa, 1837.

H. Clay, Letter to Secretary of State Richard C. Anderson, US Minister to Colombia, in W.R. Manning (ed.), *Diplomatic Correspondence of the US Concerning the Independence of the Latin-American Nations*, New York: OUP, 1925, Document 145.

M. Chevalier, 'Sobre el progreso y porvenir de la civilización,' *Revista Española de Ambos Mundos* 1, 1853.

A. De la Peña y Reyes, *El Congreso de Panamá y algunos otros proyectos de unión hispano-americana*, México: Publicaciones de la Secretaría de Relaciones Exteriores, 1926.

G.A. De la Reza, 'Simon Bolívar: Un pensamiento sobre el Congreso Anfictiónico de Panamá,' in G.A. De la Reza (ed.) *Documentos sobre el Congreso Anfictiónico de Panamá*, Carácas: Fundación Biblioteca Ayacucho/Banco Central de Venezuela, 2010.

A. De Tocqueville, *De la démocratie en Amérique*, Paris, 1836,

S. Dennison, *Joaquim Nabuco: Monarchism, Panamericanism and Nation-Building in the Brazilian Belle Epoque*, Bern: Peter Lang, 2006.

E. Dussel, E. Mendieta and C. Bohórquez, *El pensamiento filosófico latinoamericano, del Caribe y "latino" (1300–2000)*, México: Siglo XXI.

E. Dussel, 'Philosophy in Latin America in the Twentieth Century: Problems and Currents,' in E. Mendieta (ed.) *Latin American Philosophy Currents, Issues, Debates*, Bloomington: Indiana University Press, 2003.

J. Esquirol, 'Latin America,' in B. Fassbender and A. Peters (ed.) *The Oxford Handbook of the History of International Law*, Oxford: OUP, 2012.

I. González Niño, *Simón Bolívar, precursor del derecho internacional americano*, Bogotá: Instituto Colombiano de Estudios Latinoamericanos y del Caribe, 1985, pp. 1–131; M. Mackenzie, *Los ideales de Bolívar en el derecho internacional americano*, Bogotá: Biblioteca del Ministerio de Gobierno: Colección Bolivariana, 1955.

G. Hernández, *The ICJ and the Judicial Function*, Oxford: OUP, 2014

M. Hudson, 'The Central American Court of Justice,' *AJIL* 26, 1932.

W. Hull, *The Two Hague Conferences and Their Contributions to International Law*, Boston: Ginn and Cpy, 1908.

I. Jaksic, *Andrés Bello: la pasión por el orden*, Santiago de Chile: Editorial Universitaria/Imagen De Chile, 2001.

E. Jean, *La cour de justice centre-américaine*, Paris: E Sagot, 1928.

D. Kennedy, 'Toward a Historical Understanding of Legal Consciousness: The Case of Classical Legal Thought in America 1850–1940,' *Research in Law and Sociology* 3, 1980.

M. Koskenniemi, *The Gentle Civilizer of Nations: The Rise and Fall of International Law 1870–1960*, Cambridge: CUP, 2001.

E. Kuznesof, 'Ethnic and Gender Influences on Spanish Creole Society in Colonial Spanish America,' *Colonial Latin American Review* 4, 1995.

B. Lavalle, *Las promesas ambiguas: ensayos sobre el criollismo colonial en los Andes*, Lima: Pontificia Universidad Católica, 1993.

M. Lucena Giraldo, *A los cuatro vientos: las ciudades de la América hispánica*, Madrid: Marcial Pons Historia, 2006.

L. Lund, 'Barbarian Theorizing and the Limits of Latin American Exceptionalism,' *Cultural Critique* 47, 2001, 54–90.

J.A. Mazzotti (ed.), *Agencias criollas: la ambigüedad 'colonial' en las letras hispanoamericanas*, Pittsburgh: Instituto Internacional de Literatura Iberoamericana, 2000.

L. Obregón, 'Completing Civilization: Creole Consciousness and International Law in Nineteenth-Century Latin America,' in A. Orford (ed.), *International Law and Its Others*, Cambridge: CUP, 2006.

L. Obregón, 'Construyendo la región americana: Andrés Bello y el derecho internacional', in S. B. González and J. Poblete (eds), *Andrés Bello y los estudios latinoamericanos*, Pittsburgh: Instituto Internacional de Literatura Iberoamericana, 2009.

L. Obregón, 'The Civilized and the Uncivilized in the History of International Law,' in B. Fassbender and A. Peters (eds), *The Oxford Handbook of the History of International Law*, Oxford: OUP, 2012.

J.J. Malanson, 'The Congressional Debate over U.S. Participation in the Congress of Panama, 1825–1826: Washington's Farewell Address, Monroe's Doctrine, and the Fundamental Principles of U.S. Foreign Policy,' *Diplomatic History* 30, 2006.

E.R. Pérez Calvo and L.R. Pérez Calvo, *Vida y trabajos de Carlos Calvo: Los Calvo en el Río de la Plata*, Buenos Aires: Ediciones Dunken, 1996.

J.L. Phelan, *El origen de la idea de América*, Mexico: Universidad Nacional Autónoma de México, 1979.

E.A. Posner and M. de Figueiredo, 'Is the ICJ Biased?' *Journal of Legal Studies* 34 (2), 2005.

J.M. Raymond and B. J. Frischholz, 'Lawyers Who Established International Law in the US, 1776–1914,' *AJIL* 76, 1982.

A.J. Rivadeneira Vargas, *El bogotano JM Torres Caicedo (1830–1889): La multipatria latinoamericana* (*Colección Lecturas De Bogotá*), Bogotá: Academia Colombiana de Historia, Alcaldía de Bogotá, 1989.

D.J. Robinson (ed.), 'Social Fabric and Spatial Structure in Colonial Latin America,' in *Dellplain Latin American Studies* 1, Department of Geography, Syracuse University, 1979.

C. Rojas, *Civilization and Violence: Regimes of Representation in Nineteenth-Century Colombia*, Minneapolis: University of Minnesota Press, 2001.

J.M. Samper (ed.), 'Derecho público latino-americano,' in *Miscelánea o colección de artículos escogidos de costumbres, bibliografía, variedades y necrología*, Paris: Eugene Vanderhaeghen, 1869, pp. 350–59

J. Schoonhoven and C. Tymen De Jong, 'The Dutch Observer at the Congress of Panama in 1826,' *The Hispanic American Historical Review* 36.

A. Smith, 'Good Fences Make Good Neighbours?: The Wall Decision and the Troubling Rise of the ICJ as a Human Rights Court,' *Harvard HRJ* 18, 2005.

J.M. Torres-Caicedo, *Unión latino-americana, pensamiento de Bolívar para formar una liga americana. Su origen y sus desarrollos (...)*, Paris: Rosa y Bouret, 1865.

E. Vieira Posada, *La integración de América Latina:* Bogotá: Editorial Univ. Javeriana, 2004.

J.M. Yepes, *Del Congreso de Panamá a la Conferencia de Caracas, 1826–1954*, Caracas: Cromotipo, 1955.

H.F. Zamudio (ed.), 'Función del poder judicial en los sistemas constitucionales Latinoaméricanos,' in *Latino América: constitución, proceso y derechos humanos*, México: Porra, 1988.

2 The influence of the Latin American doctrine on International Law

The rise of Latin American doctrines at The Hague Academy during the early twentieth century

Ricardo Abello-Galvis and Walter Arévalo-Ramírez

The entry of Latin America into the community of nations is one of the most important facts in the history of civilization. It resulted not only in widening the field occupied by International Law but also in radically modifying its character.

(Manuel Alejandro Álvarez. 1909. Judge of the International Court of Justice (ICJ) from 1944 to 1955)

I Introduction: two meanings of 'doctrine' in International Law

'Doctrine' in International Law has two principal meanings. A doctrine is a particular form of state-practice, unilateral or collective, which the state has declared through a formal statement of its sovereign will, being political or legal interest, intention or expected behavior.[1]

Formal statements by high authorities, carrying the name of the politician that conceived them (e.g. Drago doctrine) is the usual way of expressing state doctrine. They are sovereign messages to the international community regarding the position of the state in particular matters.

On the contrary, '*the teachings of the most highly qualified publicists*'[2] (Statute of the ICJ art. 38 (1–d)), are also commonly recognized by international lawyers as doctrine (*doctrina*). This, especially in Latin America and Europe, thanks to their collective roots in civil law, is a legal system in which the scholar's interpretation is highly regarded as a subsidiary source of law.[3]

1 Rubin, A., 'International Legal Effects of Unilateral Declarations', *AJIL*, 71, 1977, 1.

2 *United Nations, Statute of the I.C.J.*, 18 April 1946.

3 Peil, M., 'Scholarly Writings as a Source of Law: A Survey of the Use of Doctrine by the International Court of Justice', *Cambridge Journal of International & Comparative Law*, 1, 2012, 136.

The quest to define the influence of the Latin American doctrine in International Law must consider both meanings of the word, and recognize the importance to those that proclaimed, developed and promoted such doctrines, they being Latin American scholars who, during the early twentieth century, yielded enormous influence in International Law conventions and academic institutions.

There are early writings by Latin American scholars who have studied the possibility of a 'Latin American international law'.[4] Such scholars have delivered their teachings at The Hague Academy of International Law (hereinafter 'The Academy'), which since its inception in 1923 is acknowledged as being the most renowned annual International Law course as well as a unique intercultural legal forum.[5]

It is from these sources that we intend to recreate the emergence of the most influential doctrines of International Law that originated in Latin America and then affected the *corpus* of International Law.

We will present a brief retrospection of the courses that Latin American scholars presented at The Academy during the first decades of the twentieth century. Their intention was to promote Latin American doctrines and regional concepts on various legal issues, coupled with the continent's proposals for the future organization of a growing international community.

Furthermore, we will provide a review of the legal practices and doctrines that emerged from Latin America during the nineteenth and twentieth centuries, which grew beyond a regional doctrine and ended up influencing the development of modern institutions of International Law.

The legal issues found in the courses presented by J. M. Yepes at The Academy in 1930[6] and 1934,[7] which are among the first legal and historical efforts of a scholar to demonstrate the influence of Latin American doctrine in the progressive development of modern International Law, inspire this review. Along with the earlier works[8] of Judge Alejandro Álvarez who was the principal promoter of the notion of 'Latin American International Law'.[9] Both, pioneers in addressing the same questions that we pose in this publication on the impact of Latin American doctrines in International Law.

4 Álvarez, A., 'Latin America and international law', *AJIL* 3.2, 1909, 270.

5 Gormley, W. P., 'The Hague Academy of International Law: A Study in Intercultural Education and Communication', *Journal of Legal Education*, 1961, 512–5.

6 Yepes, J. M., 'La contribution de l'Amérique latine au développement du droit international public et privé', *RCADI*, 32, 1930.

7 Yepes, J. M., 'Les problèmes fondamentaux du droit des gens en Amérique', *RCADI*, 47, 1934.

8 Álvarez, A., '*Le droit international américain: son fondement-sa nature d'après l'histoire diplomatique des états du nouveau monde et leur vie politique et économique*', Paris: A. Pedone, 1910.

9 Álvarez, A., 'Latin America and International Law', *AJIL* 3.2, 1909, 269–353.

II Regional doctrines at The Hague during the first half of the twentieth century

The impact following the participation of Latin American States in international conferences during the first half of the twentieth century was undeniable. A reflection of such legal influence was the presence of Latin American scholars, as well as diplomats in the first courses of The Academy. This period is considered the golden age of Latin American international legal thinking prior to the decline of the movement in the second half of the century.[10]

From 1923 to 1947, the most influential legal scholars of the region regularly visited The Hague and lectured for the Academy's annual courses. They shared the influential legal doctrines that Latin America had to offer with the world, proving to its interlocutors that their history of independence had created a regional legal consciousness of its own.[11]

A. Peaceful settlement of disputes and the need of a World Court

During the first course of The Academy in 1923, Francisco León de La Barra, former Secretary of State for Foreign Relations and former President of Mexico (1911), delivered a course on pacific settlement of disputes,[12] the first of its kind at The Academy. It reflected the views of Latin American states regarding dispute resolution, which were later included in instruments, such as The Pact of Bogotá.

Francisco León de La Barra was a pioneer in asserting that despite the relevance of arbitration for regional disputes during the nineteenth century,[13] in topics such as international boundaries and sovereign debt,[14] the progression to this was the establishment[15] of a permanent world court.

10 Lorca, A. B., 'International Law in Latin America or Latin American International Law-Rise, Fall, and Retrieval of a Tradition of Legal Thinking and Political Imagination', *Harvard International Law Journal*, 47, 2006, 283.

11 Obregón, L., 'Construyendo la región americana: Andrés Bello y el derecho internacional', *La idea de América en el pensamiento ius internacionalista del siglo XX.*, Instituto Fernando el Católico, IFC, 2010.

12 De La Barra, F., 'La médiation et la conciliation internationales', *RCADI*, 1, 1923, 553–68.

13 Summers, L., 'Arbitration and Latin America', *California Western International Law Journal*, 3, 1972, 1.

14 For example, the arbitrations regarding borders between Colombia and Venezuela, and between Guatemala and Honduras: Scott, J., 'The Swiss Decision in the Boundary Dispute between Colombia and Venezuela', *AJIL*, 16.3, 1922, 428–31.

15 De La Barra, F., 'La médiation et la conciliation internationales', *RCADI*, 1, 1923, 563–65.

B. The Monroe Doctrine, self-determination and non-intervention

In 1924, Simón Planas-Suárez, Minister Plenipotentiary of Venezuela at The Hague, Rome, Vienna and Lisbon, delivered a course at The Academy on the extension of the Monroe Doctrine in South America.[16]

Planas-Suárez explained how the Monroe Doctrine was put into practice by Latin American states primarily through diplomacy. He demonstrated how the doctrine of the United States, in relation to the rejection of foreign intervention in the region by former colonial powers during the early nineteenth century,[17] was subsequently embraced and then transformed by the Latin American countries according to their own views on the definition of independence and self-determination. These values were later acknowledged in the Pan-American Conferences that inspired a true Latin American spirit regarding regional identity in the shape of Pan-Americanism and shared legal values, which included republicanism and non-intervention.[18]

C. Diplomacy, arbitration, the urge for international organizations and Uti Possidetis Iuris

In 1925, Alberto Guani, Uruguay's Delegate at the League of Nation's Council, presented a lecture at The Academy regarding Latin American practices in foreign relations that proved their international solidarity, from the time of their independence to their participation in the League of Nations.

According to Guani, the practices that demonstrated the spirit of cooperation that characterized Latin American International Law included: (i) the early projects of an American Confederation; (ii) the ongoing Pan-American Conferences; (iii) the expansion of the Drago doctrine for the rejection of armed intervention as a means of debt collection; and (iv) the collective proclamation of arbitration as the preferred method of dispute settlement, to be included in the 1883 Conference in Caracas and promoted by the Latin American states at the Second Peace Conference at the 1907 Hague Conference, this the result of their conspicuous participation.[19]

Guani presented *uti possidetis iuris* as a Latin American principle that assisted in the avoidance of regional boundary conflicts thanks to its recognition by most countries in the region,[20] and as an effective mechanism to claim sovereignty over territories inherited from the Spanish Crown administrative divisions.[21] This principle is also accompanied by the Latin American

16 Planas-Suárez, S., 'L'extension de la doctrine de Monroe en Amérique du Sud', *RCADI*, 5, 1924, 267–366.

17 Hughes, C., 'Observations on the Monroe Doctrine', *American Bar Association Journal*, 1, 1923, 559–66.

18 Planas-Suárez, op. cit., p. 299.

19 Hicks, F. C., 'The Equality of States and the Hague Conferences', *AJIL*, 1908, 530–61.

20 Guani, A., 'La solidarité internationale dans l'Amérique Latine', *RCADI*, 8. 1925, 293–310.

21 Guani, op. cit., p. 310.

reluctance to embrace *res nullius*, which helped to avoid conflicts regarding territorial conquest among the republics in Latin America.[22]

D. Pan-Americanism, sources, harmonization and codification of International Law

In The Academy's course of 1928, Francisco José Urrutia, Colombian Diplomat and Permanent Representative to the League of Nations Assembly, who later served as a Permanent Judge of the Permanent Court of International Justice (PCIJ) at The Hague,[23] delivered a lecture regarding the codification of International Law in Latin America.

Urrutia described the region as a leader in legal codification, breaking down[24] how many principles of International Law were proclaimed by young Latin American republics over the course of six Pan-American Conferences and their resulting treaties. These included: condemnation of territorial conquest (later codified in the Organization of American States [OAS] and UN Charters),[25] the naturalization of foreign individuals, human rights, navigable international watercourses, reduction of armed forces, *ius in bellum* and the duties and rights of states in the event of civil strife. The region also advanced in terms of the codification of private International Law,[26] and the ideal of a Pan-American Union that would later lead to the OAS, the need for a permanent World Court and the call for a global effort of codification and harmonization of International Law.

E. The Latin American experience with international tribunals, collective security and the League of Nations

Former Guatemalan Minister of Foreign Affairs and Member of the Permanent Court of Arbitration at the time, José Matos, delivered a course during the 1929 sessions of The Academy, reviewing the main contributions of Latin America to the League of Nations.

He considered that the 1826 Congress of Panama, convened by Simón Bolívar, was a blueprint for the Paris Conference that led to the League of Nations as its principal organs, such as the Assembly and the Council, were

22 Kohen, M. G., 'La contribución de América Latina al desarrollo progresivo del Derecho Internacional en materia territorial', *Anuario Español de Derecho Internacional* XVII, 2001, 64.

23 Quintana, J. J., 'The Latin American Contribution to International Adjudication: The Case of the International Court of Justice', *Netherlands International Law Review*, 39.01, 1992, 127–54.

24 Urrutia, F. J., 'La codification du droit international en Amérique', *RCADI*, 22, 1928, 81–236.

25 UN Charter Article 2(4).

26 Lorenzen, E. G., 'Pan-American Code of Private International Law', *Tulane Law Review*, 4, 1929, 499.

inspired by the Panama institutions. Indeed, the use of thematic commissions to solve international disputes was a shared trait. Both conferences upheld the ideal of the implementation of a collective security system and a non-aggression pact that would bind its members.

Matos highlighted Latin America as an example of dispute settlement and judicial effectiveness thanks to the region's experience with the Central American Court of Justice (*Corte de Justicia Centroamericana*),[27] the first permanent international court in history. The Court was the result of the Washington Conference of 1907 that created a system of cooperation between Costa Rica, El Salvador, Guatemala, Honduras and Nicaragua.

This system was based on a series of conventions regarding peace and amity, extradition, communications, governmental coordination and the Tobar doctrine on the prohibition of granting recognition to governments that resulted from a *coup d'état.*

The Court's example substantially influenced the development of the practice and procedure of future international tribunals. The rulings were binding upon the members of the Washington Agreement, and the Court had jurisdiction to solve disputes involving treaty conflicts of any nature between and among its five member-states. It also had jurisdiction over claims addressed by individuals of Central American countries against any member-state for violation of treaties. This procedure envisaged an early version of the principle of exhaustion of local remedies.[28] It allowed claims involving human rights violations, as well as the possibility for individuals to submit cases against the state, not covered by a treaty, under mutual consent.[29]

Despite its short-term existence, the Court delivered rulings that still today hold influence in International Law. Examples worthy of mention include its ruling in the 1917 case involving El Salvador and Nicaragua, regarding historic bays[30] and the conditions required for a gulf to be declared as such, which was a centerpiece for the debate[31] in the 1992 ruling of the ICJ in the case El Salvador/Honduras: Nicaragua Intervening.[32]

Rodrigo-Octavio de Langgaard Menezes, President of the Brazilian Delegation at the 1st Assembly of the League of Nations, delivered a course

27 Not to be confused with the Corte Centro-Americana de Justicia (Central American Court of Justice), a Tribunal founded in 1993.

28 Hudson, M. O., 'The Central American Court of Justice', *AJIL*, 1932, 759–86.

29 Matos, J., 'L'Amérique et la Société des Nations,' *RCADI*, 28, 1929, 1–104.

30 Ruling of the Central American Court of Justice in the case between *El Salvador v. Nicaragua*. Judgment of 9 March 1917; *AJIL. Judicial Decisions Involving Questions of International Law*, 11, 1917, 703.

31 Abello-Galvis, R., 'Eaux et baies historiques en droit international', *Estudios Socio-Jurídicos*, 5, n.1, 2003, 33–76.

32 *Land, Island and Maritime Dispute (El Salvador/Honduras: Nicaragua intervening), I.C.J. Reports 1992*, p. 351. *International Court of Justice*, September 11, 1992.

at The Academy during the 1930 session. The subject was the legal traditions that were entrenched in the colonial epoch of the continent and that, based on the indigenous experience, shaped modern legal principles applied to the region, such as the rights and duties during occupation and the limits on government and civil liberties.[33]

F. Jesús María Yepes and the study of Latin American doctrines as developers of International Law

Jesús María Yepes was a Colombian diplomat, International Law professor, Senator, Delegate of Colombia at the Assembly of the League of Nations (1934–9) and Plenipotentiary of Colombia at the San Francisco Conference. As the signatory of the United Nations Charter, he proposed the second line of article 2 of the Charter, regarding good faith in the fulfillment of international obligations. Yepes was one of the most influential promoters of the Latin American ideal for an organization of states[34] and the need for a permanent court.[35] In the sessions of 1930,[36] 1934[37] and 1947,[38] he delivered a series of lectures that widely assessed the question of the contribution of Latin America to the development of International Law and the principal doctrines and legal matters that are at the forefront of such a contribution.

An invitee on three occasions to deliver courses at The Academy, Yepes was honored and internationally recognized as a leading scholar – an honor he shares with only a few internationally acclaimed legal scholars, such as Roberto Ago and Hans Kelsen, thus accorded by their number of invitations to lecture.[39] Following the thematic structure his lectures, which propelled Latin American doctrines into the forefront as true initiators of the most universal concepts of International Law, we will finally present a review of doctrines, principles and ideas that constitute the principal contribution of the region to the development of International Law.

33 Octavio, R., 'Les sauvages américains devant le droit', *RCADI*, 31, 1930, 177–292.

34 Yepes, J. M., '*Philosophie du Panamericanisme et organisation de la paix*', La Baccomnière, 1945.

35 Yepes, J. M., '*Commentaire théorique et pratique du Pacte de la Société des nations et des statuts de l'Union panaméricaine*', Vol. 2. Paris: A. Pedone, 1934, 375.

36 Yepes, J. M., 'La contribution de l'Amérique latine au développement du droit international public et privé', *RCADI* 32, 1930, 691–800.

37 Yepes, J. M., 'Les problèmes fondamentaux du droit des gens en Amérique', *RCADI*, 47, 1934, 1–144.

38 Yepes, J. M., 'Les accords régionaux et le droit international', *RCADI*, 71, 1947, 227–344.

39 Abello-Galvis, R., 'Editorial', *ACDI – Anuario Colombiano de Derecho Internacional* 1, n. 1, 2008, 7–11.

III Latin American practices and doctrines that contributed to the development of modern International Law institutions

As mentioned above, the ideal of an organization of states was always in the Latin American mindset regarding International Law. From the Congress of Panama to the region's attendance at the Second Peace Conference at the Hague 1907, their role in the League of Nations and the San Francisco Conference, the nations of the 'New World' have always been promoters of two principles that are a prerequisite for the development of many legal doctrines in public International Law. Namely, international solidarity through lawfulness and the ideal of an organized international society with permanent institutions.

A. International organizations and compulsory jurisdiction of international courts

The work of the Brazilian delegate, Raul Fernandes, is recognized as one of the most valuable contributions within the Advisory Committee appointed by the League of Nations to work on the constitution of a permanent court leading to the creation of the PCIJ (1922–46). He championed the legal equality of states before the Court and the binding effect of its rulings as a jurisdictional mechanism not restricted to a mere arbitration.[40] Regarding the scope of the Court's jurisdiction, Latin American delegations also promoted the compulsory jurisdiction of the Court for all disputes involving members of the League, but this proposal was not well received by European nations.[41]

To break this deadlock, Fernandes proposed the historic article 36 (2) of the PCIJ Statute, which would later become article 36 (2) of the ICJ Statute. This article permits states to declare at any time that they recognize the Court's jurisdiction as compulsory *ipso facto* and without special agreement, in relation to any other state accepting the same obligation. This procedure has been adopted by many modern courts and has effectively motivated inter-state judicial settlement of disputes.[42]

B. The practice of codifying International Law and the expansion of codified Latin American principles

Latin America has contributed to consolidating the codification of International Law as a desirable state practice. Multiple institutions, such as

40 Yepes, J. M., 'La contribution de l'Amérique latine au développement du droit international public et privé', *RCADI*, 32, 1930, 712.

41 Fernandes, R., '*Nonagésimo Aniversário – Conferências e Trabalhos Esparsos*', Vol. 1, Rio de Janeiro, M.R.E., 1967. 174.

42 Cançado Trindade, A. A., *The Construction of a Humanized International Law: A Collection of Individual Opinions (1991–2013)*, Leiden: Brill Nijhoff, 2014, 525.

the International Law Commission of the United Nations and the International Committee of the Red Cross, share this practice. The contribution of the continent in this matter not only reflects the aspiration of codifying International Law, but also the constant need to codify regional practices, as demonstrated by Francisco José Urrutia in his course at The Academy. The first Six Pan-American Conferences provide a shining example of this practice.

Furthermore, since the Panama Congress in 1826,[43] the region pressed for the codification of international practices. One of the goals of the congress organized by Bolívar was, indeed, the compilation of all rules and legal principles applicable to the relationship among Latin American countries, as well as that of the region with other states.

During the 1847–8 American Congress, in Lima, it was proclaimed that codification provided the means of unifying the rules applicable to Latin American states. It was considered the only way by which they could appear as a united family of peoples before the international community.[44] In the Confederation Treaty, the region codified its practices in relation to naval warfare, including collective defense, flags and the propriety of vessels, innocent passage, port blockade and other issues that are currently part of the modern law of the sea.[45]

A second codification process played a major role in private International Law: the 1888–9 Congress of Montevideo. It aimed to codify a long list of issues, as follows: conflict of laws in civil and commercial matters, intellectual property, choice of law and forum, legal personality and nationality. One of its groundbreaking contributions was the proclamation of *lex domicilii*, a common practice among Latin American states, as a legal principle. At the time, European nations preferred nationality instead of domicile as criteria to resolve conflicts of laws. The Montevideo Congress, however, helped to promote the use of *lex domicilii* worldwide.

C. The Latin American origin of the law of war

Yepes and many historians of international humanitarian law affirm that Latin American doctrine was of the utmost importance to the proclamation of many principles of *ius in bello*. They also viewed the 1820 Armistice and the Treaty for the Regularization of the War of the same year, between Simón Bolívar and the Spanish Commander-in-Chief, Pablo Morillo, as one of the first treaties, if not the first, to effectively codify and apply the rules of war to an ongoing conflict. The treaty proclaimed most humanitarian rules of war of the time and included a very specific set of rules regarding treatment and

43 Treaty of Perpetual Union, League and Confederation, Panama, 15 July, 1826.

44 Congress of Lima. 'Protocolo de la Conferencia del 20 de Diciembre de 1847', 1847.

45 Garcia-Amador, F. V., 'The Latin American contribution to the development of the law of the sea', *AJIL*, 68.1, 1974, 33–50.

exchange of prisoners of war, protection of civilians, rules applicable to occupied territories and rules of conduct for the belligerents.[46]

This treaty is said to have inspired the rules included in the 1858 Treaty of Paris ending the Crimean War. Its humanitarian formulas inspired most humanitarian rules proclaimed during the 1907 Peace Conference at The Hague, along with new proposals made by Latin American representatives, such as the ban on anchored submarine contact mines. The majority of these rules regarding armed conflict provided precious input for future conferences on the laws of war, such as the First Geneva Convention in 1864, which Latin American countries greatly participated in, the 1949 Geneva Conventions and their protocols.

D. The early Latin American condemnation of war

Latin America as a continent was pioneering in the condemnation of war by means of conventional instruments. The Panama Congress proclaimed that war was contrary to international morality, while the First Pan-American Conference, in 1889, banned the right of conquest from Latin American practice, i.e. the prohibition of offensive wars. During the Second Peace Conference at The Hague, Latin American delegations addressed reservations to the 'Porter Proposition' asserting that coercive means were not allowed for the settling of international disputes or to recover debt. They also stated that arbitration should be applied as the mandatory dispute settlement mechanism, and not just a mere prerequisite to be exhausted before war could be declared.

The efforts to ban war during the fifth Pan-American Conference[47] and the procedure introduced by the Mexican and Brazilian delegations at the sixth Pan-American Conference both proclaimed the prohibition of war in the following terms: '1) Every act of aggression is illegal, and therefore, is prohibited, 2) The American States shall apply all the peaceful meanings available to solve disputes that may arise between them'.[48]

These early formulations on the prohibition of war inspired the Paris Treaty (Briand-Kellogg Pact) and inspire modern instruments, such as the United Nations Charter and the Pact of Bogotá.[49]

46 Yepes, J. M., 'La contribution de l'Amérique latine au développement du droit international public et privé', *RCADI*, 32, 1930, 740

47 *Treaty to avoid or prevent Conflicts between the American States, signed at Santiago, Chile, on May 3, 1923.*

48 *Journal of Sessions of the VI Pan-American international Conference, La Habana. February 1928.*

49 Villalta Vizcarra, A. E, 'La Contribución de América al Derecho Internacional', *El Derecho Internacional en las Américas: 100 años del Comité Jurídico Interamericano, OAS, 2006.*

E. Other principles found in Latin American legislation regarding International Law

Finally, Yepes and Álvarez early identified many International Law principles in their municipal law, from Andres Bello Codes, to the first national constitutions of the new American republics. Some of these principles enshrined in municipal law include the abolition of slavery, the freedom of commerce in times of war, the legal equality of foreigners, the freedom of the seas, the limitation of diplomatic immunity and the reluctance to extradite individuals accused of political charges.[50]

Bibliography

Abello-Galvis, R., 'Eaux et baies historiques en droit international', *Estudios Socio-Jurídicos* 5, n.1, 2003, 33–76.

Abello-Galvis, R., 'Editorial', *ACDI – Anuario Colombiano de Derecho Internacional* 1, n. 1, 2008, 7–11.

Álvarez, A., 'Latin America and International Law', *AJIL* 3.2, 1909, 269–353.

Álvarez, A., *Le droit international américain: son fondement-sa nature d'après l'histoire diplomatique des états du nouveau monde et leur vie politique et économique*, Paris: A. Pedone, 1910.

Cançado Trindade, A. A., *The Construction of a Humanized International Law: A Collection of Individual Opinions (1991–2013)*, Leiden: Brill Nijhoff., 2014. 525.

De Lapradelle, A. G., 'Latin America at the Hague Conference', *Yale Law Journal* 17, 1907, 270.

Fernandes, R., '*Nonagésimo Aniversário – Conferências e Trabalhos Esparsos*', Vol. 1, Rio de Janeiro, M.R.E., 1967. 174.

Garcia-Amador, F. V., 'The Latin American Contribution to the Development of the Law of the Sea', *AJIL* 68.1, 1974, 33–50.

Gormley, W.P., 'The Hague Academy of International Law: A Study in Intercultural Education and Communication', *Journal of Legal Education* 1961, 512–5.

Guani, A., 'La solidarité internationale dans l'Amérique Latine', *RCADI* 8, 1925, 203–340.

Hicks, F. C., 'The equality of States and the Hague Conferences.', *AJIL* 2.3, 1908, 530–61.

Hudson, M. O., 'The Central American Court of Justice', *AJIL* 26.4, 1932, 759–86.

Hughes, C., 'Observations on the Monroe Doctrine', *American Bar Association Journal* 1, 1923, 559–66.

Kohen, M. G., 'La contribución de América Latina al desarrollo progresivo del Derecho Internacional en materia territorial', *Anuario Español de Derecho Internacional* XVII, 2001, 64.

León de La Barra, F., 'La médiation et la conciliation internationales', *RCADI* 1, 1923, 553–68.

50 Yepes, J. M., 'La contribution de l'Amérique latine au développement du droit international public et privé', *RCADI*, 32, 1930, 750.de l'Am lt international public et privé, *RCADI*, Vol. 32, 1930, 750.

Lorca, A. B., 'Law in Latin America or Latin American International Law-Rise, Fall, and Retrieval of a Tradition of Legal Thinking and Political Imagination', *Harvard International Law Journal* 47, 2006, 283.

Lorenzen, E. G., 'Pan-American Code of Private International Law', *Tulane. Law. Review* 4, 1929, 499.

Matos, J., 'L'Amérique et la Société des Nations', *RCADI* 28, 1929, 1–104.

Obregón, L., 'Construyendo la región americana: Andrés Bello y el derecho internacional', *La idea de América en el pensamiento ius internacionalista del siglo XX.* Instituto Fernando el Católico. IFC, 2010.

Obregón, L., 'Regionalism Constructed: A Short History of Latin American International Law', *European Society of International Law (ESIL) Conference Paper Series No. 5/2012.*

Octavio, R., 'Les sauvages américains devant le droit', *RCADI* 31, 1930, 177–292.

Peil, M., 'Scholarly Writings as a Source of Law: A Survey of the Use of Doctrine by the International Court of Justice', *Cambridge Journal of International & Comparative Law* 1, 2012, 136.

Planas-Suarez, S., 'L'extension de la doctrine de Monroe en Amérique du Sud', *RCADI* 5, 1924, 267–366.

Quintana, J. J., 'The Latin American Contribution to International Adjudication: The Case of the International Court of Justice', *Netherlands International Law Review* 39.01, 1992, 127–54.

Rubin, A., 'International Legal Effects of Unilateral Declarations', *AJIL* 71, 1977, 1.

Scott, J., 'The Swiss Decision in the Boundary Dispute between Colombia and Venezuela', *AJIL* 16.3, 1922, 428–31.

Summers, L., 'Arbitration and Latin America', *California Western International Law Journal* 3, 1972.

Urrutia, F. J., 'La codification du droit international en Amérique', *RCADI* 22, 1928, 81–236.

Villalta Vizcarra, A. E., 'La Contribución de América al Derecho Internacional', *'El Derecho Internacional en las Américas: 100 años del Comité Jurídico Interamericano'.* OAS. 2006.

Yepes, J. M., 'Les problèmes fondamentaux du droit des gens en Amérique', *RCADI*, 47, 1934, 1–144.

Yepes, J. M., 'La contribution de l'Amérique latine au développement du droit international public et privé', *RCADI* 32, 1930, 691–800.

Yepes, J. M., Les accords régionaux et le droit international, *RCADI* 71, 1947, 227–344.

Yepes, J. M., *Commentaire théorique et pratique du Pacte de la Société des nations et des statuts de l'Union panaméricaine* 2, Paris: A. Pedone, 1934, 375.

Yepes, J. M., *Philosophie du Panamericanisme et organisation de la paix.* Neuchâtel: La Bacconnière, 1945.

Yepes, J. M., *Philosophie du panaméricanisme et organisation de la paix: le droit panaméricain.* Éditions de la Baconnière; Paris, 1945.

Yepes, J. M., 'La contribution de l'Amérique latine au développement du droit international public et privé', *RCADI* 32, 1930.

Yepes, J. M., 'Les problèmes fondamentaux du droit des gens en Amérique', *RCADI* 47, 1934.

Jurisprudence

Land, Island and Maritime Dispute. (El Salvador/Honduras: Nicaragua intervening), Judgment, I.C.J. Reports 1992, p. 351.

Ruling of the Central American Court of Justice in the case between *El Salvador v. Nicaragua*, Judgment of 9 March 1917; *AJIL. Judicial Decisions Involving Questions of International Law.* Vol. 11, 1917, 703.

3 Latin America and the II Hague Peace Conference of 1907

Antônio Augusto Cançado Trindade

I Historical antecedents

Of the 26 states that participated in the I Hague Peace Conference of 1899, only one (Mexico) was from Latin America; in turn, of the 44 participating states at the II Hague Peace Conference of 1907, 18 were Latin American – as acknowledged by its *Acte final*[1] – which represented almost half of the participants. This substantial increase, as far as Latin American participation was concerned, can nowadays, over one century on, be fully understood and appreciated in its historical context.

On the American continent, inter-state concertation had in fact begun one decade before the first Hague Peace Conference of 1899: at the series of International Conferences of American States that commenced in 1889, and, between the I and the II Hague Peace Conferences (of 1899 and 1907). The second Conference of American states took place in Mexico City in 1901; the third in Rio de Janeiro in 1906.[2] In both, the participating states displayed their preparedness in taking part in the work of systematization of international law at a universal level.

Thus, at the II Conference of American States of Mexico City of 1901, Latin American states took note of the three Conventions adopted at the I Hague Peace Conference of 1899, particularly the one on Peaceful Settlement of International Disputes, and recognized as theirs the principles underlying it. Moreover, in order to enhance recourse to arbitration, Latin American states adopted at the aforementioned Mexico Conference a

1 Those 18 states were: Argentina, Bolivia, Brazil, Chile, Colombia, Cuba, Dominican Republic, Ecuador, El Salvador, Guatemala, Haiti, Mexico, Nicaragua, Panama, Paraguay, Peru, Uruguay and Venezuela. Cf. 'II Conférence Internationale de la Paix – Actes et documents', in vol. I: *Séances plénières de la Conférence*, La Haye: Ministère des Affaires Étrangères/Imprimerie Nationale, 1907, pp. 689–701.

2 For an account, cf., e.g., F.V. García Amador, 'Sistema Interamericano a través de Tratados, Convenciones y Otros Documentos' in vol. I: *Asuntos Jurídico-Políticos*, Washington, DC: OAS General Secretariat, 1981, pp. 67 and 133–141; César Sepúlveda, *El Sistema Interamericano*, 2nd ed., Mexico: Ed. Porrúa, 1974, pp. 23–8.

General Treaty on Arbitration, open to signature on 30 January 1902.[3] By pledging the same ideals of the states that had participated in the I Hague Peace Conference of 1899, Latin American states sent a clear message to these states, to the effect that they were quite well prepared, and willing, to participate in the forthcoming conference: the II Hague Peace Conference of 1907.

To the same effect, Latin American states began to give expression to the principle of the prohibition of the use of force in inter-state relations also before the II Hague Peace Conference of 1907. It may be recalled, in this connection, that the Drago Doctrine had been formulated five years earlier, in response to an armed attack by three European powers (Germany, Great Britain and Italy) in Puerto Cabello against Venezuela. The note, disclosed by the Foreign Minister of Argentina, L.M. Drago, became well-known in European juridical circles four years prior to the II Hague Peace Conference.[4]

The Drago Doctrine was brought to the attention of the III International Conference of American States held in Rio de Janeiro in 1906, but, as most of the delegates were representatives of debtor states, they decided to defer discussion of the matter and to schedule it for the following year, at the aforementioned II Hague Peace Conference,[5] where it would at last gain international recognition.[6]

The principle of the prohibition of the use of force or coercion in inter-state relations indeed found expression at the 1907 Hague Peace Conference, with the more concentrated support of Latin American states, which from that time on and over decades exercised considerable influence in the evolution of international law.

As the days of the II Hague Peace Conference drew closer (15 June to 18 October 1907), the decision had been taken that this Conference would be a 'universal' one, and that Latin American states were to be invited. Latin American states had rightfully earned their place in the concert of nations. Their presence alone at the Hague Conference in 1907 gave concrete

3 F.-J. Urrutia, 'La codification du Droit international en Amérique', in *Recueil des Cours de l'Académie de Droit International de La Haye* 22, 1928, p. 113, and cf. pp. 116–7.

4 C. Calvo, *Manuel de Droit international public et privé*, 3rd rev. ed., Paris: LNDJ/A. Rousseau Éd., 1892, p. 118, and cf. pp. 83 and 215; A. Bello, *Principios de Derecho Internacional*, 3rd ed., Paris, Garnier, 1873, pp. 12–4; A. de Avila Martel, 'La Filosofía Jurídica de Andres Bello', in *Congreso Internacional 'Andres Bello y el Derecho'*, Santiago: 1981, Santiago: Editorial Jurídica de Chile, 1982, pp. 41–62.

5 MRE, *Informes y Notas de la Delegación de Colombia en la II Conferencia de la Paz de La Haya, 190*, Rotterdam: (ed.) M. Wyt & Zonen Impr., 1908, pp. 38–9.

6 J.C. Arellano, 'La Doctrina Drago y Su Importancia Americanista' in *Boletín de la Academia de Ciencias Políticas y Sociales'* 36, Caracas: 1977, n. 71, pp. 157, and cf. pp. 154–5; L.M. Drago, *La República Argentina y el Caso de Venezuela*, Buenos Aires: Impr. Ed. Coni Hermanos, 1903, pp. 1–326; A.N. Vivot, *La Doctrina Drago*, Buenos Aires: Impr. Ed. Coni Hermanos, 1911, pp. 5–367.

expression to another fundamental principle of international law: that of the juridical equality of states.[7]

II Contributions to the conference

The contribution of Latin American states to the work and outcome of the II Hague Peace Conference of 1907, and to the developments thereafter, focused mainly on four issues: the recourse to arbitration and non-use of force; the basic principle of the juridical equality of States; the strengthening of international jurisdiction; and the direct access of individuals to international justice.

A. Recourse to arbitration and non-use of force

During the consideration of the matter of peaceful recovery of contract debts at the Peace Conference of 1907, the Porter proposal – of prior recourse to arbitration – represented a significant advance towards international peace. However, much of this was owed to the Drago Doctrine, which did not concede, under any circumstance, 'armed aggression' for the recovery of international contract debts.[8]

Latin American states supported the resort to arbitration to avoid the use of force for the recovery of contract debts. It is significant that the project on compulsory arbitration was supported by *all* American states participating in the II Hague Peace Conference of 1907. Indeed, while none of the other delegations was outright opposed to it, each sought to accept the outcome with greater or lesser reservation and or limitation, being finally approved on 15 October 1907.[9] As to the principle of non-use of force, the Drago Doctrine, even with the inserted Porter amendment, represented a sensible advance towards recourse to arbitration and abandonment of 'measures of violence'.[10]

The *Acte final* of the II Hague Peace Conference of 1907 included a statement of 'principles' that the Conference considered as being 'unanimously recognized', namely, 'the principle of compulsory arbitration'.[11] The *Acte final* also added a unanimous resolution whereby the 1907 Hague Peace Conference urged states to further the study of the 'limitation of military expenditures', and expressed the *voeux* of the adoption of a Convention for the establishment of an Arbitral Court of Justice, of control of military expenditures, of observance of the principles on the laws and customs of war,

7 See F.-J. Urrutia, 'La codification du Droit international en Amérique', op. cit. p. 120, and cf. pp. 116–17 and 133–4.

8 S. Pérez Triana and W.T. Stead (intr.), *Doctrina Drago – Colección de Documentos*, London: Impr. Wertheimer, Lea & Co., 1908, pp. LXXIV and XLIV.

9 F.-J. Urrutia, 'La codification du Droit international en Amérique', op. cit., pp. 134–5.

10 *Ibid*., p. 137.

11 *Ibid*., p. 700.

of peaceful settlement of inter-state disputes and of convening a III Hague Peace Conference sometime in the future.[12]

Although the projected III Peace Conference never took place, the 1928 General Treaty for the Renunciation of War as a National Policy Instrument (the so-called Briand-Kellogg Pact, in force from 24 July 1929) was concluded two decades after the II Peace Conference, and became an instrument of almost universal application. It was followed, in Latin America, by the 1933 Saavedra Lamas Pact (the so-called Anti-War Pact, adopted at the VII Conference of American States in Montevideo),[13] amid constant reassertions of the principle of prohibition of the use of force in successive regional Conferences of American States (e.g., of Havana in 1928, of Montevideo in 1933, of Buenos Aires in 1936, of Lima in 1938 and of Mexico in 1945).[14]

Whatever doubts may have existed as to the prohibition of measures 'short of war', the fact remains that the 1928 Briand-Kellogg Pact became the point of no return for traditional international law and also for the no longer warranted *jus ad bellum*, thus heralding the advent of the new *jus contra bellum*.[15] Latin American states gave a valuable contribution to this effect, as stressed by the subsequent Saavedra Lamas Pact, and this was followed by their overwhelming support for the adoption of Article 2(4) of the UN Charter.

B. The principle of the juridical equality of states

The II Hague Peace Conference of 1907, with those Latin American states present, gave concrete expression to the principle of the juridical equality of states, grouping together the most powerful with the more fragile or vulnerable states in the world. It was regarded as constituting an advance in international law since the intention was to abolish the increase of armaments, and to regulate the conduct of war, the 'most hideous cause of grief and misery'.

The principle of the *juridical equality of states* was consistently relied upon by the delegations of Latin American states in the course of the debates and work of the I Commission of the II Hague Peace Conference of 1907.[16] Over the course of the 1907 Hague Peace Conference, this principle was also

12 *Ibid*., pp. 700–1.

13 Cf. Carlos Saavedra Lamas, *Por la Paz de las Américas*, Buenos Aires: M. Gleizer Ed., 1937, pp. 91–134.

14 Cf., e.g., H. Accioly, *Tratado de Direito Internacional Público*, 2nd ed., vol. I, Rio de Janeiro: [MRE], 1956, pp. 266–268; L.A. Podestá Costa and J.M. Ruda, *Derecho Internacional Público*, vol. II, Buenos Aires: Tip. Ed. Argentina, 1985, pp. 221–4.

15 Cf. R. Kolb, *Ius contra bellum – Le droit international relativ au maintien de la paix*, Bâle/Bruxelles: Helbing & Lichtenhahn/Bruylant, 2003, pp. 39–47.

16 *In*: *II Conférence Internationale de La Paix – Actes et documents* (La Haye – 15 juin/18 octobre 1907), vol. II: *I Commission*, La Haye: Ministère des Affaires Étrangères/ Imprimerie Nationale, 1907, pp. 20–1, 145, 153–55, 618, 625 and 688.

invoked in the debates on the proposed Convention on Recourse to Arbitration for the Recovery of Contract Debts. The referred principle was thus directly related to that of the non-use of force (cf. *supra*).

While the II Hague Peace Conference of 1907 sought to consolidate and expand the achievements of the earlier I Conference of 1899, some expansion efforts were met with resistance. For example, although the idea of a permanent Court of Arbitration met general acceptance, no agreement was reached as to the structure of the proposed Court. Likewise, the proposed International Prize Court, though regarded as an 'undoubted progress' in 'eliminating a cause of conflicts among nations', gave rise to similar reservations, bearing in mind the principle of juridical equality of states.[17]

During the voting, in plenary session, of the Draft Convention on the Establishment of an International Prize Court, several delegations of Latin American states also expressed their reservations; in particular, to the proposed system of appointing judges, the Brazilian delegation being the sole to vote against the project.[18]

The work of the II Hague Peace Conference of 1907 heralded the general acknowledgement of the basic principle of the juridical equality of states,[19] as a manifestation of the progressive democratization of international law.[20]

C. *The strengthening of international jurisdiction*

Reference has already been made to the general support, expressed at the II Hague Peace Conference of 1907, for the initiative of the creation of a permanent Court of Arbitration and an International Prize Court (whereto individuals would be granted direct access). The difficulties, which proved insurmountable in those days, existed in relation to the *organization* of these Courts, in particular, the appointment of their judges (*supra*).

Despite the fact that both the proposed permanent Court of Arbitration[21]

17 MRE, *La República Argentina en la II Conferencia Internacional de la Paz – Haya 1907*, Buenos Aires: Impr. A. Pech, 1908, p. 9.

18 *In*: *II Conférence Internationale de la Paix – Actes et documents* (La Haye – 15 juin/18 octobre 1907), vol. I: *Séances plénières de la Conférence*, La Haye: Ministère des Affaires Étrangères/Imprimerie Nationale, 1907, p. 168. See J. Cabral, *Evolução do Direito Internacional*, Rio de Janeiro, Typ. Rodrigues & Cia., 1908, pp. 109–11, 284–5, 293 and 297; C. Bevilaqua, *Direito Público Internacional – A Síntese dos Princípios e a Contribuição do Brasil*, 2nd ed., vol. II, Rio de Janeiro: Livr. Ed. Freitas Bastos, 1939, pp. 169–70.

19 F.-J. Urrutia, 'La codification du Droit international en Amérique', op. cit., pp. 138 and 140; L.A. Podestá Costa and J.M. Ruda, op. cit., pp. 162–5; César Sepúlveda, *Derecho Internacional*, 15th ed., Mexico: Ed. Porrúa, 1986, pp. 265–70.

20 H. Valladão, *Democratização e Socialização do Direito Internacional – Os Impactos Latino-Americano e Afro-Asiático*, Rio de Janeiro: Livr. J. Olympio Ed., 1961, pp. 48–98.

21 D.J. Bederman, 'The Hague Peace Conferences of 1899 and 1907', in *International Courts for the Twenty-First Century* (ed. M.W. Janis), Dordrecht: Nijhoff, 1992, pp. 10–1; S. Rosenne, 'Introduction', in PCA, *The Hague Peace Conferences of 1899 and 1907 and International Arbitration – Reports and Documents* (ed. S. Rosenne), The Hague: T.M.C. Asser Press, 2001, p. XXI.

and the International Prize Court were the object of much debate at the II Hague Peace Conference of 1907, these courts did not see the light of day, as envisaged in the original proposals. They constituted, however, manifestations of recognition of the need to enhance international jurisdiction in the years to come. Over the course of the debates within the I Commission of the II Hague Peace Conference of 1907, there was support among delegations of Latin American states for the idea of compulsory arbitration, to the extent that it enhanced the principle of non-use of force.[22]

The Mexican delegation to the II Hague Peace Conference insisted on the need for observance of the principle of the 'juridical equality of states', and reported on the 'vigorous opposition' on the part of the 'main Latin American delegations' to the non-observance of that principle in the composition of the projected Court.[23] This led to a motion to approve only the 'rules pertaining to the operation of the permanent Court', and the procedures before it, leaving open for subsequent deliberation the form of constitution of the new international body.

The pacifist ideals of Latin American states proposed at the II Hague Peace Conference of 1907[24] contributed to the growing acceptance of international arbitration. In this context, the delegation of the Dominican Republic was the first to propose compulsory arbitration as a means of settlement of all possible inter-state disputes.[25]

Although by the end of the II Hague Peace Conference of 1907 recourse to arbitration had been made or remained optional,[26] the ideal of compulsory arbitration had thus stamped its presence at this Conference, and remained alive in the years to come.[27] It exerted influence on the chapter of peaceful settlement of international disputes as a whole.[28] This can be seen in particular in relation to judicial settlement, as the 'clause Raul Fernandes' (Article

22 In: *II Conférence Internationale de La Paix – Actes et documents* (La Haye – 15 juin/18 octobre 1907), vol. II: *I Commission*, La Haye: Ministère des Affaires Étrangères/ Imprimerie Nationale, 1907, pp. 13–4, 356, 915, 918–21, 924–5, 659, 791 and 9[illegible]5.

23 Secretaría de Relaciones Exteriores (SRE)/[Archivo Diplomático], *Informe a la SRE de la Delegación de México a la II Conferencia de Paz de La Haya*, del 19 October 1907, pp. 1, 3, 9, 13–5 and 17 (internal circulation – document provided to the Author of the present study by the SRE/Archivo Diplomático in Mexico City, on 26 March 2007).

24 J.-M. Yepes, 'La contribution de l'Amérique Latine au développement du Droit international public et privé', *Recueil des Cours de l'Académie de Droit International de La Haye* 32, 1930, pp. 741–2 and 764–5.

25 *Ibid.*, pp. 764–5. For an account of the memorable debates of the II Hague Peace Conference on the matter on 05 October 1907, see J.-M. Yepes, op. cit., p. 765.

26 D. Gaurier, *Histoire du Droit international*, Rennes, Presses Universitaires de Rennes, 2005, p. 429.

27 C. Wilfred Jenks, *The World beyond the Charter*, London, G. Allen & Unwin, 1969, pp. 42–3.

28 D.J. Bederman, 'The Hague Peace Conferences of 1899 and 1907', op. cit., p. 1[illegible]. And cf., on the international commissions of inquiry, e.g., C. Bevilaqua, *Direito Público Internacional*, op. cit., pp. 163–4.

36(2)) – the optional clause of compulsory jurisdiction – of the Statute of the Permanent Court of International Justice (PCIJ), and later, of the International Court of Justice (ICJ), was to bear witness from 1920 and 1945 onwards. This clause came to be recognized as a Latin American contribution to the foundation of international jurisdiction.[29]

The proposal for a permanent Court of Arbitral Justice formed the 'principal model' for the Statute of the PCIJ in 1920. Several of its provisions re-emerged, either unchanged or altered, and also appeared in the Statute of the ICJ in 1945.[30] The strengthening of international jurisdiction, as proposed and supported by Latin American delegations at the II Hague Peace Conference of 1907,[31] would thus later on see the light of day.

D. Direct access of individuals to international justice

Although the projected International Prize Court, set forth in the [XII] Convention on the Establishment of an International Prize Court, was never created, the Convention, though not having entered into force, presented issues of relevance for the evolution of international law. First, it foresaw the establishment of a jurisdiction above national jurisdictions to decide on last appeal on maritime prizes; second, it provided, in such circumstances, for the direct access of individuals to the international jurisdiction; third, it envisaged a type of international compulsory jurisdiction; and, fourth, it admitted the proposed Court's authority to decide (*compétence de la compétence*).[32]

The initiative of the aforementioned Convention further indicated that, already one century ago, there existed an awareness that 'the individual is not without standing in modern international law'.[33] At the II Hague Peace Conference of 1907, the discussions on the projected International Prize Court (international jurisdiction on appeal on prizes) focused, for the first time, on the recognition by an international diplomatic Conference, of the right of petition by individuals against other states (so as to avoid direct intervention by their own states). Although there was no unanimity on the matter, most participants deemed that it would be in the states' interests – particularly the small or weaker ones – to avoid giving to this type of case the

29 R. Fernandes, *Nonagésimo Aniversário – Conferências e Trabalhos Esparsos*, vol. I, Rio de Janeiro: MRE, 1967, pp. 174–5. And, for a recent reassessment, cf. A.A. Cançado Trindade, 'The Relevance of International Adjudication Revisited: Reflections on the Need and Quest for International Compulsory Jurisdiction', in *Towards World Constitutionalism – Issues in the Legal Ordering of the World Community* (eds R. St. John Macdonald and D.M. Johnston), Leiden: Nijhoff, 2005, pp. 515–42.

30 S. Rosenne, op. cit., p. XXI. And cf. also A. Eyffinger, 'A Highly Critical Moment: Role and Record of the 1907 Hague Peace Conference', in *Netherlands International Law Review* 54, 2007, n. 2, pp. 217 and 227.

31 Cf. A. Truyol y Serra, *Histoire du Droit international*, Paris, Economica, 1995, p. 113.

32 J. Cabral, *Evolução do Direito International*, op. cit., pp. 97–8.

33 J.B. Scott, 'The Work of the Second Hague Peace Conference' in *American Journal of International Law* 2, 1908, p. 22.

character of an inter-state dispute. The debates of the 1907 Hague Peace Conference led to the prevailing view of allowing individuals to 'personally' address their complaints before the projected International Prize Court.[34]

An initiative in a similar vein occurred in 1907. On the last day of the Central American Peace Conference, convened in Washington, DC from 14 November to 20 December 1907, five Central American states (Costa Rica, El Salvador, Guatemala, Honduras and Nicaragua) adopted the Convention for the Establishment of a Permanent Central American Court of Justice.[35] The referred Court was endowed with jurisdiction that extended not only to inter-state disputes but also to disputes opposing individuals to the states parties.

In its decade of operation, ten cases were submitted to the Court: five initiated by individuals and five inter-state cases.[36] Thus, motivated by the II Hague Peace Conference of 1907, the Central American states succeeded in establishing an international tribunal with a broad jurisdictional basis, pioneering in granting direct access not only to states but also to individuals and, thus contributing to the acknowledgement of individuals as subjects of international law.[37] It became gradually recognized – as pointed out three decades after the 1907 Hague Conference – that the ideal whereto humankind should orient itself was that of securing the right of human beings to directly access international jurisdiction.[38]

III The outcome of the conferences

As a whole, the two Hague Peace Conferences of 1899 and 1907 served clear purposes[39] – forming the *opinio juris comunis* – and projected their

34 *Ibid.*, pp. 40–1.

35 Promptly ratified, by March 1908, by the five signatory states.

36 Cf. *ibid.*, pp. 376–7; and cf. F.A. von der Heydte, 'L'individu et les tribunaux internationaux' in *Recueil des Cours de l'Académie de Droit International de La Haye* 107, 1962, p. 321.

37 C.J. Gutiérrez, *La Corte de Justicia Centroamericana*, San José de Costa Rica: Edit. Juricentro, 1978, pp. 42, 106, 150–2.

38 S. Séfériadès, 'Le problème de l'accès des particuliers à des juridictions internationales', 51 *Recueil des Cours de l'Académie de Droit International de La Haye* (1935), p. 66; and cf. p. 42. And, for a recent reassessment, cf. A.A. Cançado Trindade, *El Acceso Directo del Individuo a los Tribunales Internacionales de Derechos Humanos*, Bilbao: Universidad de Deusto, 2001, pp. 9–104; A.A. Cançado Trindade, *Los Tribunales Internacionales Contemporáneos y la Humanización del Derecho Internacional*, Buenos Aires: Ed. Ad-Hoc, 2013, pp. 7–185; A.A. Cançado Trindade, *Le Droit international pour la personne humaine*, Paris: Pédone, 2012, pp. 45–368.

39 S. Rosenne, 'Third-Party Dispute Settlement at the Turn of the Century: Some Old Problems Revisited and Some New Problems Not Foreseen at The Hague in 1899/1907, or in the Statutes of the PCIJ and the ICJ – Introductory Statement', in U.N., *International Law as a Language for International Relations – Proceedings of the U.N. Congress on Public International Law*, New York: 1995, U.N./Kluwer, The Hague: 1996, pp. 479–80; and cf. also E. Jiménez de Aréchaga, *El Derecho Internacional Contemporáneo*, Madrid: Tecnos, 1980, p. 108.

influence over time throughout the following decades. First, they sought to find ways to avoid war by developing methods for the peaceful settlement of international disputes; second, they sought 'to humanize war through the development of humanitarian law so as to spare all belligerents some of the horrors of war';[40] and third, they sought to encourage the codification of international law so as to preserve international peace and to further develop the discipline at the universal level.

The II Hague Peace Conference of 1907 fostered the codification of international law, which also occurred in Latin America at the regional level. In the years that followed, codification projects flourished in the region, such as those prepared by, e.g., Epitacio Pessôa,[41] Lafayette Rodrigues Pereira,[42] Alejandro Álvarez[43] and Gustavo Guerrero.[44] The work of codification, which nowadays is universally undertaken at the this level, appears as a manifestation of the juridical conscience.[45]

A most significant outcome from the II Hague Peace Conference of 1907 was the codification it achieved of the law of war and neutrality, with the aim of 'humanizing' the methods of combat, despite recurring claims of state sovereignty. The non-use of force and recourse to arbitration and other means of peaceful settlement of international disputes; the respect for the juridical equality of states; and the strengthening of international jurisdiction with the direct access of individuals to international justice. Such were the contributions of Latin American states to the work and outcome of the II Hague Peace Conference of 1907, and from that time to the evolution of international law. Such contributions have now become part and parcel of contemporary international law.

Bibliography

Accioly, H., *Tratado de Direito Internacional Público*, 2nd ed., vol. I, Rio de Janeiro: [MRE], 1956.

Avila Martel, A., 'La Filosofía Jurídica de Andres Bello', in *Congreso Internacional 'Andres Bello y el Derecho'*, Santiago: 1981, Santiago: Editorial Jurídica de Chile, 1982, pp. 41–62.

40 *Ibid.*, p. 108.

41 Epitacio Pessôa, *Projecto de Código de Direito Internacional Público*, Rio de Janeiro: Imprensa Nacional, 1911, pp. 5–323.

42 L.R. Pereira, *Princípios de Direito Internacional*, vols. I–II, Rio de Janeiro: J. Ribeiro dos Santos Ed., 1902–3.

43 F.-J. Urrutia, 'La codification du Droit international en Amérique', op. cit., pp. 162–163.

44 Cf. G. Guerrero, *La Codification du Droit International*, Paris: Pédone, 1930, pp. 9–10, 13, 24, 27 and 150, 182 and 175; and cf. also A.A. Cançado Trindade and A. Martínez Moreno, *Doctrina Latinoamericana del Derecho Internacional*, vol. I, San José of Costa Rica: Inter-American Court of Human Rights, 2003, pp. 5–64.

45 A. Mahiou, 'Rapport général', in *La codification du Droit international* (Colloque d'Aix-en-Provence de la Société Française pour le Droit International), Paris: Pédone, 1999, pp. 30, 41–2 and 45.

Bederman, D.J., 'The Hague Peace Conferences of 1899 and 1907', in *International Courts for the Twenty-First Century* (ed. M.W. Janis), Dordrecht: Nijhoff, 1992.

Bello, A., *Principios de Derecho Internacional*, 3rd ed., Paris: Garnier, 1873.

Bevilaqua, C., *Direito Público Internacional - A Síntese dos Princípios e a Contribuição do Brasil*, 2nd ed., vol. II, Rio de Janeiro: Livr. Ed. Freitas Bastos, 1939.

Cabral, J., *Evolução do Direito Internacional*, Rio de Janeiro, Typ. Rodrigues & Cia., 1908.

Cançado Trindade, A.A., *Los Tribunales Internacionales Contemporáneos y la Humanización del Derecho Internacional*, Buenos Aires: Ed. Ad-Hoc, 2013

Cançado Trindade, A.A., *Le Droit international pour la personne humaine*, Paris: Pédone, 2012.

Cançado Trindade, A.A., 'The Relevance of International Adjudication Revisited: Reflections on the Need and Quest for International Compulsory Jurisdiction', in *Towards World Constitutionalism - Issues in the Legal Ordering of the World Community* (eds R. St. John Macdonald and D.M. Johnston), Leiden: Nijhoff, 2005, pp. 515–42.

Cançado Trindade, A.A. and Martínez Moreno, A., *Doctrina Latinoamericana del Derecho Internacional*, vol. I, San José of Costa Rica: Inter-American Court of Human Rights, 2003.

Cançado Trindade, A.A., *El Acceso Directo del Individuo a los Tribunales Internacionales de Derechos Humanos*, Bilbao: Universidad de Deusto, 2001.

Calvo, C., *Manuel de Droit international public et privé*, 3rd rev. ed., Paris: LNDJ/A. Rousseau Éd., 1892.

Drago, L.M., 'La Cuestión de Venezuela', in *Anales de la Facultad de Derecho y Ciencias Sociales*, vol. IV, Buenos Aires: Libr. Prudent Hermanos y Moetzel, 1903, pp. 50–59.

Drago, L.M., *La República Argentina y el Caso de Venezuela*, Buenos Aires: Impr. Ed. Coni Hermanos, 1903.

Eyffinger, A., 'A Highly Critical Moment: Role and Record of the 1907 Hague Peace Conference', *Netherlands International Law Review* 54, 2007, n. 2.

Fernandes, R., *Nonagésimo Aniversário - Conferências e Trabalhos Esparsos*, vol. I, Rio de Janeiro: MRE, 1967.

Gaurier, D., *Histoire du Droit international*, Rennes: Presses Universitaires de Rennes, 2005.

Guerrero, G., *La Codification du Droit International*, Paris: Pédone, 1930.

Gutiérrez, C.J., *La Corte de Justicia Centroamericana*, San José de Costa Rica: Edit. Juricentro, 1978.

Jiménez de Aréchaga, E., *El Derecho Internacional Contemporáneo*, Madrid: Tecnos, 1980.

Kolb, R., *Ius contra bellum - Le droit international relativ au maintien de la paix*, Bâle/Bruxelles: Helbing & Lichtenhahn/Bruylant, 2003.

Pérez Triana, S. and Stead, W.T. (intr.), *Doctrina Drago - Colección de Documentos*, London: Impr. Wertheimer, Lea & Co., 1908.

Pessôa, E., *Projecto de Código de Direito Internacional Público*, Rio de Janeiro: Imprensa Nacional, 1911.

Podestá Costa, L.A. and Ruda, J.M., *Derecho Internacional Público*, vol. II, Buenos Aires: Tip. Ed. Argentina, 1985.

Rodrigues Pereira, L., *Princípios de Direito Internacional*, vols. I-II, Rio de Janeiro: J. Ribeiro dos Santos Ed., 1902–3.

Rosenne, S., 'Introduction', in PCA, *The Hague Peace Conferences of 1899 and 1907 and International Arbitration - Reports and Documents* (ed. S. Rosenne), The Hague: T.M.C. Asser Press, 2001.
Saavedra Lamas, C., *Por la Paz de las Américas*, Buenos Aires: M. Gleizer Ed., 1937.
Séfériadès, S., 'Le problème de l'accès des particuliers à des juridictions internationales', 51 *Recueil des Cours de l'Académie de Droit International de La Haye*, 1935.
Sepúlveda, C., *El Sistema Interamericano*, 2nd. ed., Mexico: Ed. Porrúa, 1974.
Sepúlveda, C., *Derecho Internacional*, 15th ed., Mexico: Ed. Porrúa, 1986.
Truyol y Serra, A., *Histoire du Droit international*, Paris: Economica, 1995.
Urrutia, F.-J., 'La codification du Droit international en Amérique', in *Recueil des Cours de l'Académie de Droit International de La Haye* 22, 1928.
Valladão, H., *Democratização e Socialização do Direito Internacional - Os Impactos Latino-Americano e Afro-Asiático*, Rio de Janeiro: Livr. J. Olympio Ed., 1961.
Vivot, A.N., *La Doctrina Drago*, Buenos Aires: Impr. Ed. Coni Hermanos, 1911.
Von der Heydte, F.A., 'L'individu et les tribunaux internationaux', in *Recueil des Cours de l'Académie de Droit International de La Haye* 107, 1962.
Wilfred Jenks, C., *The World beyond the Charter*, London: G. Allen & Unwin, 1969.
Yepes, J.-M., 'La contribution de l'Amérique Latine au développement du Droit international public et privé', in *Recueil des Cours de l'Académie de Droit International de La Haye* 32, 1930.

4 Latin America and the Central American Court of Justice

Rosa Riquelme

I Establishment: a permanent court with compulsory jurisdiction

The settlement of controversies exclusively by pacific means is inexorably linked to the idea of a compulsory jurisdiction. Despite the failure of The Hague Peace Conferences (1899 and 1907) to broaden the idea of compulsory arbitration (Martens project 1899, and projects of the United States and the Dominican Republic, 1907),[1] it provided a model for American delegates with the objective of creating organs of compulsory jurisdiction. It was generally agreed that the generalization of this obligation would lead to a proscription of war.

In 1902, four Central American States, Costa Rica, El Salvador, Honduras and Nicaragua, created a somewhat regional Permanent Arbitral Court (PCA). By means of the Treaty of Peace and Compulsory Arbitration (*Pacto de Corinto*), the contracting governments established the principle of compulsory arbitration to settle any dispute that might occur among them via the Court of Central American Arbiters.[2] Although this Court never functioned, it was the inception of the Corte de Cartago, which today is a permanent institution.[3]

1 The Conventions for the pacific settlement of international conflicts, adopted in The Hague Peace Conferences, limited themselves to recommending arbitration 'only when the circumstances enable it', as well as to celebrating treaties 'in order to extend arbitration to all cases considered pertinent to be submitted' (Art. 38, 1907 Convention). The Permanent Arbitration Court, created in 1899, and practically unaltered in 1907 (Arts 41 to 50 of the 1907 Convention), is a kind of arbiters *reserve* formed by four national names proposed by state members, with the possibility of proposing candidates, by nationals groups, to act as judges at the International Court of Justice (Art. 5, Statute of the International Court of Justice); Riquelme Cortado, R., 'The promotion of means and methods for a pacific agreement of controversies in the centennial commemoration of the First International Conference for Peace (1899–1999)', *Anuario de Derecho Internacional*, no. XV, 1999, 385–480.

2 Art. I Treaty between Central American states Providing for the Arbitration of Differences, 20 January 1902 (Martens NGR 2, 243).

3 Hudson, M. O., 'The Central American Court of Justice', *AJIL*, vol. 26, no. 4, 1932, 759–86 (759–60); Romano, C., 'Trial and error in International Adjudication', in Romano, C., Alter, K. and Shany, Y. (ed.), *The Oxford Handbook of International Adjudication*, Oxford: Oxford University Press, 2013, 127–8.

The Central American Court of Justice (CACJ) (*Corte de Justicia Centro Americana*; Court of Cartago) cemented its origin in the Conference for Central American Peace held in Washington, from 14 November to 20 December 1907, under the auspices of the presidents of the United States of America and Mexico: Theodore Roosevelt and Porfirio Díaz. Its aim was to avert war as a method of settling existing regional conflicts.[4] Influenced by the optimism that presided over the celebration of The Hague Peace Conferences, the governments of the five Central American republics signed the General Treaty of Peace and Amity ('Washington Treaty') at the Washington Conference.[5] Art. 1 of said treaty expressed the feeling of the moment and considered as one of their first duties, in their mutual relations, the maintenance of peace, and they bind themselves to always observe the most complete harmony, and decide every difference or difficulty that may arise amongst them, of whatsoever nature it may be, by means of the CACJ, created by the Convention which they have concluded for that purpose in this date,[6] i.e. the Convention for the Establishment of a CACJ ('The Convention'),[7] signed on 20 December 1907.[8]

The Convention defined in its preamble the pacifist mission with which it was entrusted[9] in order to guarantee the state parties their rights and to maintain 'peace and harmony inalterably in their relations, without being obliged to resort in any case to the employment of force'. In keeping with this objective, the High Contracting Parties agreed to constitute and maintain a permanent tribunal 'to which they bind themselves to submit all controversies or questions which may arise among them, of whatsoever nature and no matter what their origin may be' (Art. I, Convention).

The official inauguration of the court took place on 25 May 1908 in its original headquarters in the city of Cartago (Costa Rica) and was thus known as the 'Court of Cartago', a name it maintained despite the later transferral of its headquarters to the capital, San José (see Art. V, Convention).

4 Anderson, L., 'The Peace Conference of Central America', *American Journal of International Law*, vol. 2, no. 1, 1908, 144–51.

5 Scott J. B., 'The Central American Peace Conference of 1907', *AJIL*, vol. 2, no. 1, 1908, 121–43. Anderson, L., 'The Peace Conference of Central America', *AJIL*, vol. 2, no. 1, 1908, 144–51.

6 See in *AJIL*, vol. 2, no. 1/2, Supplement, 1908, 219–29; also the Additional Convention to the General Treaty.

7 See *AJIL*, vol. 2, no. 1/2, Supplement, 1908, 251–5. Other Treaties signed during the Washington Conference (1907) were: Extradition Convention; Convention on Communications; Convention for the Establishment of an International Central American Bureau; Convention for the Establishment of a Central American Pedagogical Institute; and Convention Concerning Future Central American Conferences; see *AJIL*, vol. 2, no. 1/2, Supplement, 1908.

8 The Convention was ratified by El Salvador (20 January 1908); Nicaragua (15 February 1908); Costa Rica (28 February 1908); and Guatemala (12 March 1908).

9 Marín, E., *La labor del pacifismo y la Corte de Justicia Centroamericana*, San José: Tipografía Alsina, 1908.

II Composition

Being an institution of genuinely Central American affiliation, the Court of Cartago is composed of five Justices, one appointed by each republic, selected by their legislative power (Art. VII (1), Convention) 'from among the jurists who possess the qualifications which the laws of each country prescribe for the exercise of high judicial office; and who enjoy the highest consideration, both because of their moral character and their professional ability' (Art. VI (1), Convention).

The Court was conceived to represent 'the national conscience of Central America, wherefore the Justices who compose the Tribunal shall not consider themselves barred from the discharge of their duties because of the interest which the Republics, to which they owe their appointment, may have in any case or question' (Art. XIII, Convention).

The attendance of the five Justices was considered indispensable 'in order to make a legal quorum in the decisions of the Court' (Art. VI (3), Convention). The Court had to elect from among its own members a president and vice-president (Art. XII, Convention), however, in practice, these officers were elected each year. The Justices of the court – including substitutes – are appointed for a term of five years (with the possibility of re-election), which shall begin the day they assume their duties (Art. VIII, Convention). From that moment on, they would be entitled to the same personal immunities as those granted to the magistrates of the Supreme Court of Justice of their member states under domestic law. In other contracting republics, they would be granted the privileges and immunities due to diplomatic agents (Art. IX, Convention). The salary of the Justice of the country where the Court resides has to be fixed by the respective government; furthermore, each state should then contribute to defray the expenses originated by the court in the amount established by the Convention (Art. VII (2), Convention).

III Jurisdiction and applicable law

The Court of Cartago was the first international court with jurisdiction over sovereign states, with a permanent function and compulsory authority. It was also the first occasion that individuals gained access to an international court. The Court could have attained the status of a true constitutional regional tribunal had it not been for Costa Rican reservation in recognizing its competence to adjudge 'the conflicts which may arise between the Legislative, Executive and Judicial Powers, and when as a matter of fact the judicial decisions and resolutions of the National Congress are not respected' (Annexed Article, Convention).

The court approved its Regulations,[10] and Art. 1 gives an accurate definition of its characteristics:

10 Regulation of the Central American Court of Justice, 2 December 1911; see *AJIL*, vol. 8, no. 3, Supplement, 1914, 179–94.

> The object of the Central American Court of Justice is to guarantee with its authority, based upon the honour of the States, and within the limits of the powers which have been granted to it, the rights of each of them in their reciprocal relations, and to maintain peace and harmony among them. It is, by nature, by its attribution and by the character of its jurisdiction, a Permanent Court of International Justice, with power to adjudge and decide, upon petition, all cases included in its constitutive law.

Special emphasis was thus placed on the fact that the Court could never act at the expense of a state. This is the reason why the court was unable to initiate mediation proceedings in relation to the process of revolution in Nicaragua (1910–1912).

In relation to disputes between states, the court could seize on any matter by unilateral declaration 'in case the respective Departments of Foreign Affairs had not been able to reach an understanding' (Art. I, Convention). This requirement was very broadly interpreted by the Court, which considered an impossibility by force of circumstances as sufficient, without requiring the exhaustion of all diplomatic means.

Art. II, Convention authorized the court to attempt to judge 'questions which individuals of one Central American country may raise against any of the other contracting Governments, because of the violation of treaties or conventions, and other cases of an international character' with or without the support of their government. Two alternative conditions were established for such a claim: either to 'the remedies which the laws of the respective country provide against such violation shall have been exhausted' or where the denial of justice is required to be shown.

The court would also have jurisdiction for the cases that, by common accord, the contracting governments may submit to it, no matter whether they arise between two or more of them or between one of said governments and individuals (Art. III, Convention).[11] It shall also have jurisdiction over cases arising between any of the contracting governments and individuals, when by common accord they are submitted to it (Art. III (2), Convention). The Court can likewise take cognizance of the international questions that by special agreement any one of the Central American governments and a foreign government may have determined to submit to it (Art. IV, Convention).

As judge of its own competence, the court was authorized 'to determine its jurisdiction, interpreting the Treaties and Conventions germane to the matter in dispute, and applying the principles of international law' (Art. XXII, Convention). From the moment in which any suit was filed against one or more Governments and before a final decision pronounced 'the court may at the request of any one of the parties fix the situation in which the

11 The text of this paragraph was corrected by an additional protocol of the same date, signed by the plenipotentiaries taking part in the Conference.

contending parties must remain' (Art. XVIII, Convention), i.e. the court could fix interim provisional measures of protection.

In the exercise of its judicial function, the court shall then apply 'principles of international law' in deciding matters of law, and with respect to matters of fact, the court shall be 'governed by its free judgement' (Art. XXI, Convention).[12]

The sentences of the court were final and could not be altered once the parties in dispute were notified. The court could only be requested by any involved party to state the interpretation given to its judgment (Art. XXIV, Convention). Every final decision was to 'be rendered with the concurrence of at least three of the Justices of the court' (Art. XXIII, Convention). In the event of disagreement, one of the substitute magistrates had to be called at random 'and if still a majority of three be not thus obtained in this way, other Justices shall be successively chosen by lot until three uniform votes shall have been obtained' (Art. XXIII, Convention). In any case, the definitive sentences had to be signed by all Justices and communicated in writing to the five governments of the contracting republics; the interested parties are solemnly bound by said judgment (Art. XXV, Convention).

IV The role and importance of the court for the settlement of disputes in the region

During its short-term existence, the Court of Cartago tried ten cases, the majority of which (seven) were initiated by individuals.[13] Yet, only one of the cases initiated by an individual was admitted to the court's proceedings: in *Bermudez v Costa Rica* (1914)[14] the court ruled that the plaintiff's expulsion from his country had not been arbitrary or illegal since it had been carried out in accordance with the treaties and legislation in force at that time.[15] The

12 The Ordinance of Procedure (6 November 2012) provided (Art. 72) that the Court should 'consider the facts to which the controversy refers with absolute freedom of judgement, and the question of law upon which it may depend according to the treaties and the principles of law'; see *AJIL*, vol. 8, no. 3, Supplement, 1914, 194–213.

13 Dodd, T., 'La Corte de Justicia Centroamericana, 1907–1918: Su legado, la defensa de derechos individuales', *Revista de Historia* (Costa Rica), 12–13, 1985–6, 67–82; Montiel Argüello, A.,'La Corte de Justicia Centroamericana y los derechos humanos', in *Liber Amicorum, Colección de Estudios Jurídicos en Homenaje al prof. Dr. D. José Pérez Montero* 2, Oviedo: Servicio de Publicaciones de la Universidad de Oviedo, 1988, 933–42.

14 *Alejandro Bermúdez Nuñez v Costa Rica*, Central American Court of Justice (*Judgment*, 7 April 1914); in *Anales de la Corte de Justicia Centroamericana*, vol. 4, nos 9-11, 1-119 (p. 84); see also Hudson, M. O., 'The Central American Court of Justice', op. cit., pp. 772–3.

15 Doctrine has been commented by Castro Ramírez, M., *Cinco años en la Corte de Justicia Centroamericana: Exposición a los poderes públicos de El Salvador y a Centro América en general*, San José (Costa Rica): Imprenta Lehmann, 1918, pp. 82–106; Gutiérrez C. J., *La Corte de Justicia Centroamericana* (tesis), Tegucigalpa (Honduras): Biblioteca Judicial Dr Ricardo Gallardo, 1957, pp. 35–8; Hudson, M. O., 'The Central American Court of Justice', op. cit. pp. 772–3.

Court did not admit other cases submitted by individuals. Three cases were considered to be outside the court's jurisdiction, which were: *Castillo v Costa Rica* (1908),[16] *Irías v Nicaragua* (1909)[17] and *Escalante v Costa Rica* (*Election of President Flores of Costa Rica*) (1914).[18] The cases *Díaz v Guatemala* (1909)[19] and *Molina Larios v Honduras* (1913)[20] were dismissed due to the strict application of the rule of non-exhaustion of international remedies (Art. II, Convention). The case *Cerda v Costa Rica* (1911)[21] was rejected by the court on the grounds that the legality of a Justice could only be challenged by the contracting governments.[22]

Furthermore, three inter-state disputes were submitted to the court. The first being *Honduras v Guatemala and El Salvador* (1908),[23] in which the plaintiff alleged that the respondents had promoted and protected an internecine revolt on the Honduran territory thus violating its sovereignty. The Court, however, considered that Guatemala and El Salvador had

16 *Onofre Silva Castillo v Costa Rica*, Central American Court of Justice (*Jurisdiction*, 19 August 1908), Archivo Nacional de Costa Rica, Sección Histórica, Archivo de la Corte de Justicia Centroamericana, RR.EE. caja no. 4.

17 *Laureano Irías v José Santos Zelaya (Nicaragua)*, plaintiff 25 August 1909. See Gutiérrez, C. J., *La Corte de Justicia Centroamericana*, op. cit., pp. 23–4.

18 *Escalante v Costa Rica* (*Election of President Flores of Costa Rica*), Central American Court of Justice (*Judgment*, 3 July 1914); in *Anales de la Corte de Justicia Centroamericana*, vol. 4, nos 11–13, pp. 1–12; see also Castro Ramírez, M., *Cinco años en la Corte de Justicia Centroamericana: Exposición a los poderes públicos de El Salvador y a Centro América en general*, op. cit. p. 107–10; Hudson, M. O., 'The Central American Court of Justice', op. cit., p. 773.

19 *Pedro Andrés Fornos Díaz v Guatemala*, Central American Court of Justice (*Resolution on Inadmissibility*, 6 March 1909); in *AJIL*, vol. 3, no. 3, 1909, 737–47; see also Castro Ramírez, M., *Cinco años en la Corte de Justicia Centroamericana: Exposición a los poderes públicos de El Salvador y a Centro América en general*, op. cit. p. 28–41; Hudson, M. O., 'The Central American Court of Justice', op. cit., pp. 769–70.

20 *Felipe Molina Larios v Honduras*, Central American Court of Justice (*Judgment, Admissibility of the Action*, 10 December 1913); in *Anales de la Corte de Justicia Centroamericana*, vol. 3, nos. 1–8, 26–67; see also Castro Ramírez, M.,*Cinco años en la Corte de Justicia Centroamericana: Exposición a los poderes públicos de El Salvador y a Centro América en general*, op. cit. p. 61; Hudson, M. O., 'The Central American Court of Justice', op. cit., p. 772.

21 *Salvador Cerda v Costa Rica*, Central American Court of Justice (*Judgment, Admissibility*, 14 October 1911), in *Anales de la Corte de Justicia Centroamericana* (Tipografía de Avelino Alsina, San José 1911), vol. I, no. 3, 203.

22 Although the plaintiff of Salvador Cerda aimed the restoration of his individual rights in accordance with provisions of the Washington Treaty, the most controversial point was whether the election of Justice Gutierrez Navas was valid or not; Hudson, M. O., 'The Central America Court of Justice', op. cit., p. 768.

23 *Honduras v Guatemala and El Salvador*, Central American Court of Justice (*Judgment* 19 December 1908); see *AJIL*, vol. 3, no. 3, 1909, 729–36. See also 'The First Decision of the Central American Court of Justice', ibid., vol. 3, no. 2, 1909, 434–36. Formal complaints were lodged by *Honduras and Nicaragua v Guatemala and El Salvador*, see Central American Court of Justice (*Orders* 13 July 1908), ibid., 838–40. See Brown, P. M., 'Costa Rica v Nicaragua', *AJIL*, vol. 11, no. 1, 1917, 156–60; Hudson, M. O., 'The Central American Court of Justice', op. cit., pp. 773–4.

observed the obligation of neutrality. In any case, the formal legality of this decision became a matter of dispute as it was signed by only three magistrates and not by five, as required by Art. XXIV, Convention.[24]

The signing of the Convention between Nicaragua and United States regarding the Nicaraguan Canal Route and a Naval Base on the Gulf of Fonseca (*Bryan-Chamorro Treaty*)[25] on 5 August 1914, by which the government of Nicaragua granted the United States the right to construct an inter-oceanic canal through the San Juan river and to establish a naval base in the Gulf of Fonseca was the origin of the two following cases. In its decision on *Costa Rica v Nicaragua* of 30 September 1916,[26] the court ruled that by concluding the Bryan-Chamorro Treaty, Nicaragua had violated Costa Rica's rights to prior consultation, as laid down in the 1858 Treaty of Territorial Limits between Costa Rica and Nicaragua (Art. VIII, Cañas-Jeréz Treaty);[27] it was also according to the interpretation thereof made in the *San Juan River Case* (Cleveland Award, 1888).[28] Nicaragua had also violated Costa Rica's rights laid down in Art. IX, Washington Treaty. The Court ordered Nicaragua to maintain the *status quo ante* but declined to rule on the validity of the Bryan-Chamorro Treaty.[29]

The Court confirmed its stance by the decision rendered in the case *El Salvador v Nicaragua*. Again, it failed to rule on the validity of the Bryan-Chamorro Treaty. In its decision of 9 March 1917,[30] the Court confirmed Nicaragua's obligation to maintain the *status quo ante* and declared the waters of the Gulf of Fonseca a historic bay which possesses the characteristics of a closed sea and belongs to the three States that surround them, i.e. El Salvador, Honduras and Nicaragua.[31] The International Court of Justice,

24 Castro Ramirez, M., *Cinco años en la Corte de Justicia Centroamericana: Exposición a los poderes públicos de El Salvador y a Centro América en general*, op. cit., p. 31, Hudson, M. O., 'The Central American Court of Justice', op. cit. pp. 768–9.

25 Bryan-Chamorro Treaty was signed 5 August 1914, entered into force 22 June 1916; see in *AJIL*, vol. 10, no. 4, 1916, 258–60.

26 See *AJIL*, vol. 11, no. 1, 1917, 181–229.

27 Treaty of Territorial Limits between Costa Rica and Nicaragua, signed 15 April 1958, 48 BSFP 1049.

28 *San Juan River Case* (*Costa Rica v Nicaragua*), Award 24 March 1988, 28 RIAA, 208.

29 In the 2000s, Costa Rica and Nicaragua have sought to use the Cleveland Award of 1888 and the decision of the Central American Court of Justice of 1916 as an argument to support its own case concerning *Dispute Regarding Navigational and Related Rights(Costa Rica v Nicaragua)*; however, the International Court of Justice observes that none of the points under examination in this case was settled by these previous decisions: *Dispute Regarding Navigational and Related Rights (Costa Rica v Nicaragua), Judgment 13 July, I.C.J. Reports 2009*, par. 49.

30 *El Salvador v Nicaragua*, Central American Court of Justice (*Judgment*, 9 March 1917), *Anales de la Corte de Justicia Centroamericana*, vol. 4, nos. 16–18, 96–170; translation in *AJIL*, vol. 11, no. 3, 1917, 674–730.

31 The judgment of the Court of Cartago (1917) is in part in the form of answers by the judges to questions (twenty-four in all) formulated by the Court. The answers relevant for the present purpose are about the international legal status of the Gulf of Fonseca and on the consequences of that status for the waters of the Gulf; see ninth question, ibid., p. 717.

in the *Land Island and Maritime Frontier Dispute Case*, conferred the international legal status of the Gulf of Fonseca (*El Salvador/Honduras: Nicaragua Intervening*, 1992).[32]

Nicaragua did not accept these decisions rendered by the Court of Cartago and, in March 1917, it communicated its intention to denounce the Convention. Nevertheless, since the Convention was anyway due to expire ten years after the last ratification, pursuant to Art. XXVII of the Convention, the Court continued its functioning until April 1918, after which it was then dissolved. All subsequent attempts to revive the Court were unsuccessful. The Central American Arbitration Tribunal created by the Washington Conference of 1923[33] was completely different from the former Court of Cartago, as it did not have a body of permanent judges, but instead a list of arbiters such as the Permanent Arbitration Tribunal, created under the auspices of The Hague Peace Conference.

V Conclusion

The CACJ was the first international court of its kind with obligatory jurisdiction for all controversies that could arise between states within the region. It was also the first international court to allow individuals from a Central American state to bring charges against another state without the support of their government. This was a groundbreaking measure that has taken a century to flourish in relation to other international human rights tribunals. However, the Convention did not allow individuals to take action against their own government, nor was the court able to function as a true federal tribunal due to the reservations of Costa Rica.[34] Besides, the court's compromissory jurisdiction, provided for in Arts. III and IV of the Convention, had never been activated.

Despite the fact that the Court of Cartago had to represent 'the national conscience of Central America' (Art. XIII, Convention), the method of electing its Justices (by the legislative power of each country), and the way they received their salaries (by the respective government) were elements that strongly limited its independence. The privileges and immunities enjoyed by the Justices during their terms were insufficiently clear. Furthermore, the court lacked mechanisms to enforce the execution of its

32 *Land Island and Maritime Frontier Dispute Case (El Salvador/Honduras: Nicaragua Intervening), Judgment 11 September, I.C.J. Reports 1992*, p. 350 (269). Riquelme Cortado, R., *La intervención de terceros Estados en el proceso internacional: Un caso test, la intervención de Nicaragua en la controversia marítima Honduras/El Salvador*, Madrid: Tecnos, 1993.

33 Convention for the Establishment of an International Central American Tribunal, signed 17 February 1923; see in *AJIL*, vol. 17, no. 2, Supplement, 1923, 83–106.

34 See in these regard the critical opinion of Hudson, M. O., 'The Central American Court of Justice', op. cit., p. 785.

final judgments, a limitation that still today affects the majority of international courts and tribunals.

It should be emphasized, however, that despite its limitations and imperfections, the Court of Cartago was a truly shining example of an endeavour to resolve all regional conflicts solely by peaceful means. In fact, the proceedings of the court in inter-state disputes might have been decisive in avoiding the use of force. The responsibility of the United States in the termination of the Court's functions due to the conclusion of the Bryan-Chamorro Treaty with Nicaragua should also be duly noted, as well as the fact that although this treaty was valid until 1972, no construction had been undertaken.

In short, it can be said that the creation of the Court of Cartago shaped a decisive step in the development of methods of peaceful settlement of disputes by judicial means, thus setting the precedent for the creation of the Permanent Court of International Justice fourteen years later. At the regional level, the Court of Cartago was the inception for the CACJ, which today is the main permanent legal body of the Central American Integration System (*Sistema de la Integración Centro Americana*, SICA).[35]

Bibliography

Allain, A., *A Century of International Adjudication: The Rule of Law and Its Limits*, The Hague: TMC Asser, 2000.

Alvarez, A., 'Latin America and International Law', *AJIL*, vol. 3, no. 2, 2009, 269–353. Online. Available www.jstor.org/stable/2186127 (accessed 19 June 2015).

Anderson, L., 'The Peace Conference of Central America', *AJIL*, vol. 2, no. 1 1908, 144–51. Online. Available www.jstor.org/stable/2186563 (accessed 5 July 2015).

Bonilla Castro, A. and Malavassi Aguilar, R. E., 'Los rostros de la Casa Amarilla. Proyectos arquitectónicos e iconografía de la Corte de Justicia Centroamericana. San José (Costa Rica), 1907–1918', *Revista Costarricense de Política Exterior*, vol. XI, núm. 1, 2012, 7–30. Online. Available www.academia.edu/7780804/ (accessed 11 July 2015).

Brown, P. M., 'Costa Rica v. Nicaragua', *AJIL*, vol. 11, no. 1, 1917, 156–60. Online. Available www.jstor.org/stable/2187282 (accessed 5 July 2015).

Caflisch, L., 'Cent ans de règlement pacifique des différends interétatiques', *Recueil des Cours* 288, 2001, 245–467.

Castro Ramírez, M., *Cinco años en la Corte de Justicia Centroamericana: Exposición a los poderes públicos de El Salvador y a Centro América en general*, San José (Costa Rica): Imprenta Lehmann, 1918.

Dodd, T. J., 'La Corte de Justicia Centroamericana, 1907–1918: Su legado, la defensa de derechos individuales', *Revista de Historia* (Costa Rica), nos 12–13,

35 Riquelme, R., 'Central American Integration System', *The Max Planck Encyclopaedia of Public International Law* – MPEPIL. II, Berkshire, Buckinghamshire and Oxfordshire (United Kingdom): Oxford. University Press, 2012, pp. 37–46. Online. Available: www.mpepil.com/ (accessed 20 November 2015).

1985–86, 67–82. Online. Available www.revistas.una.ac.cr/index.php/historia/article/viewFile/3190/3049 (accessed 13 August 215).

Eyma, J., *La Cour de Justice centre-américaine*, Paris: E. Sagot, 1928.

Fernández Sánchez, P. A., 'Territorial Consequences of Central-American Conflicts Before the International Court of Justice', *The Global Community*, vol. 1. no. 5, 2006, 21–49.

Gutiérrez, C. J., *La Corte de Justicia Centroamericana* (thesis), Tegucigalpa (Honduras): Biblioteca Judicial Dr. Ricardo Gallardo, 1957. Online. Available www.csj.gob.sv/BVirtual.nsf/1004b9f7434d5ff106256b3e006d8a6f/ab99fc5ec92bb40006256b3e00747ae9?OpenDocument (accessed 11July 2015).

Hudson, M. O., 'The Central American Court of Justice.' *AJIL*, vol. 26, no. 4, 1932, 759–86. Online. Available www.jstor.org/stable/2189583 (accessed 13 August 2015).

Hull, W. I., 'Obligatory Arbitration in The Hague Conferences', *AJIL*, vol. 2, no. 4, 731–42. Online. Available www.jstor.org/stable/2186642 (accessed 19 June 2015).

Karnes, T. L., *The Failure of a Union: Central America 1824–1960*, Chapel Hill: University of North Carolina Press, 1961.

León Gómez, A., *La Corte de Managua: Defensa de su institucionalidad*, Managua: Corte Centroamericana de Justicia, 1997.

López Villamil, H., *La Corte Centroamericana de Justicia en la política internacional*, Madrid: Imprenta Juan Bravo, 1960.

Marín, E., *La labor del pacifismo y la Corte de Justicia Centroamericana*, Tipografía Alsina, 1908.

Moreno, L., *Historia de las Relaciones Interestatales de Centroamérica*, Madrid: Compañía Iberoamericana de Publicaciones, 1928.

Munro, D. G., *The five Republics of Central America (their political and economic development and their relation with the United States)*, New York: Russell, 1918.

Perotti, A. D., 'Sistema de la Integración Centroamericana: La Corte Centroamericana de Justicia', *Revista de Derecho Privado y Comunitario* (Santa Fe, Argentina) 1, 2005, 597–628.

Paz Barnica, E., *Reestructuración institucional de la integración centroamericana*, Tegucigalpa: Edición Nuevo Continente, 1972.

Politis, N., 'Une Expérience the Tribunal International Permanent en Amérique Centrale', *Revue d'Histoire Diplomatique* 36, 1922, 123–37.

Riquelme Cortado, R., 'La promoción de medios y métodos de arreglo pacífico de las controversias en la conmemoración del centenario de la primera Conferencia Internacional de la Paz (1899–1999)', *Anuario de Derecho Internacional* 15, 1999, 385–480. Online. Available http://dialnet.unirioja.es/servlet/articulo?codigo=1035744 (accessed 30 July 2015).

Riquelme Cortado, R., 'Central American Court of Justice', *The Max Planck Encyclopaedia of Public International Law* – MPEPIL. II, pp. 28–33. Online (update, 2014) Available www.mpepil.com/.

Riquelme Cortado, R., 'Central American Integration System', *The Max Planck Encyclopaedia of Public International Law* – MPEPIL. II, Berkshire, Buckinghamshire and Oxfordshire (United Kingdom): Oxford University Press, 2012, pp. 37–46. Online. Available www.mpepil.com/ (accessed 11 July 2015).

Romano, C., Alter, K. and Shany, Y. (ed.), *The Oxford Handbook of International Adjudication*, Oxford: Oxford University Press, 2013.

Scott, J. B., 'The Central American Peace Conference of 1907', *AJIL*, vol. 2, no. 1, 1908, 121–43. Online. Available www.jstor.org/stable/2186562 (accessed 5 July 2015).

Scott, J. B., 'Recommendations for the Third Conference at the Hague', *AJIL*, vol. 2, no. 4, 1908, 815–22. Online. Available www.jstor.org/stable/2186647 (accessed 20 June 2015).

Tomuschat, C., 'International Courts and Tribunals with Regionally Restricted and/or Specialized Jurisdiction', in Max Planck Institute for Comparative Public Law and International Law (ed.), *Judicial Settlement of International Disputes: International Court of Justice, Other Courts and Tribunal Arbitration and Conciliation: An International Symposium*, Berlin: Springer, pp. 285–416.

VV A cien años de la Corte de Cartago (1907–2007): Justicia, integración y derechos humanos, *Memoria del Primer Encuentro de Cortes Internacionales y Regiones de Justicia del Mundo*, Managua, 4–5 de Octubre 2007, Corte Centroamericana de Justicia, 2007.

5 Latin America and the Permanent Court of International Justice

Miriam Cohen

The history of international dispute settlement, and in particular, that of the Permanent Court of International Justice (PCIJ) cannot be recited without reference to the contribution of Latin American states. In a similar vein, since the inception of the PCIJ, international peaceful dispute settlement in Latin America has also significantly evolved. From the inception of the idea of creating a permanent court to settle international disputes, Latin American states contributed significantly to the compulsory jurisdiction of that Court. Their prominent role continued long after the creation of the PCIJ, with its successor, the International Court of Justice (ICJ).

This chapter places the important role of Latin American states in the solidification of the peaceful settlement of international disputes, from the prism of their relevant role in the creation of the PCIJ and their support of that institution. It first overviews the brief history of the PCIJ and addresses the contribution of Latin American states to the inception and foundational aspects of that Court. In this sense, it focuses on the significant role of Latin American states to the development of the idea of compulsory jurisdiction, through a review of relevant statements of delegates of the region in the negotiation of the PCIJ statute. This chapter ultimately argues that Latin American states performed a cardinal role in the creation of the Court, and more broadly, in the progressive development of international law.

I Where it all began: the short history of the Permanent Court of International Justice in the broader context of international dispute settlement

While recognizing that the history of the peaceful settlement of international disputes dates back to the classical era, the *judicial* settlement of *international* disputes is intrinsically linked with the creation of the PCIJ. The first modern international tribunal was the Central American Court of Justice, which was in operation for a short period of time, from 1907 to 1917. However, the PCIJ was effectively the first World Court, since the former was a regional Court.[1]

The history of the PCIJ cannot be properly examined without mentioning the participation of Latin American states, in its formative years, and with regard to the conception of compulsory jurisdiction, which remains relevant to this day. Similarly, the (legal) history of Latin America cannot be explained without mentioning the events that led to the creation of the PCIJ. A descriptive overview will assist in establishing the rationale for the creation of the PCIJ.

Article 14 of the Covenant of the League of Nations[2] referred to the establishment of a Court – the PCIJ – which was competent to adjudicate disputes of an international nature and to provide advisory opinions on questions referred to it by the Council or Assembly of the League of Nations.

In order to give life to this provision,[3] the Council of the League of Nations appointed in 1920 an Advisory Committee of Jurists to draft the statute of what was to become the PCIJ.[4] Soon after, a report emanated from the Advisory Committee of Jurists containing the draft scheme of the statute, which the Committee would submit to the Council in August of the same year. After further examination and amendments, the Council submitted the draft for consideration by the First Assembly of the League of Nations in its meeting in Geneva in November of the same year.

This draft statute was then submitted to a Third Committee where it underwent exhaustive study by a subcommittee. The Assembly unanimously adopted a revised version of what would become the PCIJ statute in December 1920. This statute required ratification by the states members of the Assembly. Within a few months, in September 1921, the majority of the League member states signed and ratified the statute of the PCIJ, thus rendering it into force.

1 On the Central American Court of Justice, see generally: *Central American Court of Justice*, Washington, DC: Carnegie Endowment for International Peace Library, 1928. See also, Maldonado Jordison, S., 'The Central American Court of Justice: Yesterday, Today and Tomorrow?', *Connecticut Journal of International Law* 25, 2009, 183–242, and Eyma, J., *La Cour de justice Centre Americaine*, Paris: E. Sagot & cie, 1928, 171–6.

2 According to this provision: 'The Council shall formulate and submit to the Members of the League for adoption plans for the establishment of a Permanent Court of International Justice. The Court shall be competent to hear and determine any dispute of an international character which the parties thereto submit to it. The Court may also give an advisory opinion upon any dispute or question referred to it by the Council or by the Assembly', League of Nations, *Covenant of the League of Nations*, 28 April 1919, available at: www.refworld.org/docid/3dd8b9854.html (last accessed 20 September 2015).

3 For the history of the creation of the PCIJ, see the PCIJ official website, online, available at: www.icj-cij.org/court/index.php?p1=1&p2=1 (accessed 11 July 2016). See also, Spiermann, O., 'International Legal Argument in the Permanent Court of International Justice: The Rise of the International Judiciary', in *Cambridge Studies in International and Comparative Law*, Cambridge: Cambridge University Press, 2005.

4 See, generally, Comité Consultatif de Juristes, *Procès-verbaux des Séances du Comité (16 juin – 24 juillet 1920)*, La Haye: Van Langehuysen Frères, 1920.

The first election of Judges of the Court was held on 14 September 1921. The first panel of judges comprised two Latin American judges: Ruy Barbosa from Brazil, who passed away before having heard a case, was soon replaced by Epitácio da Silva Pessoa, and Antonio Sánchez de Bustamante y Sirven from Cuba. In the Peace Palace in The Hague, the preliminary session to establish the Court's Rules opened on 30 January 1922, and on 15 February 1922 the inaugural sitting was conducted.

An important question arose concerning the composition of the PCIJ and, in this respect, Latin American states provided a lasting contribution. The Brazilian project for the creation of a Court of Arbitral Justice presented in 1907 envisaged one judge per member state and advocated for a judge from the nationality of each party to a dispute.[5] The idea that every party to a dispute have a judge of its nationality in the Court remains intact to this day.[6] When deliberating on a draft statute for the PCIJ, the Advisory Committee took into account the 1907 Brazilian project and discussed at length the question of the nationality of judges.[7] Article 31 of the PCIJ statute dealt with the question at hand.[8] In 1929, a prior revision of the statute stated that the elected judges 'should represent the main forms of civilization and the principal legal systems of the world'. The revised version of the statute came into force in 1936. Thus, this new formula provided a foundation for the operation of the PCIJ and its successor.

A total of 45 states initially signed the PCIJ statute. This event was actually a fair representation as at that time only 52 states were members of the League of Nations, and particularly because the PCIJ was not part of the League of Nations.[9]

The history of the PCIJ was short-lived, with an even shorter heyday. With the onset of World War II, the Court came to an end, having conducted its last wartime session in early 1940, prior to the German occupation of The Netherlands. In spite of its short tenure, its contribution to international law cannot be underestimated.

Within the 24-year history of the Court, it developed a powerful jurisprudence[10] and planted the seed for the notion of peaceful settlement of

5 Cf. Brazil's proposal, which counted on one judge and one substitute judge appointed by each member state: *Actes et documents de la deuxième conference*, La Haye, Vol. II 1908, 620, 1047.

6 The successor of the PCIJ – the ICJ – still adopts to this day the rule that every party to a dispute has the right to have a judge of its nationality sitting at the Court.

7 *Procès-Verbaux of the Proceedings of the Committee*, 121, 168, 197, 222, 528, 576 and 614.

8 For a detailed account of the question of the nationality of Judges before the PCIJ, see Hill, N., 'National Judges at the Permanent Court of International Justice', *The American Journal of International Law*, vol. 25 no. 4, 1931, 670–83.

9 On this point, cf. Hudson, M. O., *International Tribunals – Past and Future*, Washington: Carnegie Endowment for International Peace/Brookings Institution, 1944, 76–8.

10 Concerning the significance of the PCIJ, cf. Spiermann, O., 'International Legal Argument in the Permanent Court of International Justice: The Rise of the International Judiciary', op. cit. n. 4.

international disputes by resort to international adjudication. From 1922 to 1940, the legacy of the Court accumulated a total of 27 advisory opinions and 29 contentious cases, mostly concerning European countries.[11]

II Latin American states and the creation of the PCIJ

After providing a descriptive overview of how the PCIJ came into being, this chapter will now focus on influence of Latin American states in the creation and the subsequent functioning of the PCIJ.

According to Russel and Tokatlian, between the end of the nineteenth century and the 1930s, Latin America was concerned with safeguarding its sovereignty and the legal equality of states in light of the threat of intervention by greater powers and armed conflict by the United States, against a backdrop where Latin America responded by creating doctrine and multilateral diplomacy.[12] Amidst this setting, the discussions leading to the adoption of the PCIJ statute and its entering into force relied on strong support by Latin American countries.[13] This is not surprising in view of Latin American countries pioneering the establishment of the first regional Court of Justice: the Central American Court of Justice.

A. The role of Latin American states in the conception of the compulsory jurisdiction of the PCIJ[14]

It can be said that the Latin American countries were strong supporters of the idea of compulsory jurisdiction for the PCIJ. This is also applicable to its successor, the ICJ.[15]

The Advisory Committee of Jurists discussed the notion of compulsory jurisdiction at length while deliberating and establishing the draft statute of

11 See the PCIJ official website.

12 Russel R., and Tokatlian, J. G., 'Grand Strategy', in Dominguez, J. and Covarrubias, A. (eds), *Routledge Handbook of Latin America in the World*, London: Routledge, 2014, 58–73, at 63.

13 Concerning the Latin American contribution to international law more generally, see Caminos, H., 'The Latin American Contribution to International Law', *Proceedings of the 105th Annual Meeting (American Society of International Law)* 80, 1986, 157–72; Yepes, J. M., 'La contribution de l'Amérique Latine au développement du droit international public et privé', *Recueil des Cours de l'Académie de Droit International de la Haye* 32, The Hague: Brill Nijhoff, 1930; Urrutia, F. J., 'La codification du droit international en Amérique', *Recueil des Cours de l'Académie de Droit International de la Haye* 22, The Hague: Brill Nijhoff, 1928.

14 I am grateful to the writings of Judge Cançado Trindade concerning the history of compulsory jurisdiction before the PCIJ, which contributed to the conception of the present chapter. See e.g.: *Application of the International Convention on the Elimination of All Forms of Racial Discrimination (Georgia v. Russian Federation), Preliminary Objections, Judgment, I.C.J. Reports 2011*, p. 70, Dissenting Opinion of Judge A.A. Cançado Trindade, paras. 27–36.

15 See in this regard, ibid.

the PCIJ.[16] Had it been fully adopted, the compulsory jurisdiction of the PCIJ would have meant that the Court would have jurisdiction to hear certain disputes without the states' prior agreement.[17] The Committee had reached a consensus on a draft provision to this effect; the compulsory jurisdiction would enable the Court to hear all disputes of a 'legal' nature, while other sorts of disputes would have required states' prior consent.[18]

As Judge Cançado Trindade noted: 'It was seemingly intended that the introduction of such system of compulsory jurisdiction in disputes of a legal nature, would also extend to other cases in so far as they were covered by general or specific conventions between the (contending) parties.'[19]

The question of the compulsory jurisdiction of the Court was of the utmost importance to the Advisory Committee itself, which commented that 'the majority of [the Committee] recognized that the States forming the League of Nations, in constituting the Court, must give it a competence in cases of a legal nature, without any convention other than the constituent statute of the Court.'[20]

Notwithstanding, the Council of the League of Nations did not pursue the proposal of the Advisory Committee regarding compulsory jurisdiction; the question was thus referred to the Assembly for further discussion.[21]

Latin American states also played a key role within the Assembly in the conception of a compulsory jurisdiction of the Court. Some of the same members of the Advisory Committee of Jurists were also members of the Assembly as delegates of their countries.

The road for the adoption of the compulsory clause was much bumpier in the Assembly than it had been in the Advisory Committee of Jurists.[22] This is maybe due to the fact that the Assembly was under the impression that the draft provision relating to compulsory jurisdiction conflicted with the Covenant of the League. This question was then assigned for discussion by a Sub-Committee. While acknowledging the impasse, the Sub-Committee decided not to propose changes to the Article.

16 Comité Consultatif de Juristes, op. cit., n. 5, 218.

17 For a discussion of the compulsory jurisdiction, cf. Gross, L., 'Compulsory Jurisdiction under the Optional Clause: History and Practice', in Damrosch, L. F. (ed.), *The International Court of Justice at a Crossroads*, New York ASIL: Transnational Publishers, 1987.

18 For a detailed and insightful account of the *travaux préparatoires* concerning the compulsory jurisdiction of the PCIJ, see Cançado Trindade, A. A., *The Construction of a Humanized International Law: A Collection of Individual Opinions (1991–2013)*, Vol. 2, Leiden: Nijhoff, 2015, pp. 1174–83.

19 Ibid., 1175.

20 Comité Consultatif de Juristes, op. cit. n. 5, 727–8.

21 Cf. on the matter of compulsory jurisdiction before the Advisory Committee of Jurists: Scott, J. B., *The Project of a Permanent Court of International Justice and Resolutions of the Advisory Committee of Jurists – Report and Commentary*, Washington, 1920, 98.

22 *Documents concerning the Action Taken by the Council of the League of Nations under Article 14 of the Covenant*, Geneva, 1921, 1 *et seq.*

To be further explored in the next section, the contribution of the Brazilian delegate, Raul Fernandes, was critical to reaching a diplomatic compromise.[23] His proposal, which constituted a modification of the original draft provision, stated that states could decide whether or not to sign such a clause.[24] The revised version of his proposal determined that states wishing to submit to the compulsory jurisdiction of the Court could do so by means of a declaration. This proposal eliminated the deadlock and the Assembly approved it along with the rest of the draft statute.[25]

B. Statements of delegates of Latin American states in support of the compulsory jurisdiction of the Court

While many delegates from states outside Latin America also voiced their support for the compulsory jurisdiction of the Court,[26] it is relevant, at this stage, to mention some of the points made by Latin American delegates during the debates concerning the compulsory jurisdiction of the Court. These statements illustrate the crucial role of Latin American states in supporting the idea of compulsory jurisdiction.

The Argentinean delegate, Mr Pueyrredon, stated his view that if the Court's 'jurisdiction was not obligatory, the Court would ... be merely an arbitration tribunal'.[27] The Brazilian delegate also insisted on the maintenance of the original draft proposed by the Committee of Jurists.[28] Similarly, another Latin American delegate defended the compromissory clause: 'the parties should be obliged to bring the case before the Court if they had not agreed to submit it to arbitration or to enquiry by the Council'.[29]

23 B. C. J. Loder explains this compromise in the following terms: 'Although on the one hand it was perceived that the opposition of the Council to the proposal of the Committee of Jurists should be respected, on the other hand it would not do to overlook the wishes on the great majority, for they saw in compulsory jurisdiction the only guarantee of enforcing justice. The condition for a satisfactory solution was to find a compromise between these two views. The honour of having found this is due to the Delegate from Brazil, Senhor Fernandes, one of the ten jurists, a man as sagacious as he is energetic. To cast this solution into a form acceptable to everybody was the task of the Sub-Commission of the III Committee.' Loder, B. C. J., 'The Permanent Court of International Justice and Compulsory Jurisdiction', *British Yearbook of International Law* 2, 1921–2.

24 Cf. ibid., 107 and Annex 11.

25 Ibid., 110.

26 Although the focus of the present chapter is on Latin American states, it should also be mentioned that delegates of other states, in addition to the Latin American delegates, also spoke in strong terms in favour of the compulsory. See for example the statements of the delegates of Greece (Mr Politis), of Switzerland (Mr Huber) and of Belgium (Mr Lafontaine), op. cit.

27 Cf. League of Nations, *Records of the First Assembly – Meetings of the Third Committee (PCIJ)*, Geneva, 1920, 285.

28 Ibid.

29 Ibid., 286 (delegate of Panama). See also the views of the delegate of Portugal in the same sense.

Following the same line of reasoning, the Bolivian delegate, Mr Tamayo, supported the 'just, humane and truthful [nature] of compulsory arbitration'; not acquiring compromissory jurisdiction would be tantamount to 'promising us justice for tomorrow ... not giving it to us today'.[30] In relation to the revised draft, the Cuban delegate expressed the view that:

> Although the principle of compulsory jurisdiction is not included in the resolution of the Third Committee, we will vote for that resolution. We understand the saying of the Latin philosopher: *natura non fecit saltus.* Perfection cannot be attained in a moment from that which does not as yet exist. The laws of evolution govern all things. We must begin by building a little chapel, and in the course of time the League of Nations will be able to build a cathedral.[31]

Perhaps a statement that best captures and condenses the contribution of Latin American states to the idea of compulsory jurisdiction was from the Peruvian delegate, Mr Cornejo: 'Latin America, by a very great majority, perhaps unanimously, desires compulsory jurisdiction and the reign of peace.'[32]

It became very clear, however, in the discussions of the delegates, that although there was strong support for the idea of compulsory jurisdiction, the statute of the PCIJ would not include a compromissory clause.[33] However, in order for the Court to come into existence, a compromise was necessary.

As previously stated, it was the Brazilian delegate, Mr Fernandes, who proposed a compromise solution, currently known as the optional clause. This made the draft text (of what was to become Article 36 of the PCIJ statute) agreeable, permitting negotiations to proceed and for the statute to be adopted. His proposal was basically a compromise to ensure that the draft statute would be adopted. Indeed, he stood by the original draft and strongly criticized the modifications made in the recommendations of the Advisory Committee of Jurists by the League's Council. Thus, he stated that: 'For

30 Ibid., 248.
31 Ibid., 246–7.
32 Ibid., 244.
33 See in this regard the statement of the delegate of Portugal:

> When we realize the necessity of remaining faithful to the declarations of the preamble to the Covenant and of declaring ourselves on the question of compulsory jurisdiction, we shall be forced to accept compulsion. The tribunal we are going to found will then be provided with compulsory jurisdiction, which will be admitted by all the Members of the League of Nations. Such is my wish. I accept the institution of a Permanent Court of International Justice because I have confidence in the future. If we are not to reach the end of which I have spoken, we are deceiving ourselves. The tribunal will disappear, and with it the League of Nations, if, to settle their disputes, the Members of the League are still at the liberty to resort to war.
>
> (Ibid., 246)

legal questions the Court should have jurisdiction because the decisions of the Court are the application of the law, and make law.'[34]

To sum up, it is worth quoting the words of the Norwegian delegate, Mr Hagerup, who recognised the contribution of Latin American states to the conception of the idea of compulsory jurisdiction:

> There are already in existence a large number of conventions which provide for compulsory jurisdiction. I am happy to pay tribute here to the important part played by the States of South America in this movement. They deserve a large share of the credit for the extension on the idea. ... I wish to emphasize here how highly desirable it would be for all States, which are bound by treaties providing for compulsory jurisdiction in a general way, to modify them so that this jurisdiction shall henceforth devolve upon the Court which we are about to set up. Such an attitude will greatly help to extend the Court's jurisdiction. I have already in my first speech pointed out the importance of the motion of the Brazilian Delegation, ... according to which States which desire to extend the scope of compulsory jurisdiction are permitted to do so by means of the simple declaration. ... I have been, from the outset, a champion of the principle of compulsory jurisdiction. I can accept wholeheartedly the proposal now presented.[35]

Thus, the idea of compulsory jurisdiction, as originally conceived (prompt compulsory jurisdiction for all states) for what was to become the first World Court, died before it could even mature, but not without support from various delegates, and most importantly, those from Latin American states. The compulsory jurisdiction of the Court, through the compromissory clause as it was originally intended by the draft of the Advisory Committee, yielded to the idea of the 'optional clause' proposed by the Brazilian delegate. Indeed, the optional clause was the next best option at the time and also an important victory in order to attract supporters for the PCIJ statute. Mr Fernandez himself explained that the Advisory Committee of Jurists had the difficult task of balancing the challenge of creating the PCIJ while also making sure that the principle of juridical equality of states would not be compromised.[36]

The compromise of the optional clause was ultimately successful in guaranteeing the acceptance of the PCIJ statute and, ultimately, the creation of the Court. It 'was intended to pave the way for further development towards compulsory jurisdiction, and served its purpose in the following two decades.'[37]

34 Ibid., 285.

35 Ibid., 250.

36 Cf. Fernandes, R., *Nonagésimo Aniversario – Conferencias e Trabalhos Esparsos*, Rio de Janeiro: M.R.E., 1967.

37 *Application of the International Convention on the Elimination of All Forms of Racial Discrimination (Georgia v. Russian Federation), Preliminary Objections, Judgment, I.C.J. Reports 2011*, op. cit. n. 16, para. 38.

It stems clearly from the debates and statements made during the negotiation of the statute with regards to the compulsory jurisdiction that there were divergent opinions between states that wished to have a prompt jurisdiction of the PCIJ and those 'more powerful States' who wished to have a gradual trust of the tribunal.[38]

A review of these statements adequately encapsulate and demonstrate the degree according to which Latin American delegates, representing their states, contributed to the creation of the PCIJ and to important debates concerning its functioning, especially regarding the compulsory jurisdiction.[39]

III Concluding remarks: Latin American contribution to the PCIJ as a lasting contribution to international law

The focus of the present chapter was on the contribution of Latin American states to the creation of the PCIJ, as the first international judicial settlement Court (the first World Court), and more specifically, on the conception of the idea of compulsory jurisdiction. In addition to the support for the idea of an international court to hear legal disputes between states, the historic contribution of Latin America consisted of ensuring that the statute of the Court would be accepted by states through the optional clause compromise. The contribution of Latin American states to the PCIJ, and more generally to international law, cannot be addressed only in a single chapter, as their role goes beyond that of contributing to the creation of the PCIJ, and the latter's conception of compulsory jurisdiction. Latin America as a region has played an important role in the progressive development of international law, and its contribution to the PCIJ is only but one facet of this broader contribution.[40]

The support of Latin American states meant that the Court became a reality, and thus the peaceful judicial settlement of international disputes saw the light of day. The significance of the Court itself to the development of international law is clear: it paved the way to the permanent World Court (ICJ), it contributed to avoiding escalation of conflicts by providing a judicial forum for the settlement of disputes, and it left a lasting legacy in terms of international legal doctrine and jurisprudence. Latin American states played a role in this accomplishment and thus, it can be said, that the history of the PCIJ, and of international dispute settlement itself, cannot be told without honouring the region and its longstanding commitment to international peace and justice.

38 Cf. Cançado Trindade, A. A., op. cit. n. 13.

39 Cf. Yepes, op. cit. n. 13.

40 Concerning the broader question of Latin America's contribution to international law, see Yepes, op. cit. n. 13.

Bibliography

Books

Cançado Trindade, A. A., *The Construction of a Humanized International Law: A Collection of Individual Opinions* (1991–2013), Vol. 2, Leiden: Nijhoff, 2015.

Central American Court of Justice, Washington, DC: Carnegie Endowment for International Peace Library, 1928.

Eyma, J., *La Cour de justice Centre Americaine*, Paris: E. Sagot & cie, 1928, 171–6.

Fernandes, R., *Nonagésimo Aniversario – Conferencias e Trabalhos Esparsos,* Rio de Janeiro: M.R.E., 1967.

Hudson, M. O., *International Tribunals – Past and Future,* Washington, DC: Carnegie Endowment for International Peace/Brookings Institution, 1944. 76–8.

Scott, J. B., *The Project of a Permanent Court of International Justice and Resolutions of the Advisory Committee of Jurists – Report and Commentary*, Washington, DC: 1920.

Spiermann, O., 'International Legal Argument in the Permanent Court of International Justice: The Risae of the International Judiciary', in *Cambridge Studies in International and Comparative Law*, Cambridge: Cambridge University Press, 2005.

Journal articles

Caminos, H., 'The Latin American Contribution to International Law', *Proceedings of the 105th Annual Meeting (American Society of International Law)* 80, 1986, 157–72.

Hill, N., 'National Judges at the Permanent Court of International Justice', 25 *American Journal of International Law* 670, 1931.

Loder, B. J., 'The Permanent Court of International Justice and Compulsory Jurisdiction', *British Yearbook of International Law* 2, 1921–2.

Maldonado Jordison, S., 'The Central American Court of Justice: Yesterday, Today and Tomorrow?', *Connecticut Journal of International Law* 25, 2009, 183–242.

Urrutia, F. J., 'La codification du droit international en Amérique', *Recueil des Cours de l'Académie de Droit International de la Haye* 22, The Hague: Brill Nijhoff, 1928.

Yepes, J. M., 'La contribution de l'Amérique Latine au développement du droit international public et privé', *Recueil des Cours de l'Académie de Droit International de la Haye* 32, The Hague: Brill Nijhoff, 1930.

Official documents

Actes et documents de la deuxième conference, La Haye, Vol. II 1908, 620, 1047.

Comité Consultatif de Juristes, *Procès-verbaux des Séances du Comité (16 juin – 24 juillet 1920),* La Haye, Van Langehuysen Frères, 1920.

Documents concerning the Action Taken by the Council of the League of Nations under Article 14 of the Covenant, Geneva, 1921.

League of Nations, *Covenant of the League of Nations*, 28 April 1919. Available at: www.refworld.org/docid/3dd8b9854.html (last accessed 20 September 2015).

League of Nations, *Records of the First Assembly – Meetings of the Third Committee (PCIJ)*, Geneva, 1920.

Book chapters

Gross, L., 'Compulsory Jurisdiction under the Optional Clause: History and Practice', in L. F. Damrosch (ed.), *The International Court of Justice at a Crossroads*, New York ASIL: Transnational Publishers, 1987, 19–57.

Russel, R. and Tokatlian, J. G., 'Grand Strategy', in Dominguez, J. and Covarrubias, A. (eds), *Routledge Handbook of Latin America in the World*, London: Routledge, 2014, 58–73.

Jurisprudence

Application of the International Convention on the Elimination of All Forms of Racial Discrimination (Georgia v. Russian Federation), Preliminary Objections, Judgment, I.C.J. Reports 2011, p. 70, Dissenting Opinion of Judge A. A. Cançado Trindade.

6 Latin America and the International Court of Justice

The Pact of Bogotá

María Teresa Infante Caffi

I The Pact of Bogotá as a source for the compulsory jurisdiction of the ICJ

The American Treaty on Pacific Settlement or Pact of Bogotá, adopted at the Ninth International Conference of American States in 1948, relates to the structure of the Inter-American System.[1] The Pact inherited proposals advanced by the Inter-American Juridical Committee (IAJC) from 1945, when a draft Inter-American System of Peace emerged.[2] Draft studies conducted during the years 1938–45 and the work of the IAJC to coordinate and systematize the Inter-American Peace Agreements had demonstrated preferences for a combination of procedures, including a regional court. On the whole, the concept of an Inter-American System of Peace was enshrined in the Chapultepec Conference.[3]

Later, in Bogotá, judicial and arbitral procedures became central to the discussion and this resulted in an innovative scheme of compulsory

1 A historical introduction to the System in *The Inter-American System. Treaties, Conventions & Other Documents*. Vol. I, Legal – Political Affairs, Part I. A compilation annotated by García Amador, F. V., Historical Introduction, London; New York: Oceana Publications, 1983, pp. 1–66.

2 'Anteproyecto de "Sistema Interamericano de Paz" e Informe Anexo'. Comité Jurídico Interamericano, 1945, Unión Panamericana, Washington, D.C., 6 October 1945 Also, Lleras, A., 'El Pacto de Bogotá y la OEA. Informe sobre la IX Conferencia Internacional Americana'. *Antología*. V. El Diplomático, pp.128–247, in Morales Benítez, O., Bogotá: Villegas Editores, 2006. Available at: http://jorgeorlandomelo.com/bajar/pactodebogota.pdf (accessed 1 June 2015). Also, Valencia-Ospina, E., 'The Role of the International Court of Justice in the Pact of Bogotá', in *Liber Amicorum 'In Memoriam' of Judge José María Ruda*, The Hague: Kluwer, 2000, pp. 296–9.

3 Conferencia Interamericana sobre Problemas de la Guerra y de la Paz: Informe sobre los Resultados de la Conferencia presentado por el Consejo Directivo de la Unión Panamericana, México, D.F.; febrero 21–marzo 8, 1945. Unión Panamericana, Washington, DC, 1946, serie sobre Congresos y Conferencias N° 47, Resolución XXXIX. Also Connell Smith, G., *The Inter American System*, Oxford: Oxford University Press, 1966, pp. 209–10.

jurisdiction subject to limits and exceptions.[4]

References to the Pact in distinct cases before the Court[5] revealed new facets of this instrument, where its scope and technicalities became apparent in judicial processes, and its clauses called for interpretation. Signs of this were demonstrated by the reservations made by Bolivia,[6] Ecuador,[7] and Peru,[8] and those postulated by Argentina (signatory),[9] or the positions maintained by Venezuela (signatory).[10] Fenwick's comments confirm the above conclusion,[11] which still remain valid since that time. Under Articles LV and LVI, the Pact expressly acknowledges the right of the parties to make reservations.

4 Lupinacci, J. C., 'Los Procedimientos Jurisdiccionales en el Tratado Americano de Soluciones Pacíficas (Pacto de Bogotá)', *Anuario Uruguayo de Derecho Internacional*, 1962, pp. 174–83; Jiménez de Aréchaga, E., 'Tentativas de Reforma del Pacto de Bogotá', *Anuario Jurídico Interamericano*, 1986, pp. 3–11; Infante Caffi, M. T., 'El Pacto de Bogotá. Vigencia y Propuesta de Reforma', Comité Jurídico Interamericano, Cursos de Derecho Internacional, Vol. 2 (Parte 1), *El Sistema Interamericano (1974–2001)*, pp. 1043–63; Valencia-Ospina, E., 'The Role of the International Court of Justice in the Pact of Bogotá', in *Liber Amicorum 'In Memoriam' of Judge José María Ruda*, The Hague: Kluwer, 2000, pp. 296–9.

5 *Border and Transborder Armed Action (Nicaragua v. Honduras), Jurisdiction and Admissibility, Judgment, I.C.J. Reports 1988*; *Territorial and Maritime Dispute between Nicaragua and Honduras in the Caribbean Sea (Nicaragua v. Honduras), Judgment, I.C.J. Reports 2007*; *Territorial and Maritime Dispute (Nicaragua v. Colombia), Preliminary Objections, Judgment, I.C.J. Reports 2007*; *Dispute regarding Navigational and Related Rights (Costa Rica v. Nicaragua), Judgment, I.C.J. Reports 2009*; *Aerial Herbicide Spraying (Ecuador v. Colombia), I.C.J. Reports 2008* (discontinued); *Maritime Dispute (Peru v. Chile), I.C.J. Reports 2014*; *Construction of a Road in Costa Rica along the San Juan River (Nicaragua v. Costa Rica)* (Pending); *Certain Activities Carried out by Nicaragua in the Border Area (Costa Rica v. Nicaragua)* (pending); *Obligation to Negociate Access to the Pacific Ocean (Bolivia v. Chile)*, (pending); *Maritime Delimitation in the Caribbean Sea and the Pacific Ocean (Costa Rica v. Nicaragua)* (pending); *Alleged Violations of Sovereign Rights and Maritime Spaces in the Caribbean Sea (Nicaragua v. Colombia)* (pending); *Question of the Delimitation of the Continental Shelf between Nicaragua and Colombia beyond 200 Nautical Miles from the Nicaraguan Coast (Nicaragua v. Colombia)* (pending).

6 Signatories and Ratifications, Pact of Bogotá. Available at: www.oas.org/juridico/english/sigs/a-42.html#Bolivia (accessed 30 March 2015).

7 Ecuador: Signatories and Ratifications, Pact of Bogotá. Available at: www.oas.org/juridico/english/sigs/a-42.html#Ecuador (accessed 30 March 2015).

8 Peru: Signatories and Ratifications, Pact of Bogotá. Available at: www.oas.org/juridico/english/sigs/a-42.html#Perú (accessed 30 March 2015).

9 Argentina: Signatories and Ratifications, Pact of Bogotá. Available at: www.oas.org/juridico/english/sigs/a-42.html#Argentina (accessed 30 March 2015).

10 In 1987, Venezuela gave praise in favor of the voluntary jurisdiction of the Court and of an optional clause. Consejo Permanente de la Organización de los Estados Americanos. Comisión de Asuntos Jurídicos y Políticos, OEA/Ser. G. CP/CAJP-663/87. 19 marzo 1987, Observaciones de los Gobiernos de los Estados Miembros al Nuevo Tratado Americano de Soluciones Pacíficas (AG/RES.821 (XVI-0/86), pp. 4–5.

11 Fenwick, C. G., 'The revision of the Pact of Bogota', *American Journal of International Law* 48, 1954, p. 124.

A direct consequence of the Pact is the termination of the validity of a series of treaties as enunciated by its Article LVIII. None of these treaties granted jurisdiction to the Permanent Court of International Justice (PCIJ). Although the Pact is a treaty historically associated with the Organization of American States (OAS) and corresponds to what Article 23 of its Charter indicates, its structure and functioning is autonomous from any other treaty.[12] Some OAS members have also accepted the compulsory jurisdiction of the Court under Article 36, 2 of the Statute. On the other hand, the establishment of a PCIJ and the International Court of Justice (ICJ) shows how Articles 37, 38, and 94 of the Statute of the ICJ owe much to initiatives of Latin American countries.[13]

The characteristics of the Pact of Bogotá as a dispute settlement instrument may be systematized as follows:

- It contains a complete set of mechanisms ranging from a diplomatic means including conciliation and adjudication via judicial and arbitral procedure;
- It establishes a complex relationship between the recourse to the ICJ, the competence of the Court, and arbitration, whereas the latter may be available in case the ICJ declares itself to have no jurisdiction in controversies other than those set out in Articles V, VI and VII of the Treaty (Article XXXV). It also lists a number of causes that set a limit to the competence and admissibility of cases before international tribunals as well as other mechanisms of dispute settlement; and
- It allows for reservations by the parties with the result that they can be subject to different mechanisms or obligations in respect to each and every other party.

II The structure of the pact: a test of legal logic

Article II says that:

> The High Contracting Parties recognize the obligation to settle international controversies by regional procedures before referring them to the Security Council of the United Nations.

Consequently, in the event that a controversy arises between two or more signatory states which, in the opinion of the parties, cannot be settled by

12 Quintana, J. J., 'The Latin American Contribution to International Adjudication: the Case of the International Court of Justice', *Netherlands Law Review* 39, 1992, pp. 127–54. Also, Fuentes, X., 'Latin American states and the International Court of Justice', in Klein, N. (ed.), *Litigating International Law Disputes*, Cambridge: Cambridge University Press, 2014, pp. 79–105.

13 Quintana, J. J., 'The Latin American Contribution to International Adjudication: the Case of the International Court of Justice', op. cit. n. 12, pp. 130–1.

direct negotiations through the usual diplomatic channels, the parties bind themselves to use the procedures established in the present Treaty, in the manner and under the conditions provided for in the following articles, or, alternatively, such special procedures as, in their opinion, will permit them to arrive at a solution.

For the admissibility of a case, Article II raises a treaty interpretation issue where the question is whether it is mandatory to exhaust regional mechanisms before moving towards a general procedure. The second point posed by Article II is whether the opinion to be taken into account is the one expressed by both parties or by just one of them.

This question arose in a case between Nicaragua and Honduras,[14] where the Court followed the stricter interpretation and the hypothesis of the view of both parties was upheld.

The point emerged when the IAJC analyzed possible amendments to the Pact and the proposal to adjust the phrase to state 'in the opinion of one of the Parties' was not approved.[15] Ecuador suggested in 1973 to add the term 'just' in Article II, to the search of solutions to disputes.[16]

This approach does not differ substantively from the language of Article 26 of the OAS Charter where in case of a dispute between two or more American states, which in the opinion of one of the parties cannot be settled through the usual diplomatic channels, the parties shall agree on a method of settlement. No power is granted to the Organization to impose a method to the parties.

Article IV sets another characteristic of the system when it says that: 'Once any pacific procedure has been initiated, whether by agreement between the parties or in fulfillment of the present Treaty or a previous Pact, no other procedure may be commenced until that procedure is concluded.' This clause was discussed in the jurisdiction/admissibility phase of the *Border and Transborder Armed Actions (Nicaragua v. Honduras) case*, submitted by Nicaragua simultaneously with a lawsuit against Costa Rica in 1986.[17] Honduras contended that a 'special procedure' (*Contadora* process) was in place and that according to Article IV of the Pact of Bogotá, Nicaragua was precluded by it and by 'elementary considerations of good faith' from

14 *Border and Transborder Armed Action (Nicaragua v. Honduras), Jurisdiction and Admissibility, Judgment, I.C.J. Reports 1988*, pp. 94–6, para. 63–7.

15 CJI/RES.II-13/1985. 'Dictamen Examen del Tratado de Soluciones Pacíficas ("Pacto de Bogotá") tomando en cuenta las reservas que le han formulado los Estados signatarios del mismo, así como las razones que podrían tener algunos Estados para no ratificarlo a fin de determinar si, para asegurar su viabilidad, se requiere formular reformas a dicho instrumento', pp. 22–3.

16 Organización de los Estados Americanos, Comisión Especial para Estudiar el Sistema Interamericano y Proponer Medidas para su Reestructuración. OEA/Ser.P CEESI/doc.26/73. Vol. IV Parte 2, 8 noviembre 1973, pp. 201–6

17 *Border and Transborder Armed Actions (Nicaragua v. Costa Rica), I.C.J. Reports 1986.* Discontinued in 1987.

commencing any other procedure for pacific settlement until such time as the said process had been concluded.[18]

Neither Article II nor IV was found to bar the jurisdiction of the Court. The Court concluded that the *Contadora* process did not satisfy the conditions of negotiations laid down in Article II of the Pact,[19] and that at the time of the application (28 July 1986), the process was concluded within the meaning of Article IV, so it could not be considered either as a special procedure.[20]

III The structure of compulsory jurisdiction under the Pact of Bogotá

The Pact of Bogotá is one of the treaties that established the competence of the Court following the language of Article 36, 2 of the Statute of the ICJ.[21]

The role of the ICJ is evident since Article XXXI begins with the words 'In conformity with Article 36, paragraph 2, of the Statute of the International Court of Justice' and that

> the High Contracting Parties declare that they recognize, in relation to any other American State, the jurisdiction of the Court as compulsory *ipso facto*, without the necessity of any special agreement so long as the present Treaty is in force, in all disputes of a juridical nature that arise among them concerning:
>
> a The interpretation of a treaty;
> b Any question of international law;
> c The existence of any fact which, if established, would constitute the breach of an international obligation;
> d The nature or extent of the reparation to be made for the breach of an international obligation.

This text, a sort of contractual expression[22] of Article 36, 2 of the Statute of the ICJ places the Pact under the provision of Article 36, 1 of the said Statute.

The discussion on the grounds for compulsory jurisdiction of the ICJ in a case where the relationship between Article XXXI and XXXII (failing to

18 The *Contadora* process was defined as a combination of consultation, mediation, and negotiation.

19 *Border and Transborder Armed Action (Nicaragua v. Honduras), Jurisdiction and Admissibility, Judgment, I.C.J. Reports 1988*, p. 99, para.75.

20 Ibid., p. 105, para 93.

21 Hudson, M., *International Tribunals: Past and Future*, Washington, DC: Carnegie Endowment for International Peace and the Brookings Institution, 1944, pp. 17[illegible]–9.

22 Jiménez de Aréchaga, 'The Compulsory Jurisdiction of the International Court of Justice under the Pact of Bogotá and the Optional Clause', in Dinstein, Y. (ed.), *International Law at a Time of Perplexity Essays in Honor of Shabtai Rosenne*, Dordrecht: Martinus Nijhoff Publishers, 1988, p. 356.

reach agreement in a conciliation procedure) was brought into consideration was settled in the sense that the two articles (XXXI and XXXII) were considered separate sources of jurisdiction.[23] Therefore, Article XXXII is neither subordinate nor dependent upon a conciliation procedure previously undertaken.

A short time prior to this (1985), the IAJC[24] unsuccessfully proposed to amend Article XXXII in order to render the recourse to the ICJ subject to the previous agreement of the parties whenever they failed to achieve agreement after a conciliation procedure.

The case *Nicaragua v. Honduras* (1986–1988) gave rise to another discussion related to Article XXXI of the Pact: firstly, whether the application of this provision had to be supplemented by a declaration;[25] and, secondly, whether this Article entailed a collective declaration of acceptance of the compulsory jurisdiction of the Court. In 1988, the Court did not uphold the first assumption and in respect of the second, it affirmed that Article XXXI had to be amended by the rules of the Pact itself and not by unilateral declarations of the parties.[26] On other matters, faced with two titles of jurisdiction, the Court affirmed that, firstly, the Pact of Bogotá does not necessarily prevail over and exclude a second title of jurisdiction and, secondly, the commitment under Article XXXI of the Pact does not depend upon the existence of an optional declaration on the same matter.[27]

In the *Nicaragua v. Honduras case*, where Nicaragua stated that the competence of the Court was based on successive Honduran declarations, and/or by virtue of the provisions of Article XXXI of the Pact of Bogotá and Article 36, paragraph 1, of the Statute of the Court, the Court gave full attention to Article XXXI of the Pact.[28] The Court decided that article XXXI was independent of the unilateral declaration of acceptance of the jurisdiction of the Court. 'The commitment in Article XXXI ... is an autonomous commitment, independent of any other which the parties may have undertaken or may undertake'.[29]

23 *Border and Transborder Armed Action (Nicaragua v. Honduras), Jurisdiction and Admissibility, Judgment, I.C.J. Reports 1988*, pp. 89–90, para. 47–8.

24 CJI/RES.II-13/1985. Dictamen Examen del Tratado de Soluciones Pacíficas ('Pacto de Bogotá') tomando en cuenta las reservas que le han formulado los Estados signatarios del mismo, así como las razones que podrían tener algunos Estados para no ratificarlo a fin de determinar si, para asegurar su viabilidad, se requiere formular reformas a dicho instrumento. pp.28–30.

25 United States reservation reflects this approach. Signatories and Ratifications, Pact of Bogotá. Available at: www.oas.org/juridico/english/sigs/a-42.html#United States (accessed 30 March 2015).

26 *Border and Transborder Armed Action (Nicaragua v. Honduras), Jurisdiction and Admissibility, Judgment, I.C.J. Reports 1988*, p. 84, para. 34.

27 Ibid., p. 88, para. 41.

28 Ibid., p. 90, para. 48.

29 Ibid., p. 85, para. 36.

In the *Territorial and Maritime Dispute case* between *Nicaragua v. Colombia*,[30] the Court was called to decide whether the declarations made by the parties under the optional clause (Article 36, 2 of the Statute) could provide 'a distinct and sufficient basis of jurisdiction', irrespective of the application of the Pact of Bogotá and its Article VI.

Confronted with such a specific situation in the *Nicaragua v. Colombia case*, the ICJ refined in 2007 this approach and stated that as its jurisdiction was on the basis of Article XXXI of the Pact, there was no purpose in examining the optional clause.[31]

IV Restrictions and reservations to the Pact of Bogotá: different sources, same consequences

The Pact validates restrictions to the competence of the Court. Whenever it declares to be without jurisdiction to hear the controversy by virtue of Articles V, VI and VII of the Pact, such controversy shall be declared ended, as Article XXXIV sets forth. Also, according to the wording of Article IV, 'Once any pacific procedure has been initiated, whether by agreement between the parties or in fulfillment of the present Treaty or a previous pact, no other procedure may be commenced until that procedure is concluded.'

Article VI poses a test to the functioning of the Pact. It was introduced into the Pact following the Peruvian initiative and after a substantive discussion. As the Peruvian delegate stated in 1948, 'the article must establish the principle according to which the procedures (under the Pact) do not apply to matters that have been resolved by an agreement of the parties, by an arbitral award or by the decision of a tribunal.' Then, the delegate continued: 'it would be very dangerous to attenuate the formula. In the first place, it would be very difficult to attenuate it; secondly, it would open the door to provoke a dispute, which is exactly what we want to avoid.'[32]

In the midst of the 1970's process of restudying the Inter-American System, while acknowledging that this Article guaranteed the *pacta sunt servanda* and *res judicata* principles, Ecuador[33] sought to introduce

30 *Territorial and Maritime Dispute (Nicaragua v. Colombia), Preliminary Objections, Judgment, I.C.J. Reports 2007.*

31 Ibid., p. 872, para. 132.

32 Minutes of the Third Session of the Third Commission of the Ninth Inter-American Conference, 27 April 1948, p. 135. Ministerio de Relaciones Exteriores de Colombia, *Novena Conferencia Internacional Americana, Actas y Documentos, Vol. IV*, 1953, pp. 132–6.

33 Reformas que propone la Delegación del Ecuador al Pacto de Bogotá y a la Carta de la OEA. Organización de los Estados Americanos. CEESI, OEA/Ser.P. CEESI/Subcom.I/doc.30/73 corr.1. 24 de octubre 1973. And, Exposición del Presidente de la Delegación del Ecuador, Embajador Galo Leoro F., para fundamentar el Proyecto de Reformas al Pacto de Bogotá y a la Carta de la OEA, en materia de solución de controversias. CEESI, OEA/Ser.P. CEESI/Subcom.I/doc.31/73, 23 de octubre 1973, pp.207–208 Also, Leoro, G., 'La reforma del Tratado Americano de Soluciones Pacíficas o Pacto de Bogotá', *Anuario Jurídico Interamericano 1981*, pp. 68–9.

exceptions to the rule in respect of the validity or *de facto* unenforceability of a treaty, the interpretation or revision of an award or the interpretation of a decision. This proposal was not upheld by the Special Commission set out by the OAS to restructure the system.

Reservations made by some states upon signing or ratifying the Pact shed light on the substance of the said provision. A study conducted by the General Secretariat of the OAS (1985) commented that some reservations to Article VI were 'equal to depriving such acts from their legal effectiveness if faced with the possibility that already settled disputes might be reopened'.[34]

Upon signing and ratifying (2011), Bolivia declared that 'The Delegation of Bolivia makes a reservation with regard to Article VI, inasmuch as it considers that pacific procedures may also be applied to controversies arising from matters settled by arrangement between the Parties, when the said arrangement affects the vital interests of a state.' Chile objected to this and stated that the Pact was not in force between the two countries.[35]

Peru's reservation to Article VI was also confirmatory of the jurisdictional character of its content: 'Reservation with regard to Article XXXIII and the pertinent part of Article XXXIV, inasmuch as it considers that the exceptions of *res judicata*, resolved by settlement between the parties or governed by agreements and treaties in force, determine, in virtue of their objective and peremptory nature, the exclusion of these cases from the application of every procedure.' In 2006, Peru withdrew this reservation in the midst of the preparation of a submission against Chile.[36] Bolivia did the same in 2013.[37]

In conjunction with Articles V and VII, the character of Article VI is confirmed in Article XXXIV of the Pact that says that 'If the Court, for the reasons set forth in Articles V, VI and VII of this Treaty, declares itself to be without jurisdiction to hear the controversy, such controversy shall be declared ended.' These two issues relate to the procedural stages.

Likewise, Article 90 of the OAS Charter (as reformed in 1985) states that 'In performing their functions with respect to the peaceful settlement of disputes, the Permanent Council and the respective ad hoc committee shall observe the provisions of the Charter and the principles and standards of

34 Organization of American States, Permanent Council, OEA/Ser.G CP/doc.1560/85 (Parte II), 9 April 1985. Original: Spanish, pp. 17–8.

35 Objection by the Republic of Chile to the reservation made by the Government of the Plurinational State of Bolivia at the time it ratified the American Treaty on Pacific Settlement, Pact of Bogotá. Available at: www.oas.org/dil/a-42_objection_chile_06-15-2011.pdf (accessed 30 March 2015).

36 Signatories and Ratifications, Pact of Bogotá. Available at: www.oas.org/juridico/english/sigs/a-42.html#Perú (accessed 30 March 2015).

37 www.oas.org/dil/esp/Nota%20de%20Bolivia.%20ND-039-13.pdf (accessed 30 March 2015).

international law, as well as take into account the existence of treaties in force between the parties'.[38]

The Judgment of the Court of 13 December 2007, closing the preliminary objections phase in the *Nicaragua v. Colombia case* concerning the territorial and maritime boundary dispute, illustrates the jurisdictional system of the Pact. In this case, provisions contained in Articles VI and XXXIV were invoked by Colombia to challenge the jurisdiction of the Court. The 2007 judgment followed the approach under which 'the clear purpose of this provision was to preclude the possibility of using those procedures, and in particular judicial remedies, in order to reopen such matters as were settled between the parties to the Pact, because they had been the object of an international judicial decision or a treaty'.[39]

The Court found that the matter of sovereignty over the islands of San Andrés, Providencia and Santa Catalina had been settled by the 1928 Treaty 'within the meaning of Article VI of the Pact of Bogotá', thus considering the wording of Article I of the Treaty. Therefore, there was no need to proceed further into the interpretation of the treaty and there was nothing relating to this issue 'that could be ascertained only on merits'.[40] The Court used the term 'settled' to simplify the reference to the terms 'settled' and 'governed' employed in Article VI,[41] and concluded that it had jurisdiction under Article XXXI of the Pact to entertain the dispute concerning maritime delimitation.[42]

The subtle distinction between the terms 'settled' and 'governed', becomes relevant whenever the question is regarding whether a matter is governed by a treaty negotiated irrespective of the existence of a dispute, or it has been settled by a third party decision as a conclusion to a dispute

38 Upon signing the Protocol of Amendment of the OAS Charter (1985), Peru stated that it was only an initial, albeit significant, step in the process of restructuring the Inter-American system, as provided for in resolution AG/RES. 745 (XIV-0/84):

> The Delegation of Peru states by way of a reservation that the powers conferred upon the Secretary General in Article 116 may not be exercised for matters that have already been resolved through settlement by the parties or through the decision of an arbitrator or a judgment handed down by an international court, or that are governed by agreements or treaties in force. Also, in accordance with international law, good offices are a means of peaceful settlement whose scope has been specified in international treaties, including the Pact of Bogotá. This procedure assumes the consent of the parties, and it is in this sense that the Delegation of Peru understands the powers conferred upon the Permanent Council in the new Article 84 of this Protocol.
>
> Available at: www.oas.org/dil/treaties_A-50_Protocol_of_Cartagena_de_Indias_sign.htm#Peru (accessed 1 May 2015).

39 *Territorial and Maritime Dispute (Nicaragua v. Colombia), Preliminary Objections, Judgment, I.C.J. Reports 2007*, p. 858, para. 77.

40 Ibid., p. 861, para. 88.

41 Ibid., p. 848, para. 39

42 Ibid., p. 869, para. 120.

settlement procedure. The thesis that the Pact does not aim to reopen differences already settled as well as to protect treaties from revisionist tendencies has been advanced.[43] In the case *Bolivia v. Chile*, Chile's preliminary objection to the jurisdiction of the Court,[44] invoked Article VI of the Pact as encompassing both situations, including but not limited to former disputes either governed or settled by agreements or a decision.[45]

As to the effect of Article VI in the sphere of jurisdiction, the question arises as to the meaning of what is preliminary in light of the facts exposed by the parties, taking into account Article 79, 1 of the Rules of the Court, and whether the objection has to be decided before any other proceeding on the merits.[46]

In the case *Territorial and Maritime Dispute* (*Nicaragua v. Colombia*), judges Abraham, Al-Khasawneh, Bennouna and Ranjeva were of the opinion that the objection to the Court's jurisdiction referred to a matter that did not have preliminary character as it concerned the merits of the case. Judge Abraham[47] called this to be the effect of the unusual combination of Articles VI and XXXIV that transmuted a question of merits into a question of jurisdiction.

V Jurisdiction and compliance

Compliance and post-adjudicative enforcement or a monitoring jurisdiction of the ICJ under the Pact of Bogotá is an issue currently under proceedings before the Court following the application of Nicaragua against Colombia in 2013 entitled 'Alleged Violations of Sovereign Rights and Maritime Spaces in the Caribbean Sea'.[48] The point is whether the Court retains an inherent jurisdiction over the implementation or execution of a judgment or for ensuring its compliance.[49]

43 García-Corrochano, L. El Tratado Americano de Solución Pacífica de Controversias (Pacto de Bogotá), Agenda Internacional, 1997, pp. 55–61. Available at: http://revistas.pucp.edu.pe/index.php/agendainternacional/article/view/7179/7379 (accessed 16 September 2016).

44 For Bolivia, only previously settled disputes are addressed by Article VI. *Obligation to negotiate access to the Pacific Ocean (Bolivia v. Chile), Written Statement, 7 November 2014.*

45 *Obligation to negotiate access to the Pacific Ocean (Bolivia v. Chile), Preliminary Objection of the Republic of Chile, 15 July 2015.*

46 Thirlway, H., *The Law and Procedure of the International Court of Justice: Fifty Years of Jurisprudence*, Vol.II, Oxford: Oxford University Press, 2013, pp. 1816–24.

47 *Territorial and Maritime Dispute (Nicaragua v. Colombia), Preliminary Objections, Judgment, I.C.J. Reports 2007. Separate Opinion of Judge Abraham*, p. 905, para. 7.

48 *Alleged violations of sovereign rights and maritime spaces in the Caribbean Sea (Nicaragua v. Colombia)*. Available at: www.icj-cij.org/docket/files/155/17978.pdf (accessed 12 July 2016).

49 Pillepich, A., 'Article 94', in. Cot, J. P. and Pellet, A. (eds), *La Charte des Nations Unies, Commentaire par article*, vol. II, 3rd ed., Paris: Economica, 2005, p. 1993. Also Oellers Frahm, K., 'Article 94', in Simma, B., Khan, D.E., Norte, G. and Paulus, A. (eds), *The Charter of the United Nations* (3rd edition), Oxford: Oxford University Press, 2012, p. 1961.

For the continuation of the exercise of jurisdiction, the interplay between Article 60 of the Statute and the Pact of Bogotá must come into operation. Are there situations in which the Court may retain inherent jurisdiction? So far, the Court does not appear to have admitted that it has jurisdiction to deal with the execution of its decisions.[50] The *Nuclear Test* cases[51] show that the concept of 'inherent powers' was used in a concise manner, 'to ensure that the exercise of its jurisdiction over the merits, if and when established, shall not be frustrated, and the other being, to provide for the orderly settlement of all matters in dispute, to ensure the observance of the "inherent limitations on the exercise of the judicial function" of the Court, and to "maintain its judicial character"'.

The particulars of this issue may be aligned with Article 60 of the Statute setting out that the judgment is final and without appeal, and that in the event of a dispute as to the meaning or scope of the judgment, the Court shall construe it upon the request of any party. There is also the freedom of the parties after the conclusion of the adjudicatory proceedings, to undertake negotiations that do not purport to derogate the effect of *res judicata* of a given judicial decision.[52]

Article LVI on denunciation poses other questions to the ICJ jurisdiction, which are currently submitted to the preliminary objection raised by Colombia in the case lodged by Nicaragua before the Court entitled *Question on the Delimitation of the Continental Shelf between Nicaragua and Colombia beyond 200 Nautical Miles from the Nicaraguan Coast* (*Nicaragua v. Colombia*)[53] in 2014.

VI The functioning of the Pact of Bogotá and its bearing on the jurisdiction of the International Court of Justice

In hindsight, the situation in which the Pact of Bogotá was perceived as a useless instrument seems to have been overcome.[54] Instead, El Salvador (1973) and Colombia (2012) have denounced the Pact.

50 Guillaume, G., 'De l'exécution des décisions de la Cour internationale de Justice' (1997), in *La Cour internationale de justice à l'aube du XXIe siècle, Le regard d'un juge*, Paris: Éditions A. Pedone, 2003, pp. 178–9.

51 *Nuclear Tests (New Zealand v. France), Judgment, I.C.J. Reports 1974*, p. 463, para. 23; *Nuclear Tests (Australia v. France), Judgment, I.C.J. Reports 1974*, p. 259, para. 23.

52 Wellens, K., *Negotiations in the Case Law of the International Court of Justice*, Farnham: Ashgate, 2014, pp. 292–3.

53 *Question on the Delimitation of the Continental Shelf between Nicaragua and Colombia beyond 200 Nautical Miles from the Nicaraguan Coast (Nicaragua v. Colombia)*. Available at: www.icj-cij.org/docket/files/154/17532.pdf

54 See Consejo Permanente de la Organización de los Estados Americanos. Comisión de Seguridad Hemisférica. Solución Pacífica de Controversias en la Organización de los Estados Americanos. Available at: www.oas.org/csh/spanish/ncsdoc%20soluc%20pacif.asp#5 (accessed 30 May 2015).

Initiatives such as those taken in 1954[55] and in 1971[56] to study possible amendments or a substitution of the Pact of Bogotá encountered opposition. In 1971, the IAJC decided in favour of maintaining the Pact in force instead of proposing amendments. Following this, the process, led by the Comisión Especial para Estudiar el Sistema Interamericano y Proponer Medidas para su Reestructuración (CEESI, 1972–1975), was only conducive to reform the Rio Treaty. In this ambit, Ecuador suggested to give powers to the ICJ to decide *ex aequo et bono* upon the failure to reach an agreement in a conciliation procedure according to Article XXXII.[57]

A new cycle of studies corresponded to the reform process of the Charter of the OAS and the adoption of the Cartagena Protocol in 1985.[58] Developments of 1984 and 1985, where the IAJC devoted its work to the dispute settlement systems, provide a picture of the position of states at that time.

Upon the request of the General Assembly of the OAS in 1984 (Resolution 745 (XIV-0/84)) purporting to the study of reforms to the instruments of the Inter-American System, and of the Permanent Council of the Organization in respect of the Pact of Bogotá, in particular, the IAJC undertook a study[59] whose results were submitted to the Permanent Council. Accordingly, a draft of a new pact based on the provisions of the existing Pact was presented. Colombia then submitted a draft project along the lines of the committee's project,[60] introducing substantive changes in the recourse to arbitration and the access to the ICJ upon failure to reach agreement in a conciliation procedure. Jiménez de Aréchaga criticized this approach.[61]

55 Dictamen del Comité Jurídico Interamericano sobre el fortalecimiento del sistema interamericano de paz, Organización de los Estados Americanos, Secretaría General, Recomendaciones e Informes del Comité Jurídico Interamericano, v.X (1967/1973), Washington, 1978, p. 407. Also, Comité Jurídico Interamericano, OEA/Ser.Q/IV.3. CJI-6, Trabajos realizados por el Comité Jurídico Interamericano durante el Período Ordinario de Sesiones celebrado del 16 de agosto al 13 de septiembre de 1971, Noviembre de 1971, para. 15.

56 Ibid., para. 24.

57 Organización de los Estados Americanos, Comisión Especial para Estudiar el Sistema Interamericano y Proponer Medidas para su Reestructuración. OEA/Ser.P CEESI/doc.26/73. Vol.IV Parte 2, 8 noviembre 1973, p. 211

58 Solución Pacífica de Controversias en la Organización de los Estados Americanos. Available at: www.oas.org/csh/spanish/ncsdoc%20soluc%20pacif.asp (accessed 25 April 2015).

59 CJI/RES.II-13/1985, 'Dictamen 'Examen del Tratado Americano de Soluciones Pacíficas ('Pacto de Bogotá'), tomando en cuenta las reservas que le han formulado los Estados signatarios del mismo, así como las razones que podrían tener algunos Estados para no ratificarlo a fin de determinar si, para asegurar su viabilidad, se requiere formular reformas a dicho instrumento'.

60 Nota del Embajador Representante Permanente de Colombia en relación con el tema 'Estudio del Tratado Especial a que se refiere el Artículo 26 de la Carta de la Organización de los Estados Americanos sobre los Medios Adecuados para Resolver las Controversias y Determinar los Procedimientos Pertinentes a cada uno de los Medios Pacíficos', OEA/Ser.P, AG/doc.2030/86, 1 octubre 1986, pp. 1–15.

61 Jiménez de Aréchaga, E. '*Tentativas de reforma del Pacto de Bogotá*', Anuario Jurídico Interamericano, 1986, pp. 1–11.

In sum, the ICJ 1985 and 1986 projects did not seek to derogate the powers conferred to the ICJ. The phrase contained in the 1987 Report of the Chairman of the Informal Working Group to Study the New American Treaty of Peaceful Settlement[62] reflects the situation that has prevailed until today:

> The positions taken in the Working Group have made clear the existence of a substantive disagreement in the Organization concerning the advisability of revising or replacing the Pact of Bogotá. This disagreement transcends legal considerations and can only be resolved through a political agreement among the member states.

In turn, the Pact and its Article XXXI, jointly considered with Article VI, as part of a *corpus juris* of international law in the Americas,[63] continues to be scrutinized.

62 Report of the Chairman of the Informal Working Group to Study the New American Treaty of Peaceful Settlement, Permanent Council of the Organization of American States, Committee on Juridical and Political Affairs. OEA/Ser.G. CP/CAJP-686/87, 27 July 1987, p. 3.

63 Abello, R., 'Analyse de la Compétence de la Cour Internationale de Justice selon le Pacte de Bogotá', *Revista Colombiana de Derecho Internacional*, 2005, p. 437.

Part II

Procedural contribution of Latin American cases to the development of international law

7 Obligation to negotiate

Karel Wellens

I Introduction

Negotiation 'is an institution of international law' situating itself 'at the cross-roads of international law and international diplomacy'.[1] Negotiations basically consist 'of discussions between the interested parties with a view to reconciling divergent views, or at least understanding the different positions maintained'.[2]

The distinction between negotiations and *consultations* – such as the ones provided in the 1975 Statute on the River Uruguay and in the Ramsar Convention – both aiming at the protection of the environment and of shared natural resources – should not be stretched too far as they may gradually merge without clearly moving from one stage to another.[3]

Latin American cases have played a major role in slowly, but gradually, putting flesh to the bone of the multifaceted obligation to negotiate in good faith by providing the International Court of Justice (hereinafter 'ICJ' or the 'Court') the opportunity to judicially reinforce the central role of negotiations in the present international political and legal order. The obligation to negotiate operated as a precondition for recourse to the ICJ in the *Border and Transborder Armed Actions case*,[4] it played an important role in the community interest regime of the *Pulp Mills case*[5] and it became the very subject-matter of the dispute in the case concerning the *Obligation to Negotiate Access to the Pacific Ocean*, between Bolivia and Chile (hereinafter '*Bolivia Chile* case').[6]

1 A. Watts, 'Negotiation and International Law', in P. Borba Casella (ed.), *Dimensão Internacional do Direito. Estudos em Homenagem a G. E. do Nascimento e Silva*, São Paulo: LTr, 2000, p. 519.

2 M. Shaw, 'Settling Territorial Disputes', in *Liber Amicorum Jean-Pierre Cot. Le procès international*, Bruxelles: Bruylant, 2009, at p. 258.

3 C. Fombad, 'Consultation and Negotiation in the Pacific Settlement of International Disputes', *African Journal of International and Comparative Law*, 1, 1989, p. 709 and pp. 711–712.

4 *Border and Transborder Armed Actions (Nicaragua v. Honduras), Jurisdiction and Admissibility, Judgment, I.C.J. Reports 1988.*

5 *Pulp Mills on the River Uruguay (Argentina v. Uruguay), Judgment, I.C.J. Reports 2010.*

6 *Obligation to Negotiate Access to the Pacific Ocean (Bolivia v. Chile), Application Instituting proceedings, I.C.J. 2013.*

II The legal basis of the obligation

The freedom of any state to choose and conduct its foreign policy in 'co-ordination with that of another State'[7] constitutes the very basis of any process of negotiation. Although negotiation may still 'widely (be) regarded as essentially a voluntary process'[8] and there may be *a mere need to negotiate* in case of preferential rights such as the fullest and most unrestricted right of transit granted to Bolivia pursuant to Article VI of the 1904 Treaty of Peace and Friendship with Chile,[9] *the duty to negotiate* has gradually evolved to become one of the general principles of contemporary international law – although perhaps not as a matter of customary international law[10] – a 'principle which underlies all international relations'.[11]

There is no exhaustion rule under general international law,[12] but, for instance, Articles 12 and 60 of the 1975 Statute on the River Uruguay[13] contain a legal obligation to negotiate prior to instituting proceedings before the ICJ.

Paragraph 13 of the San Isidro de Coronado Agreement of 12 December 1989 is a fine example of a *pactum de contrahendo*, as it obliged the parties to 'negotiate an extrajudicial settlement of the dispute', which was the subject of the proceedings.[14] On the other hand, it is only rarely that states find themselves under an obligation *not to negotiate* without prior consent of another state, such as, for instance, Chile and Peru pursuant to Article 1 of the Supplementary Protocol to the 1929 Lima Treaty and Nicaragua and Costa Rica pursuant to Article VIII of the 1858 Treaty of Limits.[15]

In the *Pulp Mills* case, the Court repeatedly underlined the necessity for parties to 'implement in good faith the consultation and co-operation procedures provided for 'through the use of a joint mechanism established by that treaty'.[16]

7 *Military and Paramilitary Activities in and against Nicaragua (Nicaragua v. United States of America), Merits, Judgment, I.C.J. Reports 1986*, p. 133, paragraph 265.
8 A. Watts, op. cit., p. 525.
9 Treaty between the Republics of Chile and Bolivia (Treaty of Peace and Friendship, Santiago, 20 October 1904).
10 M. Rogoff, 'The Obligation to Negotiate in International Law: Rules and Realities', *Michigan Journal of International Law*, *16*, 1979, p. 153.
11 *North Sea Continental Shelf, Judgment, I.C.J. Reports 1969*, p. 48, paragraph 86.
12 *Land and Maritime Boundary between Cameroon and Nigeria, Preliminary Objections, Judgment, I.C.J. Reports 1998*, p. 303, paragraph 56.
13 Statute of the River Uruguay (1975 Statute) (Salto, 26 February 1975, 1295 UNTS 21425).
14 Treaty of Territorial Limits between Costa Rica and Nicaragua (1858 Treaty of Limits) (San José, 15 April 1858). *Border and Transborder Armed Actions (Nicaragua v. Honduras), Order of 14 December 1989, I.C.J. Reports 1989*, p. 175.
15 *Maritime Dispute (Peru v. Chile), Judgment, I.C.J. Reports 2014*, p. 3, at p. 52, paragraph 131. *Dispute regarding Navigational and Related rights (Costa Rica v. Nicaragua), Judgment, I.C.J. Reports 2009*, p. 251, paragraph 93.
16 *Pulp Mills on the River Uruguay (Argentina v. Uruguay), Provisional Measures, Order of 13 July 2006, I.C.J. Reports 2006*, p. 134, paragraph 82 and *Order of 23 January 2007, I.C.J. Reports 2007*, p. 16, paragraph 53.

In the *Certain Activities* case, the Court by way of provisional measure and *motu proprio*, judicially imposed the obligation to negotiate on one party but, after having joined the cases, it did not accede to the request by Nicaragua to expand the scope to both parties when modifying its previous order.[17] At the merits stage, the Court reiterated 'the value of ongoing co-operation between the parties in the performance of their respective obligations in connection with the San Juan River', one of these obligations being to consult and to negotiate.[18]

According to Bolivia, the legal basis for Chile's obligation to negotiate has to be found in unilateral acts, such as an offer to negotiate and commitments to do so, and 'agreements, diplomatic practice and a series of declarations' creating legitimate expectations.[19] Judge Gaja mentioned 'the possibility *to infer* from negotiations an obligation to negotiate'.[20]

III The nature and scope of the obligation

The 'obligation *to be ready to negotiate* with a view to concluding an agreement' today still represents 'the minimum of international co-operation'.[21] In the *Border and Transborder Armed Actions* case – where the focus of the obligation was on the peaceful settlement of the dispute – the Court, irrespective of the opinion of the parties, took into account the nature of the *Contadora* process and the fact that no other negotiation was *contemplated* at the critical date to conclude that the dispute was not 'capable of being settled by direct negotiation through the usual diplomatic channels'.[22]

Conclusive material is needed to convince the Court that *participation in a process* renders unpersuasive a party's position that the matter has

17 *Certain Activities Carried out by Nicaragua in the Border Area (Costa Rica v. Nicaragua), Provisional Measures, Order of 8 March 2011, I.C.J. Reports 2011*, p. 6, at p. 27, paragraph 86 (2) and *Certain Activities Carried out by Nicaragua in the Border Area (Costa Rica v. Nicaragua): Construction of a Road in Costa Rica along the San Juan River (Nicaragua v. Costa Rica), Order of 16 July 2013, Provisional Measures, I.C.J. Reports 2013*, p. 230, at p. 233, paragraph 14 and at p. 240, paragraph 40 (1).

18 *Certain Activities Carried out by Nicaragua in the Border Area (Costa Rica v. Nicaragua), Construction of a Road in Costa Rica along the San Juan River (Nicaragua v. Costa Rica), Judgment of 16 December 2015, I.C.J. Reports 2015 (not yet reported)*, paragraph 223.

19 *Obligation to Negotiate Access to the Pacific Ocean (Bolivia v. Chile)*, Application instituting proceedings, III, paragraph 16, 17–18 and IV, 31.

20 *I.C.J., Declaration of judge Gaja, Obligation to Negotiate Access to the Pacific Ocean (Bolivia v. Chile), Preliminary Objection, Judgment of 24 September 2015, I.C.J. Reports 2015, (not yet reported)*, paragraph 4 (emphasis added).

21 *I.C.J., Dissenting Opinion of judge De Visscher, International status of South-West Africa, Advisory Opinion: I.C.J. Reports 1950*, p. 128, at p. 189 (emphasis added).

22 *Border and Transborder Armed Actions (Nicaragua v. Honduras), Jurisdiction and Admissibility, Judgment, I.C.J. Reports 1988*, p. 94, paragraph 63, p. 95, paragraph 65 and p. 99, paragraph 75 (emphasis added).

already been settled.[23] Bolivia used a similar argument in its case against Chile.[24]

From the Court's judgment in the *Pulp Mills* case it may be concluded that the obligation to negotiate – oriented towards reaching an agreement on the modalities of execution of the 1975 Statute – is a *merely procedural* one 'essential to co-operation between States'[25] but also a 'means to bring about the peaceful settlement of a dispute that is a *substantive obligation* under both general and conventional international law'.[26] In the *Bolivia v. Chile* case the nature of the alleged obligation could turn out to be more substantive than procedural.

As rightly noted by Judge *ad hoc* Arbour it is not always clear whether the judicially imposable obligation as requested by Bolivia is merely one of conduct or of result.[27] Although the Court made it clear that it is not asked 'to adjudge and declare that Bolivia has a sovereign right of access' and that accepting the existence of the obligation would not result in predetermining the outcome of the process as argued by Chile,[28] it is difficult to understand Bolivia's third submission as not implying such a right, the line of distinction being rather thin indeed.[29]

The obligation is not unlimited in time as both Articles VI and XXII of the Pact of Bogotá[30] and Article 12 of the 1975 Statute illustrate.

Subject to the Court's determination, the obligation to negotiate comes to a (temporary) end when the process has led to a successful outcome, when a particular mechanism may be considered to have been concluded and/or when according to (one of) the parties the dispute cannot be settled.

Despite the common view of the parties, the Court ruled that the *Contadora* process had been concluded: a definitive failure was not necessary, it was sufficient that there was no prospect of it being continued or resumed. The fact that the *Contadora* process played different roles at different times

23 *Territorial and Maritime Dispute (Nicaragua v. Colombia), Preliminary Objections, Judgment, I.C.J. Reports 2007*, p. 832, at p. 866, paragraph 109, at p.867, paragraph 113 and at pp. 868–869, paragraph 119.

24 *Judgment of 24 September 2015, I.C.J. Reports 2015, (not yet reported)*, paragraph 47.

25 *Pulp Mills on the River Uruguay (Argentina v. Uruguay) Judgment, I.C.J. Reports 2010*, p. 67, paragraph 145.

26 K. Wellens, *Negotiations in the Case Law of the International Court of Justice. A Functional Analysis*, Farnham: Ashgate, 2014, p. 32.

27 *I.C.J., Dissenting Opinion of Judge* ad hoc *Arbour, Obligation to Negotiate Access to the Pacific Ocean (Bolivia v. Chile), Preliminary Objection, Judgment of 24 September 2015, I.C.J. Reports 2015, (not yet reported)*, paragraphs 6–8 and 12–13.

28 *Obligation to Negotiate Access to the Pacific Ocean (Bolivia v. Chile), Preliminary Objection, Judgment of 24 September 2015, I.C.J. Reports 2015, (not yet reported)*, paragraph 32.

29 *I.C.J., Dissenting Opinion of Judge* ad hoc *Arbour, Obligation to Negotiate Access to the Pacific Ocean (Bolivia v. Chile), Preliminary Objection, Judgment of 24 September 2015, I.C.J. Reports 2015, (not yet reported)*, paragraph 24.

30 American Treaty on Pacific Settlement (Pact of Bogotá) (Bogotá, 30 April 1948, 30 UNTS 449).

and the *Contadora* Group's own assessment were decisive elements in this regard.[31]

During proceedings, parties have co-operational duties towards each other and towards the Court and these might entail 'perhaps in *statu nascendi*, a corollary duty to continue negotiations and to resume a process which has previously failed'.[32]

IV The content of the obligation

The obligation to negotiate in good faith – 'firmly rooted in general international law'[33] – implies that, in order to render the process of negotiations potentially successful, 'parties have to take the necessary measures in order to create a climate that is conducive' to the process,[34] 'to maintain a constructive atmosphere' and 'to refrain from any conduct which might undermine' the process and its progress.[35] The *Pulp Mills* case has *amply* shown the importance of this latter duty.

'[T]o *embark* on disputed activities' during the period of negotiations provided for in the 1975 Statute would deprive the relevant procedural provisions from their purpose and constitute non-compliance with the obligation to negotiate.[36]

Substantive good faith requires the parties to conduct themselves in a loyal and reasonable way by giving serious consideration to proposals tabled by the other party whereas *procedural good faith* 'requires *inter alia* the willingness to compromise and in case of deadlock further negotiation'.[37]

It is jurisprudence *constante* that the 'content of the obligation to negotiate' is that states have 'not only to enter into negotiations with a view to arriving at an agreement, and not merely go through a formal process of negotiations … [but that] they are under an obligation so to conduct

31 *Border and Transborder Armed Actions (Nicaragua v. Honduras), Jurisdiction and Admissibility, Judgment, I.C.J. Reports 1988*, p. 100, paragraph 80 and pp. 104–105, paragraphs 89–92.

32 Wellens, op. cit., p. 196.

33 *I.C.J., Separate Opinion of Judge Greenwood, Pulp Mills on the River Uruguay (Argentina v. Uruguay) Judgment, I.C.J. Reports 2010*, p. 14, at p. 226, paragraph 16.

34 Wellens, op. cit., p. 296.

35 General Assembly Resolution 53/101, Principles and Guidelines for International Negotiations, para. 2 (e).

36 *Pulp Mills on the River Uruguay (Argentina v. Uruguay), Judgment, I.C.J. Reports 2010*, p. 14, at p. 66, paragraph 143 (emphasis added) and at p. 67, paragraph 147. Individual judges disagreed with the majority on the scope of this prohibition and as to whether or not the actions had taken place during a still ongoing process of negotiations: Separate Opinion of Judge Greenwood, at pp. 225–226, paragraph 13, at p. 227, paragraph 16 and at p. 229, paragraph 22, and Separate Opinion of Judge Keith, at pp. 130–131, paragraph 25.

37 T. Hassan, 'Good Faith in Treaty Formation', *Virginia Journal of International Law* 21, 1981, 443–481, at 476, 478 and 479.

themselves that the negotiations are meaningful, which will not be the case when either of them insist upon its own position without contemplating any modification of it'. States have 'to pursue them as far as possible with a view to concluding agreements' even 'if an obligation to negotiate (does) not imply an obligation to reach agreement'.[38]

The *Pulp Mills* case provided an opportunity for reaffirmation of these fundamental requirements and for judges to add that the participation in the process 'should be with an open mind', that 'a commitment to negotiate does not imply an obligation to agree' and that the process should not necessarily 'lead to a particular outcome'. The 'commitment to negotiate does not entail an obligation to reach an agreement'.[39]

Insisting 'on a multilateral framework (such as the *Contadora* process) as the only acceptable basis for negotiating' does not only make a claim to offer to enter into a process untenable,[40] but it runs counter to the Court's requirements on good faith and it comes close to coercive diplomacy.

In order to properly manage their transboundary river regime, both states in the *Pulp Mills* case have established the institutional mechanism of CARU (*Comisión Administradora del Río Uruguay).* Their liberty to derogate from the treaty-based procedures gave rise to divergent opinions between the parties and among the judges. At the provisional measures stage, the Court ruled that consultation and co-operation procedures provided for have to be implemented in good faith and that CARU would constitute 'the envisaged forum'.[41] At the merits stage – although confirming that the parties were 'entitled to depart from the procedures laid down by the 1975 Statute' – the Court was of the opinion that 'both parties have the obligation to enable CARU' to play its role and 'to continue their co-operation through CARU'.[42] Bypassing the original institutional framework through direct negotiations in a different institutional setting such as GTAN (*Grupo Técnico de Alto Nivel)* does not necessarily lead to a successful outcome as the case illustrates.

Denial of the existence of a dispute as allegedly Chile does in the *Bolivia*

38 *North Sea Continental Shelf, Judgment, I.C.J. Reports 1969*, p. 47, paragraph 85 and pp. 47–48, paragraph 87.

39 *Pulp Mills on the River Uruguay (Argentina v. Uruguay) Judgment, I.C.J. Reports 2010*, p. 67, para. 146, and Separate Opinion of Judge *ad hoc* Torres Bernardez, p. 261, paragraph 92 and Separate Opinion of Judge Greenwood, p. 226, paragraph 16.

40 *I.C.J., Separate Opinion of Judge Shahabuddeen, Border and Transborder Armed Actions (Nicaragua v. Honduras), Jurisdiction and Admissibility, Judgment, I.C.J. Reports 1988*, p. 69, p. 151.

41 *Pulp Mills on the River Uruguay (Argentina v. Uruguay), Provisional Measures, Order of 13 July 2006, I.C.J. Reports 2006*, p. 134, paragraph 82 and *Order of 23 January 2007, I.C.J. Reports 2007*, pp.16–17, paragraph 53.

42 *Pulp Mills on the River Uruguay (Argentina v. Uruguay) Judgment, I.C.J. Reports 2010*, p. 62, para. 128 and p. 101, paragraph 266.

Chile case[43] does not *ipso facto* prove bad faith.[44] Whether such an attitude 'closes any possibility of negotiating a solution',[45] as Bolivia asserts, will be for the Court to decide.

V Assessment of the process of negotiation and claims of non-compliance

States and the Court have to make their own assessment of the process of negotiations and of claims of non-compliance.

In the *Border and Transborder Armed Actions* case, parties disagreed as to when the process really started.[46] In the *Pulp Mills* case, parties disagreed on which of the procedures on prior negotiations had not been followed, although after having changed the format of the process both parties determined 'that the negotiations undertaken within GTAN have failed'.[47]

While in the *Border and Transborder Armed Actions* case, the Court was – given the jurisdictional stage of the proceedings – '*not called to pronounce on the legal consequences* of a breakdown in the negotiations', but merely to determine the nature of the *Contadora* process, it still noted 'a more guarded attitude' of Honduras compared with a positive approach by Nicaragua towards the *Contadora* process[48] during which '*considerable progress* has been achieved on (its) *main objective*'.[49]

Although it 'is not within the *inherent jurisdiction* of the Court to make *motu proprio* a judicial assessment of the parties' performance,' there is a duty for the Court 'to assess *motu proprio* the (partial) success of those negotiations',[50] since a dispute must *continue* 'to exist at the time when the Court makes its decision'.[51]

Until the *Bolivia Chile* case, it was rather 'unlikely that issues', such as assessment of the process of negotiations and claims of non-compliance with

43 *Obligation to Negotiate Access to the Pacific Ocean (Bolivia v. Chile), Application instituting proceedings*, III, paragraph 29.

44 E. Zöller, *La bonne foi en droit international public*, Paris: Pedone, 1977, p. 131.

45 *Obligation to Negotiate Access to the Pacific Ocean (Bolivia v. Chile), Application instituting proceedings, I,* paragraph 4.

46 *Border and Transborder Armed Actions (Nicaragua v. Honduras), Jurisdiction and Admissibility, Judgment, I.C.J. Reports 1988*, p. 98, paragraph 71.

47 *Pulp Mills on the River Uruguay (Argentina v. Uruguay) Judgment, I.C.J. Reports 2010*, pp. 64–65, paragraph 135 and p. 38, paragraph 40 and p. 70, paragraph 157.

48 *Border and Transborder Armed Actions (Nicaragua v. Honduras), Jurisdiction and Admissibility, Judgment, I.C.J. Reports 1988*, p. 99, paragraph 75 (emphasis added) and p. 98, paragraph 73.

49 *Military and Paramilitary Activities in and against Nicaragua (Nicaragua v. United States of America), Merits, Judgment, I.C.J. Reports 1986*, p. 145, paragraph 291 (emphasis added).

50 Wellens, op. cit., p. 50 and p. 198.

51 *Nuclear Tests (New Zealand v. France), Judgment, I.C.J. Reports 1974*, p. 476, paragraph 58.

the obligation, would 'be resolved by a judicial body'.[52] Bolivia's application obliges the Court to make such an assessment during which it may encounter four different scenarios: 'the deadlock may still be there; there is no chance of resumed negotiations; there is no deadlock; and there is a chance of resumed negotiations'.[53] A survey of the process of negotiations will allow the Court to assess any prospect of progress.[54]

The alleged violation not to have negotiated consists in an omission. In such circumstances the Court requires 'a lower standard of proof for State responsibility to be incurred'.[55] With regard to the alleged recent change of Chile's position into a refusal to resume the negotiations, the Court will attach 'corroborative weight' to 'the conduct of the parties in negotiations' during the twentieth century and the first 15 years of the twenty-first century.[56]

Bolivia has to provide the Court with sufficient evidence to make it conclude that one party 'has consistently refused to consult' in order to attribute the failure of the negotiations solely on Chile, while it remains to be seen whether it can avoid 'to pronounce on the Parties' responsibilities for the failure of the negotiations'.[57] Pursuant to the Court's requirement, the material presented by the parties should be conclusive to allow the Court 'to evaluate the significance of the meetings held'.[58] It is clear that given Bolivia's submissions 'a cursory examination of the views, propositions and arguments consistently maintained by the two sides'[59] will not suffice.

As a result of Bolivia's third submission – which seems to imply *a contrario* a claim of Chile's bad faith – the Court's judgment might 'entail a finding of bad faith which is not explicitly expressed'.[60] In order to find 'the existence of bad faith "there is something more needed" than the failure of particular

52 Rogoff, op. cit., p. 172.

53 S. Torres Bernardez, 'Are Prior Negotiations a General Condition for Judicial Settlement by the International Court of Justice', in C. Barea et al. (eds.), *Liber Amicorum in Memoriam of Judge José Maria Ruda*, The Hague: Kluwer Law International, 2000, p. 515, paragraph 30.

54 *Border and Transborder Armed Actions (Nicaragua v. Honduras), Jurisdiction and Admissibility, Judgment, I.C.J. Reports 1988*, p.101, paragraph 81.

55 K. del Mar, 'The International Court of Justice and Standards of Proof', in K. Bannelier, T. Christakis and S. Heathcote (eds.), *The ICJ and the Evolution of International Law: The Enduring Impact of the Corfu Channel Case*, London: Routledge, 2012, p. 108.

56 *Land, Island and Maritime Frontier Dispute (El Salvador/Honduras: Nicaragua intervening), Judgment, I.C.J. Reports 1992*, p. 550, paragraph 317.

57 *Gabcikovo-Nagymaros Project (Hungary/Slovakia), Judgment, I.C.J. Reports 1997*, p. 7, at p. 66, paragraph 107 and Dissenting Opinion of Judge Herczegh, at p. 192.

58 *Territorial and Maritime Dispute (Nicaragua v. Colombia), Preliminary Objections, Judgment, I.C.J. Reports 2007*, p. 832, at pp. 868–869, paragraph 119.

59 *South West Africa cases (Ethiopia v. South Africa; Liberia v. South Africa), Preliminary Objections, Judgment of 21 December 1962: I.C.J. Reports 1962*, p. 319, at p. 345.

60 *I.C.J., Dissenting Opinion of Judge Yusuf, Whaling in the Antarctic (Australia v. Japan: New Zealand intervening), Judgment, I.C.J. Reports 2014*, p. 226, at p. 402, paragraph 54.

negotiations'. The lack of result is not the standard to measure whether 'the obligation to negotiate has been undertaken in good faith'.[61] 'Clear and convincing evidence' has to show 'that the party concerned did not *intend* to engage in meaningful negotiations'.[62]

When dealing with Bolivia's request that Chile has to comply *promptly* with its alleged obligation to negotiate, the Court will take into account that 'what period of time may be involved in the observance of duties to consult and negotiate' are 'matters which necessarily vary according to the requirements of the particular case'.[63]

VI Conclusion

Constraints of space do not allow us to touch upon another aspect of the obligation to negotiate. Indeed, *reasonableness* appears to play an important role in both the procedural and substantive faces of the obligation: the efforts to negotiate and reach an agreement should be made within a reasonable period of time – that is until and/or unless a reasonable possibility of the process becoming successful is absent – and the efforts themselves – such as tabling proposals and responding to them – should be reasonable.

The *acquis* surfacing from the cases briefly reviewed here may be summarized in the following way.

The question whether a dispute had already been finally settled, whether it could not be settled or whether it still could be settled has divided the parties in the cases and will most probably continue to divide parties confronted with similar exclusion clauses. By skilfully – but not without attracting criticism from within the Bench[64] – distinguishing the subject-matter of the dispute before it from the matter arguably already settled in the 1904 Treaty, the Court, in the *Bolivia Chile* case, avoided a decision on the question whether the matter of the sovereign access to the Pacific Ocean had indeed already been settled.

The *Border and Transborder Armed Actions* and *Pulp Mills* cases have reaffirmed that it is first of all up to the parties to make an assessment of their negotiations, subject to the Court's own evaluation of the process and which as the cases have demonstrated may differ from that of the parties.

61 *Application of the Interim Accord of 13 September 1995 (the former Yugoslav Republic of Macedonia v. Greece), Judgment of 5 December 2011, I.C.J. Reports 2011*, p. 644, at p. 685, paragraphs 132 and 134.

62 Ibid. and *I.C.J., Dissenting Opinion of Judge Greenwood, Pulp Mills on the River Uruguay (Argentina v. Uruguay) Judgment, I.C.J. Reports 2010*, p. 14, at p. 227, paragraph 16 (emphasis added).

63 *Interpretation of the Agreement of 2 March 1951 between the WHO and Egypt, Advisory Opinion, I.C.J. Reports 1980*, p. 73, at p. 96 paragraph 49.

64 *Obligation to Negotiate Access to the Pacific Ocean (Bolivia v. Chile), Preliminary Objection, Judgment of 24 September 2015, I.C.J. Reports 2015*, (not yet reported), Declaration of Judge Gaja, paragraph 4, Separate Opinion of Judge Trindade, paragraph 67, and Dissenting Opinion of Judge *ad hoc* Arbour, paragraphs 29 and 30.

In the *Border and Transborder Armed Actions* case, the Court reaffirmed that negotiations should have been pursued 'as far as possible with a view to settling the dispute'.[65]

In the *Pulp Mills* case, the Court made it clear that during a process of negotiations states should abstain from even embarking on activities that would render applicable treaty provisions without purpose and would be contrary to the procedural obligation to negotiate. Having noted the failure of both the original and alternative processes, the Court unreservedly directed the parties back to the negotiating forum as it was established by the 1975 Statute.

The *Bolivia Chile* case brings together various aspects of the obligation to negotiate as they manifested themselves in the *Border and Transborder Armed Actions* case and the *Pulp Mills* case. The Court has another opportunity – Bolivia's submissions remaining unchanged at the merits stage – to refine its judicial policy in this regard.

In the *Bolivia Chile* case, the Court will have to pronounce itself on the independent existence of the obligation to negotiate. It will have to adjudge and declare alleged state responsibility for non-compliance[66] and in doing so decide whether the alleged violation, in its procedural aspect – which given its specific content gives rise to a separate ground for responsibility[67] – is of such a gravity that it amounts to a *material breach*.[68] The Court may possibly judicially impose upon Chile – and inevitably also upon Bolivia – the obligation to negotiate in good faith without predetermining the potential outcome of the process while staying well within the boundaries of its judicial function, as it did in *Haya de la Torre*[69] and duly taking into account whether there is a reasonable possibility for a successful outcome.

In case of judicially imposed negotiations, the Court may expect to be informed 'in *particular with regard to the way in which they have been carried out, or their duration, scope or state of progress*'.[70]

65 *Border and Transborder Armed Actions (Nicaragua v. Honduras), Jurisdiction and Admissibility, Judgment, I.C.J. Reports 1988*, p. 69, at p. 95, paragraph 65.

66 *Judgment of 24 September 2015, I.C.J. Reports 2015, (not yet reported)*, paragraph 31.

67 *Pulp Mills on the River Uruguay (Argentina v. Uruguay) Judgment, I.C.J. Reports 2010*, p. 14, at p. 49, paragraph 79.

68 *I.C.J., Dissenting Opinion of Judge* ad hoc *Torres Bernardez, Pulp Mills on the River Uruguay (Argentina v. Uruguay) Judgment, I.C.J. Reports 2010*, p. 14, at p. 265, paragraph 103.

69 *Haya de la Torre (Colombia v. Peru), Judgment, I.C.J. Reports 1955.*

70 *Arrest Warrant of 11 April 2000 (Democratic Republic of the Congo v. Belgium), Provisional Measures, Order of 8 December 2000, I.C.J. Reports 2000*, p. 182, at 200, paragraph 66 (emphasis added).

Bibliography

C. Fombad, 'Consultation and Negotiation in the Pacific Settlement of International Disputes', (1989) 1, *African Journal of International and Comparative Law*, pp. 707–24.

T. Hassan, 'Good Faith in Treaty Formation', (1981), 21 *Virginia Journal of International Law* 1, 443–81.

K. del Mar, 'The International Court of Justice and Standards of Proof', in K. Bannelier, T. Christakis and S. Heathcote (eds.), *The ICJ and the Evolution of International Law: The Enduring Impact of the Corfu Channel Case*, London: Routledge, 2012, pp. 98–123.

M. Rogoff, 'The Obligation to Negotiate in International Law: Rules and Realities' (1979) 16 *Michigan Journal of International Law*, 141–85.

M. Shaw, 'Settling Territorial Disputes', in *Liber Amicorum Jean-Pierre Cot. Le procès international*, Bruxelles: Bruylant, 2009, pp. 255–81.

S. Torres Bernardez, 'Are Prior Negotiations a General Condition for Judicial Settlement by the International Court of Justice', in C. Barea, J. Barberis, J. Barboza, H. Caminos, E. Candioti, E. de la Guardia, H. Gutiérrez Posse, G. Moncayo, and E. Rey Caro (eds), *Liber Amicorum in Memoriam of Judge José Maria Ruda*, The Hague: Kluwer Law International, 2000, pp. 507–25.

A. Watts, 'Negotiation and International Law', in P. Borba Casella (ed.), *Dimensão Internacional do Direito. Estudos em Homenagem a G. E. do Nascimento e Silva*, São Paulo: LTr, 2000, pp. 519–36.

K. Wellens, *Negotiations in the Case Law of the International Court of Justice. A Functional Analysis*, Farnham: Ashgate, 2014.

E. Zôller, *La bonne foi en droit international public*, Paris: Pedone, 1977.

8 Application for revision of a judgment (Article 61, Statute of the Court)

Leonardo Nemer Caldeira Brant[1]

I The application for revision before the International Court of Justice

The revision of an international judgment has an ancient origin. In an arbitration treaty signed between Italy and Argentina back in 1898, Article 13 foresaw that the application for revision of a sentence would be allowed if made prior to its execution and before the same tribunal that delivered it.[2] Following this, the debate regarding the admissibility of an application for revision continued during the *travaux préparatoires* of the 1899 Hague Convention for the Pacific Settlement of International Disputes. Aptly quoting Abraham Lincoln's famous phrase that 'nothing is settled until is settled right', the American delegate emphasized that in particular circumstances, the principle of finality of international awards and judgments has to give way to overriding considerations of equity. He emphatically disagreed with the Russian delegate's belief in the absolute primacy of the principle of *res judicata*.[3] Despite Russia's opposition, Article 55 of the Convention was adopted and acknowledged the possibility of a party to demand a revision of the arbitration awards.[4]

1 The author would like to thank Paula Wardi D. G. Lana and Bruno de Oliveira Biazatti, both Post-Graduate students in International Law at *Instituto de Altos Estudos em Direito* (IAED), for their help in the research and editing of this piece.

2 Brant, L., *A Corte Internacional de Justiça e a construção do Direito Internacional*, Belo Horizonte: O Lutador, 2005. p. 431.

3 Zimmermann, A., *International Court of Justice: A Commentary*, New York: Oxford University Press, 2006. p. 1302.

4 Article 55 rules:

> The parties can reserve in the *compromis* the right to demand the revision of the award. In this case, and unless there be an agreement to the contrary, the demand must be addressed to the Tribunal which pronounced the award. It can only be made on the ground of the discovery of some new fact calculated to exercise a decisive influence on the award, and which, at the time the discussion was closed, was unknown to the Tribunal and to the party demanding the revision. Proceedings for revision can only be instituted by a decision of the Tribunal expressly recording the existence of the new fact, recognizing in it the character described in the foregoing paragraph, and declaring the demand admissible on this ground. The *compromis* fixes the period within which the demand for revision must be made.'

In the creation of the Permanent Court of International Justice (PCIJ), the possibility of judgment revision was established in Article 61, which substantially coincided with Article 55 of the 1899 and 1907 Hague Conventions for the Pacific Settlement of International Disputes.[5] The Statute of the International Court of Justice (hereinafter 'ICJ' or 'the Court') followed the tendency and also recognizes the application for revision with only some minor semantic changes from the PCIJ's Statute.[6]

Furthermore, Article 61 of the ICJ Statute establishes the conditions that have to be fulfilled in order to apply for revision of a sentence: the application must be based upon the discovery of some fact; the nature of this fact must be decisive; it must be unknown to the Court and also to the party claiming revision before the sentence was pronounced, provided that such ignorance was not under malice; the application must be made at latest within six months subsequent to the discovery of this new fact; and no application for revision can be made later than ten years from the date of the judgment.[7]

Historically, this procedure hasn't received the same acceptance as the application for interpretation[8] and it has been considered as merely a recourse of exceptional nature.[9] The Court was confronted with this issue in only three cases. The first one is the case regarding the Continental Shelf between Tunisia and Libya. Following this, only two applications for revision were submitted: the case concerning the *Prevention and Punishment of the Crime of Genocide (Bosnia and Herzegovina v. Yugoslavia)*,[10] and the judgment of 11 September 1992, regarding the *Land, Island and Maritime Frontier Dispute* between El Salvador and Honduras, with Nicaragua intervening.[11] The application for revision of this last case will be further explored in this chapter.

5 Zimmerman, A., *International Court of Justice: A Commentary*, op. cit. n. 3, p. 1303.

6 Statute of the International Court of Justice, 26 June 1945, art. 61 [ICJ Statute].

7 ICJ Statute, op. cit., art. 61.

8 The possibility of interpretation of a judgment finds place in the article 60 of the ICJ Statute. While it has been applied in seven cases (*Temple of Preah Vihear, Avena and Other Mexican Nationals, Land and Maritime Boundary between Cameroon and Nigeria, Aerial Incident at Lockerbie between Libya and United States, Aerial Incident at Lockerbie between Libya and United Kingdom, Continental Shelf between Tunisia and Libya, Asylum Case between Colombia and Peru)*, the procedure for revision have been applied in only three.

9 Brant, L., *A Corte Internacional de Justiça e a construção do Direito Internacional*, op. cit. n. 2, p. 431.

10 *Application for Revision of the Judgment of 11 July 1996 in the Case concerning Application of the Convention on the Prevention and Punishment of the Crime of Genocide (Bosnia and Herzegovina v. Yugoslavia), Preliminary Objections (Yugoslavia v. Bosnia and Herzegovina), Judgment, I.C.J. Reports 2003*, p. 7.

11 *Application for revision of the Judgment of 11 September 1992 in the case concerning the Land, Island and Maritime Frontier Dispute (El Salvador/Honduras: Nicaragua intervening) (El Salvador v. Honduras), Judgment, I.C.J. Reports 2003*, p. 394.

II The application for revision in the case concerning the *Land, Island and Maritime Frontier dispute between El Salvador and Honduras*

A. The original proceedings

In 2002, El Salvador submitted a request for revision of the judgment delivered on 11 September 1992 by the Court's Chamber, which was formed to deal with the case regarding *Land, Island and Maritime Frontier Dispute* between El Salvador and Honduras, with Nicaragua intervening. In this dispute, the Court analyzed three different but interconnected questions. Firstly, it decided to establish the land boundary between the parties in the six questionable sectors. The Court also resolved the juridical issues concerning the situation of several islands in Golfo de La Fonseca and the waters that belonged or were situated close to its proximity.

The third and most relevant point, being the focus of the study, is the dispute concerning the sixth sector of the land frontier. During the original proceedings, Honduras defended that this sector boundary would follow the current stream of the Goascorán River, i.e., the trajectory in which the river flows into the Gulf north-west of Islas Ramaditas in the Bay of La Unión. On the other hand, even though El Salvador had recognized the Goascorán as the limit of the Province as adopted in the time of Spanish colonization, it claimed that the frontier should be delimited by a previous course of the river, probably abandoned in the sixteenth century, which reached the Gulf at the Estero La Cutu.[12]

In relation to El Salvador's argument, the Chamber declared that it had not received any kind of document that might prove the change in the course of the river, and as a result there existed the lack of scientific evidence capable of proving that the Goascorán ever had its estuary at Estero la Cutu. According to the Chamber, El Salvador's allegation was a new claim, incompatible with the history of this controversy, as the party had never made a demand regarding the possibility of the frontier being delimited by an alleged old stream of the river, as opposed to that identified by the Court in 1821.[13] In addition, the Chamber had access to two pieces of evidence that supported the theory that the Goascorán already flowed through its current stream in 1821: the 1974 expedition record and the 'Carta Esférica'. Thus, in its 1992 final judgment, the Court decided to establish the sixth sector's boundary following the current course of the river.[14]

12 *I.C.J. Reports 2003*, op. cit., p. 397.

13 Ibid., pp. 400–401. Considered the critical date of the dispute because it constitutes the year of independence of Central America from the Spanish Crown and when Honduras and El Salvador were no longer part of the same larger administrative area.

14 *Land, Island And Maritime Frontier Dispute (El Salvador/Honduras: Nicaragua Intervening), Judgment, I.C.J Reports 1992*, p. 30.

The Submission of the Application for Revision and the Court's Decision on its Admissibility

On 10 September 2002, El Salvador submitted its application for revision of the 1992 judgment, only one day before the 'ten years after the day of judgment', which is the deadline established in Article 61 of the ICJ Statute. In its application, El Salvador cited Article 100, paragraph 1 of the Rules of Court, requiring the tribunal to proceed to form the Chamber that would hear the application for revision. The Court decided to accept the demand and to form a special chamber to deal with the case. This was the first and only occasion that an application for revision was submitted to a special chamber of the Court.

The Statute and the Rules of Court foresee a 'two-stage procedure' in face of an application for revision. The first stage is limited to the question of admissibility of the request. The second one is in relation to the analysis of the merits of the case, and only occurs if the Court declares the application admissible. Therefore, in El Salvador's case, the first step taken by the ICJ was to decide whether or not the party's request was compatible with the conditions established in Article 61 of the Statute.[15]

The first of these conditions is the discovery of a new fact related to the case, for which El Salvador pursued two lines of argumentation. The first being the recent discovery of new scientific, technical and historical evidence regarding the Goascorán River's old stream and its abrupt avulsion that occurred near 1762. The historical proof presented by Salvadoran agents consisted of a monograph published by the Sociedad de Geografía e História de Honduras and written by the Departamento de Valle, along with Ulises Meza Cálix's work entitled *Geografía de Honduras.* As technical proof, El Salvador referred to the study of soil analysis in Goascorán's estuary zone, which was undertaken earlier in 2002. Finally, on the subject of scientific evidence, the party invoked a report developed by the Coastal Environments, Inc. titled *Geologic, Hydrologic and Historic Aspects of the Goascorán Delta – A Basis for Boundary Determination.*[16]

El Salvador's second line of argument regarded the cartographic and documental proof. It argued that in 1992 the Court's decision regarding the frontier of the sixth sector was primarily based on copies of documents from the Naval Museum of Madrid, presented by Honduras: being a nautical letter and the 'El Activo' expedition record. El Salvador, however, claimed that the original version of these documents was subsequently found in the Newberry Library, in Chicago, enabling the critical questioning of the reliability of the copies presented by Honduras and accepted by the Court.[17]

15 *I.C.J. Reports 2003*, op. cit., p. 398.
16 Ibid., pp. 401 and 402.
17 Ibid., p. 407.

The second condition for admissibility of an application for revision, i.e., the decisive nature of the new fact had also been fulfilled, according to El Salvador. After all, in the Salvadoran view, the 1992 judgment was based on the assumption that there was not an avulsion of the Goascorán River and, with the acknowledgement of these new documents, El Salvador believed it was now in the position to prove its case.[18] El Salvador also made it clear that due to the civil war that had devastated its territory since 1980, and that was still occurring by the time of the 1992 judgment, the unawareness of these new facts was not under bad faith. Furthermore, the scientific and technical evidence could not have been previously obtained as the state of science and technology in 1992 was still underdeveloped.

In response to these allegations, Honduras argued that the admissibility of an application for revision must rely on the real discovery of a new fact – and not only on simple allegations or even on new evidence haphazardly documented. Considering that the exceptional nature of the revision process implies rigid requirements for its admissibility, Honduras also affirmed that any new fact to be considered as such must be in existence prior to the Court handing down its final judgment. By the 1992 judgment, both the Court and El Salvador were already aware of the Goascorán River's possible avulsion. This predicated the alleged new fact presented by El Salvador as not being new and hence unacceptable.[19]

Moreover, Honduras stated that the new documents presented by El Salvador before the Court were part of a famous collection that had been available in the Newberry Library at least since 1927. Thus, El Salvador could have easily learned of their existence. Honduras also noted that the internal conflict alleged by the other party could not be considered an impediment to the conduction of research outside the Salvadoran territory.[20]

El Salvador claimed that it reasoned its application with new discoveries concerning old facts. Although, even if these facts were already well known by the Court, they should again be analyzed in light of these new discoveries.[21] The Chamber focused on determining whether the new facts brought by El Salvador fulfilled all conditions established in Article 61, and for this purpose, it sought to place them back into context. The Chamber stressed that following this provision, the acceptance of the application relies on the existence of a new fact being capable of influencing the case in a decisive way and that was previously unknown to the Court, as well as to the requesting party before the sentence. Taking into account its definition of a new fact, the Chamber decided that the facts alleged by El Salvador could not be considered new in the sense of Article 61's text.[22]

18 Ibid., p. 402.
19 Mr Dupuy, *CR 2003/3*, p. 16. [*CR 2003/3*]
20 *I.C.J. Reports 2003*, op. cit., p. 408.
21 Mr Mendelson, *CR 2003/2*, p. 36.
22 *I.C.J. Reports 2003*, op. cit., p. 411.

The Chamber also sought to investigate if the new facts brought by El Salvador in the original versions of the 'El Activo' expedition report and the 'Carta Esférica' had a decisive nature in the Court's 1992 decision. Regarding this issue, it observed that there were no significant differences between the versions presented by El Salvador, which originated from Chicago, and those conserved in Madrid, which were the basis for the 1992 judgment. Hence, the documents presented by El Salvador only served to confirm the original judgment.[23]

Finally, the Court recalled the considerations of principle on which it relied to carry out its decision regarding the six sectors of El Salvador–Honduras boundary. During the original proceedings, the Court stated that the frontier ought to be delimited by the application of the principle generally accepted in Spanish America of the *uti possidetis juris*.[24] By that time though, the Chamber had already noted that this main position could be determined by adjudication and by treaty and, therefore, questioned whether it could be determined in another manner, such as by acquiescence or recognition, for example. This possibility was considered credible as there was sufficient evidence showing that the parties had indeed clearly accepted a variation or at least an interpretation of the *uti possidetis juris* principle.[25] Thus, when the Court denied El Salvador's request, which argued that the 1821 boundary did not follow the current stream of Rio Goascorán, it did so based on the El Salvador's actions during the 1880s, when the parties had agreed to recognize this river as the sixth sector frontier. The Chamber observed that in the terms of the 1880s negotiation, the interpretation of the words 'River Goascorán', as meaning a Spanish colonial boundary, which in 1821 followed a long-abandoned course of the river, was deemed out of question.[26] By that, the Chamber concluded that the facts presented by El Salvador were not decisive factors, and accordingly, the application for revision was found inadmissible.[27]

B. Dissenting opinion of Judge ad hoc *Paolillo*

Judge *ad hoc* Felipe Paolillo[28] presented the only dissenting opinion of the case. His main disagreement with the majority of the Chamber relied on the *ratio decidendi* of the 1992 judgment. In the original proceedings, in what relates to the sixth sector's boundary, the Chamber declared that the dispute between the parties was simple, and concerned only the conflict about

23 Ibid., p. 410.

24 Principle under which administrative colonial boundaries must be respected at the time of independence, when colonies become sovereign states. *Frontier Dispute (Burkina Faso/Republic of Mali), I.C.J. Reports 1986*, p. 565.

25 *I.C.J. Reports 2003*, op. cit., p. 404.

26 Ibid., p. 406.

27 *I.C.J. Reports 2003*, op. cit., p. 411.

28 Judge Paolillo was appointed by El Salvador.

whether or not the Goascorán River had a previous course different from that of the current stream, as alleged by El Salvador.

Judge Paolillo also believed that the Chamber did not ask itself whether or not the documentary evidence brought by El Salvador in its application for revision could be seen as new facts, since, according to its opinion, the documents had already failed to satisfy another condition established in Article 61, being the decisive nature of the fact. However, he thought that if the Chamber had analyzed the nature of the fact, it would have already implicitly considered these facts as 'new', despite the conflict regarding whether or not the new evidence could be considered as new facts.

The Judge concluded that since the *ratio decidendi* of the 1992 judgment was the lack of evidence on the previous course of the river and on its abrupt change, any evidence demonstrating the avulsion should be considered a decisive factor. Finally, Paolillo found it regrettable that the application for revision was held inadmissible in the first stage of the procedure, and thus preventing the Chamber from ruling on the merits of the application.[29] According to him, this would have allowed the Chamber to maintain or revise its 1992 judgment on the basis of more robust and reliable information than had previously been available during the original proceedings.

III Contribution of the case to the development of international law and procedural aspects of the ICJ

The application for revision of the case concerning the *Land, Island and Maritime Frontier Dispute* between El Salvador and Honduras, although considered inadmissible by the Court, brought some new debates about jurisdictional prerequisites and procedural aspects of this mechanism in the ICJ. Firstly, the discussion relied upon the possible derogation from the requirements of Article 61, since El Salvador alleged, during the revision proceedings, that Honduras had implicitly acknowledged the admissibility of the Salvadoran request. Regarding this affirmation, the Court denied the possible existence of any form of tacit acceptance from Honduras and made it clear that revision was not available simply by consent of the parties. Hence, this case emphasized the importance of the fulfillment of all the conditions established in Article 61 for the admissibility of revision proceedings and to elucidate that it can't be agreed upon the parties of the dispute.[30]

The second collaboration of this case for the ICJ arises from the fact that, until now, this is the first and only occasion that an application for revision of a judgment delivered by a Court's chamber was submitted. The Rules of Court establishes that '[i]f the judgment was given by a Chamber, the request for its revision or interpretation shall be dealt with by that

29 *Dissenting opinion of Judge* ad hoc *Paolillo. I.C.J. Reports 2003*, pp. 424 and 425.

30 Zimmerman A., *International Court of Justice: A Commentary*, op. cit. n. 3, pp. 1310 and 1311.

Chamber'[31] and, following this legal framework, a Chamber was formed to deal with El Salvador's application. While in the original proceedings, the Chamber was composed of Judges Shigeru Oda, José Sette-Camara, Robert Jennings and the Judges *ad hoc* Nicolas Valticos and Michel Virally; in the application for revision the new Chamber was composed by Judges Gilbert Guillaume, Francisco Rezek and Thomas Buergenthal and the Judges *ad hoc* Santiago Torres Bernárdez and Felipe Paolillo. Therefore, the Court didn't seem to have considered a reconstitution of the original Chamber creating, instead, a new one. Even if the maintaining of the same judges had been considered, it would be impossible to make it work, since by 2003 José Sette-Camara and Robert Jennings were no longer judges in the ICJ and Shigeru Oda was already at the end of his mandate.

Hence, this case was important for having shown that the possibility of requesting for revision for a period of up to ten years from the date of the original judgment was more than theoretical, and for having forced the Court to face the consequences brought by the use of this whole term. It's now clear to the Court that 'it might be materially impossible to reconstitute exactly the same chamber. Still, it ought to be generally advisable that the composition of the new chamber should include – if possible – those members of the Court who had already participated in the original judgment and who are still on the bench when the request for revision is brought'.[32]

Besides, the dispute between Honduras and El Salvador confirmed the position taken by the Court in all previous cases, i.e., the absence of merely one of the conditions established in Article 61 is enough for the Court to declare the petition inadmissible.[33] Such a strict standard is necessary given the exceptional nature of the procedure of revision, and, due to the fact that a decision accepting the request for revision may threaten legal security by mitigating the authority of the *res judicata* principle. Therefore, the revision of a judgment that is endowed with the authority of this principle will be acceptable only within the boundaries of the contribution brought by the new fact.

It is evident that the attenuation of the authority of the *res judicata* maxim is considered a necessary consequence of the search for the good administration of justice and the loosening of the immutability of a jurisdictional decision. Nevertheless, the Court still today hesitates to recognize and perform the mechanism of revision due to the possibility of a threat to the definitive nature of its decisions.[34] The ICJ has historically privileged legal

31 *Rules of Court*, 14 April 1978, art. 100, para. 1.

32 Zimmerman, A., *International Court of Justice: A commentary*, op. cit. n. 3, p. 1328.

33 *Application for Revision of the Judgment of 11 July 1996 in the Case concerning Application of the Convention on the Prevention and Punishment of the Crime of Genocide (Bosnia and Herzegovina v. Yugoslavia), Preliminary Objections (Yugoslavia v. Bosnia and Herzegovina), Judgment, I.C.J. Reports 2003*, p. 12.

34 Brant, L., Godinho, F., 'O processo de revisão da sentença perante a Corte Internacional de Justiça à luz de suas últimas decisões (Iugoslávia vs. Bósnia e El Salvador vs. Honduras)', in *Anuário Brasileiro de Direito Internacional v. 2*, Belo Horizonte: O Lutador, 2007, p. 98.

security to the detriment of the revision as a manner to guarantee the non-continuation of the controversy or, as it might be interpreted, in order to avoid the emergence of a new controversy associated with a previous dispute already adjudicated. As D. W. Bowett concludes, 'the respect which States show for awards would be undermined if the awards lacked finality and binding force'.[35]

This obligatory and definitive effect of judicial decisions is a universally recognized principle[36] that is materialized in the legal and social impossibility of questioning once again the points that have already been decided by the ICJ.[37] Nonetheless, the mere existence of the possibility of revision, by its very nature, leads to the conclusion that a definitive judgment is not unalterable. '[T]he history of jurisprudence evidences that regardless of how advanced a legal system may be, a miscarriage of justice is always possible'.[38] However, the decision of the Court to declare inadmissible an application for revision for the third time may eventually lead to a contrary understanding. In fact, since 2003, no other application for revision was submitted.

IV Conclusion

Despite the existence of Judge Paolillo's dissenting opinion, the Court's decision on the inadmissibility of the application for revision submitted by El Salvador maintains its understanding regarding this procedure, as already demonstrated in previous cases in which this kind of request was made. Therefore, in the conflict between legal security and good administration of justice, the Court's case-law reveals preference to the former. This choice to maintain the authority of the *res judicata* principle and to achieve legal security and predictability can be explained by the importance of the Court as an organ dedicated to perpetuate international peace and security – which must be achieved by means of a definitive sentence.

Bibliography

Doctrine

Bowett, D., 'Res Judicata and the Limits of Rectification of Decisions by International Tribunals', in *Revue Africaine de Droit International et Comparé*, September 1996, T.8, nc.3, p. 577.

35 Bowett, D., 'Res Judicata and the Limits of Rectification of Decisions by International Tribunals', *Revue africaine de droit International et comparé*, 1996, T. 8, nc 3, p. 577.

36 Brant, L., Godinho, F., 'O processo de revisão da sentença perante a Corte Internacional de Justiça à luz de suas últimas decisões (Iugoslávia vs. Bósnia e El Salvador vs. Honduras) ', op. cit. n. 34, p. 99.

37 ICJ Statute, op. cit., art. 60.

38 Zimmerman, A., *International Court of Justice: A Commentary*, op. cit. n. 3, p. 1301.

Brant, L., *A Corte Internacional de Justiça e a construção do Direito Internacional*, Belo Horizonte: O Lutador, 2005.

Brant, L., Godinho, F., 'O processo de revisão da sentença perante a Corte Internacional de Justiça à luz de suas últimas decisões (Iogusláνia vs. Bósnia e El Salvador vs. Honduras)', in *Anuário Brasileiro de Direito Internacional v. 2*, Belo Horizonte: O Lutador, 2007, pp. 132–152.

Zimmermann, A., *International Court of Justice: A Commentary*, New York: Oxford University Press, 2006.

Judicial decisions

Application for Revision and Interpretation of the Judgment of 24 February 1982 in the Case concerning the Continental Shelf (Tunisia/Libyan Arab Jamahiriya) (Tunisia v. Libyan Arab Jamahiriya), Judgment, I.C.J. Reports 1985, p. 192.

Application for Revision of the Judgment of 11 July 1996 in the Case concerning Application of the Convention on the Prevention and Punishment of the Crime of Genocide (Bosnia and Herzegovina v. Yugoslavia), Preliminary Objections (Yugoslavia v. Bosnia and Herzegovina), Judgment, I.C.J. Reports 2003, p. 7.

Application for Revision of the Judgment of 11 September 1992 in the case concerning the Land, Island and Maritime Frontier Dispute (El Salvador/Honduras: Nicaragua intervening) (El Salvador v. Honduras), Judgment, I.C.J. Reports 2003, p. 392.

Frontier Dispute (Burkina Faso/Republic of Mali), I.C.J. Reports 1986, p. 554.

Land, Island and Maritime Frontier Dispute (El Salvador/Honduras: Nicaragua Intervening), Judgment, I.C.J Reports 1992, p. 351.

Miscellaneous

Dissenting opinion of Judge ad hoc *Paolillo. I.C.J. Reports 2003*, p. 413.

I.C.J. Oral Proceedings, Application for Revision of the Judgment of 11 September 1992 concerning the Land, Island and Maritime Frontier Dispute (El Salvador/Honduras: Nicaragua intervening, CR 2003/2.

I.C.J. Oral Proceedings, Application for Revision of the Judgment of 11 September 1992 concerning the Land, Island and Maritime Frontier Dispute (El Salvador/Honduras: Nicaragua intervening, CR 2003/3.

9 Discretional intervention (Article 62, Statute of the Court)

Beatrice I. Bonafé

I Introduction

Intervention is an incidental proceeding by which the International Court of Justice (ICJ) can declare the participation of a third state admissible in a case between two *other* litigant states. The general purpose of intervention is to allow the Court to protect the interests of third states that may be affected by its decision and, therefore, improve the judicial settlement of multilateral disputes.

However, at the international level litigation is still generally perceived as a private business between the parties, and third-state intervention is mostly regarded as an intrusion rather than a means to ensure a better settlement of a common dispute. There are certainly exceptions: in particular, the participation of third parties is increasingly secured before judicial bodies established by international regional organizations in cases where one of the member states or the organization is involved.

This tension between private and collective interests largely explains why for years the rules governing intervention have been unclear and the relevant jurisprudence of the Court has been reluctant to elaborate on the requirements of intervention.[1] The compromise formula, which was originally inserted in Article 62 of the Permanent Court of International Justice's Statute and was later maintained in the ICJ Statute, was intentionally drafted in very general terms as no agreement could be found on a number of issues 'leav[ing] them to be decided as and when they occurred in practice and in the circumstances of each particular case'.[2] Inevitably, this uncertainty has not prompted third states to ask permission to intervene in cases brought before the Court; they have generally found other ways to bring to the attention of the Court their interests that might have been affected by one of its decisions. Article 62 has been invoked in ten cases by 14 states and three

1 For a general analysis, see C. Chinkin, 'Article 62', in A. Zimmermann, C. Tomuschat and K. Oellers-Frahm (eds), *The Statute of the International Court of Justice. A Commentary*, 2nd edn, Oxford: Oxford University Press, 2012, p. 1529.

2 *Continental Shelf (Tunisia/Libyan Arab Jamahiriya), Application to Intervene, Judgment, I.C.J. Reports 1981*, para. 23.

states have been granted permission to intervene. In its case law, the Court has adopted a rather restrictive approach; it has clarified some controversial aspects of intervention, but it has left others still unsettled.

After a brief overview of the basic rules governing intervention, the following analysis will focus on two cases involving Latin American states. They are particularly useful for explaining the approach taken by the Court with respect to both the aspects of intervention: those that are today no longer controversial; those that still require further clarification.

II Rules governing intervention before the court

In addition to Article 62 of the Statute, the main rules concerning so-called 'discretionary intervention' are to be found in Articles 81, 84 and 85 of the Rules of Court. The *basic* condition of intervention is set out in the Statute: the third state must show 'that it has an interest of a legal nature which may be affected by the decision in the case'. Indeed, the purpose of intervention is to offer the third state a means by which to protect itself against the possible effects of a decision of the Court in which it has not been directly involved. When comparing intervention under Article 62 of the Statute to intervention under Article 63, the latter ensures the protection of *general interests* of third states in the interpretation of multilateral treaties, whereas the protection afforded by Article 62 is limited to situations in which the *specific interests* of a third state may be affected. Article 62 cannot be invoked to seek protection of rights too remotely connected to the subject matter of the dispute, such as an interest in the legal rules and principles to be applied by the decision.[3] Thus, the existence of a qualified 'legal interest' is a crucial element of intervention, whose scope largely depends on this requirement. Article 62 does not provide a definition of the required 'legal interest'. In its jurisprudence the Court has clarified that the state seeking to intervene does not have to show that one of its 'rights' may be affected, but simply that one of its 'interests' may be affected. It must show that the interest is the object of a real and concrete claim based on international law; that it must be connected to the subject matter of the dispute brought before the Court; and that it could possibly be affected, in its content and scope, by the Court's future decision in the main proceedings.[4] More generally, the Court's case law shows that it is not easy for third states to prove that they have such a qualified legal interest.[5]

3 Ibid., paras 29–30; *Land Island and Maritime Frontier Dispute (El Salvador/Honduras), Application to Intervene, I.C.J. Reports 1990*, para. 76; *Sovereignty over Pulau Ligitan and Pulau Sipadan (Indonesia/Malaysia), Application for Permission to Intervene, Judgment, I.C.J. Reports 2001*, paras 52, 80, 83, 93.

4 See in particular *Territorial and Maritime Dispute (Nicaragua v. Colombia), Application of Costa Rica for Permission to Intervene, Judgment, I.C.J. Reports 2011*, paras 23–26.

5 For an analysis of the different types of legal interests that may justify the participation of third states in contentious cases see B. Bonafé, 'Interests of a Legal Nature Justifying Intervention before the ICJ', *Leiden Journal of International Law* 25, 2012, 739–757.

While Article 62 does not include other requirements, Article 81 of the Rules directs states wishing to intervene to specify in their request the 'object of the intervention' as well as 'any basis of jurisdiction' existing between the parties and the third state. As shall be discussed as follows, defining the 'object' of intervention had proved extremely difficult for third states at a time when the Court had not yet made up its mind as to the actual scope of protection that Article 62 was meant to afford. This issue was closely connected to that of the so-called 'jurisdictional link', which is whether intervention under Article 62 required an agreement between the parties to the case and the third state seeking to intervene in order for its participation to be admissible. This was one of the more controversial issues if not the most controversial issue about intervention, because requiring a jurisdictional link meant rendering intervention dependent on the consent of the parties. The other option was to consider that the Statute offered a sufficient legal basis to that effect, and therefore that third states could intervene notwithstanding the absence of the parties' consent.

Intervention is an incidental proceeding. Third states seeking to intervene must file an application to do so before the closure of the written proceedings in the principal case. Article 81, paragraph 1, of the Rules of Court provides that a third state can submit an application for permission to intervene 'not later than the closure of the written proceedings'. At that stage, information regarding the parties' case can only be found in the application or the notification of the special agreement. The notification of acts instituting proceedings before the Court to UN members in accordance with Article 42 of the Rules proves essential to the protection of the rights of third states that may be willing to appear before the Court. The same rationale explains the now well-established case law of the Court according to which decisions declaring counter-claims admissible are notified to third states.[6] However, acts instituting proceedings may contain insufficient information. Third states' access to the pleadings of the parties might remedy such a situation. Unfortunately, the Court has interpreted Article 53 of the Rules in a way that would exclude the release of the initial written pleadings when one of the parties has objected to it.

The application to intervene triggers an incidental proceeding that entails a decision by the Court as to the admissibility of intervention. According to Article 84 of the Rules, the decision is taken after having given the parties the opportunity to express their views on the conformity of the request with the Statute's requirements. If the request is granted, the third state can take part in the contentious proceedings and be recognized under the procedural rights provided under Article 85 of the Rules, the extent of which can be limited by the Court in accordance with what is necessary to safeguard the

6 *Application of the Convention on the Prevention and Punishment of the Crime of Genocide, Counter-claims, Order of 17 December 1997, I.C.J. Reports 1997*, para. 39.

interests of the requesting state.[7] The relevant provisions however do not specify whether other rights can be exercised by the third state, or what the legal effects would be of the Court's decision on the merits with respect to the intervening state.

III Leading cases and the gradual definition of the forms of intervention

This brief overview should sufficiently illustrate how crucial were the aspects of intervention not addressed by the drafters of the Statute. Persistent divisions among the members of the Court did not allow for the settling of those issues in the more detailed provisions of the Rules either. The Court might well had been left with the task of dealing with the individual circumstances of each intervention case, but the diverging positions of its members could hardly have produced something different than the hesitations and ambiguities of its early case law.

A. The object of 'non-party' intervention

After some unsuccessful attempts that were complicated by the uncertainty surrounding the object, scope and basic requirements of intervention, a chamber of the Court in 1990 admitted for the first time the participation of a third state, Nicaragua, in the *Land Island and Maritime Frontier Dispute (El Salvador/Honduras)* case. This groundbreaking case paved the way to what is now considered to be 'ordinary' intervention before the Court.

In the 1980s, two applications to intervene were rejected by the Court, mainly due to the absence of a jurisdictional link between the states seeking to intervene and the parties to the principal proceedings. Malta was denied permission to intervene because otherwise it would have been able to submit its own claims without being bound by the decision of the Court.[8] Despite all the efforts of the third state to show that its intention was merely to inform the Court of the interests it had that might have been affected, Italy's request was rejected because its intervention would have involved the introduction of a 'fresh dispute' between the third state and the parties and this 'would have inevitably led the Court to make findings on Italy's rights' in contrast to 'the basic principle that the jurisdiction of the Court to deal with and judge a dispute depends on the consent of the parties thereto'.[9] Therefore, one might have assumed that, for the Court, intervention necessarily implied the submission of claims related to the subject matter of the

7 *Land Island and Maritime Frontier Dispute (El Salvador/Honduras)*, supra note 3, para. 102.

8 *Continental Shelf (Tunisia/Libyan Arab Jamahiriya)*, supra note 2, para. 32.

9 *Continental Shelf (Libyan Arab Jamahiriya/Malta), Application to Intervene, Judgment, I.C.J. Reports 1984*, paras 33–34.

dispute by the third state and accordingly the need for a jurisdictional link. This would in turn have allowed the Court to make findings on such an additional dispute, and allowed the decision on the merits to be binding not only for the parties but also for the intervening state.

The 1990 decision marks a clear departure from this conception of intervention. The Chamber considered that being *informed* of the third state's interests capable of being affected is a 'perfectly proper' object for intervention.[10] Thus, intervention should be recognized as having a much more limited aim. The third state would be allowed to intervene in order to give an exhaustive account of the interests it wanted to be protected. It would not be in a position to submit new claims. In order to ensure its participation in the main proceedings, a jurisdictional link would not be required. In addition, the decision of the Chamber would not be binding for the intervening state. It was this clear position on the 'object' of intervention that allowed the Chamber to define the legal requirements and effects of intervention accordingly.

In line with prior decisions of the Court, El Salvador considered that:

> If the object of the intervention is to inform the Court of its rights or claims, Nicaragua will have a full opportunity to do so (as Italy did) in the oral proceedings to be convened in accordance with Article 84, paragraph 2, of the Rules, without any need to allow its intervention.[11]

By contrast, the new conception adopted by the Chamber meant that intervention in the main proceedings had to be kept separate from the incidental proceedings concerning the admissibility of the request to intervene:

> [I]f it were necessary for a State which considered that its legal interests might be affected by the decision in a case to give an exhaustive account of these interests in its application for permission to intervene, or at the hearings held to consider whether permission to intervene should be granted, there would be no point in the institution of intervention and in the further proceedings to which it should give rise.[12]

As a consequence of the position adopted by the Chamber, intervention was much more accessible to third states. In order to be admitted, they were just asked to show that their legal interests could possibly be affected; no jurisdictional link was required to grant permission to intervene; and intervening

10 *Land Island and Maritime Frontier Dispute (El Salvador/Honduras)*, supra note 3, para. 90. Particular attention is devoted to the 'object' of intervention by S. Torres Bernárdes, 'L'intervention dans la procédure de la Cour internationale de Justice', *Recueil des Cours*, vol. 256, 1995, pp. 363–403.

11 Ibid., para. 88.

12 Ibid., para. 89.

states were not bound by the decision of the Court. Clearly, the risk was to open the floodgates of intervention in contentious proceedings. The Chamber was well aware of that risk and of the need to balance the protection of third states with the rights of the parties. Indeed, the 1990 decision did not abandon a restrictive approach towards Article 62 intervention

First, the Chamber confirmed that a rigorous approach applies to the determination as to possible legal interests susceptible of being affected. The circumstances of the case were peculiar: the task of the Chamber was not only to delimit the frontier line between El Salvador and Honduras but also to determine the legal regime applicable to the islands and the maritime spaces in the Gulf of Fonseca. In 1917, the Central American Court of Justice recognized that the three riparian states – El Salvador, Honduras and Nicaragua – were co-owners of the waters of the Gulf and concluded that the waters remained 'undivided and in a state of community' between El Salvador and Nicaragua. El Salvador claimed that this 'objective legal regime' had consolidated over time and extended to Honduras. That claim was opposed by Honduras, which asked the Court to delimit the waters of the Gulf. What is relevant here is that Nicaragua was inevitably involved in the case because the Chamber had to establish whether or not the waters of the Gulf were subject to a *condominium*. Whatever position the Chamber had taken in that regard, it would have affected the legal interests of Nicaragua: 'the very definition of condominium points to this conclusion'.[13] Accordingly, Nicaragua was considered to have a qualified legal interest, justifying intervention with respect to the question of the legal regime governing the waters of the Gulf. The question of whether, in the absence of a jurisdictional link, Nicaragua would have been bound by the future decision of the Court was later decided by the judgment on the merits: 'this Judgment is not *res judicata* for Nicaragua'.[14] If the Chamber could hardly deny the involvement of Nicaragua in the issue of the applicable regime to the waters of the Gulf, it displayed a much more restrictive approach when it excluded that the legal interests of Nicaragua could be affected by the maritime delimitation. True, the claims of Nicaragua were quite general but it seems undeniable that Nicaragua had a legal interest in the delimitation.[15] The Chamber also confirmed that a mere interest in the applicable legal rules would not qualify as a sufficient interest under Article 62.[16] In addition, it set a relatively high standard of proof when asking the third state to demonstrate 'convincingly' the existence of its legal interest.[17]

13 Ibid., para. 72.

14 *Land Island and Maritime Frontier Dispute (El Salvador/Honduras: Nicaragua Intervening), Judgment, I.C.J. Reports 1992*, para. 424.

15 The decision shows that what is difficult to prove is rather the fact that the legal interest could be affected. See *Land Island and Maritime Frontier Dispute (El Salvador/Honduras)*, supra note 3, para. 84.

16 Ibid., para. 76.

17 Ibid., para. 61.

The second element that rendered intervention less attractive than it might have appeared at first was the limited scope of the procedural rights accorded to Nicaragua, notwithstanding the broad claims advanced by the intervening state.[18] In the 1990 decision, the Chamber made it clear that the intervening state is not a 'party' and its participation in the main proceedings should be limited to the recognition of the procedural rights provided under Article 85 of the Rules.[19] While Nicaragua was entitled to submit written and oral observations with respect to the subject matter of intervention as defined by the Court, it was entitled neither to nominate a judge *ad hoc* nor request the reformation of the Chamber as originally constituted. More generally, the third state cannot exercise all other rights pertaining to the parties.[20]

The contours of 'non-party' intervention, as defined in the 1990 decision, have been confirmed by the full Court and this is the sole form of intervention that has been successfully applied for by third states.[21] Today, non-party intervention is considered to be the 'ordinary' form of Article 62 intervention.[22]

B. The restrictive interpretation of the legal interest

Subsequent case law has adopted a similarly restrictive approach with respect to the requirement of the legal interest that may be affected. In particular, permission to intervene has been recently denied when the interest of the third state is connected to a maritime delimitation case. This is unfortunate because maritime delimitations are the typical situation in which the Court can take into account third states' interests without passing judgment on their claims. Therefore, this approach would entail the rejection of the vast majority of intervention applications,[23] and would clearly be inconsistent with

18 Ibid., para. 41.

19 Accordingly, 'the intervening State shall be supplied with copies of the pleadings and documents annexed and shall be entitled to submit a written statement within a time-limit to be fixed by the Court', and it 'shall be entitled, in the course of the oral proceedings, to submit its observations with respect to the subject-matter of the intervention'.

20 Ibid., para. 102. For a more detailed analysis of the procedural rights of the intervening state see B. Bonafé, *La protezione degli interessi di Stati terzi davanti alla Corte internazionale di giustizia*, Napoli: Editoriale Scientifica, 2014, p. 141 ff.

21 See *Land and Maritime Boundary between Cameroon and Nigeria, Application to Intervene, Order of 21 October 1999, I.C.J. Reports 1999*, p. 1029, and *Jurisdictional Immunities of the State (Germany v. Italy), Application for Permission to Intervene, Order of 4 July 2011, I.C.J. Reports 2011*, p. 494.

22 See *Territorial and Maritime Dispute (Nicaragua v. Colombia), Application of Honduras for Permission to Intervene, Judgment, I.C.J. Reports 2011*, Dissenting Opinion of Judge Abraham, para. 8.

23 Out of ten, seven Article 62 applications have been introduced in maritime delimitation cases.

the decision of the (full) Court granting Equatorial Guinea permission to intervene in the case between Cameroon and Nigeria.[24]

The 2011 judgment on the application of Costa Rica to intervene in the maritime delimitation dispute between Nicaragua and Colombia is particularly telling in that regard. Costa Rica's application to intervene – as is usual with requests to intervene in maritime delimitation cases – met all the requirements set out in 1990: its 'object' was to inform the Court as to what the third state's legal interest was that could be affected;[25] the absence of a jurisdictional link was not an obstacle to the admission of such a form of intervention;[26] and finally a legal interest could be said to exist since the third state identified an area where its interests overlapped with the rights of the parties.[27] Permission to intervene was denied because the Court excluded the possibility that Costa Rica's interests could be affected. The Court treated this aspect as a separate requirement,[28] and added that the requesting state 'must show that its interest of a legal nature in the maritime area bordering the area in dispute between [the parties] needs a protection that is not provided by the relative effect of decisions of the Court under Article 59 of the Statute'.[29] Since the parties did not ask to fix the endpoint in the area where the interests of the third state might be involved, the Court considered that it, alone, was able to protect Costa Rica's interests simply by ending the line before that area.

This approach is particularly problematic. On the one hand, the incidental proceedings concerning the admissibility of a request of intervention is transformed into a sort of *amicus curiae* procedure: 'the only way for a third State to submit information about its interest of a legal nature which may be affected by a decision of the Court would be to make an application that the Court considers inadmissible'.[30] On the other hand, the existence of 'alternative remedies' to intervention is not a requirement of Article 62 and the fact that the Court could be able to protect the interests of the third state has nothing to do with the purpose of intervention, which affords the third state the opportunity to give an exhaustive account of its involvement in the case. The approach of the Court bears a dangerous resemblance to the position of El Salvador that the Court rejected in 1990.

24 *Land and Maritime Boundary between Cameroon and Nigeria,* supra note 21, p. 1029.
25 *Territorial and Maritime Dispute (Nicaragua v. Colombia),* supra note 4, para. 35.
26 Ibid., para. 38.
27 Ibid., paras 65–66.
28 Ibid., paras 53 and 67.
29 Ibid., para. 87.
30 Ibid., Declaration of Judge Gaja, para. 4. See also Bonafé, *supra* note 5, p. 474 ff. and H. Thirlway, 'The Law and Procedure of the International Court of Justice 1960–1989. Supplement 2011: Parts Eleven, Twelve and Thirteen', *British Yearbook of International Law* 83, 2012, pp. 135–137 and 148.

C. The uncertainties still surrounding intervention as a 'party'

The parallel and similarly unsuccessful attempt of Honduras to intervene in the same maritime delimitation dispute between Nicaragua and Colombia provides an opportunity to briefly discuss another unsettled issue concerning Article 62, namely, whether other 'objects' of intervention can be included in Article 62 and more specifically whether the third state can intervene in a capacity different from the 'non-party' status already recognized under the Court's case law.

Honduras asked to intervene either as a 'party' or a 'non-party'. The Court accepted that 'a State may be permitted to intervene under Article 62 of the Statute either as a non-party or as a party',[31] and referred to a passage of the 1990 decision on Nicaragua's application to intervene which in fact was rather ambiguous.[32] In any case, the Court clarified that 'party' intervention required a jurisdictional link[33] because its 'object' was to allow the third state to 'ask for rights of its own to be recognized by the Court in its future decision, which would be binding for that State in respect of those aspects for which intervention was granted'.[34] As to the legal interest that may be affected, the Court said that this is a common requirement to all forms of intervention under Article 62 and added that the 'rights' of the state seeking to intervene have to be connected to the subject matter of the dispute put before the Court.[35]

The Court denied Honduras permission to intervene on the basis that it failed to show that it had a sufficiently qualified legal interest.[36] Therefore, the decision did not rule on a number of aspects of 'party' intervention that remain uncertain. In particular, the legal status of the intervening state and the precise definition of the 'jurisdictional link'. Both are crucial in determining the future of 'party' intervention and the willingness of third states to use it as a means of protecting their interests. As to the former, if 'party' intervention is to represent a credible alternative to the introduction of a new case and a possible joinder with the main proceedings, the Court will have to accommodate the rights of the parties with the foreseeable result being that intervening states will attempt to extend their procedural rights beyond

31 *Territorial and Maritime Dispute (Nicaragua v. Colombia)*, supra note 22, para. 27.

32 A clear position in favour of the possibility to trace back to Article 62 two different forms of intervention had already been expressed by Judge Oda in two dissenting opinions attached to the Court's judgments concerning Malta's request to intervene, supra note 8, p. 27 ff., and Italy's request to intervene, supra note 9, p. 91 ff.

33 *Territorial and Maritime Dispute (Nicaragua v. Colombia)*, supra note 22, para. 28.

34 Ibid., para. 29.

35 Ibid., para. 44. One may wonder whether with respect to 'party' intervention the third state has to show that its 'rights' rather than its 'interests' may be affected as its purpose is actually to submit claims before the Court.

36 Ibid., para. 75.

those recognized by Article 85 of the Rules.[37] On the other hand, the jurisdictional link may represent the real obstacle for third states wishing to intervene and, if this form of intervention is to have any actual application, the Court should arguably adopt a flexible approach at least under particular circumstances. Such an example would be when the participation of the third state is necessary for the Court to exercise its contentious jurisdiction.[38]

IV Concluding remarks

The case law concerning intervention reveals how difficult it can be for the Court to balance the protection of the interests of third states with the rights of the parties. A major concession to the protection of third states was accorded in 1990 when it was accepted that intervention, or at least a form of intervention, did not depend on the principle of consent. However, the jurisprudence of the Court has adopted a very restrictive approach regarding the definition of the legal interest that may justify intervention. This may have reassured the parties that the Court would not accept undue interferences but it has not prompted third states to have recourse to Article 62. Apart from the uncertainties still connected to 'party' intervention, the least that can be said is that intervention is far from being an accomplished means for the protection of third states interests.

Bibliography

R. Bernhardt, 'Judicial and Arbitral Settlement of International Disputes Involving More than Two States', *Annuaire de l'Institut de droit international – Session de Berlin*, vol. I, 1999, p. 60.

B. Bonafé, 'Interests of a Legal Nature Justifying Intervention before the ICJ', *Leiden Journal of International Law* 25, 2012, p. 739.

B. Bonafé, *La protezione degli interessi di Stati terzi davanti alla Corte internazionale di giustizia*, Napoli: Editoriale Scientifica, 2014.

C. Chinkin, 'Article 62', in A. Zimmermann, C. Tomuschat and K. Oellers-Frahm (eds), *The Statute of the International Court of Justice. A Commentary*, 2nd edn, Oxford: Oxford University Press, 2012, p. 1529.

E. Lagrange, 'Le tiers à l'instance devant le juridictions internationalles à vocation universelle (CIJ et TIDM)', in H. Ruiz-Fabri and J. M. Sorel (eds), *Le tiers à l'instance devant les juridictions internationales,* Paris: Pedone, 2005, p. 9.

37 Some authors consider, albeit for different reasons, that 'party' intervention has no place under the Statute; see R. Bernhardt, 'Judicial and Arbitral Settlement of International Disputes Involving More than Two States', *Annuaire de l'Institut de droit international – Session de Berlin*, vol. I, 1999, p. 92; E. Lagrange, 'Le tiers à l'instance devant le juridictions internationalles à vocation universelle (CIJ et TIDM)', in H. Ruiz-Fabri and J. M. Sorel (eds), *Le tiers à l'instance devant les juridictions internationales*, Paris: Pedone, 2005, pp. 68–69.

38 See B. Bonafé, supra note 20, pp. 174–181.

H. Thirlway, 'The Law and Procedure of the International Court of Justice 1960–1989. Supplement 2011: Parts Eleven, Twelve and Thirteen', *British Yearbook of International Law* 83, 2012, p. 1.

S. Torres Bernárdes, 'L'intervention dans la procédure de la Cour internationale de Justice', *Recueil des Cours*, vol. 256, 1995, p. 193.

10 Intervention by third parties under Article 63 of the Statute

Luis González García[*]

I Introduction

The right of third parties to intervene on a question of interpretation of a treaty of which they are party is well established in international law. In particular, it is expressly recognised by Article 63 of the Statute of the International Court of Justice (the 'Court') and has been subject to treatment by the referred Court. The declaration of intervention of Cuba in the *Haya de la Torre (Colombia v Peru) case*,[1] and the declaration of intervention of El Salvador in the *Military and Paramilitary Activities in and against Nicaragua (Nicaragua v United States of America) case*[2] are arguably the two most significant cases that contributed to the development of the right of intervention in international law. In fact, it would be almost impossible for commentators and academics to discuss the right to intervene under Article 63 without referencing these two Latin American cases. This chapter will explore the conditions of intervention under Article 63 of the Statute, the practical problems of interventions by third states, the impact of these two cases in the development of the right to intervene and the future of this right in international law.

II The conditions of intervention

Article 63 addresses a situation in which a state is concerned with the interpretation by the Court of a convention to which it is a party. This is a significant procedural right for third states in international law because states party to a convention have a legitimate interest in its interpretation since an interpretation rendered by the Court, which may constitute an authoritative precedent, may affect the ambit of their obligations under the convention.[3]

* The author wishes to thank Paloma Carrasco García who assisted with the research on some points of this chapter.

1 *Haya de la Torre (Colombia v Peru), Judgment, 13 June 1951, I.C.J. Reports 1951.*

2 *Military and Paramilitary Activities in and against Nicaragua (Nicaragua v United States), Order, 4 October 1984, I.C.J. Reports 1984.*

3 Palchetti, P. 'Opening the International Court of Justice to Third States', (ed.) *Max Planck Yearbook of United Nations Law* 6, 2002, p. 141.

This is so because an interpretation by the Court may impact on the manner in which the third state is implementing its rights or obligations under the treaty and potentially be affected in the future by a claim against it. Article 63 of the Statute reads as follows:

> 1 Whenever the construction of a convention to which states other than those concerned in the case are parties is in question, the Registrar shall notify all such states forthwith
> 2 Every state so notified has the right to intervene in the proceedings; but if it uses this right, the construction given by the judgment will be equally binding upon it.

Therefore, a third party can intervene in proceedings provided that the following conditions are met: (1) that a party making a Declaration of Intervention is a party to the convention in question. Problems as to whether a state is effectively a party to a particular convention will depend on whether the state consented to be bound by the treaty and for which the treaty is in force;[4] (2) that the construction of a convention must be put into question. The wording of Article 63 does not appear to refer only to situations where the construction of a treaty is disputed between the parties to the proceedings. This means that a third state may intervene even when the parties do not give different interpretations to a treaty as long as the Court is required to rule on the construction of a convention. A third state may also intervene in cases of lack of appearance in which the Court must carry out its adjudicatory powers by its own initiative;[5] and (3) the registrar must notify the states involved that the interpretation of a convention to which they are party is disputed in a particular case. This is a rather delicate function of the registrar because it is entrusted with the task of identifying *prima facie* those states that are entitled to receive notice under Article 63 (1). However, this obligation does not mean that a state that has not been notified under Article 63 would be prevented from filing a Declaration of Intervention. In fact, Article 82 (3) of the Rules of the Court permits a state party to a convention to intervene even when not in receipt of notice by the registrar. Finally it is important to note that the judgment of the Court on the legal question of interpretation of the treaty becomes binding upon the third state intervener the same way it will be for the parties to the proceedings.

4 *Vienna Convention on the Law of the Treaties*, 23 May 1969, Article 2. 1155 U.N.T.S. 331, 8 I.L.M. 679.

5 *Nuclear Tests (New Zealand v France), Judgment, ICJ Reports 1974*, and *Request for an Examination of the Situation in Accordance with Paragraph 63 of the Court's Judgment of 20 December 1974 in the Nuclear Tests (New Zealand v. France), I.C.J. Reports 1995.*

III Practical problems

The intervention of third states under Article 63 in cases before the Court gives rise to complex procedural issues. Firstly, the timing of the Declaration of Intervention may disrupt the proceedings between the disputing parties with the consequence of prolonging the proceedings. Secondly, the Declaration may also be effectively requesting the Court to reopen certain jurisdictional or substantive issues. Thirdly, a third party intervention could force a disputing party to respond to the Declaration, which may be seen as procedurally unfair for a party to the case. Fourthly, an additional problem is whether the disputing parties can challenge the admissibility of the Declaration of Intervention. Does a third party have the right to a hearing on the question of the admissibility of its Declaration? Finally, a further problem is whether a Declaration of Intervention can take place at the jurisdictional phase of a proceeding. Can an intervention be rejected because it ventures beyond the scope of the construction of a convention even though elements of the Declaration are within the scope of the construction of a convention? All these procedural problems arose in two Latin American cases: the Cuban Declaration in *Haya de la Torre* and the Salvadoran Declaration in the *Nicaragua case*.

IV The *Haya de la Torre case*

The *Haya de la Torre case* was the first time where intervention under Article 63 of the Statute has been attempted before the Court.[6]

A. Factual background

On 15 October 1949, Colombia submitted an Application to the Court against Peru concerning the dispute that arose between the two countries in connection with the asylum granted to M. Victor Raúl Haya de la Torre in the Colombian Embassy at Lima. The Court rendered its judgment on 20 November 1950 declaring that the asylum had not been granted in conformity with the Convention on Asylum signed at Havana in 1928.[7] Peru then requested Colombia to facilitate the immediate delivery of the refugee Haya de la Torre. Colombia contended *inter alia* that delivering the refugee would violate the Havana Convention.[8]

6 The first case dealing with the right to intervene involved the PCIJ in 1923 known as the SS '*Wimbledon*' *(United Kingdom and others and Poland (intervening) v Germany)*, Intervention, Judgment, (1923) PCIJ Series A no 1.

7 *Asylum* case (*Colombia v Peru*), Judgment, I.C.J. Reports 1950.

8 *Haya de la Torre case, Summary of the Summary of the Judgment, 13 June 1951*. Online. Available www.icj-cij.org/docket/index.php?sum=68&p1=3&p2=3&case=14&p3=5 (accessed 27 August 2015).

Colombia then instituted proceedings against Peru before the Court.[9] Colombia questioned *inter alia* whether it was bound to deliver M. Haya de la Torre to Peru. Peru in sequence equally requested the Court to (1) state in what manner the judgment would be executed by Colombia; (2) dismiss the submissions that considered Colombia not bound to deliver the refugee to Peru; and (3) declare that the asylum ought to have ceased immediately after the delivery of the judgment of 20 November 1950. Notice of application under Article 40 (3) of the Statute of the Court was given to the UN Members and also to other states entitled to appear before the Court. On 26 January 1951, the Court then notified the states party to the Havana Convention of their right to intervene under Article 63 of the Statute.

B. The Cuban intervention

As a party to the Havana Convention, Cuba informed the registrar of its intention to intervene in the proceedings. Cuba attached to its Declaration a memorandum containing its views in relation to the interpretation of the Havana Convention as well as its opinion regarding the right of asylum.

Colombia raised no objections to the Declaration made by Cuba. Peru, however, did so. It argued *inter alia* that the Cuban Declaration was not based on the Havana Convention and that the issues raised by Cuba were *res judicata* (in the sense that they were decided in the *Asylum case*). Peru further contended that the Cuban Declaration was not genuine. A hearing was held on the issue of the Cuban Declaration. Colombia argued on this occasion that Cuba should have the right to be heard as a party to the Havana Convention.

C. The judgment on the intervention issue

In the judgment, the Court first addressed the issue of the admissibility of the Cuban intervention under Article 63. The Court first dealt with the question of whether the right to intervene is absolute or allows for the discretion of the Court to reject Declarations. The Court stated that although Article 63 is 'as of right' it is for the Court to decide should the conditions set forth in the article be satisfied or not. Regarding the contentions made by Peru as to the inadmissibility of the intervention for being time-barred and for being, in essence, an attempt to appeal against the previous judgment, the Court observed that interventions are, in every case, incidental to the proceedings and, therefore, would only be relevant inasmuch as it is related to the proceedings. By stating this, the Court declared that the

9 *Haya de la Torre case, Pleadings, Oral Arguments, Documents: Section A: Application Instituting Proceedings*, 13 December, 1950, p. 7. Online. www.icj-cij.org/docket/index.php?p1=3&p2=3&k=d4&case=14&code=haya&p3=0 (accessed 26 August 2015).

subject matter in this case was indeed new – the surrender of the refugee Haya de la Torre, which was not addressed in the previous proceedings. It then moved on to the Cuban intervention related to the interpretation of the Havana Convention regarding the question of whether Colombia was under obligation to surrender the said refugee. The Court recognized that most of the Cuban memorandum was related to issues that had been already decided in the previous judgment and thus would have no relevance to the present case. The Court, however, found that the Cuban intervention related to a new question not decided by the judgment of 20 November 1950 – the surrender of Haya de la Torre to Peru. Thus, the Court noted, the intervention of Cuba related to a new aspect of the Havana Convention and accepted Cuba's intervention.[10]

D. The impact of the **Haya de la Torre** *case*

Perhaps one of the most important contributions of the Court in this case refers to the concept of admissibility of a Declaration of Intervention. As clarified by the Court in this case, any objection by the parties as to whether any Declaration of Intervention is admissible and must be resolved by the Court. In other words, the admission of a Declaration is not automatic. Judge Fitzmaurice summarised the point, as follows:

> Although intervention under this Article [63] is as of a right, provided the conditions stated in it are fulfilled, it is naturally for the Court to decide whether they are actually satisfied or not ... Given that these conditions are present, the Court is bound to admit the intervention, and has no discretionary power in the matter, as it would seem it must have under Article 62.[11]

The judgment is also relevant to the development of the right to intervene in international law because it clarified the conditions that must be fulfilled for a Declaration to be admissible. These include that the intervention must be adequately related to the subject matter of the current proceeding,[12] it has to be deemed 'genuine' and cannot be used as an appeal.[13] As a result of this judgment, it appears that the Court modified Article 82(2) of the Rules of Court in 1978[14] to further tighten the conditions of intervention as of right.

10 *Haya de la Torre, Judgment 13 June 1951, ICJ Reports 1951.*

11 Fitzmaurice, 'The Law and Procedure of the International Court of Justice, 1951–4', 34 *Brit Y.B. Intl'L. L.* 127, 1958.

12 Ibid. 554. See also Chinkin, C. M., 'Third-Party Intervention Before the International Court of Justice', (ed) *The American Journal of International Law* 80, 1986, 498.

13 Chinkin, op. cit., p. 509.

14 Zimmermann, A., Tomuschat, C., Oellers-Frahm, K., and Tams, C. J. (ed), *Statute of the International Court of Justice: A Commentary* (2nd ed.) n. 1, Oxford University Press, 2012, 1589.

V The *Nicaragua case*

The *Nicaragua case* is not only considered to be one of the most important cases[15] but also arguably the single most controversial case in the history of the Court.[16] Questions over the treatment by the Court of the Declaration of Intervention submitted by El Salvador contributes to the overall reputation of the *Nicaragua case*. As explained further below, the *Nicaragua case* is controversial in relation to the Court's stance concerning El Salvador's Declaration of Intervention for the following reasons: firstly, the decision of the Court not to hold a hearing; secondly, the rejection of the Declaration; and, thirdly, the question of errors contained in the Declaration.

A. Factual background and procedural history

On 9 April 1984, Nicaragua commenced proceedings alleging that the United States had committed actions constituting violation of the prohibition of the use of force and wrongful intervention in and against Nicaragua in respect of a dispute concerning the responsibility for military and paramilitary activities in Nicaragua.

The Court first dealt with issues of jurisdiction and admissibility. Two days before the closure of the written proceedings at the jurisdictional phase, El Salvador filed a Declaration of Intervention under Article 63. In its Declaration, El Salvador affirmed that 'the object of its intervention was to enable it to argue that the Court did not have jurisdiction over Nicaragua's Application.'[17] In this sense, it made reference to certain multilateral conventions on which Nicaragua based its arguments against the US, but did not mention any specific provisions. These conventions were the Statute of the International Court, the Charter of the UN, the Charter of the Organization of American States, the Convention Relative to the Duties and Rights of States in the Event of Civil Strife and the Convention on Rights and Duties of States. These treaties, it argued, gave El Salvador the right to demand Nicaragua to cease intervention in the internal affairs of the state and provide reason to intervene in the present proceedings.

Following the request for intervention by El Salvador, the parties were

15 LJIL Symposium: Discussion of the I.C.J. Nicaragua Judgment, by Dov Jacobs. Online. Available http://opiniojuris.org/2012/03/20/ljil-symposium-discussion-nicaragua (accessed 27 August 2015). Regarding the relevance of the *Nicaragua case* in the development of international law and the impact in the use of the Court by developing states see Kohen, M., 'The Principle of Non-Intervention 25 Years after the Nicaragua Judgment', (ed) *Leiden Journal of International Law* 25, March 2012, 157–64 and Fisler, L., 'The Impact of the *Nicaragua* Case on the Court and Its Role: Harmful, Helpful, or In Between?', (ed) *Leiden Journal of International Law* 25, March 2012, 135–47.

16 Llamzon, A. P., 'Jurisdiction and Compliance in Recent Decisions of the International Court of Justice', *Eur J Int Law* 18 (5), 2007, 815–52.

17 *Nicaragua case, Declaration of Intervention by El Salvador, ICJ Reports 1984.*

invited to submit written observations regarding the matter pursuant to Article 83 of the Rules of Court.

On the one hand, Nicaragua stated not to have 'any objection in principle to a proper intervention by El Salvador'[18] but referred to apparent deficiencies in the Declaration and noted that El Salvador had included in the referred document as 'a matter of record' that it felt 'under pressure of an effective armed attack'. In this line, Nicaragua contended that Article 63 'does not permit intervention for the purpose of opposing jurisdiction or to make things as a matter of record but only for the purpose of the interpretation of an identified provision of a convention'.[19] Finally, Nicaragua stated that Article 85 of the Rules of Procedure refers to time-limits in order to 'call the Court's attention'[20] to the prompt disposition of the present jurisdictional phase and the urgency of a speedy determination on the merits. On the other hand, the US stated that El Salvador's Declaration of Intervention was timely.[21]

In a further supplementary submission, El Salvador argued that Nicaragua was 'preventing El Salvador from exercising its procedural right to oral proceedings before the Court in the event of an objection'.[22]

B. The order on the intervention issue

By Order of 4 October 1984, the Court rejected El Salvador's Declaration of Intervention. It did so without holding a hearing and without providing the reasons for such denial of a hearing. The voting finalised as follows: by 9 to 6 votes: not to hold a hearing on the Salvadoran Declaration; by 14 to 1 (that of Judge Schwebel) that the Salvadoran Declaration was inadmissible 'inasmuch as it relates to the current phase of the proceedings'.[23] However, El Salvador did not make any new attempts to intervene. Several judges appended separate (including dissenting) opinions to the Order. Judge Schwebel's dissent is perhaps the strongest against the Court's ruling.

18 *Military and Paramilitary Activities in and against Nicaragua (Nicaragua v United States of America) case, Letter from the Agent of Nicaragua to the Registrar Submitting the Observations of Nicaragua on the Declaration of Intervention of El Salvador, I.C.J. Reports 1984*, p. 465. Online. Available www.icj-cij.org/docket/files/70/9623.pdf (accessed 27 August 2015).

19 Ibid. at 466.

20 Ibid.

21 *Written Observations on the Declaration of Intervention: Letter from the Agent of the United States of America to the Registrar of the International Court of Justice Submitting the Observations of the United States on the Declaration of Intervention of El Salvador I.C.J. Reports 1984*, 467.

22 Ibid., p. 473.

23 *Order, 4 October 1984, I.C.J. Reports 1984*, p. 215.

C. The impact of the Nicaragua *case*

Although it can hardly be called 'contribution', the Court's evaluation of the Salvadoran Declaration in the *Nicaragua case* has had an impact on the discussion and debate concerning the right of intervention in international law. The following are the most important controversial points of the Court's decision in this case.

D. The issue of the hearing

In his dissent, Judge Schwebel criticises the decision of the Court not to hold a hearing for El Salvador. He argued that the Court had not acted according to the observance of the due process that characterises it. This proved a valid point. In my opinion, even if the Court was right in rejecting the Salvadoran Declaration, one would expect that such decision could only be taken after granting an opportunity to El Salvador to state its position. It appears that the Court was more concerned with the protection of the procedural rights of the disputing parties than with the balance of the interest of all parties. Commentators have criticised the Court's restrictive approach towards the rights of third parties under Article 63.[24] As Professor Chinkin notes, it is not 'realistic' to think solely of the integrity of the parties and leave aside the protection of 'third party' rights.[25] Again, one may wonder why the Court failed to follow its own approach in the *Haya de la Torre case* where it granted a hearing to Cuba even though it was not obliged to do so under the rules then in force.[26] But perhaps the most striking aspect of the Court's ruling on this point is the fact that there is no further explanation on the reason why it decided 'not to hold a hearing'. As Judge Schwebel explains, if the court had granted a hearing to El Salvador to intervene rather than just literally interpreting Nicaragua's written submission, it would have been more in accordance with judicial propriety. The Court should have regarded this as an objection, and this would have presented an opportunity for El Salvador to be heard.

1. The requirements

Even should one accept the point that the Salvadoran Declaration pertained to different elements only relevant to the merits phase, it is, however, difficult to understand why it rejected the declaration *in toto*.[27] There is no

24 Greig, D. W., 'Third Party Rights and Intervention Before the International Court', *Virginia Journal of International Law* 32, 1992 (n 6), 317; Chinkin C. M., 'Third-Party Intervention Before the International Court of Justice', op. cit. n. 12, p. 511.

25 Ibid.

26 Article 66(2) of the Rules of the Court.

27 Sztucki, J., 'Intervention under Article 63 of the ICJ Statute in the Phase of Preliminary Proceedings: The Salvadoran Incident', *The American Journal of International Law*, October 1985, 1012.

apparent reason to justify the Court's decision to apply a distinct reasoning from that of the *Haya de la Torre case* where it allowed Cuba to intervene by reducing the intervention to the limits that were applicable to the proceedings ('reduced in this way and operating within these limits').[28] In the *Nicaragua case*, the Court could have acted in the same manner by separating the elements that were admissible to the case and rejecting those that were not applicable, instead of rejecting the declaration as a whole. It is also striking for the lack of reasons given by the Court in order to reject the declaration in the jurisdictional phase, as well as the remaining obscurity of certain elements that were not clarified in the separate opinions. As Judge Schwebel stated in his opinion, the Court's rejection of the declaration of El Salvador at the jurisdictional phase is questionable since there is no apparent reason why intervention should not be allowed on issues of jurisdiction and admissibility of a case.[29] The Order does not provide an answer or present elements that would lead to identify the reason why the Court decided to reject the Declaration at the jurisdictional phase.

2. *Failure to comply with Article 82 of the Rules*

The rejection of the Declaration has also been subject to criticism due to the fact that the Declaration was dismissed merely on the basis of some formal errors that could have been amended at the hearing. Some commentators have stated that the requirements outlined in Article 82 were taken too far to be used as grounds to dismiss an application that could have been otherwise perfectly valid.[30] As Professor Greig notes, the Rules of Procedure should not be contemplated as 'additional obstacles' for intervention.[31] For this reason, he argues that, by requiring the parties to identify the provisions, Article 63 was given an analogous interpretation to that of Article 62 since the Court is requiring the states to prove their legal interest in the proceedings in order to admit intervention. Greig concludes by saying that the distinction between both articles has been blurred as a result of the said interpretation of Article 63.[32]

VI The future of Article 63

Article 63 is no doubt a powerful tool for third states to ensure a more coherent interpretation of a treaty to which they are party to. But since the

28 *Haya de la Torre, Judgment, ICJ Reports 1951*, p. 77.

29 Zimmermann, A., Tomuschat, C., Oellers-Frahm, K., and Tams, C. J. (eds), *Statute of the International Court of Justice: A Commentary*, op. cit. n. 14, p. 1581.

30 Greig, D. W., 'Third Party Rights and Intervention Before the International Court', op. cit. n. 24, p. 313.

31 Ibid.

32 Ibid.

decision of the Court in the *Haya de la Torre case* to admit the Cuban intervention it has not become a routine for third states to exercise the Article 63 right to intervene in proceedings before the Court. To this point, Article 63 has been invoked in only five cases in the history of the World Court (including the Permanent Court of International Justice): (1) *S.S.* Wimbledon;[33] (2) *Haya de la Torre*;[34] (3) the *Nuclear Tests case*;[35] (4) *Nicaragua case*[36] and, more recently, (5) *Whaling in the Antarctic case*.[37] It may be argued that the restricted use of Article 63 is explained by the fact that this article was intended to protect a more limited interest of legal nature, i.e. the construction of a convention.[38] Another reason relates to the fact that there are legal consequences in case a third state decides to intervene under Article 63: the decision of the Court becomes binding on the third state. This distinctive feature of Article 63 may be a reason to dissuade any interested third party in participating in a case. But perhaps a more logical reason for the lack of use of Article 63 may relate to the rejection of the Salvadoran Declaration in the *Nicaragua case*. The Court's approach on the Salvadoran Declaration was in favour of the rights of the parties to the detriment of the rights of third states. The Court's approach in the *Nicaragua case* resulted in a positive contribution by clarifying that Article 63 provides a limited right on a question of interpretation and should not be used by states as a general right to deliver submissions on facts or law relating to particular cases. This is perhaps the most important contribution of the *Nicaragua case* to the right to intervene under Article 63. There is, however, an unintended consequence of the Court's ruling. As a result of the unfortunate treatment of El Salvador, the rejection of the Salvadoran Declaration may have caused third states to shift towards the right to intervene under Article 62 as a more applicable avenue to protect their interests. Article 62 is based on an interest in legal nature that may be affected by a ruling issued by the Court. It is intended to protect a wider set of rights, as opposed to the narrower aspect of 'construction of a convention'. One may wonder whether El Salvador would have been successful in its attempts to intervene in the *Nicaragua case* if it had submitted its interest to intervene also under Article 62 of the Statute. As Professor Chinkin notes, there seems to be no reason why a state cannot use them simultaneously.[39]

33 Supra (n 6).
34 Supra (n 1).
35 Supra (n 5).
36 Supra (n 2).
37 *Whaling in the Antarctic (Australia v Japan: New Zealand intervening)*, Judgment, 31 March 2014. Online. Available http:www.icj-cij.org/docket/files/148/18136.pdf (accessed 27 August 2015).
38 Rosenne, S. *Intervention in the International Court of Justice*, Boston: Martinus Nijhoff Publishers, 1993, p. 73.
39 Chinkin, C. M., 'Third-Party Intervention Before the International Court of Justice', *The American Journal of International Law* 80, 498.

Now, when looking forward into the future, there appears to be a glimpse of a new direction in the Court's approach to third party intervention. By granting intervention to New Zealand under Article 63 in the recent judgment in the *Whaling in the Antarctic case* – where the Court seems to have applied a similar approach as in the *Haya de la Torre case* – the Court took a positive step to the protection of third party rights and to the balance of interests in proceedings under multilateral treaties. This apparent new approach is particularly important in Latin America where states continue to take steps towards multilateralism, which is the path that international law has been increasingly following during the last decades. A fair dispute settlement system in a multilateral sphere necessarily involves the right of third states to intervene in a question of interpretation. As stated by New Zealand in its submissions before the Court, Article 63 is an integral part of the Court's framework as a forum for resolution of international disputes under multilateral treaties. By affirming New Zealand's right to intervene, the Court appears to be acknowledging the importance of the concept of equality of interests between all parties to a multilateral treaty, not only favouring the interests of the parties involved in the dispute. Doing so will promote the rule of law, which is a positive development of international law.

11 Request for interpretation of a judgment

Karin Oellers-Frahm

I Introduction

According to Art. 60 ICJ Statute '[t]he judgment is final and without appeal.' There are, however, two alternative meanings concerning 'review' of a judgment: the first one being interpretation (Art. 60 ICJ Statute); the second being revision (Art. 61 ICJ Statute). While revision may lead to a partially or totally different/new judgment, interpretation does not affect the *res judicata*; it adds nothing to the decision, but allows only for guidance to be given by the Court with regard to the meaning or scope of the judgment. Both meanings of 'review' are extremely rarely used;[1] however, interpretation has recently gained more attention, in particular in combination with a request for provisional measures.[2]

1 *Request for Interpretation of the Judgment of 20 November 1950 in the Asylum case (Colombia v. Peru), Judgment, I.C.J. Reports 1950*, p. 401; *Application for Revision and Interpretation of the Judgment of 24 February 1982 in the Case concerning the Continental Shelf* (Tunisia/Libyan Arab Jamahiriya), (*Tunisia v. Libyan Arab Jamahiriya), Judgment, I.C.J. Reports 1985*, p. 223, para. 56; *Request for Interpretation of the Judgment of 11 June 1998 in the Case concerning the Land and Maritime Boundary between Cameroon and Nigeria (Cameroon v. Nigeria), Preliminary Objections, (Nigeria v. Cameroon), I.C.J. Reports 1999*, p. 31 ss; *Request for Interpretation of the Judgment of 31 March 2004 in the Case concerning Avena and Other Mexican Nationals (Mexico v. United States of America), Order of 16 July 2008, I.C.J. Reports 2008*, p. 311 and *Judgment of 19 January 2009, I.C.J. Reports 2009*, p. 3 ss; *Request for Interpretation of the Judgment of 15 June 1962 in the Case Concerning the Temple of Preah Vihear (Cambodia v. Thailand), Order of 18 July 2011, I.C.J. Reports 2011* and *Judgment of 11 December 2013, I.C.J. Reports 2013*, p. 281 ss. Only three requests for revision have been brought before the Court which were all declared inadmissible: *Application for Revision and Interpretation of the Judgment of 24 February 1982 in the Case concerning the Continental Shelf* (Tunisia/Libyan Arab Jamahiriya), (*Tunisia v. Libyan Arab Jamahiriya), Judgment, I.C.J. Reports 1985*, p. 223, para. 56; *Application for Revision of the Judgment of 11 July 1996 in the Case concerning the Application of the Convention for the Prevention and Punishment of the Crime of Genocide (Bosnia and Herzegovina v. Yugoslavia) (Yugoslavia v. Bosnia and Herzegovina), Preliminary Objections, I.C.J. Reports 2003*, p. 7 ss.; *Application for Revision of the Judgment of 11 September 1992 in the Case concerning the Land, Island and Maritime Frontier Dispute (El Salvador v. Honduras: Nicaragua Intervening), (El Salvador v. Honduras), I.C.J. Reports 2003*, p. 392 ss.

2 Infra 3.2. and 3.3.

Interpretation proceedings[3] are rather limited in scope because the Court may only construe the judgment, but not touch upon its *res iudicata*. The advantage of an interpretation request is that it does not give rise to any question of jurisdiction because the competence to construe the judgment follows from the competence to decide the underlying case. Even if one or both parties to the original case have in the interim withdrawn their consent to the Court's jurisdiction, the competence to decide an interpretation request is unaffected, a fact that may explain the use of the interpretation proceedings in situations referring rather to the follow-up of the original judgment or to new controversial situations, which are only loosely connected to the original judgment.[4] Furthermore, there are no time-limits for a request of interpretation, so that a request for interpretation registered at the Court in 2011 in relation to a judgment delivered in 1962 is no bar to admissibility.[5] The crucial question in interpretation proceedings is whether a 'dispute' as to the meaning or scope of the judgment has arisen and what is the meaning and scope of the judgment, i.e. what was decided with *res iudicata* effect. As Latin American states have always been particularly active in the use of the various procedural means provided for in the Statute of the ICJ,[6] Latin American states also performed a significant role in bringing to life interpretation proceedings. The very first request for interpretation of a judgment concerned the *Asylum case* between Colombia and Peru,[7] and the pilot case establishing a new scenario by combining a request for interpretation with a request for provisional measures – giving rise to controversies concerning the involvement of the Court in post-judgment developments – was also brought before the Court by a Latin American state, namely Mexico in the *Avena case*.[8]

II The *Asylum case*

The case originated from the fact that a Peruvian statesman, Mr Haya de la Torre, who was accused of having plotted a (unsuccessful) military coup

3 For a comprehensive analysis see A. Zimmermann, 'Article 60', in Zimmermann, A., Tomuschat, C., Oellers-Frahm, K., and Tams, C.(eds.), *The Statute of the International Court of Justice – A Commentary,* 2nd ed., Oxford: OUP, 2012, 1469 ss.

4 Oellers-Frahm, K. 'Provisional Measures in Interpretation Proceedings – a new way to extend the Court's jurisdiction? The practice of the Court in the *Avena* and *Temple of Preah Vihear* cases', in Elias, O., and Jalloh, C. (eds.) *Shielding Humanity: Essays in Honour of Judge Abdul G. Koroma*, Leiden/Boston: Brill/Nijhoff, 2015, pp. 61–84.

5 This was the case in the *Temple of Preah Vihear case*, op. cit. n. 2.

6 Oellers-Frahm, K., 'The part of the International Court of Justice in solving disputes between American States', *L'Observateur des Nations Unies* 24, 2008, pp. 7–35.

7 *Asylum case, (Peru v. Colombia), Judgment, I.C.J. Reports 1950*, p. 266.

8 *Request for interpretation of the Judgment of 31 March 2004 in the case concerning Avena and other Mexican Nationals (Mexico v. United States of America), I.C.J. Reports 2009*, pp. 3 ss.

d'état against his government, was granted asylum in the Colombian Embassy in Lima in 1949. The dispute between Peru and Colombia concerned the question whether Colombia had the right to grant diplomatic asylum in distinction to territorial asylum and whether Mr Haya de la Torre should be granted safe conduct out of the country. The Court decided on the basis of the Havana Convention on Asylum of 1928 and customary (regional) law that it was not for Colombia to qualify the nature of the offence committed by Mr Haya de la Torre as a common crime or a political offence and that Peru was not obliged to provide safe conduct for him to leave the country. Furthermore, the Court found in favor of a counter-claim of Peru that the asylum had been irregularly granted. The very day of the delivery of the judgment, Colombia submitted a request for interpretation before the Court asking, in particular, for clarification of the question whether Mr Haya de la Torre should be surrendered to Peru.

A. Prerequisites for a request under Art. 60 ICJ Statute

Art. 60 cl. 2 provides that in the event of a *dispute* as to the meaning or scope of a judgment any party may request the Court to construe the judgment. The term 'dispute', which is used in the English version of the Statute also in Art. 36 (6) and Art. 38 (1), is reflected in both the French and the Spanish version of the Statute in different terms: 'différend' in Art. 38 (1) and 'contestation' in Art. 36 (6) and 60; and in the Spanish version 'desacuerdo' in Art. 60, 'disputa' in Art. 36 (6) and 'controversia' in Art. 38 (1). There are, thus, good reasons to hold that the general concept of the term 'dispute' as 'a disagreement on a point of law or fact, a conflict of legal views or interests'[9] is not necessarily applicable to Art. 60.[10] Furthermore, a request for interpretation requires that the dispute must relate to the scope or meaning of the judgment. The interpretation of a judgment can therefore neither put into question the *res iudicata* of the principal judgment nor broaden the scope of that judgment; it has to keep within the limits of the original judgment and to seek only clarification of what the Court has decided with binding force, and not to seek answers to questions not so decided. In this sense, it is the *dispositif* of the judgment that is decisive in the first place. However, the reasons of the Court may also have binding force if they constitute a 'condition essential to the Court's decision' being necessarily implied in the *dispositif*.[11]

9 This definition was first used by the PCIJ and later adopted by the ICJ; see *Mavrommatis Palestine Concessions, Judgment N. 2, 1924, PCIJ Series A, No. 2*, p. 11.

10 Zimmermann, op. cit. n. 3, MN 58–64.

11 Zimmermann, op. cit. n. 3 with rich case-law in footnote 167.

B. The decision of the Court

In regard to the first question, whether the parties were in dispute as to the meaning or scope of the judgment, the Court was rather short in defining the term 'dispute': 'Obviously, one cannot treat as a dispute, in the sense of that provision [Art. 60], the mere fact that one Party finds the judgment obscure when the other party considers it to be perfectly clear. A dispute requires a divergence of views; Art. 79 (2) of the Rules confirms this condition by stating that the application "shall specify the precise point or points in dispute."'[12] The Court did not need to consider this question more in detail; as the interpretation of the judgment was requested immediately after the delivery of the judgment, it rightly found that 'the very date of the request shows that such a dispute could not possibly have arisen in any way whatever'.[13]

Although the request was already inadmissible due to the non-existence of a dispute, the Court nevertheless considered also the second prerequisite, namely whether the alleged dispute related to the meaning or scope of the judgment. As this was the first case where an interpretation was requested, it was particularly important for the Court to provide some guidance to states for further use of the interpretation procedure. In the *Asylum case* the request for interpretation particularly related to the question whether Mr Haya de la Torre had to be surrendered to Colombia. Also in this regard, the answer of the Court was short and convincing. The Court referred 'to what it declared in its judgment in perfectly definite terms', namely that 'this question was completely left outside the submissions of the Parties. The Judgment in no way decided it, nor could it do so'.[14] The Court concluded that this question was a new one that could not be decided by means of interpretation, because interpretation cannot exceed the limits of the judgment. This question was the subject matter of a follow-up case brought before the Court by Colombia in 1950.[15]

In this first Art. 60 case, the Court explained the prerequisites of Art. 60 as far as it could within the context of the case. In particular, it made clear that it would be very attentive to prevent that, by the means of a request for interpretation, a decision on questions could be obtained which were not at stake in the underlying case.[16] Parties have thus to be very cautious in formulating their request for interpretation, a requirement that was of relevant significance also in the *Avena* case.[17]

12 *Request for Interpretation of the Asylum case, I.C.J. Reports 1950*, p. 395 ss at p. 403.
13 Ibid.
14 Ibid.
15 *Haya de la Torre case (Colombia v. Peru), Judgment, I.C.J. Reports 1951*, p. 71.
16 *Asylum case*, op. cit. n. 1, p. 403.
17 Infra 3.2. It may be interesting to note that also in the *Continental Shelf case (Tunisia/Libya) I.C.J. Reports 1985*, p. 217, the request for interpretation was dismissed because it was aimed at replacing the demarcation line (S.O. Oda, ibid. p. 245) and in the

III The *Avena case*

This case related to the death sentences handed down by United States courts against several Mexican nationals who were convicted in disregard of the obligations of the United States under Art. 36 of the Vienna Convention on Consular Relations, namely the obligation to inform the consular representative concerning detention and sentencing of one of its nationals. In its judgment,[18] the Court found *inter alia* that 'the appropriate reparation in this case consists in the obligation of the United States of America to provide, by means of its own choosing, review and reconsideration of the convictions and sentences of the Mexican nationals'.[19] With a view to the follow-up given to the judgment by the United States and, in particular, to decisions from courts in Texas disregarding the ICJ judgment, Mexico was of the opinion that a dispute existed between the two parties concerning the meaning of the relevant passage of the judgment and requested interpretation under Art. 60.[20] Mexico feared that on the basis of the Texas courts' decisions, its nationals faced execution in the short term and applied, together with the request for interpretation, for the indication of provisional measures aimed primarily at ordering the United States not to execute any of the Mexican nationals pending the procedure.

A. The decision of the Court under Art. 60

In this case, the only relevant question under Art. 60 was that of the existence of a dispute, not its relationship to the meaning or scope of the judgment. The dispute concerned the question whether the relevant passage of the judgment created an 'obligation of result' or, as the United States understood it according to the view of Mexico, a mere 'obligation of means'. As the United States, however, explicitly agreed that the relevant passage of the judgment created an obligation of result, there was, at first glance, no dispute so that the request would have to be dismissed. Yet in this case, the Court examined more closely than in the *Asylum case* what the term 'dispute' in Art. 60 actually means. It referred in particular to the fact that the French version of Art. 60 uses a different term for 'dispute' than Art. 36 (2), namely not the term '*différend*', but the term '*contestation*'. This led the Court to

Land and Maritime Boundary case (Cameroon v. Nigeria), op. cit. n. 1, the request for interpretation was also dismissed because the questions were already clearly decided in the judgment. Only in the *Temple of Preah Vihear case*, op. cit, n. 1, the Court admitted the request and gave an interpretation of the 1962 judgment, what was however rather controversial; see Oellers-Frahm, op. cit. n. 4; and Kulick, A., 'Article 60 I.C.J. Statute, Interpretation Proceedings, and the Competing Concepts of *Res Judicata*', *Leiden JL Int.l Law* 18, 2015, 73–89.

18 *I.C.J. Reports 2004*, p. 12 ss.

19 Ibid. Para 154 (9) of the Judgment.

20 *Request for Interpretation of the Avena case*, op. cit. n. 1.

conclude that this term was wider in scope and more flexible than the term 'dispute/*différend*', requiring only the presence of 'opposing views' as to the scope of the judgment. On this basis, the Court concluded that the parties 'apparently hold different views as to the meaning and scope of that obligation of result, namely whether that understanding is shared by all United States federal and state authorities and whether that obligation falls upon those authorities'.[21] The Court accordingly admitted the request for interpretation,[22] a decision that was highly controversial and not shared by five judges of the Court who explained their view in their dissenting opinions.[23] Their argument relied on the fact that the party to the dispute is the United States and it is the United States Executive that represents the state at the international level. Therefore, the position taken by the authorities of a state of the United States is of no relevance with regard to the obligations of the parties to the case, namely the United States and Mexico that, in fact, were in complete agreement with regard to the obligation flowing from the judgment. The dissenters rightly argued that the attitude of component organs or state authorities of the United States could only become relevant under the aspect of state responsibility should the judgment of 2004 not be complied with, irrespective of whether this would be imputable to acts of federal or state authorities. This argument had indeed also been advanced by the Court itself so that its decision on the existence of a dispute/contestation seems rather critical.

Accordingly, it may be assumed that the admission of the interpretation request has to be seen against the background that otherwise provisional measures could not be indicated, and that the Court considered the provisional protection necessary with a view to the imminent execution of the persons concerned.[24]

21 *Request for Interpretation of the Judgment of 31 March 2004 in the Case concerning Avena and Other Mexican Nationals (Mexico v. United States of America) (Mexico v. United States of America), Provisional Measures, Order of 16 July 2008, I.C.J. Reports 2008*, p. 311, para. 55.

22 Ibid., p. 311 ss.

23 *Request for Interpretation of the Judgment of 31 March 2004 in the Case concerning Avena and Other Mexican Nationals (Mexico v. United States of America) (Mexico v. United States of America), Provisional Measures, Order of 16 July 2008*, Dissenting opinion of Judge Buergenthal, joint dissenting opinion of Judges Owada, Tomka and Keith; dissenting opinion of Judge Skotnikov.

24 For comments see Charnovitz, S., 'Correcting America's continuing failure to comply with the Avena Judgment', *AJIL* 106, 2012, 572–81; Tranchant, B., 'L'arrêt rendu par la CIJ sur la demande en interprétation de l'arrêt Avena', *AFDI* 55.2009, 2010, 191–220; Arrocha Olabuenaga, Pablo, 'Case Medellin vs. Texas: comentarios al fallo de la Suprema Corte de los Estados Unidos asì come a la solicitud de interpretación del fallo Avena del 31 de marzo de 2004, presentada a la Corte internacional de Justicia por México en junio de 2008', *Anuario mexicano de derecho internacional* 9, 2009, 677–705; Requena Casanova, M., 'De nuevo el asunto Avena ante la CIJ', *Anuario español de derecho internacional* 25, 2009, 263–95.

B. The decision of the Court under Art. 41 of the Statute

According to Art. 41 ICJ Statute, the Court has the power 'to indicate, if it considers that circumstances so require, any provisional measures which ought to be taken to preserve the respective rights of either party'. These circumstances require in the first place the existence of at least *prima facie* jurisdiction – a condition that in interpretation proceedings is not problematic. Further conditions are urgency, imminent irreparable harm, plausibility of the asserted rights and a link between those rights and the measures requested.[25]

In the present case the existence of these conditions was not controversial as the fear of Mexico that one of its nationals would be executed was plausible – and, in fact proved well-founded as the execution took place on 5 August 2008. The Court indicated provisional measures, namely that no Mexican national shall be executed unless review and reconsideration of the sentence has been granted. This order repeated literally what had been decided with binding force in the original judgment of 2004. Contrary to the 2004 judgment, which is binding upon the United States without any time limit, the order granting interim protection is binding only until the judgment on the merits is delivered. It expired in fact when, by judgment of 19 January 2009,[26] the Court found that the request for interpretation had to be dismissed because it related to a matter which had not been decided in the Judgment of 2004: in the original case the Court was not asked to decide – and thus did not address – the question whether the judgment created an obligation for all United States federal and state authorities. This reasoning seems problematic: a decision imposing concrete obligations on federal or state authorities is not within the competence of the Court and would have constituted an interference within the internal system of the United States' administration. It was up to the United States Government to care for the implementation of the judgment by its territorial components – a fact that the Court had explicitly stressed in the *LaGrand case*.[27] These inconsistencies in the Court's assessment, first on the existence of a dispute, and then on the reasons for dismissing the request for interpretation on the merits both support the view that the real aim of the request (and the order of the Court)

25 For a detailed analysis of this article see Oellers-Frahm, K., 'Article 41', in *The Statute of the International Court of Justice, A Commentary*, op. cit. n. 4, p. 1027 ss.

26 *Request for Interpretation of the Judgment of 31 March 2004 in the Case concerning Avena and Other Mexican Nationals (Mexico v. United States of America), Judgment, I.C.J. Reports 2009*, p. 3.

27 *LaGrand (Germany v. United States of America), Provisional Measures, I.C.J. Reports 1999*, p. 9, where the Court explicitly stated 'whereas according to the information available to the Court, implementation of the measures indicated in the present Order falls within the jurisdiction of the Governor of Arizona; whereas the Government of the United States is consequently under the obligation to transmit the present Order to the said Governor; whereas the Governor of Arizona is under the obligation to act in conformity with the international undertakings of the United States', ibid. p. 28.

was not the interpretation of the judgment, but an appeal to the United States to comply with the 2004 judgment.[28]

C. *Interpretation proceedings and provisional measures of protection*

In the context of the limited jurisdiction under Art. 60, the rights to be protected must plausibly derive from the underlying judgment. This means that they must have been established with the force of *res judicata* in the original judgment. Although this was the first case where a request for interpretation was accompanied by a request for provisional measures, the Court failed to make any remark on whether there may be any room at all for the use of Art. 41 in a procedure for interpretation.[29] After having found that a dispute existed, the Court proceeded immediately to the examination of the circumstances requiring the indication of provisional measures. There is no doubt that a request for provisional measures is not excluded in interpretation proceedings because the dispute/contestation concerning the *res iudicata* of the judgment – in exceptional cases – may cause irreparable harm to the (newly interpreted) right at stake. However, the Court is required to be extremely attentive to keep within the framework of Art. 60. In the *Avena case,* the Court found that the rights that Mexico sought to protect had a sufficient connection with the request for interpretation, namely the meaning of para 153 (9) of the judgment.[30] Although this is true, it has to be stated that what Mexico aimed at, and what the Court finally did, was nothing else than to reconfirm the obligations of the United States stated clearly in the original judgment. This judgment keeps its binding force so that the United States would be in breach of its international obligations should one of the Mexican nationals concerned be executed without the provision of review and reconsideration of its sentence.[31] The provisional measures requested by Mexico could not have – and were not aimed to have – any other effect than to re-confirm the obligation of the United States resulting from the judgment without adding anything new to that obligation.

This situation recalls the one in the *Bosnian Genocide case*, where Bosnia-Herzegovina, after a first request for provisional measures, which was granted on 8 April 1993,[32] only some time later brought a further request for

28 Infra 3. 3.

29 This question was raised particularly in the *Temple of Preah Vihear* case with regard to provisional measures concerning the non-aggravation of the dispute, see Dissenting Opinion of Judge Donoghue, para 10, op. cit. n. 1 and Oellers-Frahm, op. cit n. 4, p.75 ss.

30 *Request for Interpretation of the Judgment of 31 March 2004 in the Case concerning Avena and Other Mexican Nationals*, op. cit. n. 23, p. 328, para 64.

31 Ibid., p. 334, paras 4–6.

32 *Application of the Convention on the Prevention and Punishment of the Crime of Genocide (Bosnia and Herzegovina v. Yugoslavia), Provisional Measures, Order of 8 April 1993, I.C.J. Reports 1993*, p. 3 ss.

provisional measures. In that case, the Court found that the 'perilous situation demands, not an indication of provisional measures additional to those indicated by the Court's Order of 8 April 1993 ... but immediate and effective implementation of those measures'.[33] In the *Avena case* the factual situation was similar; however, the procedural situation was different. The Court could only examine the question of provisional measures after having admitted the request for interpretation, at least on a *prima facie* basis. In a case where the execution of persons is at stake in breach of a binding judgment of the Court, it is comprehensible that the Court would not miss the opportunity to remind the United States of its international obligations what may be seen as a reaction to the poor means available to ensure implementation of its judgments.[34]

Furthermore, contrary to a judgment that marks the end of the involvement of the Court,[35] the situation is different in the case of provisional measures where the Court remains seized of the matter. This follows from its power to revise the measures or indicate new ones whenever circumstances so require. As provisional measures have binding force until the judgment on the merits is delivered,[36] implementation of provisional measures enjoys thus a better, although still limited supervision by the Court when compared to final judgments. This situation seems rather odd and may explain why not only parties, but also the Court, tend to use any possibility to urge the state concerned to comply with the judgment. To this end, a request for interpretation combined with a request for the indication of provisional measures is in fact a promising means because the most delicate problem in interim protection proceedings, the question of jurisdiction, does not even arise. The relevant question concerns the risk of irreparable and imminent damage to the rights at stake that must be linked, in interpretation proceedings, to the terms to be construed. As the *Avena case* demonstrates, the Court, as an organ of the United Nations charged with the peaceful settlement of disputes, would be in a dilemma when passing in silence the imminent execution of human beings in disregard of a judgment. Therefore, the 'activism' of the Court in granting provisional protection in such cases constitutes a

33 *Application of the Convention on the Prevention and Punishment of the Crime of Genocide, Provisional Measures, Order of 13 September 1993, I.C.J. Reports 1993*, p. 349, para 59.

34 Oellers-Frahm, K., 'Enforcement of the International Court of Justice Decisions', infra Chapter 12.

35 Ibid. The Court is not concerned with the implementation of the judgment; in difference to i.e. the European Court of Human Rights (ECHR) where the Committee of Ministers supervises execution (Art. 46 ECHR), there is even no institutionalised supervision mechanism since the competence of the SC under Art. 94 (2) depends on a request of the party and is only of limited effect.

36 After long-lasting controversial opinions on the binding force of provisional measures, the Court finally decided in the *LaGrand case* (*LaGrand (Germany v. United States of America), I.C.J. Reports 2001*, p. 466) that provisional measures have binding force (*I.C.J. Reports 2001*, p. 503, para 102). For details see Oellers-Frahm, K., 'Article 41', op. cit. n. 25 with bibliographical references.

sort of remedy for the lack of supervising mechanisms relating to the implementation of a judgment. The question is therefore whether the same end could be reached without stretching the prerequisites for an only *prima facie* admission of a request for interpretation in order to reach the stage of the request for provisional measures.

IV Use of interpretation procedures for lack of other means of supervision

The Court, as the principal judicial organ of the United Nations, cannot remain silent in a situation such as the one in the *Avena case*, but it could be more cautious in order not to overstretch the limits of its jurisdiction. If a request for interpretation is inadmissible, the Court could nevertheless include in the reasons or even in the operative part of the order dismissing the request a passage reminding the parties of the obligations resulting from the judgment. Such attitude would be fully in accordance with the Court's powers and would not constitute an interference with sovereign rights of the parties because it only repeats what has been decided with binding force.

The *Avena case* has thus opened a way at least to involve the Court for reminding the parties of their obligations when compliance with a judgment is lacking. Although the interpretation procedure was certainly not instituted for that aim, Mexico's use of this instrument did not amount to a misuse or abuse, but constituted only an unusual way to conduce this procedure. As Mexico's tactics in the *Avena case* were successful, it was immediately and successfully used as a model to get the Court involved in a dispute that otherwise would have encountered difficulties in reaching the Court.[37]

As international law is lacking means of enforcing not only judgments but also obligations in general, the implementation of international obligations is indeed largely dependent on the good will of the state concerned so that reminders or other soft means may be even more efficient than 'harder' means, which are, in fact, a very rare exception in international law.

Bibliography

Arrocha Olanbuenaga, P., Case Medellin vs. Texas: comentarios al fallo de la Suprema Corte de los Estados Unidos asì come a la solicitud de interpretación del fallo

37 *Request for Interpretation of the Judgment of 1962 in the Temple of Preah Vihear* Case (op. cit. n. 1). In this case the Court indicated far-reaching provisional measures that met well-founded critics by five judges of the Court, cf. supra n. 17 and K. Shulman, 'The case concerning the Temple of Preah Vihear (Cambodia v. Thailand). The I.C.J. Orders sweeping provisional measures to prevent armed conflict at the expense of sovereignty', *Tulane Journal of International and Comparative Law* 20, 2012, pp. 555–70; Tanaka, Y., 'A new phase of the Temple of Preah Vihear dispute before the I.C.J.', *Chinese Jl of Int'l Law* 11, 2012, 191–226.

Avena del 31 de marzo 2004, *Anuario mexicano de derecho internacional* 25 (2009), 677–705.

Charnowitz, S., Correcting America's continuing failure to comply with the Avena Judgment, *AJIL* 106 (2012), 572–81.

Kulick, A., Article 60 ICJ Statute, interpretation proceedings, and the competing concepts of *Res Judicata*, *Leiden JIL* 18 (2015), 73–89.

Oellers-Frahm, K., The part of the International Court of Justice in solving disputes between American States, *L'Observateur des Nations Unies* 24 (2008), 7–35.

Oellers-Frahm, K., Provisional measures in interpretation proceedings – a new way to extend the Court's jurisdiction?, in: Elias, O. and Jalloh, C. (eds.), *Shielding Humanity*, Brill 2015, 61–84.

Requena Casanova, M., De nuevo el asunto Avena ante la CIJ, *Anuario español de derecho internacional* 25 (2009), 263–95.

Shulman, K., The case concerning the Temple of Preah Vihear (*Cambodia v. Thailand*), *Tulane Jl Int. and Comp. Law* 20 (2012), 555–70.

Tanaka, Y., A new phase of the Temple of Preah Vihear dispute before the ICJ, *Chinese Jl Int'l Law* 11 (2012), 191–226.

Tranchant, B., L'arrêt rendu par la CIJ sur la demande en interprétation de l'arrêt Avena, *AFDI* 55 (2009), 191–220.

Zimmermann, A., Tomuschat, C., Oellers-Frahm, K., and Tams, C., (eds.), *The Statute of the International Court of Justice – A Commentary*, 2nd ed. 2012, OUP.

12 Enforcement of the International Court of Justice decisions

Karin Oellers-Frahm

I Introduction

In accordance with Art. 94 (1) of the UN Charter (UNC) 'Each Member of the United Nations undertakes to comply with the decision of the International Court of Justice [ICJ] in any case to which it is a party'.[1] Should all states parties to cases before the ICJ be law-abiding, there would be no room for questions of enforcement. However, the authors of the UNC were well aware of the fact that states – like individuals – do not always act as they should and therefore also included a provision for the case in which a state failed to comply with a decision of the Court. Art. 94 (2) UNC provides that 'if any party to a case fails to perform the obligation incumbent upon it under a judgment rendered by the Court, the other party may have recourse to the Security Council, which may, if it deems necessary, make recommendations or decide upon measures to be taken to give effect to the judgment'.

Under the League of Nations, compliance with court decisions, not only just those of the Permanent Court of International Justice (PCIJ) but also arbitral awards, was entrusted to the Council that 'in the event of any failure to carry out such an award or decision … shall propose what steps should be taken to give effect thereto'.[2] In contrast to Art. 94 (2) UNC, the Council could act on its own initiative, while under the Charter the Security Council (SC) takes action only upon recourse of the party that derives a title against the other party from the judgment. Other members or organs of the UN cannot request action under Art. 94 (2), but if in their view the situation resulting from the non-compliance of a judgment could lead to 'international friction or give rise to a dispute' (Art. 34 UNC), they may address the SC under Chapter VI or, as the case may be, Chapter VII of the Charter. Against this backdrop it becomes evident that Art. 94 (2) UNC constitutes

1 For a detailed analysis of Art. 94 see Oellers-Frahm, K., 'Article 94', in Simma, Bruno et al. (eds), *The Charter of the United Nations, A Commentary*, 3rd ed., Oxford: OUP, 2012, p. 1957 ff.

2 Art. 13 (4) of the Covenant; cf. Vité, S., 'Article 13', in Kolb, R. (ed.), *Commentaire sur le Pacte de la Société des Nations*, Bruxelles: Bruylant, 2015, 571 et seq.

a special basis for *seizing* the SC, namely one that does not presuppose that the non-compliance with the judgment is – already – endangering peace and security or leading to a friction or dispute referred to in Art. 34 UNC.[3]

While not all judgments of the ICJ were complied with properly,[4] recourse to Art. 94 (2) UNC was taken only once,[5] in the *Nicaragua case*,[6] which exposed the shortcomings and ambiguities inherent in this provision, which certainly has been one of the reasons for the paucity of use of the provision.[7]

II The *Nicaragua case*

This case originated from the activities of the Contras, opponents of the Nicaraguan Sandinista Government, who initiated a guerilla insurgency campaign operating from bases within neighboring states and being funded and assisted both covertly and overtly by the US.

On 9 April 1984, Nicaragua instituted proceedings before the ICJ against the US claiming that the US support for the Contras constituted an unlawful use of force against Nicaragua and an unlawful intervention in her internal affairs. As El Salvador, Honduras and Costa Rica claimed that Nicaraguan forces had engaged in military activities and assisted rebels inside their territory, the US argued to have acted in collective self-defense.

The US was in a difficult situation because support for the Contras was a matter of public knowledge and as it seemed difficult to justify its stance as collective self-defense, they were keen that the case did not progress to the merits stage. However, their strategy to deny or at least amend *ad hoc* the Court's jurisdiction failed[8] when by its judgment of 26 November 1984 the

3 Cf. Oellers-Frahm K., 'Article 94', op. cit. n.1, p. 1966 MN 18.

4 Cf. Paulson, C., 'Compliance with Final Judgments of the International Court of Justice since 1987', *AJIL* 98, 2004, 434 et seq.; Llamzon, A., 'Jurisdiction and Compliance in Recent Decisions of the International Court of Justice', *EJIL* 18, 2007, 815–52; Azar, A., *L'exécution des decisions de la Cour internationale de Justice*, Bruxelles: Bruylant, 2003; Schulte, C., *Compliance with Decisions of the International Court of Justice*, Oxford: OUP, 2004.

5 It has, however, to be noted that also in the case *Land, Island and Maritime Frontier Dispute (El Salvador v. Honduras; Nicaragua Intervening), I.C.J. Reports 1992*, p. 351 a formal accusation of non-compliance under Art. 94 (2) UNC was made in 2002 (UN Doc. No. S/2002/108). In its response to the SC of October 2002, El Salvador claimed that it had repeatedly declared its intention to request revision of the ICJ judgment and that the dispute over compliance was thus non-existent (UN Doc. S/2002/1102). El Salvador filed, one day short of the 10-year limit for revision, a request for revision to the ICJ, so that action of the SC was not necessary. After rejection of the application for revision, both states finally reached an agreement and began to demarcate the areas.

6 *Military and Paramilitary Activities in and against Nicaragua (Nicaragua v. United States), I.C.J. Reports 1986*, 14.

7 Llamzon, op. cit. n.4, 847.

8 Gill, T. G., *Litigation Strategy at the World Court*, Leiden/Boston: Martinus Nijhoff, 1989, pp. 135–40; see also Letter of the United States of 13 April 1984 to the ICJ Registrar, *I.C.J. Pleadings, Military and Paramilitary Activities*, Vol. V, 359–60. See

Court decided in favor of its jurisdiction and dismissed all preliminary objections or declared that they did not possess an exclusively preliminary character. As a consequence of this judgment the US decided to abstain from participating in the proceedings (and not to accept the final judgment) and withdrew its declaration under Art. 36 (2) of the Statute. Without delving further into details of the proceedings[9] it is sufficient in the present context to recall that in its judgment on the merits of June 1986 the Court found that the US, by supporting the Contras in Nicaragua had violated its international legal obligations not to intervene in the affairs of another state and not to use force against another state. These acts could not be justified, according to the Court's findings, as acts of collective self-defense because the support of the rebels in El Salvador by Nicaragua did not amount to an armed attack in the sense of Art. 51 UNC – reflecting customary law[10] – and that in any case the use of force as well as the level of intervention by the US was disproportionate and thus unlawful. Furthermore, the Court found that the US was required to make reparation to Nicaragua, the form and amount of which would be settled by the Court failing an agreement between the parties.

III The post-judgment developments

This judgment constituted a landmark in international law: for the first time a small state in a politically weak position for reaching a negotiated solution to the dispute gained traction before the ICJ against the most powerful state of the world, which was found to be responsible for grave violations of international law. The judgment proved to be, however, only one, although a highly pertinent step in the 'settlement' of the dispute, because the dispute would only be settled when the violations of international law were

further the Shultz letter by which the United States tried to amend their declaration of acceptance of the ICJ's jurisdiction under Art. 36 (2) of the Statute in the sense that disputes with Central American States are excepted from the Court's jurisdiction (*I.C.J. Reports, 1984*, p. 398. *Military and Paramilitary Activities in and against Nicaragua (Nicaragua v. United States of America), Jurisdiction and Admissibility, Judgment, I.C.J. Reports 1984*, p. 392).

9 Crawford, J. R., 'Military and Paramilitary Activities in and against Nicaragua (*Nicaragua v. United States of America*)', in *MPEPIL*. Available at http://opil.ouplaw.com/ view/ 10.1093/law:epil/9780199231690/law-9780199231690-e170?rskey=mJSUdR&result=1&prd=OPIL (access date January 2016), with bibliographical indications; and Schulte, op. cit. n.4, pp. 184–211.

10 Due to a reservation in the declaration of the US under the optional clause the Court had no jurisdiction to decide disputes arising under a multilateral treaty unless all parties to the treaty affected by the decision are also parties to the case before the Court (so-called Vandenberg reservation), so that the Court, in deciding the merits, could not apply the UN Charter that, in Art. 2 and Art. 51, regulates questions of use of force and self-defense, but had to rely on customary law that, in this respect, does not differ in substance from the Charter provisions.

terminated and a reparation agreement reached. As the US had declared already in 1984, after the judgment on admissibility and jurisdiction was delivered, that they would 'not accept' the judgment – although, this again, constituted a violation of international law – and as Nicaragua had no means to reach implementation of the judgment by negotiation or other bilateral means, the only possible alternative was recourse to Art. 94 (2) UNC.

Immediately after the delivery of the judgment, Nicaragua requested the SC to prevent the US Congress from according again US$100 million to the Contras. This step, however, was taken under Chapter VI, not under Art. 94 (2) UNC, which was used only some time later when President Ronald Reagan, in October 1986, ratified the Act of support for the Contras.[11] On 28 October 1986, five non-aligned states: Congo, Ghana, Madagascar, Trinidad and Tobago and the United Arab Emirates, presented the text of a resolution that only 'called for full and immediate compliance with the judgment ... in conformity with the relevant provisions of the Charter' and 'requests the Secretary-General to keep the Council informed on the implementation of this resolution'.[12] Only one state voted against the resolution, not surprisingly the US, the state concerned and a veto-power, while eleven countries voted in favor, with three abstaining.[13]

It thereby became clear that Nicaragua would not garner support from the SC for enforcing the judgment. Indeed, if a resolution in such general terms as the one presented was vetoed by the US, any resolution providing for concrete means of implementation would share the same fate. Nicaragua, therefore, turned to the General Assembly (GA), which adopted several resolutions urging the US 'to abide by its obligations resulting from the UNC'.[14] However, the action of the GA proved unpromising as, opposed to the SC, it lacks the power to take binding decisions.

Although the US did, at a later time, comply with the judgment in an 'indirect' manner,[15] the experience made by Nicaragua shows that the enforcement of a judgment against a veto-power (or another state closely related to a veto-power) would not work, because – strangely enough – the state against which implementation means are requested participates in the voting. The problem of the voting system in the SC was not addressed in the

11 Cf. Gill, op. cit. n.8, p. 285 ff.

12 SR Doc. S/18428 of 28 October 1986.

13 The affirmative votes came from Australia, Bulgaria, China, Congo, Denmark, Ghana, Madagascar, Trinidad and Tobago, USSR, United Arab Emirates and Venezuela; abstentions came from France, Thailand and the United Kingdom, S/PV.2704.

14 GA Res A/RES/41/31 of 3 November 1986, which was identical to the resolution presented to the Security Council; A/RES/42/18 of 12 November 1087 and A/RES/44/217 of 22 December 1989.

15 The US adopted finally a comprehensive program of economic assistance for Nicaragua, which induced Nicaragua to withdraw its claim brought before the ICJ requesting the determination of the amount of compensation owed to Nicaragua, *I.C.J. Reports 1991*, p. 47 ff.

Nicaragua case, what does not mean that the procedure followed in that case was uncontroversial. As no further recourse was ever made to the SC under Art. 94 (2) UNC,[16] the following considerations are still, even thirty years after the Nicaraguan experience, of theoretical character.[17]

IV The voting procedure in the Security Council

Art. 27 UNC concerning the voting procedure in the SC makes a distinction between decisions on procedural matters, which 'shall be made by an affirmative vote of nine members' (Art. 27 (2)), and those referred to in Art. 27 (3) UNC: 'Decisions ... on all other matters shall be made by an affirmative vote of nine members including the concurring votes of the permanent members; provided that, in decisions under Chapter VI, and under paragraph 3 of Article 52, a party to the dispute shall abstain from voting.'

A. Procedural matters and other matters

The Charter as well as the Rules of Procedure of the SC lack details concerning the distinction between procedural and other matters, although Arts 28 to 32 UNC address procedural questions. However, these articles fail to provide an exhaustive enumeration of what procedural matters are. In a case where the SC acts under Art. 94 (2) UNC – and that being the only aspect of Art. 27 UNC which is relevant here[18] – the decisive question is whether such action is considered as (i) a 'procedural matter', thereby not requiring the affirmative vote of the five permanent members; or as (ii) 'another matter', thus requiring the consent of all permanent members, with the exception that the party concerned is prevented from voting when the matter relates to Chapter VI of the Charter. Chapter VI concerns the pacific settlement of disputes, which allows for action of the SC in the form of recommendations, excluding binding decisions that are admissible only in situations covered by Chapter VII. In order to exclude a permanent member against which compliance with a judgment is sought from voting (and vetoing) a resolution in the SC, it would therefore be helpful to consider action of the SC in the framework of Art. 94 (2) UNC as a procedural matter, or as action under Chapter VI, unless the SC explicitly declares that it is acting under Chapter VII.

16 See, however, the attempt made by Honduras in the *Land, Island and Maritime Frontier Dispute (El Salvador v. Honduras: Nicaragua Intervening)*, *I.C.J. Reports 1992*, p. 351, supra n.5.

17 Oellers-Frahm, K., 'Souveräne Gleichheit der Staaten in der internationalen gerichtlichen Streitbeilegung? Überlegungen zu Art. 94 Abs. 2 und Art. 27 UN-Charta', in Frowein J. A., Scharioth K., Winkelmann I., and Wolfrum R. (eds), *Verhandeln für den Frieden Liber Amicorum Tono Eitel*, Heidelberg: Springer, 2003, pp. 169, 178 ss.

18 For a general analysis of Art. 27 see Zimmermann, A., 'Article 27', in B. Simma, op. cit. n.1, vol. I, 871 ss., MN 76 ss.

The Statute of the ICJ contains procedural rules on the voting within the SC only in relation to the election of judges, Art. 10(2) Statute, so that guidance has been sought in General Assembly Resolution 267 (III) of 14 April 1949, which contains a list of matters to be considered as procedural, including 'decisions to remind members of their obligations under the Charter'.[19] Whether this rather vague formulation might prove helpful in cases of Art. 94 (2) UNC, which requires more than a mere reminder of the obligations under the Charter, remains rather problematic. The question whether the General Assembly has at all the power to regulate items concerning the voting procedure in the SC is equally problematic. The SC has adopted its own Rules on Procedure[20] that, with regard to the question at hand, are, however, silent.

B. The double veto

Most authors who tackle the question of voting in the framework of Art. 94 (2) UNC, do not consider decision-making under Art. 94 (2) UNC as a 'procedural matter',[21] although expressing concern with regard to the use of the veto power.[22] As the question is not uncontroversial, the SC would, in a concrete case, have to decide as a first step whether it is concerned with a procedural matter or not. This question clearly is no procedural matter so that Art. 27 (3) UNC would be applicable admitting the veto of a permanent member that, already at this point, could and certainly would prevent the adoption of a resolution qualifying the matter as procedural with the consequence that a decision on the substance would be governed by Art. 27 (3) UNC allowing again the use of the veto.

C. The Chapter VI exception

The situation would be different only should the second part of Art. 27 (3) UNC apply, which provides for the abstention from voting by a state 'party to the dispute'. In the *Nicaragua case*, the Office of Legal Affairs of the UN, while preparing the relevant meeting of the SC, drew the attention not only to the so-called double veto that would be applicable if the question should arise, but also to the 'abstention clause' in the following terms:

19 Tanzi A., 'Problems of Enforcement of Decisions of the International Court of Justice and the Law of the United Nations', *EJIL* 6, 1995, 539 ss, 551.

20 UN Doc. S/96.

21 Vulcan, C., 'L'exécution des décisions de la Cour internationale de Justice d'après la Charte des Nations Unies', *RGDIP*, 1947, 201; Kelsen, H., 'Settlement of Disputes by the Security Council', *The Int.l Law Quarterly* 1948, 211; Azar, A., *L'exécution des décisions de la Cour internationale de Justice*, Bruxelles: Bruylant, 2003; Kamto, M., 'Considérations actuelles sur l'inexécution des décisions de la Cour internationale de Justice', in Ndiaye, T. M., and Wolfrum, R. (eds), *Liber Amicorum Judge Thomas A. Mensah*, Leiden/Boston: Nijhoff 1992, 301.

22 Oellers-Frahm, K., 'Article 94', op. cit. n.1, MN 26.

> If the matter is brought before the SC under Article 94, it could be argued that the issue of giving effect to the judgment of the Court is not in itself a dispute, and that therefore no question of the application of Art. 27, paragraph 3, arises. Article 94 is considered by several jurists as providing the SC with powers and responsibilities independent of those envisaged in Chapter VI, thus reinforcing an argument that Article 27, paragraph 3, has no application when the Council is considering a matter under Article 94, paragraph 2. Thus no member of the Council would be debarred from voting, even though it is a party in the case which has been adjudicated by the Court.[23]

This opinion has manifestly been followed by the SC and also by numerous legal scholars.[24] The SC was certainly right in qualifying – although incidentally – a procedure under Art. 94 (2) UNC not as a 'procedural matter'. Whether this implies the inapplicability of the Chapter VI exception seems more problematic and depends on the relationship between Art. 94 (2) UNC and the rest of the Charter.

V Systemic interpretation of Article 94 (2) UNC

A. Historical background of Art. 94 (2) UNC

As opposed to Art. 13 (4) of the Covenant of the League of Nations,[25] Art. 94 (2) is limited to the implementation of judgments of the ICJ. Access to the SC should be open even if international peace and security are not – yet – endangered.[26] This interpretation follows not only from the text of the Charter, but also from the reports of the San Francisco Conference.[27] The particularity of Art. 94 (2) UNC concerns the fact that it is of a preventive character, i.e. that the SC can be addressed although international peace and security are not endangered, but might be endangered at any moment by the creditor of the judgment seeking enforcement. This view reflects Art. 13 (4) of the Covenant, the provision that inspired Art. 94 (2) UNC, which explicitly provided that members of the League of Nations 'will not resort to war against a Member of the League that complies' with the judgment, creating reciprocal obligations.[28] However, this does not mean that Art. 94 (2) UNC

23 UNJYB 1986, 283. See in this context Rosenne, S., *The Law and Practice of the International Court of Justice 1920–2005*, 4th ed., Leiden/Boston: Nijhoff, 2006, vol. I, p. 246 ss.

24 Schulte, op. cit. n.4, p. 55 with bibliographical references in footnote 176.

25 Vité, S., 'Article 13', in Kolb, R. (ed.), op. cit. n.2, p. 571 ff, 582.

26 Tunçel, E., *L'exécution des décisions de la Cour internationale de Justice selon la Charte des Nations Unies*, Messeiller, 1960, p. 72.

27 Schachter, 'The Enforcement of International Judicial and Arbitral Awards', *AJIL* 54,1960, 1 ss., at p. 19.

28 For more details see Vité, S., 'Article 13', op. cit. n.2, p. 582.

provides a special competence for the SC that is completely independent from its other competences,[29] or that it confers a special competence to apply Chapter VII.[30] This article only offers a particular way to set in motion the exercise by the SC of its usual competences under the other Charter provisions. In the words of Rosenne, Art. 94 (2) UNC has to be considered as a 'particularization of powers' in contrast to 'complementary powers'[31] in the sense that it provides a particular basis for the access to, not the powers of, the SC.[32] Compliance with a judgment is considered as a political matter rather than a legal one;[33] it is entrusted to the SC as part of its function to maintain international peace and security.

B. Relationship between Chapter VI and Art. 94(2) UNC

Art. 94 (2) UNC is not the only means of recourse to the SC for reaching the implementation of a judgment. The creditor of the judgment could also proceed under Chapter VI, which relates to disputes 'the continuance of which is likely to endanger the maintenance of international peace and security' (Art. 33 UNC) or 'which might lead to international friction' (Art. 34 UNC). This situation is always present if a judgment of the ICJ is not complied with,[34] because as long as the judgment is not implemented the underlying dispute is not settled and the SC retains the competence – notwithstanding the judgment of the ICJ – to deal with it.[35] The requirement for action of the SC under Chapter VI, namely the danger for international peace and security, is no serious obstacle as it 'would evidently be a direct incentive for that state to claim that it may be compelled to resort to force or other acts endangering international peace – a consequence that was almost surely not intended and which cannot be considered desirable'.[36] This scenario might be preferred by a state with regard to the voting procedure, namely the obligation incumbent upon the state party to the dispute to abstain from voting when the SC acts under Chapter VI, but not when it acts under Art. 94 (2) UNC. This situation reveals the inconsistency of interpreting Art. 94(2) UNC isolated from the rest of the Charter and seems hardly compatible with a provision explicitly concerning peaceful compliance with a judgment and which in a narrow and formal understanding would leave only little, if any room at all, for the application of Art. 94(2) UNC.[37]

29 Oellers-Frahm, op. cit. n.17, p. 184.
30 Proposals in this sense have explicitly been dismissed in the San Francisco Conference, see Tanzi, op. cit. n.19, p. 555.
31 Rosenne, op. cit. n.23, p. 245.
32 Schachter, op. cit. note 27, p. 19 ss.
33 Llamzon, op. cit. note 4, p. 847.
34 Zimmermann, A., 'Article 27', op. cit. n.18, p. 921, MN 216.
35 Tunçel, op. cit. note 26, p 69; Tanzi, op. cit. note 19, 572; Schachter, op. cit. n.27, p. 21.
36 Schachter, op. cit. note 27, p. 20.
37 Reisman, 'Enforcement of International Judgments', *AJIL* 63,1969, 1, 14–15.

What proves decisive in cases where compliance with a judgment is at stake is the type of action, action under Chapter VI or Chapter VII irrespective of the basis of seizing the SC, Art. 94 (2) UNC or another basis.[38] In any case, the SC cannot touch upon the judgment, which is *res judicata* and unimpeachable for the SC, for the rest the SC is free to decide on the means of action.

C. The role of the Security Council under Art. 94 (2) UNC

This understanding of Art. 94 (2) UNC is further supported by the fact that under Art. 94 (2) UNC the SC 'may, *if it deems necessary*' (emphasis added) take action. It does not function as a subsidiary organ of the ICJ, but exercises its functions as one of the principal organs of the UN independently from other organs, in this case the ICJ. The fact that also in cases of Art. 94 (2) UNC the SC is acting in complete independence means that its action is governed by the question of whether the non-implementation of the judgment may endanger international peace and security. The action of the SC must be aimed 'to give effect to the judgment',[39] a wording that is larger than the term 'implementation of the judgment'. The SC thus freely assesses the situation resulting from the non-implementation of the judgment. In doing so, the judgment is an essential, but not the sole factor to be taken into account, because the SC addresses the question on a broader basis than the Court.[40] The SC, the political organ of the UN, 'ne prête pas son bras séculier à la justice; il tire simplement parti de l'activité parallèle du juge pour renforcer la légitimité de ses propres mesures'.[41] Hence, Art. 94 (2) UNC cannot be compared to national provisions concerning enforcement of judgments, but constitutes part of the provisions on the settlement of disputes.[42] Drawing from this conclusion, it is clear that the voting procedure differs according to whether decisions are taken under Chapter VI or Chapter VII, thereby preventing or permitting the concerned party to vote. To make the voting procedure in the SC dependent on the way in which the SC was seized leading to different procedures for decisions that, by their very nature,

38 Tanzi, op. cit. note 19, p. 554.

39 The *Nicaragua case* is a prominent example of non-compliance with, but giving effect to a judgment: after the failure of an agreement on compensation, Nicaragua brought the question of compensation back to the Court. When in 1990 in Nicaragua, a US-backed President, Violeta Chamorro, was elected, the US terminated its economic sanctions against Nicaragua. On 12 September 1991, Nicaragua discontinued the case before the Court on the basis of 'agreements [with the United States] aimed at enhancing Nicaragua's economic, commercial and technical development to the maximum extent possible'. See for more details Schulte, op. cit. n.4, pp. 205/206.

40 Rosenne, op. cit. note 23, p. 246.

41 Weckel, P., 'Les suites des décisions de la Cour internationale de Justice', *AFDI* 1996, 428 ff, 442.

42 Oellers-Frahm, op. cit. n.17, p. 188.

are Chapter VI decisions is inconsistent with the fact that also under Art. 94 (2) UNC the SC is exercising its usual function as the political organ of the UN. This understanding of Art. 94 (2) UNC also answers the question whether Art. 94 (2) UNC represents a special case for the application of Chapter VII powers,[43] which it does not. Only if the prerequisites of Chapter VII are fulfilled, binding measures including use of force can be taken also in cases where the SC acts under Art. 94 (2) UNC. And, finally, the above understanding of Art. 94 (2) UNC is fully in line with Art. 27 (3) UNC, which refers to 'decisions under Chapter VI'; this qualifies the decisions, not the way in which the SC was seized.

In the *Nicaragua case*, this interpretation of Art. 94 (2) UNC would have required that the US abstain from voting since the decision taken by the SC was evidently a Chapter VI decision.

VI Concluding remarks

The above analysis shows that action of the SC under Art. 94 (2) UNC is, like action under Chapter VI, of only limited extent: as the *Nicaragua case* demonstrates, the resolutions that were vetoed by the US did not even provide for concrete recommendations for complying with the judgment. Also, the resolutions taken by the GA – which does not even have the power to take binding decisions – were of a mere hortatory character.[44] International law is, thus, only efficient in adjudging a dispute against a permanent member of the SC – as long as there is a basis for jurisdiction – but it is weak with regard to enforcing a judgment against a non-complying state, in particular a permanent member. In the case that the non-compliance with a judgment would endanger peace and security, the situation is even worse because the concerned party always participates in a vote under Chapter VII, so that not even a hortatory decision could be taken should a permanent member or a state with strong political ties to a permanent member be concerned.

But this is hardly groundbreaking news in international law where actions against the will of a state remain the most delicate problem. This should, however, not lead to resignation – at least in the context of the implementation of ICJ judgments – because practice shows that the compliance record for ICJ judgments is rather positive. Also, in cases of non-compliance 'diplomatic pressure' or the possibility of recourse to the SC more often than not lead to the desired result. The *Nicaragua case* is an impressive example in the sense that questions of power are irrelevant before the Court, but that the post-judgment phase may be burdensome and disappointing. However, in the end, even if the judgment was not literally implemented, it was at least 'given effect' so that the underlying dispute was settled peacefully.

43 Ibid., p. 184. See also Tanzi, op. cit. n.19, p. 55 and Oellers-Frahm, op. cit. n.17, p. 184.
44 Schulte, op. cit. n.4, p. 204.

Bibliography

Azar, A. *L'exécution des décisions de la Cour internationale de justice*, Bruxelles: Bruylant, 2003.

Crawford, J.R. *Military and Paramilitary Activities in and against Nicaragua (Nicaragua v. United States of America)*, *MPEPIL*.

Gill, T.D. *Litigation Strategy at the International Court: A Case Study of the Nicaragua v. United States Dispute*, Dordrecht/Boston/London: Martinus Nijhoff Publishers, 1989.

Kamto, M. Considérations actuelles sur l'inexécution des décisions de la Cour internationale de Justice. In: Ndiaye, T.M. and Wolfrum, R. (eds), *Liber Amicorum Judge Thomas Mensah*, Martinus Nijhoff Publishers: Dordrecht/Boston/London, 1992, 215–33.

Kelsen, H. Settlement of Disputes by the Security Council, *The Int'l Law Quarterly*, 1948, 173–213.

Llamzon, A.P. Jurisdiction and Compliance in Recent Decisions of the International Court of Justice, *EJIL* 18 (2007), 815–52.

Oellers-Frahm, K. Article 94. In: Simma, B., Khan, D.-E., Nolte, G., and Paulus, A. (eds), *The Charter of the United Nations, A Commentary,* 3rd ed., Oxford: OUP, 2012, 1957–71.

Oellers-Frahm, K. Souveräne Gleichheit der Staaten in der internationalen gerichtlichen Streitbeilegung? Überlegungen zu Art. 94 Abs. 2 und Art. 27 UN-Charta, Frowein, J. A., Scharioth, K., Winkelmann, I., and Wolfrum, R., (eds), *Verhandeln für den Frieden*, Heidelberg: Springer, 2003, 169–91.

Paulson, C. Compliance with Final Judgments of the International Court of Justice since 1987, *AJIL* 98 (2004), 434–61.

Reisman, W.M.N. Enforcement of International Judgments, *AJIL* 63 (1969), 1–27.

Rosenne, S. *The Law and Practice of the International Court of Justice 1920–2005*, 4th ed. Martinus Nijhoff Publishers: Leiden/Boston, 2006, vol. I.

Schachter, O. The Enforcement of International Judicial and Arbitral Awards, *AJIL* 54 (1960), 1–24.

Schulte, C. *Compliance with Decisions of the International Court of Justice*, Oxford: OUP, 2004.

Tanzi, A. Problems of Enforcement of Decisions of the International Court of Justice and the Law of the United Nations, *EJIL* 6 (1995), 539–72.

Tunçel, E. *L'exécution des décisions de la Cour internationale de Justice selon la Charte des Nations Unies*, Neuenburg: Messeiller, 1960.

Vité, S. Article 13. In: Kolb, R., *Commentaire sur le Pacte de la Société des Nations*, Bruxelles: Bruylant, 2015, 571–86.

Vulcan, C. L'exécution des décisions de la Cour internationale de Justice d'après la Charte des Nations Unies, *RGDIP* 57 (1947), 187–205.

Weckel, P. Les suites des décisions de la Cour internationale de Justice, *AFDI* 1996, 428–42.

Zimmermann, A. Article 27. In: Simma, B., Khan, D.-E., Nolte, G., and Paulus, A., (eds), *The Charter of the United Nations, A Commentary*, 3rd ed., Oxford: OUP, 2012, 871–938.

Part III

Material contribution of Latin American cases to the development of international law

Diplomatic and consular protection

13 *Vienna Convention on Consular Relations (Paraguay v. United States of America), 1998*

Cristina Hoss

I Introduction and factual background

On 3 April 1998, Paraguay instituted proceedings and requested the indication of provisional measures against the US before the International Court of Justice (hereinafter 'the ICJ' or 'the Court').[1] The dispute related to alleged breaches of the Vienna Convention on Consular Relations of 1963[2] (hereinafter 'the Vienna Convention') and concerned a Paraguayan national, Mr Angel Francisco Breard, who was sentenced to the death without being informed of his rights under Article 36, paragraph 1 lit.b), of the Vienna Convention.[3] Two more prominent cases involving very similar matters, *LaGrand (Germany v. United States)*[4] and *Avena (Mexico v. United States)*,[5] were still to come before the Court. Those two cases proceeded to the merits and therefore 'superseded' the *Breard* precedent. Nevertheless, this latter case should not be forgotten for it opened the path for important legal findings in the *LaGrand* and *Avena* cases.

When determining the influence of the short-lived *Breard case*, two factors deserve particular attention: the debate surrounding the binding force of provisional measures ordered by the ICJ and the form of reparation requested by Paraguay, namely guarantees of non-repetition. These two issues were present in the 'forgotten' case of Mr Breard, but were only addressed by the Court in the cases *LaGrand* and *Avena* a few years later for the simple reason

1 *Vienna Convention on Consular Relations (Paraguay v. United States of America), Provisional Measures, Order of 9 April 1998, I.C.J. Reports 1998*, p. 248. Also available on the Court's website: www.icj-cij.org (accessed 11 March 2016).

2 *Vienna Convention on Consular Relations,* UNTS, vol. 596, p. 487. Full text available at: https://treaties.un.org/doc/Publication/UNTS/Volume%20596/volume-596-I-8638-English.pdf (accessed 11 March 2016).

3 The full history of domestic court proceedings in the US is set out in the last decision of the United States Supreme Court in that case, dated 14 April 1998: *Breard v. Greene* 523 US 371 (1998). Confirmed by *Breard v. Commonwealth*, 445 S.E.2d 670.

4 *LaGrand (Germany v. United States of America), Judgment, I.C.J. Reports 2001*, p. 466.

5 *Avena and Other Mexican nationals (Mexico v. United States of America), Judgment, I.C.J. Reports 2004*, p. 12.

that the *Breard case* did not proceed to the merits. The case was removed from the Court's list just one month after the filing of Paraguay's memorial, which occurred seven months after the execution of Mr Breard.

Paraguay proceeded to file an application before the ICJ on 3 April 1998, only days before the execution of Mr Breard, scheduled to take place on 14 April 1998. The Court held hearings on 7 April 1998 and, unanimously, adopted an order on 9 April 1998, requesting the US to take all measures to ensure that Mr Breard was not executed pending the final decision in the proceedings.[6] In reference to the order, the Secretary of State, Ms Madeleine Albright, wrote a letter to the Governor of Virginia on 13 April 1998 requesting him to exercise his powers as governor and stay Mr Breard's execution.[7] On the morning of 14 April 1998, the day scheduled for his execution, the US Supreme Court denied the last application for stay of execution on the basis of the 'procedural default' rule. This rule essentially bars any claim brought at a late stage of criminal proceedings from being examined unless established that prejudice was caused to the defendant.[8]

The Supreme Court nevertheless recalled that it was now the prerogative of the Governor of Virginia to allow the stay of execution on the basis of the letter from the Secretary of State.[9] The Governor of Virginia, however, chose not to exercise his power and Mr Breard was put to death by lethal injection on 14 April 1998.[10]

Paraguay filed a memorial on 9 October 1998[11] requesting the Court to make declarations of wrongfulness with regard to the violation of the Vienna Convention and of the order indicating provisional measures.[12]

Paraguay further requested the Court to order guarantees of non-repetition and, given that *restitutio in integrum* had become impossible after the execution of Mr Breard, it sought to obtain alternative reparation, namely compensation and moral damages.[13] On the same day, 9 October 1998, Paraguay filed a document entitled 'Conditional Request of Paraguay for an order conclusively establishing facts'. In that document, Paraguay essentially set out the prejudice it had allegedly suffered due to the execution of Mr Breard.[14]

6 *I.C.J. Reports 1998*, p. 258.

7 Letter of the Secretary of State to the Governor of the State of Virginia, dated 13 April 1998, available at: www.state.gov/documents/organization/65744.pdf (accessed on 11 March 2016).

8 *Breard v. Greene* 523 US 371 (1998).

9 Ibid. and note 7 above.

10 www.nytimes.com/1998/04/15/us/clemency-denied-paraguayan-is-executed.html (accessed on 11 March 2016).

11 *I.C.J. Pleadings, Vienna Convention on Consular Relations (Paraguay v. USA), 1998, Memorial submitted by Paraguay on 9 October 1998*, pp. 140–187.

12 Ibid., p. 185.

13 Ibid., p. 186.

14 The procedure before the Court does not provide for any such order, however, the document was admitted to the case file *I.C.J. Pleadings, Vienna Convention on Consular Relations (Paraguay v. USA)*, pp. 85–88.

Only one month following the filing of Paraguay's memorial, the Court issued an order removing the case from the list at the request of Paraguay. The Government of Paraguay had written to the Court expressing its intention to 'discontinue the proceedings with prejudice'.[15] The US agreed, and the Court issued an order to that effect on 10 November 1998. No particular reason was given for the discontinuance of the proceedings but in the meantime the US had made an official apology with the assurance of better future compliance with the Vienna Convention.[16]

Regardless of its early demise, the *Breard case* arguably had a greater impact on the Court and its jurisprudence than one would have expected from the very brief history of the proceedings before the ICJ.

Apart from the protracted debate that this case has generated among US scholars,[17] and as far as international law is concerned, the impact is most visible in the context of the determination of the nature of provisional measures issued by the ICJ (II) as well as in the emergence of guarantees of non-repetition as remedy in the law of state responsibility (III).[18]

II On binding orders and 'non-binding language'

At the time of Paraguay's filing of the case regarding the Vienna Convention on Consular Relations, the binding nature of provisional measures under the Court's Statute was still much debated.[19] In fact, Paraguay's memorial was the first in the history of the Court to request an explicit finding on the violation of an order indicating provisional measures, and henceforth a declaration on the binding nature of provisional measures as such.[20]

Different lines of argument can be identified in Paraguay's approach, as set out in its memorial of 9 October 1998.[21] Paraguay first argued that orders of

15 The expression 'with prejudice' means that the State renounces the possibility of taking any further action in the matter, *I.C.J. Vienna Convention of Consular Relations (Paraguay v. USA), Order of 10 November 1998, I.C.J. Reports 1998*, p. 426 (p. 427).

16 Statement made in Asunción (Paraguay) on 1 November 1998, in: Quigley, J., Aceves, W. J., and Shank, A., *The Law of Consular Access: A documentary guide*, London: Routledge, 2010, p. 139. See full text below, pp. 14–15.

17 For an extensive debate on that issue see notably: Agora Breard, *American Journal of International Law*, No. 4 (1998), pp. 666–712.

18 See also Hoss, C., *Satisfaction*, in: MPEPIL, Oxford: OUP, 2014.

19 See in particular: Bernhardt, R. (ed.), *Interim Measures Indicated by International Courts*, Berlin/Heidelberg/New York: Springer, 1994; see, in particular, Thirlway, H., The Indication of Provisional Measures by the I.C.J., in Bernhardt, R. (ed.), *Interim Measures Indicated by International Courts*, Berlin/Heidelberg/New York: Springer, 1994, p. 1 et seq. Kolb, R., *The International Court of Justice*, Oxford: Hart, 2013, pp. 638–650. Hersch Lauterpacht even suggested to amend the Court's Statute. Lauterpacht, H., *The Development of International Law by the International Court*, London, 1958, reprinted edition published by Cambridge: CUP, 1982, pp. 112–113.

20 CR 2000/27, p. 47, par. 19 (Dupuy).

21 *I.C.J. Pleadings, Vienna Convention on Consular Relations (Paraguay v. USA)*, Memorial submitted by Paraguay on 9 October 1998, paras 5.15–5.51, pp. 165–176.

the Court were 'decisions' within the meaning of Article 94, paragraph 1 of the UN Charter in which member states of the United Nations had vowed to comply.[22] It also stated that the ordinary meaning of Article 41 of the Statute, when read in its context, could only lead to the conclusion that provisional measures were binding.[23] Finally, the object and purpose of Article 41 of the Statute require that provisional measures be binding.[24] Paraguay referred to the rules of interpretation of the Vienna Convention on the Law of Treaties,[25] placing only little emphasis on the supplementary means of interpretation the *travaux préparatoires,* since they appear not to be clearly pointing towards the binding force of provisional measures, at least as thus conceived by the drafters and state parties to the Statute of 1920.[26]

Other arguments, such as the general principle of effectiveness of judicial orders or even arguments based on the Court's inherent powers, could have been submitted. However, they were not addressed by Paraguay in this case.[27]

The position of the US in relation to the binding nature of provisional measures proved to be rather nuanced at the time: In the case concerning *United States Diplomatic and Consular Staff in Tehran (United States of America v. Iran),* the Court had indicated provisional measures for the immediate release of hostages. In its memorial, the US stated that, by disregarding the order of the Court, Iran had violated its international 'obligations'.[28] This would indicate that the US considered that the order of the Court indicating provisional measures could be the source of obligations applicable to Iran.

Fast-forwarding eighteen years, in the course of the *Breard* proceedings, some elements indicate that the US still viewed orders indicating provisional measures as binding. During the Court's hearings held on 7 April 1998, the US argued:

22 Ibid., paras 5.16 to 5.17.

23 Ibid., paras 5.18 to 5.39.

24 Ibid., paras 5.40 to 5.50.

25 *Vienna Convention on the Law of Treaties* of 23 May 1969, UNTS, vol. 1155, p. 331.

26 See Thirlway, H., *The indication of provisional measures by the International Court of Justice*, in: Bernhardt R. (ed.), *Interim measures indicated by International Courts,* op. cit. n. 19, p. 1 (29). The ICJ in *LaGrand*, however, pointed out that the *travaux préparatoires* did 'not preclude the conclusion that orders under Article 41 have binding force.' *LaGrand (Germany v. United States), I.C.J. Reports 2001, Judgment*, p. 504.

27 See Kolb, R., *The International Court of Justice*, Oxford & Portland: Hart, 2013, pp. 643–645.

28 'Iran has formally undertaken *pursuant to Article 94 paragraph 1 of the Charter ... to comply with the decision of this Court in any case to which Iran might be a party ...* [including] *any and all Orders and Judgments* which might be entered by this Court in the course of the present litigation.' *United States Diplomatic and Consular Staff in Tehran (United States v. Iran), Pleadings, I.C.J. Reports 1980,* p. 266 (Statement of Roberts Owen) Emphasis added. See also Memorial of the US, *Pleadings, I.C.J. Reports 1980,* p. 139.

> 1.6. As this Court knows, the indication of provisional measures is a serious matter which the Court is cautious in exercising. That is especially true in the case where the Court is being asked to take action that would severely intrude upon national criminal jurisdiction of a State in a matter of violent crime.[29]

This, to any diligent reader, does not imply that the US, at the time of the hearings, were convinced of the non-binding character of provisional measures indicated by the Court.

The plain text of the wording from the Court's order of 9 April 1998 could also be interpreted as intending to create binding obligations, as the operative paragraph applied the term 'should' ('*doivent*' in the French version is even more robust) rather than 'ought', which is the language used in Article 41 of the Statute. However, when Secretary of State Madeleine Albright wrote to the Governor of Virginia to request that he stay the execution of Francisco Breard, she stated:

> Using *non-binding language*, the Court said that the United States should 'take all measures at its disposal to ensure that Angel Francisco Breard is not executed pending the final decision in these proceedings'.[30]

Clearly, Secretary of State Albright did not suggest that orders of the Court as such were non-binding, instead, she had insisted on the 'non-binding language' used by the Court.

This would indicate that initially, the US Government did not contest the binding nature of the Court's orders as such, but would rather dissect the language used by the Court in a particular order. However, the State Department when appearing before the US Supreme Court, stated that an ICJ order should not be considered binding by the US Supreme Court.[31]

Once the provisional measures in both the *Breard* and *LaGrand* proceedings had been breached, the US chose to argue that orders on provisional measures were not binding all together. Referring to the text of Article 41 of the Statute and also to the *travaux préparatoires*, the US held that measures indicated under Article 41 of the Statute were intended to be non-binding.[32] Contrary to its position taken in the *Consular staff-case*, the US argued that

29 Verbatim Record CR 98/7, p. 25 (Intervention of the US Department of State Legal Adviser, David R. Andrews).

30 Letter of the Secretary of State to the Governor of the State of Virginia, dated 13 April 1998, available at: www.state.gov/documents/organization/65744.pdf (accessed 3 September 2015), emphasis added.

31 See Quigley, J., Aceves, W. J., and Shank, A., *The Law of Consular Access: A documentary guide*, London: Routledge, 2010, p. 222; Amicus curiae brief of 14 April 1998, in *Breard v. Greene*.

32 Counter-Memorial of the United States in the *LaGrand* case, para. 142–152.

Article 94, paragraph 1 of the UN Charter related solely to judgments of the Court, not to its orders.[33]

The arguments initially applied by Paraguay in its memorial were revived and developed by Counsel for Germany, when pleading the *LaGrand case* before the Court, after the execution of the LaGrand brothers in violation of the Court's provisional measures.[34] With regard to the binding nature of provisional measures, Germany made extensive reference to the *Breard* precedent.[35] In its memorial, Germany notably pressed the point that during the proceedings in the *Breard case*, the US had not argued that the Court's orders were non-binding.[36] Germany also underlined the need for the Court to make a finding on the binding nature of provisional measures in view of the conduct of US authorities in the aftermath of the issuance of the Court's orders in *Breard* and in *LaGrand*.[37]

In the *LaGrand* proceedings it became clear that the breach of the Court's order was no minor setback, or an '*incident de parcours*', but rather that it was owed to what had become a legal position of the US on the nature of orders on provisional measures.[38] This might well have prompted the Court to make a finding on the binding nature of orders indicated under Article 41 of the Statute. As one commentator fittingly observed in 1998, after the Breard proceedings, the Court now 'needed to bite the bullet',[39] and determine that its orders on provisional measures should be respected as a matter of law, not as a matter of courtesy.[40]

Arguably, the non-observance of the order in the *Breard* precedent might well have contributed to the Court's finding in the *LaGrand case* on the binding nature of its orders indicating provisional measures.[41]

33 Counter-Memorial of the United States, para. 155–158.

34 *LaGrand (Germany v. United States of America), Provisional Measures, Order of 3 March 1999, I.C.J. Reports, 1999*, p. 9.

35 Mainly in its Memorial of Germany paras 4.123, 4.156, 4.163 to 4.165, 4.170 to 4.147.

36 Memorial of Germany, paras 1.463–1.465.

37 Memorial of Germany, paras 4.170–4.171. See also CR 2000/27, pp. 40–50 (Dupuy).

38 See in particular CR 2000/27, pp. 57–58 (Dupuy) and CR 2000/30, pp. 6–47 (Dupuy).

39 Highet, K., 'The Emperor's new clothes, death row appeals to the World Court? The Breard case as a miscarriage of (international) justice', in: *Liber amicorum Jose Maria Ruda*, Kluwer 2000, pp. 435–456 at p. 449.

40 See, for a quick overview of later cases before US courts, Dupuy, P.-M. and Hoss C., 'Avena and other Mexican nationals', in: *Max Planck Encyclopedia of Public International Law*, Oxford: OUP, 2014; also available online: http://opil.ouplaw.com/home/EPIL (accessed 11 March 2016).

41 One of the two dissenting judges, Judge Buergenthal only dissented because he considered the late filing of Germany's request for the indication of provisional measures to amount to procedural misconduct. *I.C.J. LaGrand, Judgment, dissenting opinion Judge Buergenthal, I.C.J. Reports, 2000*, pp. 556–557.

III The silent revolution: emergence of guarantees of non-repetition

At the time the *Breard case* was brought before the Court, restitution was still possible. In its application of 3 April 1998, Paraguay had therefore requested a series of measures under the heading of 'restoration'.[42] It requested in particular that the Court instruct the US not to apply the doctrine of 'procedural default'; to determine that the US was under an international obligation to undertake any future detention of Paraguayan nationals in conformity with its international obligation; to declare that any criminal liability of Mr Breard imposed in violation of the international legal obligations is void; and, to order the US to re-establish the situation that existed prior to the detention of, proceedings against, and conviction and sentencing of Mr Breard. Finally, Paraguay requested the US to provide a guarantee of non-repetition of the illegal acts perpetrated.

Paraguay's latter request appears to be rather elusive, compared with the detailed requests for 'restoration', most of which are in fact very much future-oriented rather than measures of restitution. In all likelihood, Paraguay was contemplating measures of general assurances to be undertaken for the violations not to occur again, as was usually sought in state practice. Paraguay's focus was to restore the situation and, ultimately, to overturn the death sentence of Mr Breard. Hence, the general terms of Paraguay's request for guarantees of non-repetition.

Both the amended application[43] and the memorial of Paraguay submitted on the same day, on 9 October 1998, remedied this lack of precision. By that time, 'restoration' had become impossible and appropriate alternative means of reparation had to be sought. Paraguay not only requested 'compensation' and 'satisfaction' for the US' violation of the Court's order on provisional measures, but also requested, separately, that measures be taken in order to prevent further violations.

In fact, Paraguay transposed the measures initially requested as measures of 'restoration' into its request for guarantees of non-repetition.[44]

Instead of the general assurances that would have traditionally been given verbally, the government of Paraguay now requested the Court to declare that the US was:

> under a legal obligation to provide Paraguay a guarantee that the United States will not repeat its illegal acts, but will carry out in conformity with [its] international obligations any future detention of or criminal proceedings against any Paraguayan national in its territory, whether by a constituent, legislative, executive, judicial or other power.

42 Application of Paraguay 3 April 1998, *I.C.J. Pleadings 1998*, para. 25.

43 Amended Application of 9 October 1998, *I.C.J. Pleadings 1998*, p. 73.

44 Ibid., p. 81.

In its memorial, the submission in relation to guarantees of non-repetition was phrased in the same manner.[45]

This shift from one remedy (restoration) to another (guarantees of non-repetition) was possibly the starting point of what could be called a silent revolution of the law of state responsibility:

> In 1998, state practice in the field of 'guarantees and assurances of non-repetition' consisted of linking those assurances to formal apologies, i.e. approaching these as a form of satisfaction. Apologies would be complemented by verbal assurances according to which the responsible state would commit not to violate its obligations in the future. There were no formal requirements discernible in state practice and there was no precedent in the Court's jurisprudence to serve as a basis for a request to order guarantees of non-repetition.[46]

Hence, the section on guarantees of non-repetition in the memorial of Paraguay is remarkably brief.[47] The basis of Paraguay's request appears to be Article 46 of the Draft Articles of the International Law Commission (ILC) on state responsibility, in particular, at that time, the Draft from 1996 adopted in the first reading.[48]

Given the precision of Paraguay's request for guarantees of non-repetition, it is somewhat surprising that it accepted as sufficient the rather general assurances made by the State Department in November 1998. The relevant text of the statement released by the Department of State in Asunción (Paraguay) on 3 November 1998 reads as follows:

> Recognizing that United States' compliance with the requirements of the Vienna Convention must improve, the Government of the United States has undertaken efforts to better educate officials throughout the United States of the consular notification requirements. The United States intends through these efforts to ensure that the consular rights of foreign nationals in the United States are respected, and that Paraguayan and other foreign nationals in the United States are properly notified of their right to request consular assistance if they are arrested or detained. Consular notification is no less important to Paraguayan and other foreign nationals in the United States than to US nationals outside the

45 See Application instituting proceedings, 3 April 1998, para. 25 (4); Amended Application instituting proceedings, para. 45 (e) and Memorial of Paraguay para. 6.21 (e), *I.C.J. Pleadings, 1998*, p. 186.

46 Palmisano, G., 'Les garanties de non-répétition entre codification et réalisation juridictionnelle du droit: à propos de l'affaire LaGrand', *Revue générale de droit international public (RGDIP)*, vol. 106, issue 4, 2002, pp. 753–790.

47 *Memorial of Paraguay, ICJ Pleadings 1998*, pp. 178–179.

48 *Yearbook of the International Law Commission* 1996, vol. II (2), p. 301.

> United States. We fully appreciate that the United States must see to it that foreign nationals in the United States receive the same treatment that we expect for our citizens overseas.[49]

While the text of the official statement issued by the US in Asunción (Paraguay) on 3 November 1998 goes beyond a simple apology, it does not mirror the requests of Paraguay made in its memorial concerning guarantees of non-repetition.

The statement seemed to be motivated by considerations of protection of US citizens abroad and reciprocity, rather than constituting a proper apology. As to the measures that the State Department vowed to take in order to prevent future violations, they were rather limited in scope. The measures actually taken in the aftermath of the *Breard case* included the dissemination of a booklet containing information on consular rights, as well as the organisation of seminars and training.[50] In fact, as later established in *LaGrand* and *Avena*, the measures envisaged by the statement proved to be inadequate to prevent further violations of consular rights.

All in all, the statement certainly fulfilled its purpose, namely to settle the dispute with Paraguay, but it failed to satisfy the request of Paraguay, as set out in its memorial. The reason why Paraguay accepted these assurances and settled the dispute remains unclear. It had already rejected similar statements made earlier by US authorities, which it considered insufficient, such as the apologies offered by the State Department.[51]

One might, however, venture out to assess the chances of Paraguay obtaining the requested measures: Paraguay's request for guarantees relied on little more than a few examples taken from state practice and arbitral awards, as set out in the ILC commentary on Article 46 of the ILC Draft articles on state responsibility; none of the precedents was as detailed as Paraguay's request. Assuming that the case had gone to the merits, it is unlikely that the Court would have granted such guarantees in the *Breard case*.

To a certain extent, Germany had the easier stand requesting guarantees of non-repetition as it relied on the *Breard* precedent to establish that repetition had already occurred and would in all likelihood occur again. In the factual part of its memorial, Germany consequently referred to the *Breard* precedent in order to justify its request for guarantees of non-repetition and in order to demonstrate that there was a real risk of repetition.[52]

49 Aceves, J. and Shank, A., *The Law of Consular Access: A documentary guide*, London: Routledge, 2010, p. 139.

50 *LaGrand I.C.J. Pleadings 2000*, Counter-memorial of the United States, pp. 17–18

51 1997 US Briefs 1390, Statement made by the Solicitor General before the US Supreme Court; quoted in the Memorial of Germany, par. 6.54. and reproduced in Annex 34 to the memorial.

52 *LaGrand I.C.J.*, Memorial of Germany, 16 September 2000, paras 1.06–1.07 and para 3.42.

It appears that Germany, when preparing the *LaGrand* proceedings, saw a window of opportunity for requesting and obtaining guarantees of non-repetition as a separate form of reparation. Similar to the request formulated by Paraguay, the request for guarantees from Germany was substantially more detailed than a traditional, general assurance. As did Paraguay, the German memorial had to rely on the work of the ILC, with the sparse precedents referred to therein.[53]

Over the course of the hearings, Counsel for Germany referred to *Breard*, notably to bolster Germany's case in convincing the Court that the measures taken by the US in order to improve compliance with the Vienna Convention (and arguably constituting guarantees of non-repetition but far from those requested by Paraguay) were insufficient to preserve the rights of foreign nationals under the Vienna Convention.[54] The success of this strategy can be seen in the operative paragraph of the *LaGrand* judgment, read together with the declaration of President Guillaume.[55]

The ILC took account of these developments in its preparation of the Draft articles on state responsibility. In its commentary to the Articles adopted in 2001, the ILC refers to the *Breard case* with reference to Article 31 (Satisfaction), as an example for the diplomatic practice of offering apologies as a means of reparation for immaterial damages.[56] The ILC did not see fit to add the *Breard* precedent to its commentary on Article 30 (Cessation and guarantees of non-repetition), being that the *LaGrand case* was the most relevant precedent.[57]

However, it appears that the Court has since been rather reluctant to grant guarantees in subsequent cases. As the Court has held:

> while the Court may order, as it has done in the past, a State responsible for internationally wrongful conduct to provide the injured State with assurances and guarantees of non-repetition, it will only do so if the circumstances so warrant, which it is for the Court to assess. As a general rule, there is no reason to suppose that a State whose act or conduct has been declared wrongful by the Court will repeat that act or conduct in the future, since its good faith must be presumed.[58]

For the Court to grant a request for guarantees and assurances of non-repetition, it almost requires a systematic, manifest breach or unlawful conduct on the part of the responsible state.

53 *LaGrand I.C.J.*, Memorial of Germany, paras 6.60–6.77.

54 CR 2000/26, pp. 21 et seq. (Simma); CR 2000/27, pp. 34–35 (Simma). See also the question raised by Judge Higgins CR 2000/29, p. 57 and the written answer from Germany (available on the Court's website).

55 *LaGrand, Judgment, I.C.J. Reports 2011*, pp. 516–517.

56 ILC Report 2001, *YbILC*, 2001, vol II, Part Two, p. 92.

57 ILC Report 2001, *YbILC*, 2001, vol II, Part Two, p. 90.

58 *Dispute regarding Navigational and Related Rights (Costa Rica v. Nicaragua), Judgment, I.C.J. Reports 2009*, p. 213 (p. 267).

Guarantees and assurances of non-repetition remain an extraordinary remedy. They are considered to be exceptional in as much as they require specific justification of a real risk – not only a possibility – of repetition, and in so far as they can be more intrusive into the sovereign domain of a state than other forms of reparation. The concept of guarantees of non-repetition as it now stands in the Court's jurisprudence appears as rather remote from the traditional form of guarantees, which consisted of complementary verbal assurances accompanying apologies rather than a sanction for systematic wrongful conduct.

IV Conclusion

The particular circumstances surrounding the 'forgotten' *Breard case* most probably contributed to the findings made in the *LaGrand* judgment and, hence, contributed to considerable developments in the jurisprudence of the Court.

It is all the more deplorable that the underlying tensions between domestic courts and international obligations remain ultimately unsolved, revealing once again the pressing need for a constructive dialogue between all courts and tribunals, domestic and international, in the best interest of the rule of law.[59]

Bibliography

Agora Breard, *American Journal of International Law*, No. 4 (1998), pp. 666–712.

Aceves, J. and Shank, A., *The Law of Consular Access: A documentary guide*, London: Routledge, 2010.

Dupuy, P.-M. and Hoss C., 'Avena and other Mexican nationals', in: *Max Planck Encyclopedia of Public International Law*, Oxford: OUP, 2014.

Frowein, J., 'Provisional measures by the International Court of Justice – the LaGrand Case', *Zeitschrift für ausländisches öffentliches Recht und Völkerrecht* (ZaöRV), 62, issue 1–2, 2003, pp. 55–60.

Hambro, E., 'The binding character of the provisional measures of protection indicated by the I.C.J'., in: *Rechtsfragen der internationalen Organisation, Festschrift für Hans Wehberg*, Frankfurt am Main, 1956, p. 152 et seq.

Highet, K., 'The Emperor's new clothes, death row appeals to the World Court? The Breard case as a miscarriage of (international) justice', in: *Liber amicorum Jose Maria Ruda*, Kluwer 2000, pp. 435–456.

59 See in particular the decision of the US Supreme Court in *Medellin v. Texas 128 S.Ct 1346 (2008)* and the proceedings before the I.C.J. Request for interpretation of the Judgment of 31 March 2004 in the case concerning *Avena and Other Mexican nationals (Mexico v. United States of America) (Mexico v. United States of America*. For an overview: Dupuy, P.-M., Hoss, C., 'Avena and other Mexican nationals', *Max Planck Encyclopedia of Public International Law*, Oxford: OUP, 2014.

Hoss, C., *Satisfaction*, in: *Max Planck Encyclopedia of Public International Law*, Oxford: OUP, 2014.

Kolb, R., *The International Court of Justice*, Oxord: Hart, 2013, p. 1222.

Lauterpacht, H., *The Development of International Law by the International Court*, London Stevens & Sons Limited, 1958, reprinted edition published by Cambridge: CUP, 1982.

Mendelssohn, M., 'Provisional measures of protection in cases of contested jurisdiction', *BYIL* 46 (1972–1973), p. 259 et seq.

Oellers-Frahm, K., *Die einstweilige Anordnung in der Internationalen Gerichtsbarkeit*, Berlin Heidelberg New York: Springer, 1975.

Palmisano, G., 'Les garanties de non-répétition entre codification et réalisation juridictionnelle du droit: à propos de l'affaire LaGrand', *Revue générale de droit international public (RGDIP)*, vol. 106, issue 4, 2002, pp. 753–790.

Pellet, A., 'Remarques sur la jurisprudence récente de la C.I.J. dans le domaine de la responsabilité internationale', in: Kohen, M., and Kolb, R., Tehindrazanarivelo D. (eds), *Perspectives of International Law in the 21st Century, Liber amicorum Dominicé*, Leiden: Nijhof, 2012, pp. 321–345 (341–342).

Thirlway, H., 'The indication of provisional measures by the I.C.J.', in: Bernhardt, R. (ed.), *Interim Measures Indicated by International Courts*, Berlin/Heidelberg/New York: Springer, 1994, p. 1 et seq.

14 *Haya de la Torre (Colombia v. Peru), 1950*

Pablo Sandonato de León

I Presentation

Very few topics are as intrinsically linked to Latin America as diplomatic asylum.[1] Latin America's history of civil wars and insurrections in the nineteenth and twentieth centuries demonstrated the need to expeditiously protect the life and physical integrity of those persecuted for political reasons. In such circumstances, the jurisdictional immunity enjoyed under diplomatic premise can constitute a useful tool. The practice of some Latin American states allowing persons persecuted for political reasons to enter the premises of their embassies or the private residence of their chiefs of diplomatic mission abroad in order to be protected by the jurisdictional immunity of the diplomatic premise (hereinafter the 'asylum state') raises a series of problems in international law (i.e. who may benefit from it? and under what conditions? what state or states determine whether the requester can effectively benefit from it and whether the conditions are actually met? when does it actually start and until when does it last? how does it come to an end?). A number of treaties were concluded at different times by Latin American states[2] to provide an uncontroverted legal basis for this practice while attempting to solve those problems. Among those treaties is the 1928 Havana Convention on Political Asylum (hereinafter the 'Convention').

Mr Victor Raul Haya de la Torre was a Peruvian politician and head of the American People's Revolutionary Alliance. On 3 October 1948, a military rebellion broke out in Peru and was suppressed the very same day. Criminal proceedings were instituted against Mr Haya de la Torre for instigating and directing the rebellion.

On the night between 3 and 4 January 1949, Mr Haya de la Torre felt he

1 See, among others: Perez Gomar, *Curso elemental de derecho de gentes*, Montevideo: 1864, p. 177 and A. Álvarez, *Le droit international américain*, Paris: Pedone, 1910, p. 74.

2 These are: the *1889 Montevideo Treaty on International Penal Law*, 23 January, 1889, Organisation of American States Official Records (OEA/Ser.X/7), Treaty Series 34; the *1928 Havana Convention on Political Asylum*, 20 February, 1928, 132 League of Nations Treaty Series 323; and the *1954 Caracas Convention on Diplomatic Asylum*, 28 March, 1954, United Nations Treaty Series, vol. 1438, I-24377.

was persecuted for political reasons and fearing for his life and physical integrity, requested and was permitted by the Colombian ambassador to Peru to enter the premises of the Colombian Embassy. Colombia later requested a safe-conduct to Peru to allow Mr Haya de la Torre (hereinafter the 'refugee') to leave the country. Peru denied this on the ground that the refugee was not politically persecuted but was accused of common crimes and was therefore not entitled to benefit from diplomatic asylum or, as it is also known, 'political asylum'.

The dispute was ultimately submitted to the International Court of Justice (hereinafter 'the Court') that rendered its judgment on 20 November 1950 (hereinafter the 'Asylum Judgment'). The Court ruled that Colombia was not entitled to unilaterally qualify the nature of the offence attributed to the refugee and that Peru was not bound to deliver safe-conduct. The Court also found that the asylum was not granted in conformity with the Convention because the granting of asylum cannot be an obstacle to proceedings instituted by legal authorities operating in accordance with the law.

On the same day of the Asylum Judgment, Colombia requested the Court for an interpretation of the judgment (hereinafter the 'request for interpretation'). One week later, on 27 November 1950, the Court rendered its judgment declaring the request for interpretation inadmissible.

Following the Asylum Judgment and the decision on the request for interpretation, Peru requested Colombia to surrender the refugee. However, in Colombia's view, to surrender the refugee would have been contrary to the Asylum Judgment and the Convention. The parties were unable to reach an agreement on the manner in which effect was to be given to the Asylum Judgment. On 13 December 1950, less than a month after the Asylum Judgement and some fifteen days after the Court's decision on the request for interpretation, Colombia instituted new proceedings against Peru. This new case is known as the *Haya de la Torre case.*

II – The proceedings before the Court

As a preliminary remark, it is worth mentioning that the Court counted on its bench[3] two Latin American judges: Mr José Gustavo Guerrero and Mr Alejandro Álvarez. Judge Guerrero, from El Salvador, was a distinguished Latin American internationalist of his time.[4] He was the last President of the

3 At the time the Court was composed by President Basdevant, Vice-President Guerrero and Judges Álvarez, Hackworth, Winiarski, Zoričić, De Visscher, McNair, Klaestad, Badawi Pacha, Read and Hsu Mo.

4 Among his works, see: *La responsabilidad internacional de los estados por los daños causados en su territorio a la persona o bienes de los extranjeros*, Paris: Lahure, 1926; *La codification du droit international: la première conférence (La Haye, 13 mars – 12 avril 1930)*, Paris: Pedone, 1930; *L'Union panaméricaine et la Société des nations*, Leyde: Sijthoff, 1937; *El orden internacional*, Universidad Autónoma de El Salvador, 1945; and *L'ordre international, hier, aujourd'hui, demain*, Neuchâtel: La Baconnière, 1945.

Permanent Court of International Justice and the first President of the International Court of Justice and who at the time was the Vice-President of the Court. Judge Álvarez, from Chile, was also a distinguished Latin American internationalist, well known for his stance in championing the existence of a Latin American approach to international law.[5]

In its memorial, Colombia submitted both principal and alternative claims. In its principal claim, Colombia requested the Court to determine the manner in which effect was to be given to the Asylum Judgment and to state whether Colombia was bound to surrender the refugee to Peru. Should the principal claim not be granted, Colombia, in its alternative claim, requested the Court to adjudge, in the exercise of its ordinary competence, whether in accordance with the law in force between the parties 'and particularly American international law'[6] Colombia was bound to surrender the refugee to Peru.

In its counter-memorial, Peru also submitted both principal and alternative claims. As a principal claim, Peru requested the Court to determine in what manner the Asylum Judgment was to be executed by Colombia and to dismiss Colombia's submission that it was not bound to surrender the refugee to Peru. Should the principal claim not be granted, Peru requested the Court, in its alternative claim, to declare that the asylum granted to the refugee ought to have ceased immediately after the delivery of the Asylum Judgment and must cease 'in order that the Peruvian justice may resume its normal course'.[7]

As the Court's bench did not include any judge of the nationality of the parties, Colombia appointed Mr José Joaquin Caicedo Castilla, while Peru appointed Mr Luis Alayza y Paz Soldan both to act as judges *ad hoc*.

Colombia also informed the Registrar that it relied on the Convention as the applicable law and requested the Registrar to give effect to Article 63 of the Statute of the Court and inform all states parties to the Convention. In response to the Registrar's information, Cuba addressed a letter and a memorandum to the Court. The Court considered Cuba's letter as a declaration of intervention. The memorandum was in relation to the construction of the Convention and was based almost entirely on the Asylum Judgment. It also contained Cuba's

5 Among his works, see: *Le droit international américain: son fondement, sa nature d'après l'histoire des états du nouveau monde et leur vie politique et économique*, Paris: A. Pedone, 1910; *La codification du droit international, ses tendance – ses bases*, Paris: A. Pedone, 1912; *Le droit international de l'avenir*, Washington, DC, 1916; *Le nouveau droit international public et sa codification en Amérique*, Paris: A. Rousseau, 1924; *Considérations générales sur la codification du droit international américain: (code de droit international des états américains)*, Rio de Janeiro: Impr. Nacional, 1927; and *Le droit international nouveau dans ses rapports avec la vie actuelle des peuples*, Paris: Librairie Pedone, 1959.

6 *I.C.J. Pleadings, Oral Arguments, Documents, Haya de la Torre Case (Colombia v. Peru), Judgment of June 13th, 1951, Part I*, p. 10.

7 *I.C.J. Pleadings, Oral Arguments, Documents, Haya de la Torre Case (Colombia v. Peru), Judgment of June 13th, 1951, Part I*, p. 98.

general attitude in regard to asylum. Colombia did not object to Cuba's intervention, while Peru considered that the time of the declaration had expired and was an attempt by Cuba to appeal the Asylum Judgment.

The Court admitted the intervention of Cuba during its public hearing held on 16 May 1951.

III The judgment

A. Presentation

The Court rendered its judgment in the *Haya de la Torre case* (hereinafter the 'Judgment') on 13 June 1951,[8] exactly six months after the institution of proceedings, and less than a month after the hearings. While it is impossible to draw any conclusion from the short time in which the Court delivered the judgment, it is also difficult not to take this aspect into account when reflecting on its general outcome. This is particularly so, if one recalls that the Judgment was unanimously adopted on all claims, with the exception of the claim to surrender the refugee, for which the judge *ad hoc* appointed by Peru could not concur. While the legal value of a judgment adopted by unanimity is not greater than the one adopted by the casting vote of the President, the former holds a higher moral authority, or is at least more persuasive, and reflects a general consensus within the bench as to the clarity of a given situation or the interpretation of a given norm. Certainly, in the internal dynamics of the Court, as in every human context, personalities count, and the personality of the President of the Court is certainly a factor not to be underestimated. In other words, even if the deliberations of the Court remain secret, it does not seem unreasonable to advance the hypothesis that the esteemed intellectual authority of President Jules Basdevant and Vice-President José Gustavo Guerrero both played an important role for reaching unanimity. However, the elliptical approach adopted by the Court in the Judgment, coupled with its previous decisions in the Asylum Judgment and the request for intervention, makes this judgment one of the most incomprehensible, and perhaps also most severally criticized,[9] judgments rendered by the Court. This is however not our contention in this article. As we shall see, their Judgment contributed to clarify a number of issues in international law. In addition, there is no contradiction in the Court's finding that asylum was to cease, but Colombia was under no obligation to bring that about by surrendering the refugee to the Peruvian authorities.

8 *Haya de la Torre Case, Judgment of June 13th, 1951: I.C.J. Reports*, p. 71.

9 See, for instance: Barcia Trelles, C., 'El Derecho de Asilo Diplomático y el caso Haya de la Torre', *Revista Española de Derecho Internacional* III, 3, 1950, 59–66, and: Vieira, M. A., *Derecho de Asilo Diplomático (asilo político)*, Montevideo: Biblioteca de Publicaciones Oficiales de la Facultad de Derecho y Ciencias Sociales de la Universidad de la República, 1961, p. 123.

B. Jurisdiction

The first issue considered by the Court was that of its jurisdiction. In the view of the parties consenting to its jurisdiction, there was no need for the Court to make any special analysis of this issue. In fact, the Court simply stated that the parties after having consented to its jurisdiction without objection '[t]his conduct ... is sufficient to confer jurisdiction on the Court'.[10]

A necessary word of caution is however required. It is not because the parties consented to its jurisdiction and that no objection thereon was raised that the Court could simply avoid analysing its jurisdiction. In fact, the judgment confirms that, even in circumstances where the parties consent to its jurisdiction and no objection is raised, the Court will nevertheless evaluate whether or not it has jurisdiction to deal with a case.

C. Intervention

The second issue considered by the Court was the declaration of intervention filed by Cuba. While it is clear from the Rules of the Court[11] that intervention is an incidental proceeding within a principal proceeding,[12] the Court nevertheless insisted ***ad abundantiorem cautelam*** that 'every intervention is incidental to the proceedings in a case'.[13]

The *Haya de la Torre case* was the first case where a declaration of intervention under Article 63 of the Statute was made before the Court.[14] It is therefore comprehensible that the Court felt the need to insist on one of the conditions that such a declaration must satisfy. In fact, a declaration under Article 63 becomes an intervention 'if it actually relates to the subject-matter of the pending proceedings'.[15] Once this link between the declaration and the subject-matter of the case has been established, it is solely for the Court to accept the intervention. Cuba's intervention, as a party to the Convention, was an intervention as of right.[16]

10 *Haya de la Torre Case, Judgment of June 13th, 1951: I.C.J. Reports*, p. 78.

11 Official document: Rules of Court. See: www.icj-cij.org/documents/index.php?p1=4&p2=3&p3=0 (accessed 13 July 2016).

12 Rules of Court, Part III 'Proceedings in Contentious Cases', Section 'D' Incidental Proceedings, subsection 4 'Intervention'.

13 *Haya de la Torre Case, Judgment of June 13th, 1951: I.C.J. Reports*, p. 76.

14 A precedent exists in the jurisprudence of the Permanent Court of International Justice: *Case of the S.S. "Wimbledon", Judgment of 28 June 1923 (Question of Intervention by Poland), P.C.I.J., series A, n°. 1*, p. 13.

15 *Haya de la Torre Case*, p. 76.

16 Ruda, J. M., 'Intervention before the International Court of Justice', in *Fifty Years of the International Court of Justice; Essays in honour of Sir Robert Jennings*, Vaughan Lowe and Malgosia Fitzmaurice (eds), Cambridge: Grotius Publications/Cambridge University Press, 2007, pp. 487–502.

Another issue that the Court considered when analyzing Cuba's declaration was the nature and content of Cuba's memorandum. It will be recalled that the memorandum almost exclusively concerned questions discussed in the Asylum Judgment and not specifically the interpretation of the Convention, which was actually the instrument granting Cuba the right to intervene. In fact, to the extent the memorandum discussed matters already dealt with in the Asylum Judgment, the intervention should have been rejected, as these matters were immune from review under the authority of *res judicata*. This risk was, however, avoided by the Cuban agent who, during the hearings, stated that the intervention was based on the interpretation of a new aspect of the Convention not previously decided in the Asylum Judgment, namely the surrender of the refugee.[17]

D. The principal submissions[18]

Colombia's principal submission consisted of two parts. The first part being the manner in which effect was to be given to the Asylum Judgment; the second in relation to the surrender of the refugee.

1. First part: the manner in which effect was to be given to the Asylum Judgment

In the first part of its principal submission, Colombia requested the Court to state the manner in which effect was to be given to the Asylum Judgment. Peru made the same request in its first submission. The Court rejected the submission by unanimous vote.

The parties clearly wanted the Court to make a choice on the various courses available to terminate diplomatic asylum and determine exactly what was to be done to terminate the same asylum the Court previously found to be granted not in conformity with the Convention. The Court refused to be placed at such crossroads and found that the choice among different options is 'conditioned by facts and by possibilities which to a very certain extent the Parties are uniquely positioned to appreciate'.[19] On the one hand, and taking into account the parties' expectations, such a reply can be seen as largely unsatisfactory, as it may be seen as not replying to the parties' request. On the other hand, however, the reply does not appear as unsatisfactory as one might initially perceive. In fact, once the conditions for granting diplomatic asylum are met, the actual decision whether to grant it or not is based on political evaluation.

17 *Haya de la Torre Case (Colombia v. Peru), Judgment of June 13th, 1951, Part II: Oral proceedings*, p. 172.

18 Principal and alternative submissions refers to Colombia's claims as applicant in the proceedings. Reference to respondent's claims is made against applicant's claims.

19 *Haya de la Torre Case, Judgment of June 13th, 1951: I.C.J. Reports*, p. 79.

The same could be said in relation to the decision on how to terminate diplomatic asylum. The Court perhaps had this political evaluation in mind when it noted that a choice among the different courses to terminate diplomatic asylum 'could not be based on legal considerations but only on considerations of practicability or of political expediency'.[20] In such circumstances, since the decision in question is basically a political decision, the Court's finding that 'it is not part of the Court's judicial function to make such a choice',[21] assumes its full meaning.

This is not to say that the Court will not adjudicate on legal questions having political aspects. Nor is it to say that diplomatic asylum is a non-justiciable matter. Indeed, the Court can and had already declared that the asylum had not been granted in accordance with the requirements of the Convention and had ordered it to be terminated. The fact that the Court found that the actual choice of the manner of terminating the asylum is a political choice, does not affect the conclusion that the asylum was not illegal as it was not granted in accordance with the Convention.

A distinction should then be made between the refugee seeking to protect his life or physical integrity and the decision to grant asylum once all necessary conditions are met. The latter is clearly a political decision with legal consequences, even if 'its moral justification is to be found in considerations of humanity'.[22]

2. *Second part: the surrender of the refugee*

In the second part of its principal submission, Colombia requested the Court to determine whether 'in execution' of the Asylum Judgment[23] it was bound to surrender the refugee to Peru. Peru in turn requested the Court, in its second submission, to dismiss the second part of Colombia's principal submission. In other words, Peru was seeking a declaration that Colombia was bound to surrender the refugee to Peru.

The Court replied that it could not say whether or not Colombia was bound to surrender the refugee to Peru 'in execution' of the Asylum Judgment simply because Peru had not requested the surrender of the refugee in the Asylum Judgment.

One may be disappointed by the Court's finding in not taking an upfront decision on the surrender of the refugee. However, one must also bear in mind principles of procedural law, such as *res judicata* and *ne ultra petita*, as well as the limits of the international judicial function in its current state. In this regard, it is worth recalling that the Court cannot surpass the requests of the parties. Article 36, paragraph 1 of the Statute provides, precisely, that

20 Ibid.
21 Ibid.
22 Institut de droit international, *Annuaire*, session of Bath, vol. 43, 1950, no.1, p. 138.
23 *Haya de la Torre Case*, p. 79.

the 'jurisdiction of the Court comprises all cases which the parties *refer* to it' (emphasis added). Similarly, Article 38, paragraph 1 of the Statute provides that the function of the Court is 'to decide in accordance with international law such disputes *as* are submitted to it' ('les différends *qui* lui sont soumis') (emphasis added). Like it or not, the Court cannot be blamed for what the parties did not refer or submit to it.

E. The alternative submission

In its alternative submission, Colombia requested the Court, in the exercise of its ordinary competence, to adjudge whether in accordance with the law in force between the parties 'and particularly American international law', Colombia was bound to surrender the refugee to Peru. Peru, in its second submission, requested the Court to dismiss Colombia's alternative submission that it was not bound to surrender the refugee to Peru.

The Court recalled that Article 1 of the Convention provides one option to terminate diplomatic asylum, namely the issue of safe-conduct for the departure of the refugee from the state where the diplomatic premise is located (hereinafter the 'territorial state'). Under the terms of the Asylum Judgment, safe-conduct could only be claimed under the Convention if the asylum had been regularly granted and if the territorial state had requested the departure of the refugee. The problem was that none of those requirements was met in the case at hand and the Convention remained silent as to how to terminate a diplomatic asylum when neither of those requirements was met.

Short of an express provision in the Convention, the Court could not conclude the existence of a positive obligation incumbent upon Colombia to surrender the refugee to Peru. In doing so, the Court confirmed that the existence of such an obligation cannot be presumed.

In the Court's judgment, the absence of an obligation to surrender the refugee can be also explained because interpreting the Convention's silence as to impose an obligation as such would be contrary to the Latin American tradition that inspired the Convention and according to which a political refugee ought not to be surrendered. The Court's reference to 'Latin American tradition'[24] is curious but comprehensible. Tradition is certainly not a source of international law, but is used by the Court as a pre-existing practice to interpret a subsequent treaty. This is the opposite of perhaps more common situations, as will be later foreseen by the Vienna Convention on the Law of the Law of Treaties, whereby provisions of a treaty are interpreted according to subsequent practice. Be that as it may, the Court's Judgment appears to confirm that a pre-existing practice can be an element in the interpretation of the provisions of a treaty.

24 *Haya de la Torre Case*, p. 81.

In addition, recognizing that Colombia should surrender the refugee would be, in the view of the Court, equivalent to recognizing an obligation to render positive assistance to the authorities of the territorial state in the prosecution of a political refugee. The Court concluded that such an obligation 'could not be recognized without an express provision to that effect in the Convention'.[25]

The Court also recalled that diplomatic asylum should not become a permanent situation and should remain, according to Article 2, paragraph 2 of the Convention, only 'for the period of time strictly indispensable for the person who has sought asylum to ensure in some other way his safety'.[26]

Three things now seem clear. First, once all conditions are met, the granting of diplomatic asylum is a political decision. Second, the Convention does not impose any obligation to surrender the refugee. Third, diplomatic asylum should remain for the time strictly necessary to ensure the safety of the refugee. In light of these three factors, the silence of the Convention in providing a solution on how to terminate an irregularly granted diplomatic asylum appears to suggest the parties' intention 'to leave the adjustment of the consequences [of granting diplomatic asylum]... to decisions inspired by considerations of convenience or of simple political expediency'.[27]

It should also be noted that, despite Colombia's reference to an 'American international law' in its submission as a legal ground for its claims, the Court made no reference whatsoever to it in the Judgment. It seems that by avoiding any reference to an 'American international law' the Court attempted to implicitly downplay its controversial reference in the Asylum Judgment to a regional or local custom peculiar to Latin American states.[28]

The Court certainly bore in mind the inherent tension that diplomatic asylum generates between, on the one hand, the refugee's safety and, on the other hand, the regular application of the territorial state's law and the exercise of its jurisdiction by legally constituted tribunals. In this sense, a word of caution seems to be addressed to states when granting diplomatic asylum with the Court emphasizing that asylum 'would not authorize the diplomatic agent to obstruct the application of the laws of the [territorial] country, whereas it is his duty to respect them'.[29] The issue remains open as to the options available to the territorial state should the granting of diplomatic asylum actually obstruct the regular application of its laws and the exercise of its jurisdiction by legally constituted tribunals. Similarly, the issue remains

25 Ibid.

26 *Haya de la Torre Case*, p. 80.

27 The paragraph is particularly revealing: 'to infer from this silence that there is an obligation to surrender a person to whom asylum has been irregularly granted would be to disregard both the role of these extra-legal factors in the development of asylum in Latin America, and the spirit of the Havana Convention itself'. *Haya de la Torre Case*, p. 81.

28 *Colombian-Peruvian Asylum case, Judgment on November 20th, 1950: I.C.J. Reports 1950*, p. 276.

29 *Haya de la Torre Case*, p. 81.

open as to the options available to the territorial state should the granting of diplomatic asylum constitute a threat to its national security, public order or other essential interest, and this notwithstanding the self-contained character of the rules of diplomatic law.

For all the above reasons, the Court concluded that Colombia was under no obligation to surrender the refugee.

This brings us to our last point, Peru's third and alternative submission should the Court not be able to state in which manner the asylum should conclude. In the first part of its alternative submission, Peru requested the Court to declare that the asylum was contrary to the Convention and ought to have ceased immediately after delivery of the Asylum Judgment and must cease after delivery of the judgment. In the second part of its alternative submission, Peru added that the asylum should cease 'in order that Peruvian justice may resume its normal course'.[30]

The Court replied unanimously to the first part of Peru's alternative submission. It recalled that in the Asylum Judgment it held that the asylum was not granted in conformity with the Convention. Such finding 'entails a legal consequence, namely that of putting an end to an illegal situation'.[31] In other words, 'Colombia, which had granted the asylum irregularly is bound to terminate it'.[32] In making such a declaration, the Court says nothing new, but recalls one of the principles of the international responsibility of states according to which a state that has breached an international obligation is bound to cease the wrongful conduct. Accordingly, Peru could legally demand the termination of a situation established in violation of the Convention, and Colombia would be 'bound to terminate it'.

What follows is a conclusion that has raised significant objection, at least in Latin America: 'asylum must cease, but … Colombia is under no obligation to bring this about by surrendering the refugee to the Peruvian authorities'.[33] It has been submitted that this is contradictory and of no practical use.[34] The objections raised against the Court's conclusion are within themselves objectionable.

Indeed, the conclusion of the Court may at first glance appear contradictory. The Court affirms: '[t]here is no contradiction between these two findings, surrender is not the only way of terminating asylum'.[35] There are indeed a broad range of ways to terminate diplomatic asylum, such as: (i) the refugee freely surrendering himself; (ii) the asylum state surrendering the refugee; (iii) the surrender of the refugee or by the refugee himself to a third state; (iv) the withdrawal of the charges against the refugee; (v) the refugee

30 *Haya de la Torre Case*, p. 82.
31 Ibid.
32 Ibid.
33 *Haya de la Torre Case*, p. 83.
34 See supra footnote 11.
35 *Haya de la Torre Case*, p. 82.

being acquitted of all charges; (vi) the territorial state requesting that the refugee leaves its territory; (vii) the refugee being authorized to leave the territorial state and being granted a safe-conduct; (viii) the refugee dying; (ix) a change of government in the territorial or in the asylum state allowing for the refugee to leave the diplomatic premises; (x) the refugee being pardoned; (xi) the asylum and the territorial state breaking diplomatic relations; (xii) or, even more extreme, the territorial state considering the presence of the refugee a threat to its national security, public order or other essential interest and, with prior notice and deadline, suspending the diplomatic status of the premises of the mission with the sole purpose of arresting the refugee. It was however not for the Court to make any such or alternative proposals. The Court does not make proposals but adjudges the dispute that is submitted to it, in accordance with international law.

As for the second part of Peru's alternative question, namely that diplomatic asylum should cease in order for Peruvian justice to resume its normal course, the Court did not agree for the reasons already explained. Indeed, no such obligation could be said to exist without an express provision and such an obligation would also be contrary to the Latin American tradition that inspired the Convention and, accordingly, to which a political refugee ought not to be surrendered.

Finally, it is worth noting that the last paragraph of the judgment encourages the parties, 'now that their mutual legal relations have been made clear',[36] to find a practical and satisfactory solution. In fact, once the Court has made the mutual legal relations of the parties to a dispute clear, they are presumed to be able to find a practical and satisfactory solution. Such was in fact the case, even if it still took some years for the refugee to leave the Colombian embassy in Peru.

IV Conclusion

Latin American authors of international law have traditionally identified diplomatic asylum as a Latin America legal institution. It is however not only a legal institution, but also one that lies at the crossroads of political expediency, diplomatic immunities and elementary considerations of humanity. While diplomatic asylum is, at least for those states party to the Convention, a legal notion, the decision to grant or not diplomatic asylum is political in nature and inspired by elementary considerations of humanity. The Judgment clearly shows this tension.

The Judgement, and through it two Latin American countries, contributed to clarifying the rules governing diplomatic asylum. In particular, that the decision to grant or not diplomatic asylum is a political decision, so much as it is the decision on how to terminate it. The Judgment also

36 *Haya de la Torre Case*, p. 83.

clarified that under the Convention there is no obligation to surrender the refugee to the territorial state, that the surrender of the refugee is not the only way of terminating asylum, that diplomatic asylum should not become a permanent situation and should remain only for the time strictly indispensable for the refugee to ensure his safety in some other way and that, despite the considerations of humanity that inspires it, diplomatic asylum does not authorize an obstruction of the application of the laws of the territorial state.

The Judgment, delivered less than five years after the coming into existence of the Court, also contributed to clarifying some specific aspects of the Court's procedure and of international law in general. Indeed, the Judgment confirmed that the Court will evaluate its jurisdiction to deal with a specific case even if the parties have consented to its jurisdiction and raised no objections to it. The Judgment also settled the criteria that an intervention under Article 63 of the Statute of the Court shall relate to the subject matter of the proceedings. In terms of international law in general, the Judgment legitimized the use of a pre-existing practice to interpret a subsequent treaty, assimilated a tradition to a practice, confirmed that an obligation is not to be presumed and also confirmed that a State that has breached an international obligation is bound to cease the wrongful act.

Sadly, the Judgment is remembered for what it failed to do, namely to provide a manner in order to terminate a diplomatic asylum irregularly granted. It was however not for the Court to fashion a solution when the applicable law failed to provide for one. In this regard, the criticism against the Judgment is largely unfounded.

Bibliography

A. Álvarez, *Le droit international américain: son fondement, sa nature d'après l'histoire des états du nouveau monde et leur vie politique et économique*, Paris: A. Pedone, 1910.

——, *La codification du droit international, ses tendance – ses bases*, Paris: A. Pedone, 1912.

——, *Le droit international de l'avenir*, Washington, DC: American Institute of International Law, 1916.

——, *Le nouveau droit international public et sa codification en Amérique*, Paris: A. Rousseau, 1924.

——, *Considérations générales sur la codification du droit international américain: (code de droit international des états américains)*, Rio de Janeiro: Impr. Nacional, 1927.

——, *Le droit international nouveau dans ses rapports avec la vie actuelle des peuples*, Paris: Librairie Pedone, 1959.

N. Bailey, 'Asylum and Haya de la Torre', *Journal of International Affairs* 9, 1955, 82–86.

C. Barcia Trelles, 'El Derecho de Asilo Diplomático y el caso Haya de la Torre', *Revista Española de Derecho Internacional* vol. III, 3, 1950, 59–66.

P. Behrens, 'The Law of Diplomatic Asylum – A Contextual Approach', *Michigan*

Journal of International Law 35, no. 2, 2014, 319–367.
J.J. Caicedo Castilla, 'El derecho de asilo', *Revista española de derecho internacional* 10, 1957, 446–464.
C. Diaz Cisneros, 'Las sentencias de la Corte Internacional de Justicia en el caso Haya de la Torre', *Revista jurídica argentina La Ley* 78, 1955, 831.
H.J. Gobbi, 'Ensayo de una crítica del asilo diplomático', *Revista española de derecho internacional*', 15, 1962, 413–435.
J.G. Guerrero, *La responsabilidad internacional de los estados por los daños causados en su territorio a la persona o bienes de los extranjeros*, Paris: Lahure, 1926.
——, *La codification du droit international: la première conférence (La Haye, 13 mars - 12 avril 1930)*, Paris: Pedone, 1930.
——, *L'Union panaméricaine et la Société des nations*, Leyde: Sijthoff, 1937.
——, *El orden internacional*, s/l: Universidad Autónoma de El Salvador, 1945
——, *L'ordre international, hier, aujourd'hui, demain*, Neuchâtel: La Baconnière, 1945.
L. Jully, 'L'asile diplomatique devant la Cour internationale de Justice', *Die Friedens-Warte*, Bd. 51, 1951, 20–58.
Institut de droit international, *Yearbook*, session of Bath, vol. 43, 1950, no. 1.
G. Perez Gomar, *Curso elemental de derecho de gentes*, Montevideo, 1864.
J.M. Ruda, 'Intervention before the International Court of Justice', *Fifty Years of the International Court of Justice: Essays in honour of Sir Robert Jennings*, Vaughan Lowe et Malgosia Fitzmaurice (eds), Cambridge: Grotius Publications/ Cambridge University Press, 2007, 487–502.
J.L.F. Van Essen, 'Some Reflections on the Judgements of the International Court of Justice in the Asylum and Haya de la Torre Case', *International and Comparative Law Quarterly* 1, 1952, 533–539.
M.A. Vieira, *Derecho de Asilo Diplomático (asilo político)*. Montevideo: Biblioteca de Publicaciones Oficiales de la Facultad de Derecho y Ciencias Sociales de la Universidad de la República, 1961.

15 *Asylum (Columbia v. Peru), 1949* and request for interpretation of the judgment of 20 November 1950 in the *Asylum Case (Colombia v. Peru), 1950*

Guy S. Goodwin-Gill[1]

I The *Asylum cases* in brief

On 3 January 1949, Víctor Raúl Haya de la Torre, the Peruvian founder and leader of *Allianza Popular Revolucionaria Americana* (APRA), was granted asylum in the Embassy of Colombia in Lima.[2] It would be five years, four months and three judgments of the International Court of Justice (hereinafter 'the Court') before he was able to leave.

The Colombian Ambassador requested safe-conduct for Haya de la Torre, but the government of Peru declined, arguing that the 1928 Havana Convention,[3] to which both states were party, did not oblige it to comply. Unable to agree on a *compromis*, the two governments nevertheless signed the 'Act of Lima' on 31 August 1949, following which Colombia instituted proceedings before the Court. Basing its case on the 1911 Bolivarian Agreement on Extradition, the 1928 Havana Convention on Asylum,[4] and 'American international law', Colombia argued that it had the right unilaterally and definitively to 'qualify' the nature of the offence for which diplomatic asylum had been granted; and that Peru, as the territorial state, was obliged to grant safe-conduct to the refugee.[5]

In its counter-claim, Peru argued that asylum had been granted in violation of Article 1, paragraph 1 of the Havana Convention, which prohibited the grant of asylum for common crimes, and contrary to Article 2, paragraph

1 The excellent research assistance of Sotirios-Ioannis Lekkas in the preparation of this chapter is gratefully acknowledged.
2 *Asylum Case (Colombia v. Peru), Judgment, I.C.J. Reports 1950*, pp. 266, 272.
3 Bolivarian Agreement on Extradition, Caracas, 11 July 1911, OAS Off. Rec. (OEA/SER.x.1), Treaty Series, No. 34.
4 Convention on Asylum, Havana, 20 February 1928, OAS Off. Rec. (OEA/SER.X/1), Treaty Series, No. 34.
5 *Asylum Case (Colombia v. Peru)*, op. cit., p. 269.

2, which provided further that asylum could not be granted except in urgent cases.

By 14-2, the Court rejected Colombia's submission that it had a right of unilateral and definitive qualification; and by 15-1, that it was entitled to demand safe-conduct. Also by 15-1, the Court rejected the first part of Peru's counter-claim, and held that the grant of asylum had not violated Article 1, paragraph 2 of the Havana Convention; but by 10-6, the Court upheld the Peru counter-claim that there was no urgency.[6]

On the day the judgment was handed down (20 November 1950), Colombia lodged a request for interpretation, asking whether the decision meant that legal effects were to be attributed to the Colombian Ambassador's qualification of the offence imputed to Haya de la Torre; that Peru was not entitled to demand surrender, and that Colombia was not bound to hand over the refugee; or that, on the contrary, Colombia was bound to deliver him up. On 27 November 1950, the Court held 12-1 that the request was inadmissible, there being (at that time) no dispute between the parties as to the meaning and scope of the judgment. Moreover, the first question had not been among those originally submitted by the parties, while the remaining two questions were simply alternative ways of approaching the issue of surrender – again, a point that had not been put to the Court, and on which the Court could make no decision without hearing both sides.

A dispute was not long in the making. The next day, Peru requested the immediate surrender of Haya de la Torre, which Colombia rejected on 6 December, on the ground that nothing in the decision obliged it to hand over the refugee.[7] It then instituted fresh proceedings on 13 December 1950, asking the Court how effect was to be given to the original judgment. As explained elsewhere in this volume, the Court unanimously rejected Colombia's principal submission and also part of Peru's case.

By 13-1, however, the Court upheld Colombia's alternative argument, finding 'in the exercise of its ordinary competence', rather than in terms of the first judgment,[8] that Colombia was not bound to deliver Haya de la Torre to the Peruvian authorities, as he was not accused of a common crime.

6 Ibid., pp. 282–4. Judges Alvarez, Badawi Pasha, Read, Azevedo and Caicedo Castilla dissented; Judge Zoričić voted with the majority, save with regard to the issue of urgency, on which he shared the views of Judge Read.

7 Bishop, W. W., 'Judicial Decisions: *Haya de la Torre Case* (Colombia/Peru)', *American Journal of International Law* 45, 1951, 781, 783–4.

8 *Haya de la Torre Case, Judgment, I.C.J. Reports 1951*, Convention on Asylum, Havana, 20 February 1928, OAS Off. Rec. (OEA/SER.X/1), Treaty Series, No. 3471, 79.

II The general approach of the court in the *Asylum cases*

A. Customary international law

The judgment of the Court in the first case made a clear contribution to the doctrine of international law, and it continues to be widely cited, particularly on evidence and proof of custom. Much contemporaneous criticism, however, was often directed at the perceived failure of the Court to appreciate the significance of 'American international law', the importance of diplomatic asylum in regional context and beyond, and the need for a less formalistic approach to treaty interpretation.[9] The Court was also criticized for having simply handed down decisions, rather than a solution, which some characterized as verging on a *non liquet*.[10]

More positively, leading commentators, such as Brownlie, have commended the Court's clarity and its 'leading pronouncements' on the specific elements of local or regional custom,[11] quoting extensively what has become a seminal part of the judgment dealing with local or regional custom:

> The Party which relies on a custom of this kind must prove that this custom is established in such a manner that it has become binding on the other Party. The Colombian Government must prove that the rule invoked by it is in accordance with a constant and uniform usage practised by the States in question, and that this usage is the expression of a right appertaining to the State granting asylum and a duty incumbent on the territorial State. This follows from Article 38 of the Statute of the

9 See generally, Barcía Trelles, C., 'El derecho de asilo diplomático y el caso Haya de la Torre: Glosas a una sentencia', *Revista española de derecho internacional* 3, 1950, 753; van Essen, J. L. F., 'Some Reflections on the Judgments of the International Court of Justice in the Asylum and Haya de la Torre Cases', *International and Comparative Law Quarterly* 1, 1952, 533.

10 See generally, Jully, L., 'L'asile diplomatique devant la Cour internationale de justice', *Die Friedens-Warte* 51, 1951–3, 20; Green, L. C., 'International Court of Justice: Right of Asylum Case (Colombia/Peru) ICJ Reports 1950, p.266', *International Law Quarterly* 4, 1951, 229. See also, Lauterpacht, H., *The Development of International Law by the International Court*, Cambridge: CUP, 1958, 145; Rosenne, S., *The Law and Practice of the International Court of Justice*, Leiden: Martinus Nijhoff, 4th edn, 2006, 201.

11 Brownlie, I., *Principles of Public International Law*, Oxford: Clarendon Press, 1st edn, 1966, 5–6, 8–9; 7th edn, Oxford: OUP, 2008, 7, 10, 12; Crawford, J., *Brownlie's Principles of Public International Law*, Oxford: OUP, 8th edn, 2012, 30. See also Shaw, M., *International Law*, Cambridge: CUP, 2nd edn, 1986, 63–4; 6th edn, Cambridge: CUP, 2008, 76–7, 93; Kolb, R., *The International Court of Justice*, Oxford: Hart Publications, 2013, 935; Lauterpacht, H., *Oppenheim's International Law*, London: Longmans, 8th edn, 1955, 26–7; Combacau, J., and Sur, S., *Droit international public*, 9th edn, Paris: Montchrestien, 2010, 68–9, 72–3; Rosenne, S., op. cit., 1041.

> Court, which refers to international custom 'as evidence of a general practice accepted as law'.[12]

Colombia's argument for 'American international law' was also treated by the Court as an argument for custom, which thus assimilated it to the *general* standard of Article 38, so denying its potentially special character.[13]

B. Asylum

Comments on the Court's approach to the issue of asylum and specifically to the claim of a 'unilateral right of qualification' have tended to be subordinated to questions related to customary international law.[14] Although the analysis was largely conditioned by assumptions regarding sovereignty and non-intervention, it has nevertheless added clarity to diplomatic asylum, not least by separating it from an exclusive connection to the inviolability of diplomatic premises. This, in turn, helps to show how the practice may operate beyond Latin America, despite the opposition of many states to its legal character, and thereby contribute to a better understanding of asylum generally and the rationale for protection.

III Specific aspects of the judgments

A. The customary international law arguments

How the Court characterized the institution and practice of diplomatic

12 *Asylum Case (Colombia v. Peru)*, op. cit., pp. 276–7. The precise context – regional custom as an *exception* to fundamental principles of general application – is not always appreciated, however.

13 For an alternative view, see *I.C.J., Dissenting Opinion of Judge Alvarez, Asylum Case (Colombia v. Peru)*, pp. 293–4, on 'American international law' as a 'complex of principles, conventions, customs, practices, institutions and doctrines peculiar to the region', being a continental system of law not subordinate to general international law, 'but *correlated* to it'; Alvarez, A., *Le droit international nouveau*, Paris: Pedone, 1959, 145, suggesting, perhaps not surprisingly, that 'American international law' had been accepted by the Court, 'au moins implicitement'. Judge Read, dissenting, referred to 'a body of conventional and customary law complementary to universal international law, and governing inter-State relations in the Pan-American world': *I.C.J., Dissenting Opinion of Judge Read, Asylum Case (Colombia v. Peru)*, p. 316. See also, Ronning, C. N., *Diplomatic Asylum – Legal Norms and Political Reality in Latin American Relations*, Leiden: Martinus Nijhoff, 1965, 35–41.

14 Earlier commentaries tended to pay more attention to the asylum issue; see, for example, Fitzmaurice, G. G., 'The Law and Procedure of the International Court of Justice: General Principles and Substantive Law', *British Yearbook of International Law* 27, 1, 1950. 31–40; Gonidec, P. F., 'L'affaire du droit d'asile', *Revue Générale de Droit International Public* 55, 1951, 547, 590; Jully, op. cit., 57–8; Morgenstern, F., 'Diplomatic Asylum', *Law Quarterly Review* 67, 1951, 362, 376–81; Green, op. cit., 238–9; Evans, A. E., 'The Colombian-Peruvian Asylum Case: Termination of the Judicial Phase', *American Journal of International Law* 45, 1951, 755.

asylum largely determined its approach to customary international law and its strict treatment of 'American international law'. The Court thus placed its analysis firmly in the context of sovereignty, jurisdiction, and non-intervention, and its assumption that diplomatic asylum involves a 'derogation' from sovereignty, understood as the normally exclusive jurisdiction of the territorial state, necessarily had an impact on the evidence and standard of proof.[15] Once the Court had opted for a 'general' international law approach favouring the principle of non-intervention then, as Lauterpacht observed, a restrictive interpretation of any argument for exceptions would naturally follow.[16] Having formulated the conditions for a customary rule, 'the Court proceeded to apply them in an exacting manner'.[17] Colombia's claim to a unilateral right of qualification was thus destined to stumble in the face of such fundamental principles of international law, while an alternative and more nuanced approach to the underlying premises might well have led more readily to a solution.

B. The law and practice of diplomatic asylum

Diplomatic asylum has long enjoyed a special place in the law and practice of Latin American States, but it can also be seen with some regularity in other regions.[18] However, whereas Latin American States have frequently sought to translate the practice into a treaty-based system of protection, states elsewhere have tended to deny its legal character, all the while admitting nonetheless that refuge may well be accorded in egregious cases.[19]

15 Brownlie, I., op. cit., 291; Combacau, J., and Sur, S., op. cit., 360; Crawford, J., op. cit., 451.

16 Lauterpacht, H., *Oppenheim's International Law*, op. cit., 52–4; Lauterpacht, H., *The Development of International Law*, op. cit., 381–2 ('decisive weight' given to intervention).

17 Lauterpacht, H., *The Development of International Law*, op. cit., 375. See also Jennings, R. and Watts, A., *Oppenheim's International Law*, London: Longmans; 1992; Oxford: OUP, 9th edition, repr. 2008, 30 (so far as regional or particular customary international law involves departure from a generally applicable rule, its existence is a matter of strict proof). According to Verdross and Simma, this also explains why the consent of *all* states is required: Verdross A. and Simma, B., *Universelles Völkerrecht – Theorie und Praxis*, Berlin: Duncker und Humblot 1976, 289.

18 See generally, Ronning, C. N., *Diplomatic Asylum – Legal Norms and Political Reality in Latin American Relations*, Leiden: Martinus Nijhoff, 1965.

19 Brownlie notes that the Vienna Convention on Diplomatic Relations makes no provision for diplomatic asylum, apart from an oblique reference to 'special agreements' in Article 41: Brownlie, I., op. cit., 357. Lauterpacht accepted that there was a 'legal right' to grant temporary asylum, on grounds of humanity, in the case of violence or disorder, but that otherwise diplomatic asylum must be based on treaty: Lauterpacht, H., *Oppenheim's International Law*, op. cit., 796–8. See also Jennings, R. and Watts, A., *Oppenheim's International Law*, op. cit. 1082–6 (on 'So-called diplomatic asylum'); Ronning, C. N., op. cit. 21–3, 66–96, 122–4.

Colombia argued that it enjoyed the right unilaterally to determine the political or non-political nature of the offence with which the putative asylee was charged or convicted.[20] Moreover, that its unilateral decision in favour of the asylee triggered an obligation on the part of the territorial state to accord the necessary safe-conduct. The Court held that the general principles of international law referred to in the 1911 Bolivarian Agreement,[21] 'do not recognize any rule of unilateral and definitive qualification by the State granting diplomatic asylum'.[22] It detected some confusion in the pleadings between diplomatic and territorial asylum, on the one hand, and extradition, on the other; its explanation strongly emphasizes the controlling aspect in a diplomatic asylum scenario of 'derogation' and the principle of non-interference:

> In the case of extradition, the refugee is within the territory of the State of refuge. A decision with regard to extradition implies only the normal exercise of the territorial sovereignty. The refugee is outside the territory of the State where the offence was committed, and a decision to grant him asylum in no way derogates from the sovereignty of that State
>
> In the case of diplomatic asylum, the refugee is within the territory of the State where the offence was committed. A decision to grant diplomatic asylum involves a derogation from the sovereignty of that State. It withdraws the offender from the jurisdiction of the territorial State and constitutes an intervention in matters which are exclusively within the competence of that State. Such a derogation from territorial sovereignty cannot be recognized unless its legal basis is established in each particular case.
>
> For these reasons, it is not possible to deduce from the provisions of agreements concerning extradition any conclusion which would apply to the question now under consideration.[23]

The 1928 Havana Convention also made no provision for the unilateral right claimed by Colombia, and no such right could be implied in the Convention or considered as inherent in the institution of asylum. Such an exceptional competence would itself further derogate from the 'equal rights' of qualification attributable to other states concerned, and 'thus *aggravate* the derogation from territorial sovereignty' (emphasis added), even if such a rule might make the institution more effective.[24]

The question of effectiveness was taken up in the dissenting opinions, with Judge Alvarez noting pointedly that a right of appreciation, while its exercise might be questioned, 'must naturally appertain to the State granting asylum;

20 On 'qualification', see also Ronning, C. N., op. cit., 148–57, 187–213.

21 Above, note 3.

22 *Asylum Case (Colombia v. Peru),* op. cit., p. 274.

23 Ibid., pp. 274–5.

24 Ibid., p. 275; on practice, see also at 277.

if it appertained to the territorial State, the institution of asylum would be rendered nugatory'.[25] Judge Caicedo Castilla, the judge *ad hoc* for Colombia, agreed: a unilateral right of qualification was 'inherent in the very nature' of diplomatic asylum, as also in territorial asylum, for otherwise the institution could not exist.[26]

Although Colombia relied on a considerable body of practice, it did not meet the Court's stringent evidence and proof requirements.[27] Moreover, Peru's non-ratification of the 1933 and 1939 Montevideo Conventions (on which Colombia relied as evidence of a pre-existing rule) was perceived as 'rejection' of the unilateral qualification rule, notwithstanding certain official documents appearing to accept it.[28]

The Court further held that any obligation to grant safe-conduct required that asylum must have been regularly granted and maintained; that is, to political offenders, not common criminals, only in urgent cases and for the time strictly indispensable for the safety of the refugee.[29] It interpreted the Havana Convention to mean 'that the territorial State may require that the refugee be sent out of the country, and that only "after" such a demand can the State granting asylum require the necessary guarantees as a condition of his being sent out.'[30] Peru had not asked that Haya de la Torre leave; on the contrary, it argued in its counter-claim that asylum had been granted in breach of the Havana Convention, and that its maintenance constituted a violation of the treaty.[31]

The Court had no problem in finding that Haya de la Torre was an 'accused', or that Peru had failed to discharge its burden of showing that he was accused of a common crime. The evidence conclusively showed that the offence in question – military rebellion – was a political offence, and was viewed as such even by Peruvian law.[32]

On the question of urgency, however, the Court was divided. Once again, the basic premises led the way and the Court thought it 'inconceivable' that the Havana Convention had been intended to provide protection against

25 *I.C.J., Dissenting Opinion of Judge Alvarez, Asylum Case (Colombia v. Peru)*, p. 297.

26 *I.C.J., Dissenting Opinion of Judge Caicedo Castilla, Asylum Case (Colombia v. Peru)*, pp. 363–6.

27 *Asylum Case (Colombia v. Peru)*, op. cit., p. 277.

28 Ibid., p. 278. Compare Judge Read's dissenting opinion at 321, in which he cites numerous instances of relevant practice and that of Judge Caicedo Castilla at 363. See also Lauterpacht, H., *The Development of International Law*, op. cit., 376 (non-ratification can be due to reasons other than opposition).

29 *Asylum Case (Colombia v. Peru)*, op. cit., p. 278 (Article 1, paragraph 1, 1928 Havana Convention).

30 Ibid., p. 279 (emphasis added). Even if the usual practice was for safe-conduct to be requested and granted at once, the Court considered that this did not alter the legal situation.

31 Ibid., pp. 279–80. The Court had found the counter-claim admissible, being directly connected with the subject matter of the proceedings, namely, the regularity of the asylum granted.

32 Ibid., pp. 280–2.

regular legal proceedings: 'In principle ... asylum cannot be opposed to the operation of justice.'[33] The Convention did not protect political offenders generally, for this would lead to 'foreign interference of a particularly offensive nature'.[34] Given that the essential justification for asylum lay in the 'imminence or persistence of a danger to the person of the refugee',[35] and that over three months had passed before Haya de la Torre sought and was granted asylum, Colombia had not discharged its burden of showing that the case was urgent, when considered against 'the objective existence of the facts'.[36]

Among the dissenting judges, Judge Alvarez applied the same approach to that of the 'qualification' issue: it was for the state of refuge to appreciate whether or not there was urgency, and strict compliance with the letter of the Havana Convention, 'would be tantamount to a disregard of the nature of asylum'.[37] Judge Read paid detailed attention to the urgency question, but also to the widespread practice of diplomatic asylum, citing numerous instances involving some 244 individuals and at least 17 Latin American States.[38] He could find no occasion where a political offender had ever been surrendered at the request of the territorial state, and no evidence to show that diplomatic asylum was limited to occasions of pursuit by angry mobs; on the contrary, asylum was freely granted during unsettled, post-revolutionary conditions.[39] It was unthinkable, he said, that in referring to 'urgent cases', the drafters of the Havana Convention had intended to bring to an end an institution based on ninety years of tradition.[40] As he saw it, the essential distinction was between political offenders and common criminals, and the 'urgent cases' qualification simply limited diplomatic asylum for political offenders to times of political disturbance of a revolutionary character.[41] In

33 Ibid., p. 284.

34 Ibid., p. 286. The view of the Court contrasts directly with that of Judge Alvarez who, while emphasizing the 'ex-territoriality' of the premises in which asylum was granted and taking account of Latin American practice, argued that, 'asylum in no way constitutes an intervention or a limitation of the sovereignty of the territorial State, but rather ... the legitimate exercise of a prerogative': *I.C.J., Dissenting Opinion of Judge Alvarez, Asylum Case (Colombia v. Peru)*, p. 292.

35 *Asylum Case (Colombia v. Peru)*, op. cit., p. 282.

36 Ibid., p. 287.

37 *I.C.J., Dissenting Opinion of Judge Alvarez, Asylum Case (Colombia v. Peru)*, pp. 298–9. Cf. Fitzmaurice, G. G., op. cit., 39–40, minimizing the differences between the Court and the dissents.

38 Ibid., *Dissenting Opinion of Judge Read*, pp. 317, 328; also *Dissenting Opinion of Judge Caicedo Castilla*, pp. 375–7; Ronning, op. cit., 66–96, 148–57, 187–213.

39 Ibid., *Dissenting Opinion of Judge Read*, p. 321: 'The evidence shows that asylum was granted, as a matter of course, to political offenders who were seeking to escape from ordinary judicial proceedings under the laws of the territorial States.'

40 Ibid., p. 322.

41 Ibid., Convention on Asylum, Havana, 20 February 1928, OAS Off. Rec. (OEA/SER.X/1), Treaty Series, No. 34326.

an interesting, but undeveloped remark, he also referred to Peru's submissions: 'In saying that "diplomatic asylum is based on a desire for *humanitarian protection* in times of political disturbance", the Government of Peru was furnishing the key to the solution of the problem.'[42]

C. Impact and influence

Although much has been made of the Court's pronouncements on custom, evidence, and proof, little attention has been given, particularly more recently, to what was said about diplomatic asylum and, directly and indirectly, about asylum in general. Significantly, the Court first distanced its analysis and reasoning from the 'extra-territoriality' of diplomatic premises,[43] and instead emphasized asylum as a 'state of protection':

> The grant of asylum is not an instantaneous act which terminates with the admission, at a given moment, of a refugee to an embassy or a legation. Any grant of asylum results in, and in consequence logically implies, a state of protection; the asylum is granted as long as the continued presence of the refugee in the embassy prolongs this protection.[44]

The Court's reference also to 'humanitarian grounds'[45] echoes the 'elementary considerations of humanity', which it evoked in the *Corfu Channel Case*.[46] Fitzmaurice already noted this in 1950,[47] as did O'Connell in 1970,[48] and it has resonated in a line of later decisions.[49] Secondly, as highlighted above, the Court stressed that the 'essential justification' for asylum is the 'imminence or persistence of a danger for the person of the refugee',[50] and while asylum cannot be opposed to the operation of justice, it does protect

42 Ibid. (emphasis supplied).

43 The status of the premises in which asylum is granted nevertheless remains a relevant consideration overall, however, and clearly constrains the options open to the territorial state.

44 *Asylum Case (Colombia v. Peru)*, op. cit., p. 281. Cf. '*non-refoulement* through time', as described in Goodwin-Gill, G. S. and McAdam, J., *The Refugee in International Law*, Oxford: Oxford University Press, 3rd edn, 2007, 207–8, 343–5.

45 'It has not been disputed by the Parties that asylum may be granted on humanitarian grounds in order to protect political offenders against the violent and disorderly action of irresponsible sections of the population': *Asylum Case (Colombia v. Peru)*, op. cit., pp. 282–3.

46 *Corfu Channel Case (United Kingdom v. Albania), Judgment, I.C.J. Reports* 1949, p. 4.

47 Fitzmaurice, G. G., op. cit., 33–5.

48 O'Connell, D. P., *International Law*, 2nd edn, London: Stevens, 1970, 737–8.

49 For discussion, see Goodwin-Gill, G. S., 'Europe: A place to seek, to be granted, and to enjoy asylum?', in Gortázar, C., Parra, M-C., Segaert, B. and Timmerman, C. (eds), *European Migration and Asylum Policies: Coherence or Contradiction?*, Bruxelles: Bruylant, 2012, 33.

50 *Asylum Case (Colombia v. Peru)*, op. cit., p. 282.

where 'arbitrary action is substituted for the rule of law', and 'against any measures of a manifestly extra-legal character'.[51] In this way, the judgment has helped to forge and strengthen an alternative, human rights-based rationale for asylum, which derives from the various duties of the state to provide protection to those at risk of relevant serious harm.[52]

Where diplomatic asylum is granted, the problem of what to do next can be very real,[53] and similar issues can also arise in territorial asylum. For example, in cases where the asylee is accused of a crime of international concern and may be liable to extradition: here, the possibility of local prosecution can help to avoid return to persecution while maintaining the integrity of the international justice system.[54] Diplomatic asylum nevertheless adds a sharper edge to an ongoing legal relationship between the two states;[55] and the sheltering state's non-exclusive competence raises near insurmountable obstacles to a solution in purely legal terms. Ironically, it was only in its third judgment that the Court appears to have recognized the normative quality of regional practice, and to have accepted that, while the Havana Convention provided for the surrender of common criminals, its silence on what to do with political offenders 'cannot be interpreted as imposing an obligation to surrender the refugee', even where asylum had been granted inconsistently with the Convention, for this would be contrary to the spirit of Latin American tradition, which only an express provision could achieve.[56] It is hardly surprising, therefore, that the Court turned to regional practices and traditions of good neighbourliness when it urged the parties to find the solution it could not

51 Ibid., p. 284.

52 Cf. *B v Secretary of State for Foreign and Commonwealth Affairs* [2004] EWCA Civ 1344.

53 The cases of Haya de la Torre and, more recently, Julian Assange, clearly demonstrate the practical problems. The Australian Wikileaks founder and editor entered the Ecuadorian Embassy in London on 19 June 2012 and claimed diplomatic asylum, which was granted on 16 August 2012: Neuman, W. and Ayala, M., 'Ecuador Grants Asylum to Assange, Defying Britain', *New York Times*, 16 August 2012: www.nytimes.com/2012/08/17/world/americas/ecuador-to-let-assange-stay-in-its-embassy.html (accessed 13 October 2015); BBC News Online, 'Julian Assange: Ecuador grants Wikileaks founder asylum', 16 August 2012: www.bbc.co.uk/news/uk-19281492 (accessed 13 October 2015). Over three years later, the issue had still not been resolved when the police guard on the Embassy was withdrawn in October 2015, by which time it had cost over £12 million: BBC News Online, 'Julian Assange: Police end guard at Wikileaks founder's embassy refuge', 12 October 2015: www.bbc.co.uk/news/uk-34508500 (accessed 13 October 2015)

54 Goodwin-Gill, G. S., 'Crimes in International Law: Obligations *Erga Omnes* and the Duty to Prosecute', in Goodwin-Gill G. S. and Talmon, S. (eds), *The Reality of International Law: Essays in Honour of Ian Brownlie*, Oxford: Clarendon Press, 1999, 199. See also, *Questions relating to the Obligation to Prosecute or Extradite (Belgium v. Senegal)*, *I.C.J. Reports 2012*, p. 422, on possible limits to the right of the state to grant asylum to individuals accused of serious human rights violations.

55 O'Connell, D. P., op. cit., 736.

56 *Haya de la Torre Case*, op. cit., pp. 81–2; see also *I.C.J., Dissenting Opinion of Judge Alvarez, Asylum Case (Colombia v. Peru)*, p. 322.

provide itself.[57] However, the Court's recognition of the principle of non-return (*non-refoulement*) remains critically important to this day.

The impact of the litigation was soon felt. Just four years later, as Haya de la Torre was about to leave the Colombian Embassy in Lima, Organization of American States (OAS) member states concluded two new conventions: one on territorial and the other on diplomatic asylum.[58] The Territorial Asylum Convention commences with general recognition of every state's right, 'in the exercise of its sovereignty, to admit into its territory such persons as it deems advisable' (Article 1), and specifically provides that, 'No State is under the obligation to surrender to another State, or to expel from its territory, persons persecuted for political reasons or offenses' (Article 3).

The Diplomatic Asylum Convention draws a clearer line to the litigation. Article 2 declares that, 'Every State has the right to grant asylum, but it is not obligated to do so or to state its reasons for refusing it.' There then follow a series of articles clearly intended to provide a corrective to the Havana Convention, as interpreted by the International Court of Justice in the *Asylum Case*. In particular, Article 4 provides that, 'It shall rest with the State granting asylum to determine the nature of the offense or the motives for the persecution'; Article 6, that 'urgent cases' include those where the individual is sought, either by mobs or the authorities themselves, and faces the risk of deprivation of life or liberty; Article 7, that it shall rest with the state granting asylum to determine the degree of urgency; and Article 12, that once asylum has been granted, the asylum state may request that the asylee be allowed to depart, 'and the territorial State is under obligation to grant immediately … the necessary guarantees … and the corresponding safe-conduct.'[59]

Although treaty ratification may be patchy, the underlying premises have been taken forward. During the 1950s and 1960s, asylum was extensively discussed in the UN Commission on Human Rights, the Third Committee, the Sixth Committee, and also, if briefly, in the International Law Commission.[60] Although it dealt initially with both diplomatic and territorial

57 *Haya de la Torre Case*, op. cit., p. 83; see also *I.C.J., Dissenting Opinion of Judge Alvarez, Asylum Case (Colombia v. Peru)*, pp. 293–4.

58 1954 Caracas Convention on Territorial Asylum: OAS Official Records, OEA/Ser.X/1; OAS Treaty Series No. 19; 1954 Caracas Convention on Diplomatic Asylum: OAS Official Records, OEA/Ser.X/1; OAS Treaty Series No. 18. Both treaties were adopted on 28 March 1954 and entered into force on 29 December 1954. Twelve states have ratified the Convention on Territorial Asylum, and fourteen on Diplomatic Asylum. Interestingly, Colombia signed the Diplomatic Asylum Convention on the day of adoption, but has not ratified, while Peru signed in 1960, and ratified in 1962. Colombia signed the Territorial Asylum Convention the same day, and ratified in 1968; Peru also signed on that day, but has not ratified.

59 See also Articles 15 and 20, which emphasize the protection dimension.

60 For a summary of discussions, see Goodwin-Gill, G. S., 'Introduction to the 1967 United Nations Declaration on Territorial Asylum', UN Audio-Visual Library of International Law, Historic Archives, 2012: www.un.org/law/avl/ (accessed 13 October 2015).

asylum, the 1967 Declaration adopted by the General Assembly was ultimately limited to the latter,[61] but nonetheless endorsed the proposition that, 'It shall rest with the State granting asylum to evaluate the grounds for the grant of asylum.'[62] Subject to the emerging *ordre public* on international crimes and crimes of international concern – such as terrorism – the basic principle argued for in the *Asylum Case* is clearly and implicitly accepted today in the practice of states.

Diplomatic asylum as a formal legal institution more or less clearly regulated by rules may be regionally limited, but the practice still occurs from time to time in other jurisdictions. The *Asylum Case* clarified its humanitarian foundations, and as Fitzmaurice suggested, prepared the ground for a general right, capable of being universalized.[63] In that regard, Latin American practice and tradition have played a no less formative role, even if states in other regions remain hesitant and the classic clash of jurisdictions remains unresolved. Lauterpacht noted the difficulties inherent in bringing asylum within the confines of law and policy,[64] and that one of the unfortunate consequences of the *Asylum Case* (and of incidents since and elsewhere), 'was that a person was exposed for a prolonged period to a condition of uncertainty and suspense in a manner unacceptable to human dignity'.[65] Today, countless refugees may have the 'protection' of territorial asylum, but their situation remains precarious and characterized by extreme hardship; they are in need of solutions that the law cannot as yet supply.

Bibliography

Alvarez, A., *Le droit international nouveau*, Paris: Pedone, 1959.

Barcía Trelles, C., 'El derecho de asilo diplomático y el caso Haya de la Torre: Glosas a una sentencia', *Revista española de derecho internacional* 3, 1950, 753.

Bishop, W. W., 'Judicial Decisions: *Haya de la Torre Case* (Colombia/Peru)', *American Journal of International Law* 45, 1951, 781.

Brownlie, I., *Principles of Public International Law*, Oxford: Clarendon Press, 1st edn, 1966; 7th edn, Oxford: OUP, 2008.

61 UNGA res. 2312 (XXII), 'Declaration on Territorial Asylum', 14 December 1967 (adopted without a vote). In 1974, on an Australian initiative, the General Assembly requested the Secretary-General to prepare and circulate a report on the practice of diplomatic asylum and invited member states to make known their views: UNGA res 3321 (XXIX), 14 Dec. 1974. The report (UN doc. A/10139) confirmed the regional nature of the practice; of twenty-five states which made known their views, only seven favoured drawing up an international convention on the matter.

62 Article 1(3).

63 Fitzmaurice, G. G., op. cit., 34–5.

64 See also, Jully, L., op. cit., 57.

65 Lauterpacht, H., *The Development of International Law*, op.cit., 147. As Jully puts it, 'Voilà un asile qui rassemble fort à une prison!', op. cit., 58, note 102.

Combacau, J. and Sur, S., *Droit international public*, Paris: Montchrestien, 9th edn, 2010.
Crawford, J., *Brownlie's Principles of Public International Law*, Oxford: OUP, 8th edn, 2012.
Evans, A. E., 'The Colombian-Peruvian Asylum Case: Termination of the Judicial Phase', *American Journal of International Law* 45, 1951, 755.
Fitzmaurice, G. G., 'The Law and Procedure of the International Court of Justice: General Principles and Substantive Law', *British Yearbook of International Law* 27, 1950, 1.
Gonidec, P. F., 'L'affaire du droit d'asile', *Revue Générale de Droit International Public* 55, 1951, 547.
Goodwin-Gill, G. S., 'Europe: A place to seek, to be granted, and to enjoy asylum?', in Gortázar, C., Parra, M-C., Segaert, B. and Timmerman, C. (eds), *European Migration and Asylum Policies: Coherence or Contradiction?* Bruxelles: Bruylant, 2012, 33.
Goodwin-Gill, G. S., 'Crimes in International Law: Obligations *Erga Omnes* and the Duty to Prosecute', in Goodwin-Gill, G. S., and Talmon, S. (eds), *The Reality of International Law: Essays in Honour of Ian Brownlie,* Oxford: Clarendon Press, 1999, 199.
Goodwin-Gill, G. S. and McAdam, J., *The Refugee in International Law*, Oxford: OUP, 3rd edn, 2007.
Green, L. C., 'International Court of Justice: Right of Asylum Case (Colombia/Peru) ICJ Reports 1950, p. 266', *International Law Quarterly* 4, 1951, 229.
Jennings, R. and Watts, A., *Oppenheim's International Law*, London: Longmans; 9th edn, 1992; Oxford: OUP, repr. 2008.
Jully, L., 'L'asile diplomatique devant la Cour internationale de Justice', *Die Friedens-Warte* 51, 1951–3, 20.
Kolb, R., *The International Court of Justice*, Oxford: Hart Publications, 2013.
Lauterpacht, H., *Oppenheim's International Law*, London: Longmans, 8th edn, 1955.
Lauterpacht, H., *The Development of International Law by the International Court,* Cambridge: CUP, 1958.
Morgenstern, F., 'Diplomatic Asylum', *Law Quarterly Review* 67, 1951, 362.
O'Connell, D. P., *International Law*, London: Stevens, 2nd edn, 1970.
Ronning, C. N., *Diplomatic Asylum – Legal Norms and Political Reality in Latin American Relations*, Leiden: Martinus Nijhoff, 1965.
Rosenne, S., *The Law and Practice of the International Court of Justice*, Leiden: Martinus Nijhoff, 4th edn, 2006.
Shaw, M., *International Law*, Cambridge: CUP, 2nd edn, 1986; 6th edn, Cambridge: CUP, 2008.
Van Essen, J. L. F., 'Some Reflections on the Judgments of the International Court of Justice in the Asylum and Haya de la Torre Cases', *International and Comparative Law Quarterly* 1, 1952, 533.
Verdross A. and Simma, B., *Universelles Völkerrecht – Theorie und Praxis*, Berlin: Duncker und Humblot, 1976.

16 *Avena and Other Mexican Nationals (Mexico v. United States of America), 2003* and request for interpretation of the judgment of 31 March 2004 in the case concerning *Avena and Other Mexican Nationals (Mexico v. United States of America), 2008*

Cesare P.R. Romano[1]

I Introduction

Between 1998 and 2008, the International Court of Justice (ICJ) heard a trio of cases on the application of the 1963 Vienna Convention on Consular Relations (VCCR),[2] in the context of the arrest, trial and sentencing of foreign nationals by the US: the *Breard case* brought in 1998 by Paraguay,[3] the *LaGrand case* brought in 1999 by Germany,[4] and the *Avena case* brought in 2003 by Mexico.[5]

The *Avena case* presented the ICJ with the opportunity to clarify and expand on points of law that it had previously made in the *LaGrand case*. But, unlike the *LaGrand case*, it unleashed a torrent of litigation in the US, resulting in rulings by the US Supreme Court that redefined the way the US incorporates its international legal obligations into its domestic legal system.

1 The author would like to thank Nadine Kheshen for her help in the research and editing of this piece.

2 *Vienna Convention on Consular Relations*, 24 April 1963, 596 U.N.T.S. 261, *entered into force* 19 March 1967.

3 *Case concerning the Vienna Convention on Consular Relations (Paraguay v. United States of America), Order of 10 November 1998, I.C.J. Reports 1998*, p. 426.

4 *LaGrand (Germany v. United States of America), Judgment, I.C.J. Reports 2001*, p. 466.

5 *Avena and Other Mexican Nationals (Mexico v. United States of America), Judgment, I.C.J. Reports 2004*, p. 12.

This chapter focuses on Mexico's reasons for taking the momentous step of filing a case against the US, the consequences of this decision, both legal and political, and the contribution that the *Avena* decision made to international law.

II Background

The VCCR has long been one of the centrepieces of the international legal architecture. To date, it has achieved quasi-universal acceptance, having been ratified by 177 states.[6] Yet, despite its importance, it rarely grabs headlines. Only when it is violated do states realize how crucial compliance is to ensure that the rights of their nationals abroad are respected, as well as the rights of non-nationals under their jurisdiction.

The VCCR consists of 79 articles, most of which provide for the operation of consulates, the privileges and immunities of consular officials posted to foreign countries, and the functions of consular agents. Consular officials act as a cultural bridge for nationals who face foreign criminal procedure, by providing translation, arranging legal representation, helping gather mitigating evidence, and hiring defence counsel.

These functions are regulated by Article 36 (1) of the VCCR, which provides that 'consular officers shall be free to communicate with nationals of the sending State and to have access to them' and nationals of the sending state will, likewise, have the same right to communicate and access their consular officers.[7] If a national of a sending state is detained and the national 'so requests, the competent authorities of the receiving State shall, without delay, inform the consular post of the sending State.'[8] The authorities of the receiving state also have a responsibility to 'inform the person concerned without delay of his rights.'[9] These rights include allowing a consular officer to 'visit a national of the sending State who is in prison, custody or detention, to converse and correspond with him and to arrange for his legal representation.'[10] Furthermore, while paragraph 2 states that these rights 'shall be exercised in conformity with the laws and regulations of the receiving State,' it maintains the exception that 'said laws and regulations must enable full effect to be given to the purposes for which the rights accorded under this article are intended.'[11]

The US has, by a large margin, the highest number of detainees in the world – almost 2.3 million – and approximately 5% are foreigners (thus,

6 United Nations Treaty Collection, status as at 9 August 2015. Available at: https://treaties.un.org/pages/ViewDetails.aspx?src=TREATY&mtdsg_no=III-6&chapter=3&lang=en (accessed 1 June 2016).

7 *Vienna Convention on Consular Relations*, art. 36.1(a).

8 Ibid., art. 36.1(b).

9 Ibid.

10 Ibid., art. 36.1(c).

11 Ibid., art. 36. 2.

about 120,000).[12] In theory, each of these should have been informed, either during their arrest or arraignment, of their right to notify their consular authorities. However, in a country as large as the US, with multiple levels of law enforcement in each state, each with its own criminal justice system, as well as a federal justice system, compliance with the requirements of Article 36 of the VCCR is far from universal and automatic. It should be no surprise then, that over the years several states have objected to the occasional denial of consular assistance to their nationals. In the late 1990s, Paraguay, Germany and Mexico decided to challenge the non-compliance with the VCCR of the US before the ICJ.

Mexico had the greatest interest in the issue and the highest stakes. Because of the close ties with the US, the prominence of Mexican nationals in the US territory, and the importance of their remittances to the Mexican economy, Mexico considers treatment of its nationals, particularly when they have been arrested, a priority of governmental concern. At the time of the *Avena* case, there were about 10 million Mexican nationals living in the US.[13] A good portion of the roughly 120,000 foreigners detained in US prisons around that time were likely to be Mexican citizens. Mexico has more consulates and consular officers in the US than any other state, with over forty-five consulates and hundreds of consular officers trained, specifically, to intervene on behalf of Mexican nationals.[14]

For years before taking the momentous step of filing a case before the ICJ, Mexico had been contesting the non-compliance of the US with the VCCR to no avail. Mexico closely observed when, in 1998, Paraguay brought a case to the ICJ claiming that the US had violated the VCCR by not providing Angel Breard, a Paraguayan national, the opportunity to benefit from consular assistance. Eventually, Paraguay caved under diplomatic pressure from the US and withdrew its application, preventing the ICJ from ruling on the merits.[15] However, the following year, Germany brought a similar case after the denial of consular assistance to two German nationals (the LaGrand brothers), who had been arrested, tried, and sentenced to death in Arizona.[16] This time, the ICJ did rule on the merits and found the US in violation of

12 R. Walmsley, 'World Prison Population List', International Centre for Prison Studies, 9th Ed., 2010. Available at: www.idcr.org.uk/wp-content/uploads/2010/09/WPPL-9-22.pdf (accessed 1 June 2016).

13 J. Zong and J. Batalova, *Frequently Requested Statistics on Immigrants and Immigration in the United States*, Migration Policy Institute (26 February 2015). Available at: www.migrationpolicy.org/article/frequently-requested-statistics-immigrants-and-immigration-united-states#MexicanImmigrants (accessed 1 June 2016).

14 C. Amirfar, 'AALS Panel – Mexico v. U.S.A. (Avena) – Arguments of Mexico', *European and International Law* 5, 2004, 376.

15 *Case concerning the Vienna Convention on Consular Relations (Paraguay v. United States of America), Order of 10 November 1998*, supra note 2.

16 *LaGrand, I.C.J. Reports 2001*, supra note 4, p. 466, para. 14.

both the VCCR and an order to stay the execution that it had issued while the case was being decided on the merits.[17]

III The ICJ judgment in the *Avena case*

The ICJ ruling in the *LaGrand case*, coupled with the continuing non-compliance by the US with the VCCR and the uniquely high stakes for Mexican nationals, convinced Mexico to take the bull by the horns. On 9 January 2003, it instituted proceedings before the ICJ alleging violation of all paragraphs and sub-paragraphs of Article 36 of the VCCR to the detriment of 52 Mexican nationals who had been tried and were awaiting execution in US prisons. Mexico also filed a request before the ICJ for the indication of provisional measures asking the Court to request that the US, *inter alia*, take 'all measures necessary to ensure that no Mexican national be executed.'[18]

Yet there were important differences between the facts of the *LaGrand* and *Avena* cases that warranted Mexico's departure from Germany's script. For instance, in the *LaGrand case*, the two brothers had been executed before the ICJ could reach a decision on the merits, whereas, at the time of the *Avena* proceedings, all Mexican nationals in question were still alive. Thus, while Germany could only seek reparations and non-repetition, Mexico insisted on restitution. As a result, Mexico built and expanded upon the *LaGrand* script; it argued that the US had failed to provide meaningful 'review and reconsideration' of the conviction and sentencing by neglecting to account for the violations of the VCCR, as required by the *LaGrand* decision.[19]

The ICJ took the *LaGrand case* as a point of departure for its reasoning on the merits, too, and went further. Although the Court adopted provisional measures to stay the execution of the detainees in both the *LaGrand* and *Avena* cases, the wording of the Court's order in the *Avena case* was stronger than that of the *LaGrand case*. For example, in the *Avena case*, the Court indicated that the US 'shall' – rather than 'should' – take necessary measures to stay the execution of three Mexican nationals, whose execution were imminent, pending the final judgment of the ICJ.[20] This was an obvious reaction to the failure of the US to ensure that the Governor of Arizona stayed the execution of the *LaGrand* brothers and a reminder that provisional measures are binding orders. Indeed, if one was to look for a tangible contribution from the *LaGrand/Avena* litigation to international law, this

17 Ibid., at para. 115.

18 *Avena and Other Mexican Nationals (Mexico v. United States of America), Order of 5 February 2003, I.C.J. Reports 2003*, p. 77, para. 18.

19 *LaGrand, I.C.J. Reports 2001*, supra note 4, p. 466, para. 128(7).

20 *Avena and Other Mexican Nationals, Order of 5 February 2003*, supra note 18, p. 77, para. 59(a).

would be the resolution of the debate on whether ICJ provisional measures are binding.

The decision of the ICJ in the *Avena case* has given rise to a very large number of scholarly commentaries, and we defer to those for a detailed account. Here, it suffices to say that the main points of contention in the *Avena case* were the obligation to provide consular information 'without delay' in accordance with VCCR Article 36 (1)(b) and the failure to provide meaningful 'review and reconsideration' of the conviction and sentencing, given the violation. The Court declared that the term 'without delay' does not necessarily mean 'immediately upon arrest', rather, the duty to provide consular information exists once there are grounds to believe that the individual being detained is a foreign national.[21] Accordingly, the Court found that, with the exception of one detainee, the US had violated its obligation to provide consular notification 'without delay'.[22] The Court further found that, in 49 cases, the US had violated its obligations, to enable Mexican consular officers to communicate with, have access to, and visit their nationals, and in 34 cases, the US had violated its obligation to enable Mexican consular officers to arrange for legal representation of their nationals.[23]

Mexico claimed that the US had also violated Article 36(2) by failing to provide 'meaningful and effective review and reconsideration of the convictions and sentences'.[24] The US argued that in the American legal system courts are barred from doing so by the so-called 'procedural default rule', which is a doctrine applied by American federal courts that requires a state prisoner seeking a writ of *habeas corpus* in federal court to have first presented his or her federal law argument to the state courts in compliance with state procedural rules.[25] Failure to do so bars any subsequent attempt to present the same argument to the federal courts on collateral review. This argument led the Court to take a jab at the US, by observing that the procedural default rule had not been revised despite the ICJ's judgment in the *LaGrand case*, which called attention to the problems regarding the application of this rule caused for defendants relying on violations of the VCCR in appeal proceedings.[26]

However, the Court also gave the US a break by holding that, although Article 36(2) of the VCCR had been violated by the US in the cases of three Mexican nationals, judicial re-examination was still possible in the other 49 cases.[27] Accordingly, the Court found that Mexico's request for *restitutio in*

21 *Avena and Other Mexican Nationals (Mexico v. United States of America), Judgment*, supra note 5, p. 12, para. 85–8.
22 Ibid., at para. 89.
23 Ibid., at para. 90–91.
24 Ibid., at para. 107.
25 Ibid., at para. 110.
26 Ibid., at para. 112–113.
27 Ibid., at para. 20.

integrum – meaning partial or total annulment of the conviction and sentencing of the state court – would not be an appropriate remedy; rather, adequate reparations would be a 'review and reconsideration' of the convictions and sentencing of the Mexican nationals by courts in the US.[28] The Court also indicated that the methods employed for 'review and reconsideration' should be left to the US to determine, so long as the violations of rights guaranteed under the VCCR were taken into account.[29] The Court found no evidence indicating a 'regular and continuing' pattern of breaches by the US of the VCCR, and, in fact, it acknowledged the efforts undertaken by the US to encourage implementation of its obligations under the VCCR, and considered this as a demonstration of commitment by the US in satisfaction of Mexico's request.[30]

Finally, fully aware of its role in the architecture of the international community and the precedential value of its judgments, the Court observed that, while the *Avena case* concerned only the 52 Mexican nationals in question, its judgments could not be taken to imply that the Court's conclusions would not be extended to other foreign nationals finding themselves in similar situations in the US.[31] This statement not only put the US on notice as to how the Court would rule should it be seized for the fourth time by another state raising these same violations, but it also signalled to the rest of the world, both that the Court considers itself the ultimate judge for the international community, and that its rulings extend beyond the specific case and become part of the international legal regime.

Other than that and the clarification of what 'without delay' means under the VCCR, the *Avena* decision did not add much substance to the body of international law regulating consular relations. The paradigm shift had already happened with *LaGrand*. The *Avena case* was, as it has been poetically framed, 'a fight for the soul of *LaGrand*.'[32] However, the *Avena case* will go down in history as a turning point in the discussions of how international law and domestic law interface, and what place international law has within the US legal system.

IV The legal fallout in the US and the aftermath

The ICJ issued its ruling on 31 March 2004. Within a few days, both the US government and the attorneys representing the Mexican nationals took steps to ensure implementation of the ICJ decision. The ICJ had asked US courts, both state and federal, to 'review and reconsider' the convictions and sentences of the Mexican nationals who had not been provided consular

28 Ibid., at para. 123 and 148.
29 Ibid., at para. 122.
30 Ibid.
31 *Avena and Other Mexican Nationals, Judgment*, supra note 5, p. 12, para. 151.
32 C. Amirfar, op. cit., p. 378.

assistance without delay. A judicial melee ensued, one that left a major casualty on the battlefield: the reputation of the US as an international law-abiding nation.

The first case was that of Osvaldo Torres, one of the individuals named in the *Avena case*. His attorneys filed an application for post-conviction relief. On 13 May 2004, the Oklahoma Court of Criminal Appeals decided to grant an indefinite stay of execution and remanded an evidentiary hearing to assess whether Torres had suffered prejudice by the violations of his consular rights.[33] On the same day, and after having been requested to do so by the US Department of State, the Governor of Oklahoma decided to grant clemency and commute the death sentence to life without parole.[34] That rendered subsequent litigation of the case moot and it was accordingly dismissed by the Oklahoma Court of Criminal Appeals.[35]

In February 2005, President George W. Bush attempted to ensure the compliance of state courts with the ICJ decision by issuing a memorandum for the Attorney General 'determining that the US will discharge its international obligations under the decision of the International Court of Justice … by having State courts give effect to the decision in accordance with the general principles of comity in cases filed against 51 Mexican nationals addressed in that decision.'[36] While the US Government was taking steps to ensure compliance with the ICJ ruling, it also decided to shield itself from future VCCR litigation before the ICJ by formally withdrawing from the Optional Protocol to the VCCR concerning the Compulsory Settlement of Disputes, which had been the basis of the ICJ jurisdiction in the *Breard*, *LaGrand*, and *Avena* cases.[37]

Inevitably, the US Supreme Court was bound to weigh in on the issue, which it did in several rounds commencing with the *Sanchez-Llamas v. Oregon case*.[38] Moises Sanchez-Llamas, a Mexican national, had been convicted of attempted murder in Oregon. Mario Bustillo, a Honduran national, had been convicted of murder in Virginia. Both had not been sentenced to death, but to lengthy detention periods, and in neither case had the detainees' consulates been informed. Both Sanchez-Llamas and Bustillo filed *habeas* petitions arguing that their right to consular notification had been violated. However, the courts in Oregon and Virginia both ruled that, because these claims had not been argued at the trial court level, they were

33 *Torres v. State, No. PCD-04-442, (Okla. Crim. App. May 13, 2004).*

34 *Torres v. State*, 2005 OK CR 17, para. 2, 120 P.3d 1184, 1186.

35 Ibid.

36 Memorandum for the Attorney General from the President of the United States of America on Compliance with the Decision of the ICJ in Avena (28 February 2005).

37 S. Charnovitz, 'Correcting America's Continuing Failure to Comply with the Avena Judgment', *The American Journal of International Law* 106, 2012, 579.

38 *Sanchez-Llamas v. Oregon*, 548 U.S. 331 (2006).

procedurally barred. The two cases were then consolidated and brought before the US Supreme Court.[39]

On 28 June 2006, the Supreme Court decided that the ICJ's rulings in *LaGrand* and *Avena* were not conclusive for courts in the US: faced with the interpretation of an international agreement by an international court, all American courts needed to do was to give the decision 'respectful consideration', but by no means were they bound by it.[40] The Supreme Court also found that state courts could admit evidence against defendants even if the evidence was obtained in violation of the VCCR.[41] Article 36 claims, if not timely brought, could be procedurally barred by state procedural default rules.[42] Finally, the ICJ's interpretation of Article 36 of the VCCR, requiring that the rule of procedural default be set aside, was inconsistent with the adversarial nature of the American criminal justice system, so the US could not have intended to accept that interpretation when it negotiated and ratified the VCCR.[43] However, at this time the Supreme Court did not rule as to whether Article 36 created individual rights that require honouring in state criminal proceedings.[44]

Meanwhile, the case of one of the individuals named in *Avena*, José Ernesto Medellin Rojas, was pending before the Fifth Circuit Court. After the ICJ handed down its decision in *Avena*, the Fifth Circuit Court rejected the request of Medellin's attorneys for revision of his case, holding that it could not overrule the Supreme Court's *Sanchez-Llamas* finding that the ICJ's rulings have no binding effect on courts in the US, and adding that Article 36 of the VCCR does not create individually enforceable rights.[45]

Medellin asked Texas' courts to implement the memorandum issued by President Bush.[46] On 15 November 2006, the Texas Court of Criminal Appeals found that it was not bound by the memorandum as the President had no constitutional authority to order state courts to follow a decision of the ICJ.[47] The case was then referred to the US Supreme Court, which, by a vote of six to three, upheld the ruling, finding that neither the ICJ judgment in *Avena* nor the memorandum constituted directly enforceable federal law.[48] While the Supreme Court acknowledged that the *Avena* judgment was binding as a matter of international law,[49] it found, nonetheless, that as a matter

39 *State v. Sanchez-Llamas*, 338 Or. 267, 108 P.3d 573 (2005); *Bustillo v. Johnson*, 546 U.S. 1002 (2005) opinion after grant of cert. sub nom. *Sanchez-Llamas v. Oregon*, 548 U.S. 331 (2006).

40 *Sanchez-Llamas v. Oregon*, 548 U.S. 331, 333.

41 Ibid., at 350.

42 Ibid., at 359.

43 Ibid., at 334.

44 Ibid., at 342.

45 *Medellin v. Dretke*, 371 F.3d 270, 280 (5th Cir. 2004).

46 Ibid., at 663.

47 *Ex parte Medellin*, 223 S.W.3d 315, 335 (Tex. Crim. App. 2006).

48 Ibid., at 518 and 532.

49 Ibid., at 535.

of US domestic law, a treaty has no domestic effect unless Congress gave it such force, or it is self-executing.[50]

Article 94(1) of the UN Charter obligates each member of the United Nations 'to comply with the decision of the International Court of Justice in any case to which it is a party.' To add insult to injury, the US Supreme Court excused the US of this obligation by finding that Article 94(1) was neither self-executing nor had Congress implemented legislation to give ICJ decisions effect in US courts.[51] By issuing the memorandum President Bush had overstepped the limits of his powers, encroached on the prerogatives of Congress, and the rights of states of the Union.[52] The Court interpreted Article 94 of the UN Charter as a commitment of the US to ensure implementation, but one not intended to vest ICJ decisions with immediate legal effect in domestic courts. The Court based its finding on the fact that Article 94(2) contains a mechanism to ensure compliance with decisions of the ICJ: referral to the UN Security Council. The US Supreme Court conveniently glossed over the fact that self-referral to the Security Council by the US Government would be absurd, and referral by another state, such as Mexico, would be blocked by US veto power.

Mexico made a last-ditch effort to stop the execution of Medellin and four other nationals by returning to the ICJ with a request for interpretation of the *Avena* ruling (eventually rejected by the Court), and new provisional measures.[53] It asked the Court to request that the US undertake all necessary measures to ensure that the five nationals were not executed, and that the US inform the Court of all measures taken to that effect.[54] On 16 July 2008, the ICJ found that it had *prima facie* jurisdiction to adopt the provisional measures requested, and that there was enough urgency and risk of irreparable harm to warrant their adoption.[55] The order put the US Government between the proverbial rock and a hard place. On the one hand, it faced an order from an increasingly irritated ICJ, peeved by the repeated lack of compliance with its orders and decisions, and on the other, the Supreme Court's Medellin judgment had emasculated the White House, denying it a legal basis to request governors to stay executions. All the executive branch could do was appeal to state governments to graciously agree to stay executions. To add a further ironic twist to an already grotesque imbroglio, it is worth recalling that President George W. Bush had been governor of Texas before Governor Rick Perry, and he himself had refused to stay the execution

50 Ibid., at 505.
51 Ibid., at 508.
52 Ibid., at 528.
53 Request for interpretation of the Judgment of 31 March 2004 in the case concerning *Avena and Other Mexican Nationals (Mexico v. United States of America) (Mexico v United States of America), Provisional Measures, Order of 16 July 2008, I.C.J. Reports 2008* p. 311, para. 5–6.
54 Ibid., at para. 13.
55 Ibid., at para. 73–74.

of Mexicans who had not benefitted from consular assistance. Likewise, Governor Perry rejected calls from Mexico and President Bush. José Medellin was executed on 5 August 2008.

The decision of the ICJ on the Request of Interpretation of the Avena ruling, coupled with the withdrawal of the US from the Optional Protocol to the VCCR, barred any subsequent litigation of the dispute before the ICJ. However, the tussle between the executive, legislative, and judiciary branches, and between the federal government and state authorities, continued.

President Barack Obama was sworn in on 20 January 2009, the day after the ICJ decision. He came to power on a promise of radical departure from the policies of the Bush administration and with a broad mandate from the American electorate to effect change. In the judgment interpreting *Avena*, the ICJ had left the US '[a] choice of means … in the implementation of its obligation and, failing success within a reasonable period of time through the means chosen, it must rapidly turn to alternative effective means of attaining that result.'[56] Observers of the VCCR mix-up would have bet on the Obama administration to seize the moment to seek legislation from Congress to resolve the problem once and for all. However, facing a financial crisis of unprecedented magnitude, and entangled in two costly and bloody wars, the White House did not recognize this issue as a high priority. As a result, the problem dragged on.

In June 2011, Senator Patrick Leahy (D- Vermont), the Chairman of the Senate Judicial Committee, introduced legislation regarding US compliance with its obligations under the VCCR.[57] The Consular Notification Compliance Act intended to give jurisdiction to federal courts to review the cases of foreign nationals on death row in the US who had not received consular assistance as required.[58] It included those individuals named in the *Avena case*, who had not yet been executed, as well as future similar cases. But, while Leahy's bill languished in the Senate, the legal battle to stop executions raged on. On 7 July 2011, the US Supreme Court handed down another ruling in the case of Humberto Leal Garcia, a Mexican national who had not been given consular assistance and who had been sentenced to death. Leal claimed that his conviction had been vitiated by a violation of the VCCR.[59] Leal's attorneys, as well as the US Government, appearing as *amicus curiae*, asked the Supreme Court to stay the execution so that Congress could have the time to consider Leahy's bill.[60] In a 5 to 4 decision, the Court denied Leal's application for stay of execution.[61] It stood fast to its

56 Ibid., at para. 47.
57 Charnovitz, op. cit. n. 41, p. 577.
58 Ibid.
59 *Garcia v. Texas*, 131 S. Ct. 2866, 2867 (2011).
60 Ibid.
61 Ibid., at 2868.

Medellin ruling that international legal obligations are not binding unless Congress enacts them in a statute. Expressing scepticism that such legislation would ever be enacted, it dryly observed that its role is 'ruling with the law at present, not what it might become in the future.'[62] Despite calls to Texas from the President, the State Department, and Mexico, all asking for a last-minute reprieve, Leal was duly executed within just a few hours of the Supreme Court's ruling.

Finally, after years of wrangling, a first step was taken towards ensuring proper implementation of the VCCR in 2014 when the US Supreme Court issued an amendment to the Federal Rules of Criminal Procedure. Since 1 December 2014, under Title II, Rule 5 (d) 'If the defendant is charged with a felony, the judge must inform the defendant of the following: (F) that a defendant who is not a US citizen may request that an attorney for the government or a federal law enforcement official notify a consular officer from the defendant's country of nationality that the defendant has been arrested — but that even without the defendant's request, a treaty or other international agreement may require consular notification.'[63] As a result, notification must be provided to any defendant at their initial appearance, without attempting to determine the defendant's citizenship.[64]

Whether Article 36 of the VCCR creates individual rights that may be invoked in judicial proceedings, and what, if any, remedy may exist for their violation, remains an open question as far as the US legal system is concerned.[65] The amendment issued by the Supreme Court applies only to federal courts, while the various states of the Union approach the matter differently. Indeed, one of the consequences of the Medellin and Leal Garcia decisions by the US Supreme Court is that 'the United States is no longer one nation when it comes to honouring consular commitments because the rights received by a foreign national often depend on the state in which the individual is apprehended.'[66] The states of the Union most resistant to the reach of the VCCR happen to also be retentionist on capital punishment. Furthermore, they are predominantly located in the south, where most Mexican nationals tend to live. It is clear that Mexico has still a long way to

62 Ibid., at 2867–8; *Medellin I,* 552 U.S. at 523–24.

63 Fed. R. Crim. P. 5.

64 Ibid., see Committee Notes on Rules-2014 Amendment.

65 Ibid.

66 Charnovitz, op. cit. n. 41, p. 575. Charnovitz cites, as example, the Supreme Court of Massachusetts, which, in 2011, held that if an alien did not receive the notification required by Article 36 of the Vienna Convention, a challenge to the conviction may be made in a motion for a new trial. In so holding, the court 'acknowledge[d] and accept[ed] the conclusion of the ICJ regarding the obligation that art. 36 creates when clear violations of its notice protocols have been established, that is, to provide some process by which the soundness of a subsequent conviction can be reviewed in light of the violation.' *Commonwealth v. Gautreaux*, 458 Mass. 741, 751 (2011).

go before it can claim it obtained its desired outcome throughout the course of this protracted legal struggle.

As it has been aptly remarked, '[w]e have always known about the possibility that a [State of the Union] could cause a US treaty violation by refusing to comply with America's international obligations. What is new about the *Avena* affair is that such misbehaviour can persist even in the face of an ICJ judgment against the United States. As a result, the reputation of the United States for being a law-abiding nation has been undermined.'[67]

Eight years after the Medellin ruling, the Supreme Court remains unswayable. Congress, rather than the President or the federal judiciary, has the role of effectuating compliance with the ICJ when states' laws or courts place the US in violation of a treaty. Considering how divided and ineffective the US Congress has been since the beginning of the 2000s, it is unlikely that the international and constitutional crisis triggered by the *Avena case* and cases that succeeded it will be resolved any time soon.

Bibliography

Amirfar, Catherine, 'AALS Panel – Mexico v. U.S.A. (Avena) – Arguments of Mexico,' *European and International Law* 5, 2004, 376.

Bradley, Curtis A., 'Sanchez-Llamas v. Oregon – U.S. Supreme Court Decision on the Effect of Procedural Default on the Availability of Remedies for Violation of the Vienna Convention on Consular Relations, in Light of the ICJ Opinion in Avena, and on Suppression of Evidence as a Possible Remedy,' *American Journal of International Law* 100, 2006, 882.

Charnovitz, Steve, 'Correcting America's Continuing Failure to Comply with the Avena Judgment,' *American Journal of International Law* 106, 2012, 572.

Olabuenga, Pablo Arrocha, 'Caso Medellín vs. Texas: comentarios al fallo de la Suprema Corte de los Estados Unidos así como a la solicitud de interpretación del fallo Avena del 31 marzo de 2004, presentada a la Corte Internacional de Justicia por México en junio de 2008', *Anuario mexicano de derecho internacional* 9, 2009, 677–705.

67 Charnovitz, op. cit. n. 41, p. 576.

17 *Nottebohm (Liechtenstein v. Guatemala), 1951*

Alfred M. Boll[1]

The *Nottebohm Case (Liechtenstein v. Guatemala)*[2] involved issues related to a state's right to attribute its nationality by naturalization, the relationship between nationality in municipal and international law, and a state's obligation to recognize another state's attribution of nationality and the right of diplomatic protection and international claims. The majority held Liechtenstein's application inadmissible as Friedrich Nottebohm could not be considered its national for the purposes of international law, preventing the International Court of Justice (hereinafter 'ICJ' or 'Court') from considering the case's merits and whether Guatemala had committed wrongful acts against Liechtenstein and Nottebohm.

The case is perhaps best known for the criticism that has been levied against the judgment. Most authors draw its relevance narrowly and qualify its significance as limited due to what has been called 'unconvincing' reasoning.[3] Weis calls the case's circumstances 'quite exceptional', appending the adage 'bad cases make bad law'.[4] The Court made several sweeping, arguably questionable, pronouncements on the nature of nationality and naturalization, and the case is, however, still frequently cited by authors as indicative of broad rules of international law with respect to nationality.[5] Egües, for

1 The author would like to thank Joan Sherer of the State Department's Ralph J. Bunche Library and Lisa Chiapputo, University of Helsinki, for their assistance. The views expressed herein are the author's and do not necessarily represent those of his employer.

2 *Nottebohm Case (Liechtenstein v. Guatemala) (Second Phase), Judgment, I.C.J. Reports 1955*. During the case's first phase the Court had held that it was entitled to hear the case, Guatemala having argued the Court's jurisdiction had expired. *Nottebohm Case (Liechtenstein v Guatemala) (Preliminary Objections), I.C.J. Reports 1953.*

3 O. Dörr, 'Nottebohm case', in *Max Planck Encyclopedia of Public International Law*, Heidelberg: Max Planck Foundation for International Peace and the Rule of Law, 2015. For a list of critical commentary, see J. Kunz, 'The Nottebohm judgment', *American Journal of International Law* 54, 1960, 537–8, note 2.

4 Weis, P., *Nationality and Statelessness in International Law*, Alphen aan den Rijn: Sijthoff & Noordhoff, 1979, p. 180.

5 Although Sloane's characterization of the case as having become 'a kind of doctrinal mantra' appears overstated. R. Sloane, 'Breaking the genuine link: The contemporary international regulation of nationality', *Harvard International Law Journal* 50, 2009, 3–4.

example, cites *Nottebohm*'s definition of nationality to defend Argentina's attribution of nationality by exercise of option to children born abroad of Argentine parents.[6] In Peru, press reports cited the case in relation to former President Alberto Fujimori's Peruvian nationality following his flight to Japan in 2000 amidst allegations of corruption and human rights violations.[7] Aleinikoff and Klusmeyer argue that 'the selling of nationality unquestionably cheapens its value as a form of allegiance', and oppose the practice on the basis that:

> the emerging consensus in international law reflected in the Nottebohm case suggests that a substantial connection should tie the individual to the state of which he or she is a national. Nationality should not become a commodity that individuals can purchase to further their business or personal interests.[8]

Parts of the judgment related to states' right to attribute their nationality are broadly approved of, as well as the distinction the Court drew between nationality on the planes of municipal and international law.[9] Whether *Nottebohm* can in fact be said to reflect an emerging consensus in international law that individuals should have a substantial connection to their states of nationality as postulated by Aleinikoff and Klusmeyer is, however, doubtful.[10]

The case is often cited and confused as dealing with issues of multiple nationality, although it does not. This is because the Court applied a test that is commonly used by governments and courts in cases involving multiple nationality, that of ***genuine or effective nationality***, to facts involving someone with only one nationality. An example of such confusion is the Brazilian Supreme Federal Court's analysis citing *Nottebohm* in denying a petition in favour of Andréa Ciaccio in 2004, a Brazilian–Italian dual national.[11]

6 A. Egües, 'El principio de la "nacionalidad natural"', *La Ley* C, 1990, 706.

7 C. Romero, *Alberto Fujimori es peruano y no japonés*, Lima: La Republica, 20 June 2001. Online. Available http://larepublica.pe/20-06-2001/alberto-fujimori-es-peruano-y-no-japones (accessed 1 October 2015).

8 T. Aleinikoff and D. Klusmeyer, 'Plural nationality: facing the future in a migratory world', in T. Aleinikoff and D. Klusmeyer, (eds), *Citizenship today. Global perspectives and practices*, Washington, DC: Carnegie Endowment for International Peace, 2001, p. 84. See also R. Koslowski, 'Challenges of international cooperation in a world of increasing dual nationality', in D. Martin and K. Hailbronner (eds), *Rights and duties of dual nationals. Evolution and prospects*, The Hague: Kluwer, 2003, pp. 157–82.

9 Dörr, op. cit.

10 Ibid. See also Weis, op. cit., pp. 178–80; A. Randelzhofer, 'Nationality', in R. Bernhardt (ed.), *Encyclopedia of Public International Law* 8, Amsterdam: Elsevier, 1985, p. 421.

11 Supremo Tribunal Federal, 'Arquivado HC de ítalo–brasileiro', Brasilia: 26 August 2004. Online. Available www.stf.jus.br/portal/cms/verNoticiaDetalhe.asp?idConteudo=63327&caixaBusca=N (accessed 25 August 2015).

I The background and circumstances of the case

Nottebohm, a German national by birth in 1881, went to Guatemala when he was about 24 years old to join his brothers' businesses in trade, banking, and plantations.[12] The business prospered and Nottebohm became head of the firm in 1937. He lived in Guatemala as a permanent resident German national for 34 years, occasionally traveling to Germany on vacation, including to Liechtenstein to visit a brother who lived there and had become a national. In 1939, before the outbreak of the Second World War, he travelled to Liechtenstein where he acquired its nationality by naturalization, thereby losing his German nationality under German law. He returned to Guatemala in January of 1940 on a Liechtenstein passport.

Guatemala declared its neutrality but shortly thereafter joined the allies, declaring war on Germany on 11 December 1941. Nottebohm was arrested as an enemy alien in 1943, transferred to a US military base, and subsequently interned in the US. Buys recounts the history of what she calls the US-Latin American Detention Program under which various Latin American countries and the US cooperated to deport over 4,000 German nationals or persons of German ancestry to the US and intern them during the war, for security, economic, and other reasons.[13]

According to press reports, Nottebohm's assets were estimated at US$1.5 million,[14] or some US$20.6 million today.[15] During his detention, Guatemala used Nottebohm's status as an enemy alien to take custody of his property. His company's assets in the US were also blocked. The US Government later concluded it did not have credible evidence of Nottebohm's alleged Nazi sympathies or activities, and released him in 1946.[16] When Guatemala refused him permission to return, he went to Liechtenstein. In 1950, the US Government settled a civil suit brought by the Nottebohms, returning about half the value of what it had seized.[17] Guatemala, however, kept his property, although it returned 16 coffee plantations to Nottebohm's family in 1962 after his death.[18] In 1951, Liechtenstein brought suit against Guatemala in the ICJ for the purportedly illegal confiscation of its national's property, and the treatment he suffered.

12 *Nottebohm Case*, op. cit. p. 13.

13 C. Buys, 'Nottebohm's nightmare: Have we exorcised the ghosts of WWII detention programs or do they still haunt Guantanamo?', *Chicago-Kent Journal of International and Comparative Law* 11, 2011, 15.

14 'World Court hears war seizure case', *New York Times*, 11 February 1955, 9.

15 United States Department of Labor, Bureau of Labor Statistics, CPI Inflation Calculator. Online. Available www.bls.gov/data/inflation_calculator.htm (accessed 4 October 2015).

16 Buys, op. cit., p. 11.

17 Ibid., p. 13.

18 Ibid., p. 14.

Liechtenstein claimed reparations for Nottebohm's arrest, detention, expulsion, non-readmission, and retention of his property without compensation. It asked the Court to declare that its naturalization of Nottebohm was not contrary to international law, and that its claim on Nottebohm's behalf was admissible.[19] Guatemala argued that Nottebohm's acquisition of Liechtenstein nationality had not accorded with that country's municipal laws; his naturalization had not been in accordance with general principles of law on nationality; and that his intent in acquiring Liechtenstein nationality had been to obtain the status of a neutral, not to establish a durable link with Liechtenstein and to break his ties to Germany, and thus fraudulent.[20] The conditions of Nottebohm's naturalization in Liechtenstein included swearing an oath of allegiance, and payment of a substantial fee and an annual tax, including a security deposit. A provision was applied that waived the requirement of at least three years' previous residence. Nottebohm never resided permanently in Liechtenstein before his return to Guatemala in 1940, and did not spend any substantial amount of time there. Most of his property was confiscated in 1949 or later, after he had lived permanently in Liechtenstein for some three years.

II The judgment

Eleven of the Court's 14 judges held that Liechtenstein's claim was inadmissible, ruling that Guatemala was not obligated to recognize Nottebohm's Liechtenstein nationality given the circumstances under which it had been granted, and that as a consequence Liechtenstein was not entitled to extend its protection vis-à-vis Guatemala.

A. States' right to attribute their nationality

The Court first rejected arguments that Guatemala had recognized Liechtenstein's naturalization of Nottebohm and thus should be estopped from arguing it was invalid. Guatemala had issued Nottebohm a visa in his Liechtenstein passport; the Guatemalan Ministry of External Affairs had approved Nottebohm's request that his nationality be amended in the Register of Aliens and in his identity document; the Guatemalan Civil Registry had issued him a certificate confirming the change.[21] The Court rejected these arguments as related to control of aliens rather than recognition of nationality.[22]

The Court then proceeded to issue its first sweeping pronouncement on nationality, related to states' right to attribute their nationality:

19 *Nottebohm Case*, op. cit., pp. 6–9 and 16.
20 Ibid., p. 11.
21 Ibid., p. 17.
22 Ibid., p. 19.

> It is for Liechtenstein, as it is for every sovereign State, to settle by its own legislation the rules relating to the acquisition of its nationality and to confer that nationality by naturalisation granted by its own organs in accordance with that legislation.[23]

It then qualifies this statement by saying 'It is not necessary to determine whether international law imposes any limitations on its freedom of decision in this domain.'[24] The Court then points out that nationality's most important and relevant effects are purely domestic in nature, identifying which individuals enjoy rights and are bound by obligations under municipal law. The Court thus both appears to reject Guatemala's argument that Nottebohm's naturalization was invalid under general principles of law, while holding open the possibility that international law poses limits on states' freedom in the matter. Although Weis interprets this as an explicit refusal to consider the validity of Nottebohm's naturalization according to the laws of Liechtenstein, the wording does not indicate the naturalization was invalid domestically.[25] In that sense it is doubtful that the ruling can be cited for arguments that naturalization practices such as Liechtenstein's at the time are *per se* invalid. This affects a considerable number of states, whose conditions for naturalization fall along a wide spectrum.[26]

B. Nationality in municipal and international law

Having accepted Liechtenstein's presentation of Nottebohm as a national by naturalization, the Court then addressed whether this entitled Liechtenstein to exercise diplomatic protection with respect to Guatemala. Here, its holding was unambiguous. Nationality's effects on the international plane depend on international law, not municipal law.

> International practice provides many examples of acts performed by States in the exercise of their domestic jurisdiction which do not necessarily or automatically have international effect, which are not necessarily and automatically binding on other States or which are binding on them only subject to certain conditions.[27]

23 Ibid., p. 20.
24 Ibid., p. 20.
25 Weis, op. cit. p. 179.
26 For various examples, see Kälin, C. *The global residence and citizenship handbook*, Zürich: Ideos, 2011. Koslowski warns that many such practices amount to a 'commodification' of citizenship. Koslowski, op. cit., p. 174. On the question whether international law imposes limits on states' regulation of nationality see Donner, R., *The regulation of nationality in international law*, Irvington-on-Hudson: Transnational Publishers, 1994, pp. 119–20.
27 *Nottebohm Case*, op. cit., p. 21.

The Court also recalled the specific context of diplomatic protection and international claims as measures related to states' rights vis-à-vis one another.[28]

C. The application of the principle of the genuine or effective link

The Court then turned to the central issue of how to determine 'whether the naturalisation conferred on Nottebohm can be successfully invoked against Guatemala'.[29] The Court cites the principles applied by international arbiters confronted with a nationality claim by one state that is contested by another, to assess whether the nationality in question gives rise to an international obligation, citing one of the standards applied in cases of dual nationality:

> They have given their preference to the real and effective nationality, that which accorded with the facts, that based on stronger factual ties between the person concerned and one of the States whose nationality is involved.[30]

It posited that a 'real and effective' nationality with effect on the international plane is one based on strong personal ties to a state compared with other states. At another point in the judgment the Court says that nationality as a legal bond must 'accord with the individual's genuine connection with the State'.[31]

Translating this to Nottebohm's particular circumstances, the Court asked:

> At the time of his naturalisation does Nottebohm appear to have been more closely attached by his tradition, his establishment, his interests, his activities, his family ties, his intentions for the near future to Liechtenstein than to any other State?[32]

The Court's answer was a resounding 'no', describing Nottebohm's ties to Liechtenstein as 'extremely tenuous'. Its reasoning turned on three points:

28 Ibid., p. 24. Santulli concludes that absent a determination of valid nationality by the state concerned, the nationality is invalid on the international plane. Santulli, C., *Irrégularités internes et efficacité internationale de la nationalité*, Paris: Université Panthéon-Assas, 1995.

29 *Nottebohm Case*, op. cit., p. 21.

30 Ibid., p. 22.

31 Ibid., p. 23. Aghahosseini emphasizes the importance of terminology, defining a nationality as 'valid' if it has been regularly acquired in municipal law; 'effective' if it should be given international effect for a specific purpose; and 'dominant' if it represents stronger ties between two 'valid' and 'effective' nationalities. Aghahosseini, M., *Claims of dual nationals and the development of customary international law. Issues before the Iran-United States Claims Tribunal*, Leiden: Martinus Nijhoff, 2007, p 80.

32 *Nottebohm Case*, op. cit., p. 24.

Nottebohm had always been a German national and had maintained ties to Germany – nothing in Nottebohm's application for naturalization demonstrated a desire to 'disassociate himself from the Government of his country'; Nottebohm's ties to Guatemala had been the same for 34 years and his naturalization did not change their strong nature; Nottebohm had no intention to settle, remain, or locate business interests in Liechtenstein.[33] The Court then contrasts the absence of 'any bond of attachment between Nottebohm and Liechtenstein' to 'a long-standing and close connection between him and Guatemala', to conclude that his naturalization was 'lacking in the genuineness requisite to an act of such importance, if it is to be entitled to be respected by a State in the position of Guatemala.'[34] The Court comes close to saying Nottebohm's intent was fraudulent, but does not, saying that his intent was to substitute the nationality of a belligerent state for that of a neutral state, while not 'becoming wedded to its traditions, its interests, its way of life'.[35] The Court appears to allude to Guatemala's allegations that Nottebohm had been a Nazi sympathizer, which later writers have questioned.[36] In any case, the judgment extended the principle of the genuine or effective link from the context of multiple nationality to that of a single nationality.

III The dissenting opinions and criticism of the judgment

The substantial criticism that has been levied at the Court's reasoning largely follows the careful and sharp dissenting opinions of Judges Guggenheim, Read, and Klaestad. All posit that the Court should not have gone further than to ask whether Nottebohm's naturalization had been valid under the laws of Liechtenstein, with the exception of fraudulent intent in acquisition, such as in order to conceal enemy property in wartime.[37]

Judge Read highlights that states are entitled to great deference in determining who their nationals are.

> I do not think that international law, apart from abuse of right and fraud, permits the consideration of the motives which led to naturalisation as determining its effects.[38] …
>
> [T]he practice of States has steadfastly rejected vague and subjective tests for the right to confer nationality – sincerity, fidelity, durability, lack of substantial connection.[39]

33 Ibid., p. 25.
34 Ibid., p. 26.
35 Ibid.
36 See generally Buys, op. cit.
37 *Nottebohm Case*, op. cit., pp. 57 and 65.
38 Ibid., p. 42.
39 Ibid., p. 46.

While all dissenting opinions follow the Court's reasoning that nationality's effects differ on the municipal and international planes, the dissent's emphasis that the status of nationality is one of law unrelated to subjective or emotional standards, highlights its argument that '[t]he general rule of international law is that nationality gives rise to a right of diplomatic protection'.[40] Judge Guggenheim emphasizes that cases in which nationality might be invalid on the international plane and for purposes of diplomatic protection are limited to cases of multiple nationality, when nationality was imposed without the individual's consent, or if a state still validly considers an individual its national after he or she has acquired another nationality.[41] Guggenheim calls the majority's use of the standard of a genuine or effective link an attempt to replace objective criteria with subjective considerations with no basis in international practice and that can lead to arbitrary decisions. He points out that under international law states are free to attribute their nationality to descendants of nationals through unlimited generations, even those who have lived abroad for centuries and who have no other link to the state.

The Court's emphasis that its holding was limited to Liechtenstein's protection of Nottebohm vis-à-vis Guatemala, appears questionable. What was Nottebohm's nationality for purposes of international law, if not that of Liechtenstein? It was uncontroverted that he had lost his German nationality under German law upon his naturalization in Liechtenstein, although he could have applied to retain it.[42] Judges Read and Guggenheim point out that the ruling meant Nottebohm had no state to protect him, leaving him stateless for the purposes of international law. 'Such a dissociation of nationality from diplomatic protection is not supported by any customary rule nor by any general principle of law'.[43]

Judge Guggenheim points out that the Court's holding places the whole system of diplomatic protection in jeopardy for individuals whose nationality might be disputed. He cites this as contrary to the Universal Declaration of Human Rights' provisions on the right to a nationality, and efforts to combat statelessness.[44] Far from dismissing diplomatic protection as less important than in the past, the International Law Commission's (ILC) Special Rapporteur on diplomatic protection declared in 2000 that 'as long as the State remained the dominant actor in international relations, the espousal of claims by States for violations of the rights of their nationals remained the most effective remedy for human rights protection'.[45]

40 Ibid., p. 46.

41 Ibid., p. 54.

42 Weis, op. cit. p. 179.

43 *Nottebohm Case*, op. cit., p. 60.

44 Ibid., p. 63.

45 International Law Commission, *Yearbook of the International Law Commission,* 2000, p. 38. Online. Available http://legal.un.org/ilc/publications/yearbooks/english/ilc_2000_v1.pdf (accessed 1 October 2015).

Whereas the majority expressed the issue in terms of states' rights vis-à-vis one another, Judge Read contemplates Nottebohm's position as an individual seeking redress. He emphasizes the circumstances of Nottebohm's arrest in Guatemala, his deportation and internment in the US without a trial in either country or provision for him to defend himself, and the severity and extent of the legal proceedings commenced against him to expropriate his property without compensation.[46] Alluding to the above-mentioned settlement of Nottebohm's civil suit against the US, Read implicitly contrasts Nottebohm's treatment in the US ('Since his release, he has been accorded full civil rights in the United States and has exercised them freely')[47] with his treatment in Guatemala ('He was allowed no opportunity to give evidence of the charges made against him, or to confront his accusers in open court').[48]

The dissent highlighted that Nottebohm's claims stemmed from the period after 1946 when Nottebohm was in Liechtenstein, and most from the period after 1951 when Liechtenstein brought its case against Guatemala.[49] Following the Court's reasoning, was Nottebohm's nationality still not effective at that point in time? The Court did not discuss various facts that appear relevant, among them that it appears Nottebohm was ineligible to apply for Guatemalan citizenship.[50] While Nottebohm's nephews and business partners took advantage of new legislation in 1938 that allowed Guatemalan-born German citizens to be naturalized, Nottebohm, who was born in Germany, was excluded.[51] His nephews' Guatemalan nationality, however, did not protect them. Both were arrested in December 1942 and transferred to the US as enemy aliens.[52]

IV The court's pronouncements on the nature of nationality and naturalization

The majority's pronouncements on the nature of nationality and naturalization have been widely cited, arguably incorrectly, as positing rules of international law. The Court appeared to define nationality as:

> a legal bond having as its basis a social fact of attachment, a genuine connection of existence, interests and sentiments, together with the existence of reciprocal rights and duties. It may be said to constitute the juridical expression of the fact that the individual upon whom it is

46 *Nottebohm Case*, op. cit., p. 34.
47 Ibid., p. 41.
48 Ibid., p. 35.
49 Ibid., pp. 31 and 61.
50 Buys, op. cit., p. 5. Leigh is unaware of this possibility but sees the fact as relevant. G. Leigh, 'Nationality and diplomatic protection', *International and Comparative Law Quarterly* 20, 1971, 468.
51 Buys, op. cit., p. 5.
52 Ibid., p. 8.

> conferred, either directly by the law or as the result of an act of the authorities, is in fact more closely connected with the population of the State conferring nationality than with that of any other State.[53]

Randelzhofer does not consider that the ICJ intended this to be a general definition of nationality, and says it is limited to the issue of 'diplomatic protection in the particular case of conferment of nationality by naturalisation'.[54] The author has previously argued that state practice that tolerates and even incorporates multiple nationality means the majority's definition above can only be a definition of nationality in municipal law.[55] The Court's definition may be contrasted with Weis' definition of nationality in international law, which corresponds to that laid out in the dissenting opinions, and in the author's opinion is evidenced by state practice, as:

> a technical term denoting the allocation of individuals, termed nationals, to a specific state – the state of nationality – as members of that State, a relationship which confers upon the State of nationality … rights and duties in relation to other states.[56]

The majority's characterization of naturalization as involving the 'breaking of a bond of allegiance and [the] establishment of a new bond of allegiance'[57] is also clearly faulty, given clear and widespread practice to the contrary.[58]

V The case's reception and application

The 1958 Flegenheimer Claim before the United States-Italian Conciliation Commission rejected the reasoning in *Nottebohm*, limiting the case to its facts and holding that the effective link test should not be applied to persons possessing a single nationality given the large number of persons whose legal nationality does not correspond to their domicile, economic, or personal interests or ties.

> But when a person is vested with only one nationality … the theory of effective nationality cannot be applied without the risk of causing confusion. … There does not in fact exist any criterion of proven effectiveness for disclosing the effectiveness of a bond with a political collectivity.[59]

53 *Nottebohm Case*, op. cit., p. 23.
54 Randelzhofer, op. cit., p. 421.
55 Boll, A., *Multiple nationality and international law*, Leiden: Martinus Nijhoff, 2007, p. 297.
56 Weis, op. cit. p. 59.
57 *Nottebohm Case*, op. cit., p. 24.
58 Boll, op. cit.
59 *Flegenheimer Claim, (Italian-United States Conciliation Commission), International Law Reports 1958 25*, p. 150.

Like the Flegenheimer Commission and Judge Guggenheim, most commentators maintain that a generalization of the ICJ's test in *Nottebohm* would lead to uncertainty in issues surrounding nationality of claims and diplomatic protection.[60] Applying a test of effective nationality to naturalized persons only belies the fact that even native-born nationals may have no 'genuine link' to their state of nationality.

The Iran-United States Claims Tribunal cites *Nottebohm* in its 1984 judgment holding it would apply the rule of dominant and effective nationality in cases involving dual nationality, including when one of the nationalities is that of the respondent state.[61] Aghahosseini argues it is unclear whether the Court intended to extend its judgment to such cases.[62] Interestingly, the three dissenting judges also cited *Nottebohm* for their argument that the Tribunal should have applied the rule of non-responsibility instead. They argue that *Nottebohm*'s requirement of proof of a genuine link even in cases of a single nationality reinforces the notion that states should not be allowed to bring claims against one another when they are equally placed, that is, when a claimant possesses both their nationalities.[63]

The ILC also specifically rejected *Nottebohm* in its draft rules on diplomatic protection. The Commentary specifies that

> Draft article 4 does not require a State to prove an effective or genuine link between itself and its national, along the lines suggested in the Nottebohm case ... If the genuine link requirement proposed by Nottebohm was strictly applied it would exclude millions of persons from the benefit of diplomatic protection.[64]

It should also be noted that commentators do not interpret *Nottebohm* as changing the standards applicable to corporate nationality.[65]

VI Conclusion

The *Nottebohm* case is perhaps most significant in terms of the confusion it has sown. Notwithstanding attempts to explain the Court's reasoning, its proposal that a 'genuine' link between a national and his or her state be a

60 Weis, op. cit. pp. 180–1.

61 *Iran v. the United States, Judgment, Iran-U.S. C.T.R. 5, 251.* See Aghahosseini, op. cit., p. 39.

62 Ibid., pp. 98–100.

63 Ibid., pp. 48 and 96.

64 International Law Commission, *Draft articles on diplomatic protection with commentaries*, 2006, pp. 31 and 33. Online. Available http://legal.un.org/ilc/texts/instruments/english/commentaries/9_8_2006.pdf (accessed 1 October 2015).

65 D. Harris, 'The protection of companies in international law in the light of the Nottebohm case', *International and Comparative Law Quarterly* 18, 1969, 285.

condition of diplomatic protection at international law has not been followed in practice. Still, its clear separation of nationality in municipal and international law has met with definite approval, as has its confirmation of the rule that states have wide latitude in attributing their nationality. Its description of nationality, which can only be a definition of nationality in municipal law, is unfortunately likely to continue to cause confusion in relation to norms of international law. The case's political significance also appears limited, although the Federal Republic of Germany refused to establish diplomatic relations with Guatemala as a consequence, only reopening its embassy in 1960.[66] The rejection by the ILC and in subsequent practice of the genuine link in relation to a single nationality reflects the need for international law to respond to individuals' basic rights and to ensure the access to justice that Mr Nottebohm was arguably denied.

Bibliography

Aghahosseini, M., *Claims of dual nationals and the development of customary international law. Issues before the Iran-United States Claims Tribunal*, Leiden: Martinus Nijhoff, 2007.

Aleinikoff, T. and Klusmeyer, D., 'Plural nationality: facing the future in a migratory world,' in T. Aleinikoff and D. Klusmeyer (eds), *Citizenship today. Global perspectives and practices,* Washington, DC: Carnegie Endowment for International Peace, 2001, pp. 63–88.

Boll, A., *Multiple nationality and international law*, Leiden: Martinus Nijhoff, 2007.

Buys, C., 'Nottebohm's nightmare: Have we exorcised the ghosts of WWII detention programs or do they still haunt Guantanamo?', *Chicago-Kent Journal of International and Comparative Law* 11, 2011, 1–73.

Donner, R., *The regulation of nationality in international law*, Irvington-on-Hudson: Transnational Publishers, 1994.

Dörr, O., 'Nottebohm case', in *Max Planck Encyclopedia of Public International Law*, Heidelberg: Max Planck Foundation for International Peace and the Rule of Law, 2015.

Egües, A., 'El principio de la "nacionalidad natural"', *La Ley* C, 1990, 706.

Flegenheimer Claim, (Italian-United States Conciliation Commission), International Law Reports 1958 25, p. 91.

Harris, D., 'The protection of companies in international law in the light of the Nottebohm case', *International and Comparative Law Quarterly* 18, 1969, 275–317.

International Law Commission, *Draft articles on diplomatic protection with commentaries*, 2006, pp. 31 and 33. Online. Available http://legal.un.org/ilc texts/instruments/english/commentaries/9_8_2006.pdf (accessed 1 October 2015).

66 Buys, op. cit. p. 14.

International Law Commission, *Yearbook of the International Law Commission*, 2000, p. 38. Online. Available http://legal.un.org/ilc/publications/yearbooks/english/ilc_2000_v1.pdf (accessed 1 October 2015).

Iran v. the United States, Judgment, Iran-U.S. C.T.R. 5, 251.

Kälin, C. *The global residence and citizenship handbook*, Zürich: Ideos, 2011.

Koslowski, R., 'Challenges of international cooperation in a world of increasing dual nationality', in D. Martin and K. Hailbronner (eds), *Rights and duties of dual nationals. Evolution and prospects*, The Hague: Kluwer, 2003, pp. 157–82.

Kunz, J., 'The Nottebohm judgment', *American Journal of International Law* 54, 1960, 536–71.

Leigh, G., 'Nationality and diplomatic protection', *International and Comparative Law Quarterly* 20, 1971, 453–75.

Nottebohm Case (Liechtenstein v Guatemala) (Preliminary Objections), I.C.J. Reports 1953, p. 111.

Nottebohm Case (Liechtenstein v Guatemala) (Second Phase), I.C.J. Reports 1955, p. 4.

Randelzhofer, A., 'Nationality', in R. Bernhardt (ed.), *Encyclopedia of Public International Law* 8, Amsterdam: Elsevier, 1985, pp. 416–24.

Romero, C., *Alberto Fujimori es peruano y no japonés*, Lima: La Republica, 20 June 2001. Online. Available http://larepublica.pe/20-06-2001/alberto-fujimori-es-peruano-y-no-japones (accessed 1 October 2015).

Santulli, C., *Irrégularités internes et efficacité internationale de la nationalité*, Paris: Université Panthéon-Assas, 1995.

Sloane, R., 'Breaking the genuine link: The contemporary international regulation of nationality', *Harvard International Law Journal* 50, 2009, 1–60.

Supremo Tribunal Federal, 'Arquivado HC de ítalo-brasileiro', Brasilia: 26 August 2004. Online. Available www.stf.jus.br/portal/cms/verNoticiaDetalhe.asp?idConteudo=63327&caixaBusca=N (accessed 25 August 2015).

Weis, P., *Nationality and statelessness in international law*, Alphen aan den Rijn: Sijthoff & Noordhoff, 1979.

'World Court hears war seizure case', *New York Times*, 11 February 1955, 9.

Part IV

Material contribution of Latin American cases to the development of international law

International peace and security

18 *Military and Paramilitary Activities in and against Nicaragua (Nicaragua v. United States of America), 1986*

Efthymios Papastavridis[1]

I Introduction

If there is a case in public international law that every law student remembers or is called to remember, it is the *Nicaragua case*. The Judgment of the International Court of Justice (hereinafter 'Court' or 'ICJ') in the case of *Military and Paramilitary Activities in and against Nicaragua* (1986)[2] has been monumental in the history of international law in many facets and certainly insofar as the development of the law of use of force is applied. Much ink has been spilled and even more discussion occasioned in the wake of this judgment,[3] while many subsequent international judgments have referred to it as an authoritative source of law.[4] This comes as no surprise bearing in mind that the *Nicaragua case* did address quite significant issues, including *inter alia* issues in relation to the sources of international law and, in particular, the formation of international customary law,[5] questions of

1 The author would like to thank Dr Panos Merkouris (Groningen University, The Netherlands) for reading and commenting on a previous draft of this chapter. The usual disclaimer applies.

2 *Military and Paramilitary Activities in and against Nicaragua (Nicaragua v. United States of America). Merits, Judgment. I.C.J. Reports 1986*, p. 14 (hereinafter: 'Nicaragua case').

3 See *inter alia* McDonald, R. St., 'The Nicaragua case: New Answers to Old Questions?' *Canadian Yearbook of International Law* 24, 1986, 127; Moore, N., 'The Nicaragua case and the Deterioration of World Order', *American Journal of International Law* 81, 1987, 151.

4 See most notably *Legality of the Threat or Use of Nuclear Weapons, Advisory Opinion, I.C.J. Reports 1996*, p. 226 [hereinafter: 'Nuclear Weapons Opinion']; *Oil Platforms (Islamic Republic of Iran v. United States of America), Judgment, I.C.J. Reports 2003*, p. 161 [hereinafter: 'Oil Platforms case']; *Armed Activities on the Territory of the Congo (Democratic Republic of the Congo v. Uganda), Judgment, I.C.J. Reports 2005*, p. 168 [hereinafter: 'Armed Activities case']; *Case Concerning the Application of the Convention on the Prevention and Punishment of the Crime of Genocide (Bosnia and Herzegovina v. Serbia and Montenegro), Merits Judgment, I.C.J. Reports 2007*, p. 43 (hereinafter: 'Genocide Convention case').

5 See *Nicaragua case*, paras 98–101. See comments in Kirgis, F. L., 'Custom on a Sliding Scale' *American Journal of International Law* 81, 1987, 146.

international responsibility[6] and of the law of the sea.[7]

However, *jus ad bellum* is the field of international law that, beyond any doubt, has been dramatically influenced by the *Nicaragua case*. Being the first dispute concerning use of force in the post-Charter era, the *Nicaragua case* presented a unique opportunity to the ICJ to break ground in many respects. In particular, the Court acknowledged for the first time that the use of inter-state force has been proscribed not only in the Charter but also in customary international law.[8]

Secondly, it authoritatively set forth that self-defence would warrant only measures that are proportional to the armed attack and necessary to respond to it as a rule well established in customary international law, even though not mentioned in Article 51 of the UN Charter.[9] The two conditions of necessity and proportionality were reaffirmed by the Court in its *Nuclear Weapons Opinion*[10] and in its judgments in the *Oil Platforms case*[11] and in the *Armed Activities case*,[12] while the *Institut de Droit International* (IDI) in its Santiago Resolution of 2007 declared that 'necessity and proportionality are essential components of the normative framework of self-defence'.[13]

In addition, the Court also addressed the right of collective self-defence and held that it is well established both in the Charter and in customary international law.[14] It did also set out the conditions for its lawful invocation and exercise.[15] Furthermore, the Court underscored the importance of reporting the exercise of self-defence to the Security Council, as required by Article 51 of the UN Charter. It held that the absence of a report might be one of the factors indicating whether the state in question was itself convinced that it was acting in self-defence.[16]

Besides the aforementioned general and, more or less, undisputed propositions of the Court regarding *jus ad bellum,* the Court did touch upon some other issues that still elicit significant discussion, namely the case of indirect armed attack, the permissibility of counter or any other measures in response to forcible action short of armed attack and the question of intervention

6 See, for example, the celebrated 'effective control' test regarding the attribution of the conduct of private persons to a state, which was posited by the Court in the *Nicaragua case*, para. 112 and reiterated in the *Genocide Convention case*, at 388, 394.

7 With regard to the question whether there is a customary right of access to ports the Court enunciated that 'It is also by virtue of its sovereignty that the coastal State may regulate access to its port', *Nicaragua case*, para. 213.

8 Ibid., para. 99.

9 Ibid., para. 176.

10 See *Nuclear Weapons Opinion*, para. 41.

11 See *Oil Platforms case*, para. 73.

12 See *Armed Activities case*, para. 147.

13 *Institut de Droit International*, Resolution, 'Self-Defence', *Annuaire IDI* 72, 2007, 233.

14 See *Nicaragua case*, para. 194.

15 Ibid., para. 195.

16 Ibid., para. 200.

upon invitation. Although the Court's pronouncements on these issues have been widely alluded to by international scholarly writings as authoritative statements of the relevant law, there is certainly room for different opinion, especially in light of subsequent developments. More importantly, these Court's pronouncements strike at the heart of contemporary international and non-international conflicts, such as those currently taking place in Syria and Iraq and thus certainly invite discussion.

It is the purpose of the present chapter to revisit these *dicta* of the Court in Nicaragua in view of the current international doctrine and state practice. It goes without saying that an exhaustive treatment of these issues is beyond the spatial confines of the present chapter; however, there will be an attempt to encapsulate the main arguments in relation to these matters. In any event, it is incontrovertible that the *Nicaragua case* is and will remain the *locus classicus* of law of use of force.

II The indirect armed attack and fight against terrorism

The concept of armed attack was central to the ICJ's judgment in the *Nicaragua case*; the USA claimed that its use of force against Nicaragua was justified as collective self-defence of Costa Rica, Honduras and El Salvador in response to armed attacks on those states by Nicaragua. However, the Court rejected this argument as it found that there was no armed attack by Nicaragua. While analysing the notion of armed attack, the Court considered whether an armed attack had to be conducted by a regular army. It used Article 3(g) of the 1974 General Assembly's Definition of Aggression to bolster its view that 'the sending by or on behalf of a state of armed bands, groups, irregulars or mercenaries, which carry out acts of armed force against another State of such gravity as to amount to (inter alia) an actual armed attack conducted by regular forces, or its substantial involvement therein' could be considered an armed attack.[17] The Court thus took Article 3(g) of the 1974 Definition of Aggression to reflect customary international law, while the Court relied on this clause again in the Armed Activities case [18]

On the other hand, the Court distinguished instances falling under Article 3(g) with other modes of assistance to rebels. In its words, 'the Court does not believe the concept of *armed attack* includes not only acts by armed bands where such acts occur on a significant scale but also assistance to rebels in the form of the provision of weapons or logistical or other support. Such assistance may be regarded as a threat or use of force, or amount to intervention in the internal or external affairs of other States'.[19] This was strongly criticized by Judge Schwebel and Judge Jennings in their Dissenting

17 See *Nicaragua case*, para. 195.
18 See *Armed Activities case*, para. 146.
19 *Nicaragua case*, para. 195 (emphasis added).

Opinions as very restrictive and unrealistic.[20] Be that as it may, states do not challenge the view that actions by irregular forces can constitute armed attack; the point of contention is rather the gravity of the attack itself and the degree of the state involvement that it is necessary to make the actions attributable to the state and justify action in self-defence.[21] On the face of the *Nicaragua case*, it readily appears that the Court adopted a twofold test in this regard: an 'objective' one, namely that the attack must have certain 'scale and effects' and a 'subjective' one, i.e. that the perpetrator of the attack must have a 'substantial involvement therein'.

This test however is called into question in view of the Global War on Terror in the wake of 9/11. It is commonplace that following 9/11, the legal contours of the right of self-defence and generally of the *jus ad bellum* have been significantly challenged, especially with regard to the threat posed by non-state actors and international terrorism. It is beyond the ambit of the present enquiry to address *in extenso* these perplexing issues of self-defence and the contemporary law on use of force.[22] The cardinal question to be addressed in such cases is whether the terrorist act qualifies as an 'armed attack' under Article 51 of the UN Charter as well as when it is permissible to use force in response to terrorist acts.[23]

It is true that the Nicaragua test sets the threshold high, especially as far as the need to prove the substantial involvement of a state in each and every case. According to various authors, however, a possible modification of the pertinent rules in the aftermath of 9/11 would be the shift of the threshold required from 'substantial involvement' to the more lax one of 'aiding and abetting'; in other words, acts like providing shelter or abetting terrorists *ex post facto* would be very likely to qualify as 'substantial involvement'.[24] Analogous to the 'aiding and abetting' theory and even more flexible is the argument that self-defence is justified in the case that a state is 'unable and unwilling' to counter terrorism.[25] Telling is that in launching air strikes

20 See Schwebel, Dissenting Opinion, p. 349 and Jennings, Dissenting Opinion, p. 543.

21 See Gray, C., *International Law and the Use of Force,* Cambridge: Cambridge University Press, 2000, p. 97.

22 See *inter alia* Lubell, N., *Extraterritorial Use of Force against Non-State Actors*, Oxford: Oxford University Press, 2010. Ruys, T., *Armed Attack and Article 51 of the UN Charter: Evolutions in Customary Law and Practice,* Cambridge: Cambridge University Press, 2010.

23 On the issue of the use of force in response to international terrorism post-9/11, see Tams, C., 'The Use of Force against Terrorists' *European Journal of International Law* 20, 2009, 359; Antonopoulos, C., 'Force by Armed Groups as Armed Attack and the Broadening of Self-Defence' *Netherlands International Law Review* 5, 2008, 159; K. Trapp, *State Responsibility for International Terrorism: Problems and Prospects*, Oxford: Oxford University Press, 2011.

24 Cf. Article 3(g) of the Definition of Aggression in GA Res 3314/1974. See also in this respect the remarks of Randelzhofer 'Art 51', in Simma B. (ed.), *The Charter of the United Nations. A Commentary,* Oxford: Oxford University Press, 2002, 802.

25 Deeks, A., 'Unable or Unwilling: Toward a Normative Framework for Extraterritorial Self-Defense', *Virginia Journal of International Law* 52, 2012, 483. For a fervent opponent see Corten, O., *The Law against War*, Oxford: Hart, 2012, 188.

against ISIS in Syria in 2014, the USA officially justified its action in relying on a combination of Article 51 of the UN Charter and the 'unwilling or unable' standard.[26]

A more cautious and conservative approach was adopted by the IDI in its 2007 Session:

> (i) If non-State actors launch an armed attack at the instructions, direction or control of a State, the latter can become the object of action in self-defence by the target State.[27]

The present author aligns himself with the Nicaragua test, in the sense that there must be always a serious involvement of the state that would be the target of the self-defence in response of the terrorist attack. This is without prejudice, however, to the potential invocation of other circumstances precluding wrongfulness under the law of international responsibility, such as distress, or necessity, in order to justify less grave responses to terrorist attacks. And this holds true notwithstanding Article 26 of the International Law Commission (ILC) Articles on the Responsibility of States for Internationally Wrongful Acts (hereinafter 'ARSIWA'), according to which when the internationally wrongful act of a state involves the violation of *jus cogens*, as here the use of force, the excuses in question cannot be invoked.[28] This prima facie encumbrance to the use of force in such operations premised upon necessity or distress might be superseded on account of the thesis advanced by Roberto Ago, one of the former special rapporteurs on the issue of state responsibility, who distinguished between the most serious uses of force and others less grave. Although the latter are still prohibited by Article 2(4), Ago was doubtful whether they fell under the scope of the *jus cogens* prohibition.[29] Accordingly, there is room for the invocation of the circumstances precluding wrongfulness in some exceptional cases of counterterrorism action; for example, the plea of necessity in the 1976 Entebbe incident.[30]

26 See the US Letter and further comments available at http://opiniojuris.org/2014/09/23/unwilling-unable-doctrine-comes-life/ (accessed 3 October 2015).

27 See IDI Resolution op. cit, para 10.

28 See Article 26 of the ILC Articles on Responsibility of States for Internationally Wrongful Acts, UN General Assembly Official Records; 56th Session, Supp. No. 10 at UN. Doc A/56/10; at 31 (hereinafter: 'ARSIWA').

29 See Addendum to the Eighth Report on State Responsibility by Mr Roberto Ago YbILC (1980-II), 40.

30 See more analysis in Paust, J., 'Entebbe and Self-help: Israeli Response to Terrorism', *Fletcher Forum* 2, 1978, 86.

III Forcible (counter-)measures

The above points relating to Ago's thesis on the *jus cogens* nature of the prohibition of the use of force and the potential permissibility of circumstances precluding the wrongfulness as justifications of forcible action are also relevant to another question left open by the *Nicaragua case*. In this case, the Court referred to 'measures, which do not constitute an armed attack but may nevertheless involve a use of force'[31] and found it 'necessary to distinguish the most grave forms of the use of force (those constituting an armed attack) from other less grave forms'.[32] The Court in the *Oil Platforms case* reiterated this differentiation.[33] It was also endorsed by the Eritrea-Ethiopia Claims Commission in its Partial Award on Jus ad Bellum: '[l]ocalized border encounters between small infantry units, even those involving the loss of life, do not constitute an armed attack for purposes of the Charter'.[34]

At first sight, it might appear that the distinction drawn in a sketchy way between armed attacks and lesser incursions, such as frontier incidents, was illogical and unnecessary and it was heavily criticized by many authors.[35] As it is observed, however, by Gray, 'despite the criticisms of the Nicaragua case, there were nevertheless serious reasons for the Court's distinction between armed attacks and mere frontier incidents. Its concern was with collective self-defence; it wanted to limit third state involvement. Its insistence on a high threshold for armed attack would serve to limit third party involvement'.[36]

In any event, the distinction between 'armed attack' and 'use of force' derives also from the wording of the relevant provisions of the Charter, which substantiate the existence of a logical and pragmatic gap between armed attack and the 'use of force'. On the one hand, Article 2(4) of the UN Charter forbids in general terms the 'use of force' in international relations and, on the other, Article 51 permits taking self-defence measures only in the specific case of an 'armed attack'. This thus leaves open the question on how states may respond to a use of force simpliciter, i.e. a forcible action not amounting to an armed attack, such as 'mere frontier incidents'.[37]

According to one strand of thought, given that all self-defence, whether individual or collective, must be necessary and proportionate, a mere frontier

31 *Nicaragua case*, para. 209.

32 Ibid., para. 191.

33 See *Oil Platforms case*, para. 51.

34 Eritrea-Ethiopia Claims Commission, *Partial Award on Jus ad Bellum* (Ethiopia's Claims 1–8), 2005, 45 *ILM* 430, at 433.

35 See *inter alia*, Dinstein, Y., *War, Aggression and Self-Defense*, Cambridge: Cambridge University Press, 5th edn, 2012, at 96–97; Schacter, O., 'In Defense of International Rules on the Use of Force' *University of Chicago Law Review* 56, 1986, 113.

36 Gray, C., *International Law and the Use of Force*, op. cit n. 22, 133.

37 *Nicaragua case*, para. 195.

incursion, for instance, would justify only a very limited response. The necessity and proportionality requirements would provide adequate safeguards against excessive use of force.[38] For example, Dinstein speaks of 'on-the spot reaction as the case in which a small-scale armed attack elicits at once, and in situ, the employment of counterforce by those under attack or present nearby'.[39] In a similar vein, the United States Standing Rules of Engagement coin this as 'unit self-defence'.[40]

From another viewpoint, since, arguably, the response against mere frontier incidents cannot fall under the scope of Article 51 of the UN Charter, the only possible justification seems to be proportionate measures, analogous to but less grave than self-defence. The *Nicaragua case* itself did allude to this possibility of lawful individual forcible measures, albeit not explicitly,[41] whereas it did disavow such collective measures; in the Court's words, 'the lawfulness of the use of force by a State in response to a wrongful act of which itself is not the victim, is not admitted when this wrongful act is not an armed attack'.[42]

Judge Simma in his Separate Opinion in the *Oil Platforms case* famously introduced the notion of measures involving the minimum use of force as a response to forcible action short of armed attack. Building on the ambivalent dictum of the *Nicaragua case*, he propounded the thesis that a distinction should be drawn between full-scale self-defence against an 'armed attack' within the meaning of Article 51 of the Charter, on the one hand and, on the other, the case of hostile action below the level of Article 51, 'justifying proportionate defensive measures on the part of the victim, equally short of the quantity and quality of action in self-defence expressly reserved in the United Nations Charter'.[43]

It readily appears that Judge Simma was very careful not to apply the term 'countermeasures', aligning himself thus with the orthodox view expressed by the General Assembly in the Declaration on Principles of International Law concerning *Friendly Relations and Cooperation among States in*

38 See *per* this view e.g. Higgins, R., *Problems and Processes*, Oxford: Oxford University Press, 1994, 251.

39 Dinstein, op. cit, p. 242.

40 United States Army, Judge Advocate General's Legal Center and School, *Operational Law Handbook* (2007), at 85.

41 '[I]t is not for the Court here to determine what direct reactions are lawfully open to a State which considers itself the victim of another State's acts of intervention, possibly involving the use of force. … It might however be suggested that, in such a situation, the United States might have been permitted to intervene in Nicaragua in the exercise of some right analogous to the right of collective self-defense, one which might be resorted to in a case of intervention short of armed attack', *Nicaragua case*, para. 210.

42 Ibid., para. 211.

43 See Separate Opinion of Judge Simma, para 12. See also *Armed Activities case*, paras 126–27 and Antonopoulos, C., 'Force by Armed Groups as Armed Attack and the Broadening of Self-Defence' op. cit. n. 24, 166.

accordance with the Charter of the United Nations,[44] as well as by the ILC in its work on state responsibility, namely that forcible countermeasures are prohibited. Indeed, Article 50 of ARSIWA (2001) sets forth that '[c]ountermeasures shall not affect: (*a*) the obligation to refrain from the threat or use of force as embodied in the Charter of the United Nations'. It is in this regard that Judge Simma's opinion may be read as proclaiming the permissibility of forcible measures short of self-defence within the remit of the primary rules on the use of force, as evolved by state practice, rather than within the ambit of secondary rules of international responsibility, i.e. as countermeasures under Article 22 of ARSIWA.[45]

The latter option would mean that it is accepted that such defensive measures are not part of the primary rule of self-defence, as reflected in Article 51 of the UN Charter, but either as part of customary law complementing Article 51 or countermeasures under the law of the international responsibility. Mindful of the restrictions posed not from Article 26 of ARSIWA,[46] but from Article 50 of ARSIWA, the author is of the view that forcible countermeasures taken in strict accordance with the principles of necessity and proportionality in order to defend a state from a single low-scale use force, like a frontier incident, may be excused under international law.[47]

In concluding, the ICJ's cryptic statement on the permissibility of forcible measures short of self-defence sparked a very interesting debate, which still holds today. These measures may be called 'on the spot self-defence' or 'defensive reprisals', the truth is that states do resort to such measures and the yardstick for the legality at the end of the day is solely their adherence to the overarching principles of necessity and proportionality.

IV Intervention upon invitation

Finally, reference should be made to the right of a government to invite outside intervention, which was addressed by the ICJ in the *Nicaragua case* and still incites controversy. This right attains even more importance today in view of the contemporary intervention of third states in Iraq, although mainly in Syria to combat ISIS.

44 'States have a duty to refrain from acts of reprisal involving the use of force', General Assembly resolution 2625 (XXV), annex, first principle (hereinafter: '1970 Declaration').

45 On countermeasures in general see Sicilianos, L.-A., *Les réactions décentralisées a l'illicite: Des contre-mesures a la légitime defense*, Paris: Librairie générale de droit et de jurisprudence, 1990.

46 See Ago, 'Addendum to the Eighth Report on State Responsibility', op. cit. n. 30. For a modern view see Green, J., 'Questioning the Peremptory Status of the Prohibition of the Use of Force', *Michigan Journal of International Law* 32, 2011, 215.

47 Yoram Dinstein refers to 'defensive armed reprisals', which however according to him may also be future orientated, rather than simply defensive and on the spot, as it is contended here; see Dinstein, op. cit, pp. 249–255.

The Court did examine the possible argument that the USA was justified in using force against Nicaragua to help the *contras* in their forcible opposition to the government. The ICJ said that a government may invite outside help, but a third state may not forcibly help the opposition to overthrow the government. According to the Court, 'no such general right of intervention, in support of an opposition within another State, exists in contemporary international law'.[48] This statement of the Court has authoritatively underpinned the so-called right of intervention upon invitation, which, according to the Court however, is limited to intervention by the official government of a state.[49]

In any event, the Court's pronouncement in relation to intervention upon invitation raises a host of legal issues, which have been extensively debated in scholarly writings. Firstly, the issue has been whether the legality of forcible intervention is merely limited to local unrest or it may apply also to civil wars.[50] The preponderant view seems to be that any form or assistance to other states was prohibited when a civil war was taking place and control of the state's territory was divided between warring parties.[51] Thus the permissibility of intervention becomes contingent on the classification of a certain conflict between civil war and local unrest.[52] However, admittedly, this line between unrest and civil wars has proven highly controversial and both jurisprudence and state practice do not preclude intervention in cases of civil conflicts.[53] For example, in the *Armed Activities case*, the ICJ did not question the right of Uganda to send troops to Congo – during a non-international armed conflict – as long as the deployment was done with the consent of the Congolese government.[54] Also, recent state practice, such as the intervention of France in the Ivory Coast (2010) and Mali (2013) further substantiate that the non-permissibility of intervention in cases of civil wars is not without contestation.[55]

However, when the IDI revisited this issue after its 1975 Wisbaden Resolution in its Rhodes Session, it reaffirmed in Article 2 (1) that 'military

48 Ibid., para 209.

49 See Gray, C., *International Law and the Use of Force*, op. cit. n. 22, pp. 57–58. See also Doswald-Beck, L., 'The Legal Validity of Military Intervention upon Invitation of the Government', *British Yearbook of International Law* 56, 1985, 189.

50 See *inter alia* Farer, T., 'The Regulation of Foreign Intervention in Civil Armed Conflict', *Recueil des Cours* 142, 1974-II, 291.

51 Ibid., p. 57. See also the authoritative Resolution of Institut de Droit International in its Wisbaden Session (1975); Institute de Droit International, 'Resolution, The Principle of Non-Intervention in Civil Wars', *Annuaire IDI* 56, 545 (Wisbaden, 1975).

52 On classification of conflicts, see *inter alia* Green, L., *The Contemporary Law of Armed Conflict*, 2nd edn, Manchester: Manchester University Press, 2000, 54.

53 See Gray, op. cit., 58 and Dinstein, Y., *Non-International Armed Conflicts in International Law*, Cambridge: Cambridge University Press, 2014, 78–79.

54 *Armed Activities case*, para. 190.

55 See also Gara, Y., 'The Arab Uprisings under the Light of Intervention', *German Yearbook of International Law* 55, 2012, 11.

assistance on request' should be given in respect of 'situations of internal disturbances and tensions, such as riots, isolated and sporadic acts of violence and other acts of a similar nature, including acts of terrorism, below the threshold of a non-international armed conflict in the sense of Article 1 of APII'.[56] Military assistance is allowed to the government when the opposition forces in the civil war are themselves receiving external assistance, provides consistency in international legal doctrine.

An additional issue of concern is how the invitation or the consent of the state should be given. Here the rules of international law appear to be clear: according to Article 20 of the ARSIWA, which enunciates 'consent' as a circumstance precluding wrongfulness under international law and the ILC's Commentary, 'consent by a State to particular conduct by another State precludes the wrongfulness of that act in relation to the consenting State, provided the consent is *valid and to the extent that the conduct remains within the limits of the consent given*'.[57] Accordingly, the consent or the invitation for military request by a state should be prior to this request; it should be given validly and the assistance should remain within the bounds of international law, whereas it should end when the requesting state withdraws its consent. The ICJ in the *Armed Activities case* has reaffirmed this.[58]

A final issue is whether there is any change since *Nicaragua case* in respect of the opposition requesting military assistance to combat formal governmental forces. For example, would the argument of Russia with regard to the intervention in Ukraine be tenable, i.e. that it offered military assistance to the former President Yanukovych?[59] It is highly unlikely that such a transformation of the customary rule has occurred.

The aforementioned issues have been raised recently in the current conflict in Syria and Iraq. The intervention of the USA and its allies in Iraq has been officially based on the request of the Iraqi government,[60] while in Syria, Russia is officially assisting the government upon its request.[61] Very importantly, it seems that these interventions have been authoritatively legitimized by the UN Security Council Resolution 2249(2015).[62]

56 Institute de Droit International, 'Resolution on Military Assistance on Request', *Annuaire IDI* 74, 359 (Rhodes, 2011).

57 ILC, Commentary to Article 20 ARSIWA, at p. 71.

58 *Armed Activities case*, paras 42–54.

59 Further discussion is available at www.ejiltalk.org/intervention-with-the-consent-of-a-deposed-but-legitimate-government-playing-the-sierra-leone-card/ (accessed 6 October 2015).

60 See Official Letter dated 20 September 2014 of the Permanent Representative of Iraq to the United Nations addressed to the President of Security Council, S/2014/691, available at www.securitycouncilreport.org/atf/cf/%7B65BFCF9B-6D27-4E9C-8CD3-CF6E4FF96FF9%7D/s_2014_691.pdf (accessed 6 October 2015).

61 See in this regard www.ejiltalk.org/russias-intervention-in-syria/ (accessed 5 March 2016).

62 See UN Security Council Resolution 2249 of 20 November 2015 and further comments in www.ejiltalk.org/the-constructive-ambiguity-of-the-security-councils-isis-resolution/ (accessed 5 March 2016).

V Concluding remarks

The judgment of the ICJ in the *Nicaragua case* has undoubtedly been a watershed moment in international law. States, especially from Latin America, as well as states of the Non-Aligned Movement, regained their lost trust – after the 1966 *South-West African cases* – in the rule of law on the international stage. It is telling that after this decision, the ICJ has received numerous applications by Latin American states in order to resolve their disputes with the champion being, not surprisingly, Nicaragua!

The *Nicaragua case*, however, is, and will remain, monumental also due to its findings in respect of many fields of international law, but more importantly in respect of the legal framework of the use of force. Its pronouncements in this regard are still debated and still inform state practice and international jurisprudence. Issues such as indirect aggression and self-defence, forcible countermeasures and intervention upon invitation still haunt international lawyers and the *Nicaragua case* remains the first port of call for every scholarly analysis. It has been the most important legacy of Latin America in the case law of the ICJ and is rightly triumphed as the *locus classicus* of the *jus ad bellum*.

Bibliography

Antonopoulos, C., 2008 'Force by Armed Groups as Armed Attack and the Broadening of Self-Defence', *Netherlands International Law Review* 5, 159–180.

Corten, O., 2012 *The Law against War*, Oxford: Hart Publishing.

Deeks, A., 2012 'Unable or Unwilling: Toward a Normative Framework for Extraterritorial Self-Defense', *Virginia Journal of International Law* 52, 483–550.

Dinstein, Y., 2012 *War, Aggression and Self-Defence*, 5th edn, Cambridge: Cambridge University Press.

Dinstein, Y., 2014 *Non-International Armed Conflicts in International Law*, Cambridge: Cambridge University Press.

Doswald-Beck, L., 1985 'The Legal Validity of Military Intervention upon Invitation of the Government', *British Yearbook of International Law* 56, 189.

Farer, T., 1974 'The Regulation of Foreign Intervention in Civil Armed Conflict', *Recueil des Cours* 142-II, 291.

Gara, Y., 2012 'The Arab Uprisings under the Light of Intervention', *German Yearbook of International Law* 55, 11.

Gray, C., 2000 *International Law and the Use of Force*, Cambridge: Cambridge University Press.

Green, J., 2011 'Questioning the Peremptory Status of the Prohibition of the Use of Force' *Michigan Journal of International Law* 32, 2011, 215–257.

Green, L., 2000 *The Contemporary Law of Armed Conflict*, 2nd edn, Manchester: Manchester University Press.

Higgins, R., 1994 *Problems and Processes*, Oxford: Oxford University Press, 1994.

Kirgis, F. L., 1987 'Custom on a Sliding Scale' *American Journal of International Law* 81, 146.

Lubell, N., 2010 *Extraterritorial Use of Force against Non-State Actors*, Oxford: Oxford University Press.

Paust, J., 1978 'Entebbe and Self-help: Israeli Response to Terrorism', *Fletcher Forum* 2, 86–92.

Randelzhofer, A., 2002 'Art 51', in Simma B. (ed), *The Charter of the United Nations: A Commentary*, Oxford: Oxford University Press 802.

Ruys, T., 2010 *Armed Attack and Article 51 of the UN Charter: Evolutions in Customary Law and Practice*, Cambridge: Cambridge University Press

Sicilianos, L.-A., 1990, *Les réactions décentralisées a l'illicite: Des contre-mesures a la légitime defense*, Paris: Librairie générale de droit et de jurisprudence.

Tams, C., 2009 'The Use of Force against Terrorists', *European Journal of International Law* 20, 359–397.

Trapp, K., 2011, *State Responsibility for International Terrorism: Problems and Prospects*, Oxford: Oxford University Press

Part V

Material contribution of Latin American cases to the development of international law

Territorial and maritime disputes

19 *Territorial and Maritime Dispute (Nicaragua v. Colombia), 2001*

Virginie J.M. Tassin

I Introduction

The long-standing territorial and maritime dispute between Nicaragua and Colombia is enshrined in a complex political geography. The western Caribbean's geostrategic location and the richness of its mineral and biological resources, including lagoons and reefs, makes it a fragile yet quintessential place for any sovereign battle. Hence, it is with no surprise that the tension among its neighbours has been palpable for years.

On the basis of the Pact of Bogota,[1] Nicaragua instituted proceedings against Colombia in 2001, requesting the International Court of Justice (hereinafter 'ICJ') to determine the sovereignty over the islands of Providencia, San Andrés, and Santa Catalina as well as all the appurtenant islands and keys and over the Roncador, Serrana, Serranilla, and Quitasueño keys. It further requested the Court to determine the course of a single maritime boundary separating the exclusive economic zones and the continental shelves of Colombia and Nicaragua.

By means of its judgment from 13 December 2007, the Court confirmed its jurisdiction to adjudicate the dispute. Interested to intervene, Costa Rica and Honduras filed permissions to the ICJ in 2010, which were refused in a judgment rendered on 4 May 2011. Public hearings were held shortly after and another judgment, dated 19 November 2012, (hereinafter '*Nicaragua/Colombia*')[2] finally adjudicated the claims introduced in 2001. It unanimously confirmed Colombia's sovereignty over all the seven islands, unanimously drew the single maritime boundary in favour of Nicaragua, and unanimously rejected Nicaragua's claim on the extended continental shelf.

This long-awaited judgment is of great interest to the international community because of its subject matter and its regional impact. Following the 2012 judgment, Nicaragua introduced a new claim against Colombia to

1 *American Treaty on Pacific Settlement*, 30 April 1948, UNTS 30-I-449, p. 85.

2 *Territorial and Maritime Dispute (Nicaragua v. Colombia), Judgment, I.C.J. Reports 2012*, p. 624.

the ICJ on the issue of the delimitation of the extended continental shelf[3] shortly before the effective denunciation of the Pact of Bogota by Colombia.[4] This ICJ claim is based on the Nicaraguan extended continental shelf submission that exacerbated the disputes with Colombia and its neighbours (namely Costa Rica, Panama, and Jamaica). Indeed, Columbia, Costa Rica and Panama all declared in various communications sent to the Commission on the Limits of the Continental Shelf[5] (hereinafter 'CLCS') that the new Nicaraguan claim violated their ocean spaces and threatened the regional peace and security of the Caribbean.

In light of these recent developments, only one aspect of the 2012 judgment will be analyzed. While the ICJ rejected Nicaragua's claim on the delimitation of its extended continental shelf, it did address and analyze, for the first time in ICJ's history and the first time in the history of the international settlement of disputes, the applicable law to the delimitation of the extended continental shelf between a State Party (Nicaragua) to the United Nations Convention on the Law of the Sea (hereinafter 'UNCLOS') and a non-Party State (Colombia). This task was therefore delicate and difficult given the youth of the implementation of continental shelf regime of UNCLOS dedicated to the extended continental shelf area.

II The change in Nicaragua's claim and the principle of single continental shelf

When Nicaragua instituted the proceedings on 6 December 2001, it asked the Court:

> to determine the course of the single maritime boundary between the areas of continental shelf and exclusive economic zone appertaining respectively to Nicaragua and Colombia.[6]

3 *I.C.J., Nicaragua institutes proceedings against Colombia asking the Court to "definitively determine the question of the delimitation of the continental shelf between Nicaragua and Colombia in the area beyond 200 nautical miles from the Nicaraguan coast"*, Press Release, No. 2013/21, 17 September 2013.

4 Colombia denounced the Pact of Bogota on 27 November 2012 (with effect on 27 November 2013). Nicaragua instituted the new proceedings at the ICJ on 16 September 2013. See Ibid.

5 CLCS, Communication from Jamaica in reply to the submission of Nicaragua, 12 September 2013; CLCS, Communication from Colombia, Costa Rica, and Panama in reply to the submission of Nicaragua, 23 September 2013 and 5 February 2014; CLCS, Communication of Costa Rica in reply to the submission of Nicaragua, 20 January 2014; CLCS, Communication of Panama in reply to the submission of Nicaragua, 30 September 2013 and 3 February 2014; CLCS, Communication of Colombia in reply to the submission of Nicaragua, 24 September 2013 and 6 February 2014.

6 *I.C.J., Application Instituting Proceedings filed in the Registry of the Court on 6 December 2001*, para. 8, p. 8.

This request was nonetheless transformed during the course of the proceedings. In its final submission on 1 May 2012, Nicaragua requested the Court to divide 'by equal parts the overlapping entitlements to a continental shelf of both Parties.'[7]

This transformation did not appear out of thin air. Nicaragua pursued a strategy. According to judge Owada, Nicaragua 'attempted to replace the original formulation of the claim.'[8] It is no surprise that this change exacted the wrath of Colombia, which declared it to be 'radical'[9] and 'fundamentally' changing the subject matter of the dispute.[10]

Although it is obvious that Nicaragua filed a new claim,[11] the change of subject matter is less obvious. According to the criteria of admissibility of a new claim, one of the two alternative tests must be satisfied: the new claim should be either implicit on the application,[12] or it must arise directly out of the question which is the subject matter of the application.[13] In the present case, the ICJ considered that the new claim of Nicaragua fell within the definition of the dispute in the original request and thus arose directly out of the delimitation dispute.[14]

Judge Owada criticized this conclusion. Indeed, none of the cases applied by the parties[15] were tackling the issue of a transformation of a claim during the course of proceedings.[16] He also rightly pointed out the notable inconsistency of Nicaragua's position. Indeed, Nicaragua proclaimed the role of geology and geomorphology in its memorial by stating that they have 'no relevance for the delimitation of a single maritime boundary within the delimitation area.'[17] Since geology and geomorphology are inextricably linked to the delineation of the extended continental shelf, such an affirmation highlights, as a matter of fact, a radical change in the claim.

When instituting the proceedings, Nicaragua was thus not asking the

7 *Territorial and Maritime Dispute (Nicaragua v. Colombia)*, op. cit., para. 17, (3), p. 636.

8 *I.C.J., Dissenting Opinion of Judge Owada, Territorial and Maritime Dispute (Nicaragua v. Colombia)*, para. 6, p. 722.

9 *I.C.J., Public sitting held on Friday 27 April 2012, at 10 a.m., at the Peace Palace, President Tomka presiding, in the case concerning the Territorial and Maritime Dispute (Nicaragua v. Colombia)*, CR 2012/12, para. 2, p. 44.

10 Article 40 of the Statute of the ICJ.

11 CR 2012/12, Ibid.

12 *Case concerning the Temple of Preah Vihear (Cambodia v. Thailand), Merits, Judgment of 15 June 1962: I.C.J. Reports 1962*, p. 36.

13 See *Fisheries Jurisdiction (Federal Republic of Germany v. Iceland), Merits, I.C.J. Reports 1974*, para. 72, p. 203 and *Certain Phosphate Lands in Nauru (Nauru v. Australia), Preliminary Objections, Judgment, I.C.J. Reports 1992*, para. 67, p. 266.

14 *Nicaragua/Colombia*, para. 111, p. 665.

15 *Certain Phosphate Lands in Nauru*, op. cit., p. 240 and *Ahmadou Sadio Diallo (Republic of Guinea v. Democratic Republic of the Congo), Judgment, I.C.J. Reports 2010*, p. 639.

16 *I.C.J., Dissenting Opinion of Judge Owada, Territorial and Maritime Dispute (Nicaragua v. Colombia)*, para. 6, p. 722.

17 *I.C.J., Memorial of Nicaragua, Territorial and Maritime Dispute (Nicaragua v. Colombia)*, para. 3.58, p. 215.

Court to officiate the extended continental shelf's delimitation. It was only much later, either in light of the implementation of the delineation procedure, or, according to judge Owada in connection with the 2007 *Nicaragua/Colombia* judgment on the preliminary objections,[18] that Nicaragua decided to change its claim.

As to the change in the subject matter, the conclusion of the ICJ is consistent with the jurisprudence on the extended continental shelf but it lacks explanations. All of the previous cases relating to the delimitation of the extended continental shelf clearly distinguished the delineation procedure from the delimitation procedure, and therefore the underlying subject-matter issue. The *Barbados/Trinidad and Tobago* arbitration[19] highlighted, meanwhile, the principle of unity of Article 76 of UNCLOS.[20] As a consequence, the ICJ should have delimited the extended continental shelf during the course of Nicaragua's first claim and it is regrettable that the Court omitted mention of such an important principle.

III The tension between sovereign rights and the definition/delineation of the extended continental shelf

In the present dispute, the parties agreed that the relevant provisions of UNCLOS dedicated to baselines, entitlements to maritime spaces, and the definition and delimitation procedures for an exclusive economic zone and continental shelf are customary international law. They nevertheless disagreed on the existence of any extended continental shelf in the Caribbean[21] and on the nature and content of the rules governing the entitlements of coastal states to an extended continental shelf.

When facing the question of identification of customary international law within Article 76, the ICJ stated, too briefly, that it 'considers that the definition of the continental shelf set out in Article 76, paragraph 1, of UNCLOS forms part of customary international law.'[22] In providing neither details on its analysis of the relevant states' practices nor on the interpretation exercise made according to Article 76, the ICJ evaded the question and simply declared that its task was only to determine if it was in a position to carry out a continental shelf delimitation.[23] The ICJ therefore focused its

18 *I.C.J., Dissenting Opinion of Judge Owada, Territorial and Maritime Dispute (Nicaragua v. Colombia)*, para. 23, p. 728.

19 *Barbados/Trinidad and Tobago, Award of 11 April 2006*, RIAA, Vol. XXVII, p. 147, para. 213, pp. 208–9.

20 It has been confirmed in *Dispute concerning delimitation of the maritime boundary between Bangladesh and Myanmar in the Bay of Bengal (Bangladesh/Myanmar), Judgment, ITLOS, 14 March 2012*, para. 363, p. 108 and in *Bay of Bengal Maritime Boundary Arbitration between Bangladesh and India, Award, 7 July 2014*, para. 77, p. 21.

21 Colombia objects the existence of any extended continental shelf in the disputed area.

22 *Nicaragua/Colombia*, para. 118, p. 666.

23 *Territorial and Maritime Dispute (Nicaragua v. Colombia)*, op. cit., para. 118, p. 666.

analysis on the entitlement of coastal states and did not analyze the legal nature of the procedure for establishing or delineating the extended continental shelf and its relationship with the definition.[24] This is perhaps the reason why Article 76, paras. 2 and 3, were not included in the definition of the continental shelf. Indeed, paragraph 2 makes direct reference to the delineation criteria of Article 76, paras. 4 to 6.[25]

Such restrictive interpretation appears to be problematic for future interpretation. When interpreting the meaning of natural prolongation, the International Tribunal for the Law of the Sea (hereinafter 'ITLOS') recognized that the notion of 'natural prolongation' and 'outer edge of the continental margin' in paragraphs 1 and 4 are 'closely interrelated .. the object and purpose of Article 76 of the Convention is to define the precise outer limits of the continental shelf.'[26] Recognizing only paragraph 1 of Article 76 as customary international law means cutting out the definition of the extended continental shelf to its precise limits. Although the *Bangladesh/Myanmar* case was a case between two UNCLOS state parties, the interpretation of ITLOS is in line with the object and purpose of UNCLOS. Hence, if the definition of the extended continental shelf is dissected, is it correct to affirm the customary status of its definition? How should a non-party state define the outer limits of its extended continental shelf without taking paragraphs 2–6 into consideration?

This dispute also highlights the possibility of a disjunction between the sovereign rights of the coastal states and the definition/delineation of the continental shelf. It reveals therefore a tension between the differing natures of Articles 77 and 76. In relation to this point, judge *ad hoc* Mensah rightly criticized the reference made by the ICJ to the *Nicaragua v. Honduras* case.[27] This reference wrongly casts a doubt into the question of whether non-party states are entitled to a continental shelf beyond 200 nautical miles.[28] Beyond any doubt, all coastal states, be they UNCLOS parties or not, have *ipso facto* and *ab initio* rights that cannot be challenged, they are customary international law rights.[29]

24 A.G. Oude Elferink 'Article 76 of the LOSC on the definition of the continental shelf: Questions concerning its interpretation from a legal perspective,' *The International Journal of Marine and Coastal Law* 21, 3, 269–85.

25 As outlined by Nicaragua. *I.C.J., Comments in writing of the Republic of Nicaragua on the written reply by the Colombian Government to the question put by Judge Bennouna at the public sitting held on the afternoon of 4 May 2012*, 18 May 2012, para. 12, p. 3.

26 *Bangladesh/Myanmar*, paras. 434–5, p. 127.

27 *Territorial and Maritime Dispute between Nicaragua and Honduras in the Caribbean Sea*, op. cit., para. 110, p. 695.

28 *I.C.J., Declaration of Judge ad hoc Mensah, Territorial and Maritime Dispute (Nicaragua v. Colombia)*, para. 7, p. 764.

29 V.J.M. Tassin, 'L'équilibre à trouver entre l'interaction des articles 76 et 77 et la procédure de définition du plateau continental étendu', *Les défis de l'extension du plateau continental: consécration d'un nouveau rapport de l'Etat à son territoire*, Paris: Pedone, 2013, pp. 226–57.

IV The delineation procedure and customary international law

Both Colombia and Nicaragua stated in their written proceedings that 'Article 76 reflects "well-established principles of customary international law"'.[30] These affirmations place the cart before the horse! The lack of convincing demonstrations from both parties on the delicate question of the formation of customary law led judge Bennouna to put forth a question during the oral proceedings in 2012 on the customary status of article 76.[31]

In its first brief reply, Colombia declared that implementation of Article 76, paras. 4 to 9, is undertaken only pursuant to treaty obligations. Colombia therefore included the delineation criteria and the delineation procedure (coastal states and CLCS) into the treaty obligations.

Such a position has of course been criticized by Nicaragua.[32] In contrast to the written reply of Colombia, that of Nicaragua is greatly detailed. But it does not improve its argument. Indeed, in order to demonstrate that Article 76, paras. 4 to 7, have the status of customary international law, Nicaragua used the doctrine of the 'automatic appurtenance of the continental shelf' deriving from the famous *North Sea Continental Shelf* cases.[33] It therefore used the description of the spatial extent of the continental shelf to bind this doctrine to the continental shelf delineation rather than to the rights of the coastal states. This interpretation seems to confuse the history and legal value of Article 76 with the one of Article 77.

Nicaragua proceeds by providing statistics to prove the general support of the international community for the principles of delineation of the continental shelf. It concluded that 'more than 80 States of the 90 that have continental shelf legislation appear to accept the definition in Article 76 (4)–(7) either explicitly in their laws or implicitly by their acceptance of the UNCLOS.'[34]

Although statistics are always very illustrative, Nicaragua created unclear categories between the various states.[35] Rather than observing a clear cut between states parties and non-party states, Nicaragua created a category of

30 *I.C.J., Reply of Nicaragua, Territorial and Maritime Dispute (Nicaragua v. Colombia)*, 18 September 2009, para. 2.5, p. 64.

31 'Les règles posées à l'article 76 de la convention des Nations Unies de 1982 sur le droit de la mer, pour la détermination de la limite extérieure du plateau continental au-delà des 200 milles marins, peuvent-elles être considérées aujourd'hui comme ayant le caractère de règles de droit international coutumier?'.

32 Nicaragua excludes from customary international law paragraphs 8 and 9 of Article 76.

33 *North Sea Continental Shelf, Judgment, I.C.J. Reports 1969*, para. 19, p. 22.

34 *I.C.J., Written reply of the Republic of Nicaragua to the question put by Judge Bennouna at the public sitting held on the afternoon of 4 May 2012*, para. 14, p. 5.

35 *I.C.J., Comments in writing of the Republic of Colombia on the written reply by the Nicaraguan Government to the question put by Judge Bennouna at the public sitting held on the afternoon on 4 May 2012*, para. 8, p. 2.

states having 'signed or ratified UNCLOS,' thus mixing the two into the same basket![36]

Was this nebulosity a strategy by Nicaragua to hide the weakness of its arguments? It might well be. Indeed, apart from these statistics, Nicaragua provided examples of non-party states (Ecuador and the United States) before annexing a table of legislation again mixing state parties and non-party states. Interestingly, most of the legislation of non-party states of this table follows Article 76, para. 1. Apart from Ecuador's legislation mentioning the isobath rule of Article 76, para. 5, none of them make reference to Article 76, paras. 4 to 7. Some of them, such as those from Israel, Peru, and Venezuela, even follow the model of the Geneva Convention of 1958. The examples provided by Nicaragua thus fail to prove any uniformed, extensive, and representative state practice.[37] Furthermore, as pointed out by Colombia, the inclusion of provisions within customary international law requires that the action of a state be undertaken because of a belief of compliance with a legal obligation, the *opinio juris sive necessitatis*.[38]

Nicaragua finally raised an interesting argument according to which treaties that are almost universally ratified, could in fact, lay the foundation for the formation of customary international law.[39]

The ICJ has indeed previously recognized that representative participation, including states whose interests are specially affected, might suffice to provide the basis for customary international law.[40] The quasi-universal ratification of UNCLOS could be indeed an *indicator* of the formation of customary international law.[41] Nevertheless, in the context of a dispute between a party and a non-party state, Nicaragua should have studied in detail the practice of the non-party states and countered it with the practice of state parties in order to provide for the formation of customary law. Moreover when considering the technicality of Article 76, paras. 4 to 6, and the great disparity between implementation and interpretation of these paragraphs by the CLCS and by the states, such study should go beyond the simple assertion of transposition of UNCLOS into the domestic legislations of states and focus on their practice of implementation.

36 'Non-ratification may sometimes be due to factors other than active disapproval of the convention concerned can hardly constitute a basis on which positive acceptance of its principle can be implied: the reasons are speculative, but the facts remain.' *North Sea Continental Shelf*, ibid., para. 73, p. 43.

37 Ibid., para. 74, p. 43.

38 *North Sea Continental Shelf*, ibid., para. 77, p. 44.

39 *I.C.J., Written Reply of the Republic of Nicaragua to the question put by Judge Bennouna at the public sitting held on the afternoon on 4 May 2012*, 18 May 2012, para. 11, p. 3.

40 See *North Sea Continental Shelf*, op. cit., para. 74, p. 42. See also *Case concerning Military and Paramilitary Activities in and against Nicaragua (Nicaragua v. United States), Merits, Judgment, ICJ Reports* 1986, para. 188, pp. 99–100.

41 The technique of the 'package deal' should not be confused with an accelerated way of forming customary international law. Many provisions of UNCLOS remain not yet implemented.

V The interaction between the delineation and the delimitation procedures

The ICJ rejected the claim of Nicaragua for delimitation of its continental shelf boundary with Colombia mainly because of the lack of proof detailing the outer limits of Nicaragua's continental shelf. In light of the pending case on the delimitation of the extended continental shelf between the same two parties currently before the ICJ,[42] an analysis of some of the Court and parties' arguments will be conducted in the following developments.

A. Principles of interaction

According to a core principle of Article 76, the actions of the CLCS shall not prejudice matters relating to the delimitation of boundaries between states with opposite or adjacent coasts.[43]

Pursuant to Annex I of the Rules of Procedure of the CLCS, in the case of a dispute over the delimitation of the continental shelf or in 'other cases' of unresolved land or maritime disputes, the coastal states submitting the extension claim shall inform the CLCS and assure it that the submission will not prejudice matters relating to the delimitation of boundaries. Moreover, when a land or maritime dispute exists, the prior consent of 'all states that are parties to such dispute' is required in order to proceed with the delineation exercise. This broad wording allows non-party states to participate in the delineation procedure. It therefore renders a variety of dispute scenarios possible: between state parties to UNCLOS, and between state parties and non-party states to UNCLOS.

In some disputes, states will give their prior consent despite the running dispute. This was the case for Norway when Russia submitted its extension claim in 2001.[44] In other disputes, a state might refuse to give its prior consent and will therefore block the delineation procedure and the work of the CLCS. This is the case in the delineation procedure from Myanmar, which, despite the judgment of ITLOS in 2012, is still blocked because of the lack of prior consent given by Bangladesh to the CLCS.[45]

This is not dissimilar to the current case. Following Nicaragua's submission in 2013, the heads of states and governments of Colombia, Costa Rica, and Panama expressed concerns regarding Nicaragua's submission dated 30

42 *I.C.J., Question of the Delimitation of the Continental Shelf between Nicaragua and Colombia beyond 200 Nautical Miles from the Nicaraguan Coast (Nicaragua v. Colombia)*, pending.

43 Article 76, para. 10, Article 46 and Annex I of the Rules of Procedure of the CLCS.

44 CLCS, Communication of Norway in reply to the submission of Russia, 20 March 2002.

45 Following *Bangladesh/Myanmar* in 2012, Myanmar submitted a revised submission to the CLCS on 30 July 2015. On 22 October 2015, Bangladesh reacted to this revised submission accusing Myanmar of not complying with UNCLOS and with the Rules of Procedure of the CLCS.

July 2015 claiming it constitutes 'a clear threat to regional peace and security.' By applying this argument, they rejected Nicaragua's claims and did not authorize the CLCS to proceed with examination of the delineation.[46] Colombia also reacted individually by giving an express reservation to the 'entire document' filed by Nicaragua and stressing that an action or inaction from the CLCS would not affect the sovereign rights of Colombia on its continental shelf.[47] The delineation procedure is therefore currently blocked and it is unlikely that the CLCS will examine Nicaragua's submission in such a tense context.

B. Which 'information' to take into consideration for the determination of the outer limits of the extended continental shelf?

In filing only preliminary information for the delineation of its extended continental shelf, Nicaragua asserted that this filing was in conformity with the requirements of Article 76, para. 8. This argument has been of course objected by Colombia claiming that Nicaragua's preliminary information was based on 'rudimentary and incomplete technical information' and was thus 'woefully insufficient.'[48]

Before starting its analysis, the Court noted the difference between the delimitation exercise and the delineation procedure. Pursuant to Article 76, functions of the CLCS are limited to the interpretation of data and materials for the issuance of a recommendation and to the scientific and technical advising.[49] The ICJ for its part has jurisdiction over 'any dispute concerning the interpretation or application' of UNCLOS.[50] The Court's task is therefore not to examine and interpret the data of coastal state and to determine the outer limits of the continental shelf, but rather to delimit the continental shelf between two countries. These two tasks are inextricably related yet very different.[51]

The insufficient data provided by Nicaragua led the Court to rightly conclude that the preliminary information was not sufficient to prove the location of the outer limits of the continental shelf.[52] Judge *ad hoc* Mensah disagreed with this position and warned that such a requirement could lead to an automatic exclusion of all preliminary information. According to him,

46 CLCS, Communication from Colombia, Costa Rica, and Panama in reply to the submission of Nicaragua, 23 September 2013 and 5 February 2014.
47 CLCS, Communication of Colombia in reply to the submission of Nicaragua, 24 September 2013 and 6 February 2014.
48 *Territorial and Maritime Dispute (Nicaragua v. Colombia)*, op. cit., para. 122, p. 667.
49 Article 3, Annex II, UNCLOS.
50 Article 286, UNCLOS.
51 *Bangladesh/Myanmar*, para. 379, p. 112.
52 *Territorial and Maritime Dispute (Nicaragua v. Colombia)*, op. cit., para. 129, p. 669. See also in the same case *I.C.J., Separate Opinion of Judge Donoghue*, para. 12, p. 754

a submission should 'not necessarily be required in every case to enable the Court or Tribunal to delimit a continental shelf beyond 200 miles.'[53]

I could not agree more with this assertion. All preliminary information files are different: while some of them are well furnished, others constitute just a communication declaring the intention to file a submission in the future. A well-furnished preliminary information could thus legitimately serve as a basis for a court or a tribunal. The clear distinction between the delineation and the delimitation procedures should nevertheless be kept in mind. The interpretation and application of Article 76 requiring 'both legal and scientific expertise'[54] indicate that international courts and tribunals should work together with the experts from the CLCS to resolve issues during the examination of any preliminary information or submission. The concerns of judge Donoghue with regards to a potential blockage of both CLCS and ICJ seem not to be justified. Inaction from the ICJ would go against its role and the current case law.[55] The work of both institutions is complementary and their cooperation should be praised and reinforced to ensure implementation in the spirit of UNCLOS. A lot has yet to be done.

C. The delineation procedure and the public order of oceans

While assessing the suitability of the preliminary information's data and materials for the delineation of the contested outer limits of the extended continental shelf, the ICJ also examined the question of whether Nicaragua should comply with the delineation procedure of the CLCS considering that Colombia is a non-party state.

Judge *ad hoc* Cot and judge *ad hoc* Mensah supported the argument that the requirements of Article 76, para. 8, should not apply to non-party states. According to Judge *ad hoc* Mensah, when a dispute involves two non-party States, 'Article 76 of UNCLOS should not apply as between them and may, in any event, not be available to them'.[56] Judge *ad hoc* Cot underlined for his part that Article 76, para. 8, does not represent customary international law: 'Article 76, paragraph 8, is thus *res inter alios acta* for Colombia.'[57]

53 *I.C.J., Declaration of Judge ad hoc Mensah, Territorial and Maritime Dispute (Nicaragua v. Colombia),* paras. 11–12, p. 766.

54 *Bangladesh/Myanmar*, para. 411, p. 120.

55 *I.C.J., Separate Opinion of Judge Donoghue, Territorial and Maritime Dispute (Nicaragua v. Colombia),* paras. 29–30, p. 759. See also *Bangladesh/Myanmar*, para. 392, p. 115 and T. Treves, 'Conflict between the International Tribunal for the Law of the Sea and the International Court of Justice', *New York University Journal of International Law & Policy* 31, 1999, pp. 809–21.

56 *I.C.J., Declaration of Judge ad hoc Mensah, Territorial and Maritime Dispute (Nicaragua v. Colombia),* para. 12, p. 766.

57 *I.C.J., Declaration of Judge ad hoc Cot, Territorial and Maritime Dispute (Nicaragua v. Colombia),* para. 19, p. 771.

Should this theory be applied to state parties, some extreme scenarios might emerge. Let us imagine that Nicaragua is engaged in various delimitation disputes with Colombia, Costa Rica, and Panama on matters relating to the delimitation of its extended continental shelf. According to the theory of Judge *ad hoc* Cot and Judge *ad hoc* Mensah, for cases posing Nicaragua against Costa Rica and Panama, therefore being claims between state parties, Nicaragua would be required to implement Article 76, para. 8. It would be the contrary for a dispute between Nicaragua and Colombia. In such case, what should Nicaragua do?

The difference in requirements between these disputes has the potential to create complete chaos! It would lead 'to a result which is manifestly absurd or unreasonable.'[58] Interpreting Article 76, para. 8, as non-legally binding for Nicaragua because of a dispute against a non-party state would result in ignoring the bilateral nature of the delineation procedure thereby creating a dialogue (only) between the coastal state, willing to delineate its continental shelf, and the CLCS. It would encourage a two-speed reading of UNCLOS endangering the public order of oceans.

To tackle this issue, the ICJ recalled the Preamble of UNCLOS.[59] It is thus commendable that the Court considered that the fact that 'Colombia is not a party thereto does not relieve Nicaragua of its obligation under Article 76 of the Convention.'[60] By quoting the importance of the legal order of the oceans, the Court took into consideration the letter and spirit of UNCLOS and seems to have applied the rule of effectiveness.[61] Such interpretation is greatly welcome. It stresses the key role of international courts and tribunals as guardians of the peaceful uses of oceans among all states, be they UNCLOS parties or not.

Bibliography

Magnusson, B.M., *The continental shelf beyond 200 nautical miles: delineation, delimitation and dispute settlement*, Leiden & Boston: Brill Nijhoff, 2015, p. 319.

Maw, C.E., 'The United States and the Law of the Sea after UNCLOS III: Comment', in *Law and Contemporary Problems*, Vol. 46, N°2, 1983, pp. 55–60.

Meese, R., 'Le plateau continental au-delà des 200 milles marins', Chronique, in *Annuaire du droit de la mer*, Paris: Pedone, 2005–14.

Oude Eferink, A.G., 'Article 76 of the LOSC on the definition of the continental shelf: Questions concerning its interpretation from a legal perspective', *The International Journal of Marine and Coastal Law*, 21, 3, 2006, pp. 269–85.

58 Article 32, *Vienna Convention on the Law of Treaties*.

59 *Territorial and Maritime Dispute (Nicaragua v. Colombia)*, op. cit., para. 126, p. 669.

60 Ibid.

61 *Interpretation of Peace Treaties (second phase), Advisory Opinion: I.C.J. Reports 1950*, p. 229.

Koh, S.A., 'Constitution Court of Colombia Finds that Implementation of ICJ Decision Requires Treaty (May, 2, 2014),' *ASIL*, 9 May 2014.

Tassin, V.J.M., *Les défis de l'extension du plateau continental: consécration d'un nouveau rapport de l'Etat à son territoire*, Paris: Pedone, 2013, p. 494.

Treves, T., 'Conflict between the International Tribunal for the Law of the Sea and the International Court of Justice', *New York University Journal of International Law Policy* 31, 1999, pp. 809–21.

20 *Maritime Dispute (Peru v. Chile)*, *2008*

Tullio Scovazzi

I Introduction

On 27 January 2014 the International Court of Justice (hereinafter 'ICJ') rendered the judgment on the *Maritime Dispute (Peru v. Chile) case*.[1] The jurisdiction of the Court was based on Art. XXXI of the American Treaty on Pacific Settlement (Bogotá, 1948).

The main question submitted to the ICJ related to the maritime boundary between Chile and Peru. Chile argued that the Declaration on the Maritime Zone (Santiago, 1952), adopted by Chile, Ecuador and Peru (hereinafter 'Santiago Declaration') had already established a maritime boundary along the parallel of latitude passing through the starting point of the land boundary between the two countries and extending to 200 NM, as also evidenced by subsequent agreements and practice. Peru argued that no agreed maritime boundaries existed between the parties and requested the Court to determine the boundary line according to equidistance in order to achieve an equitable result.[2]

The ICJ decided, by fifteen votes to one, that the initial segment of the single maritime boundary[3] follows the parallel of latitude passing through the starting point of the land boundary; by ten votes to six, that this initial segment runs up to point A, situated at a distance of 80 NM from the starting point, that the boundary continues along the equidistance line until its intersection with the 200 mile limit from the Chilean baselines (point B) and that it then continues along that limit until the point of intersection of the 200 mile limits measured from the Peruvian and Chilean baselines (point C) (see Maps 20.1 and 20.2).[4]

1 This comment will focus only on the main aspects of the decision.

2 *Maritime Dispute (Peru* v. *Chile), Judgment, I.C.J. Reports 2014*, p. 3, para. 22.

3 As the Parties never drew any distinction between the waters and the seabed, the ICJ concluded that the boundary is an all-purpose one (*Maritime Dispute, Judgment* para. 102).

4 See the maps that show the claims of the parties (after para. 23 of the judgment) and the boundary decided by the ICJ (after para. 190 of the judgment). The two maps are also reproduced at p. 238 of this volume.

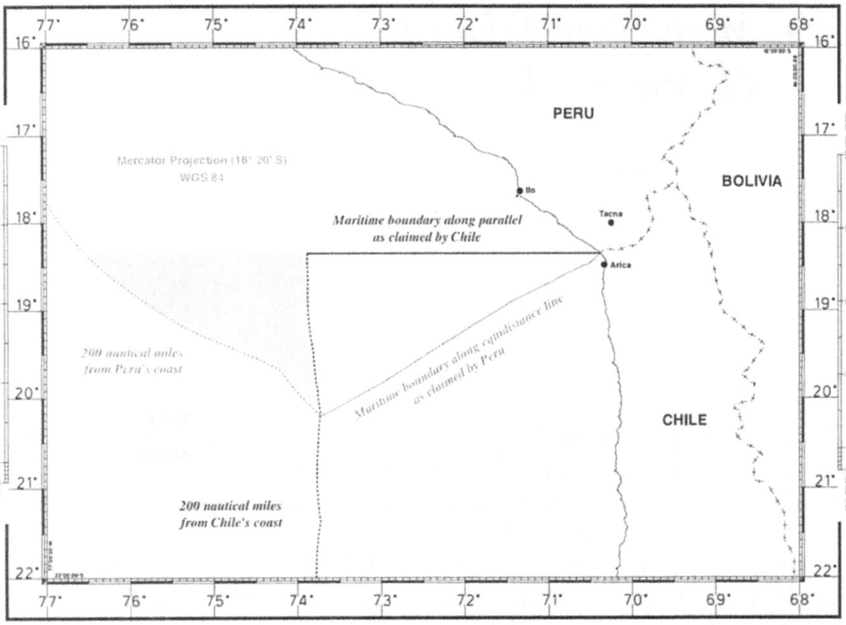

Map 20.1 The maritime boundary lines claimed by Peru and Chile, respectively

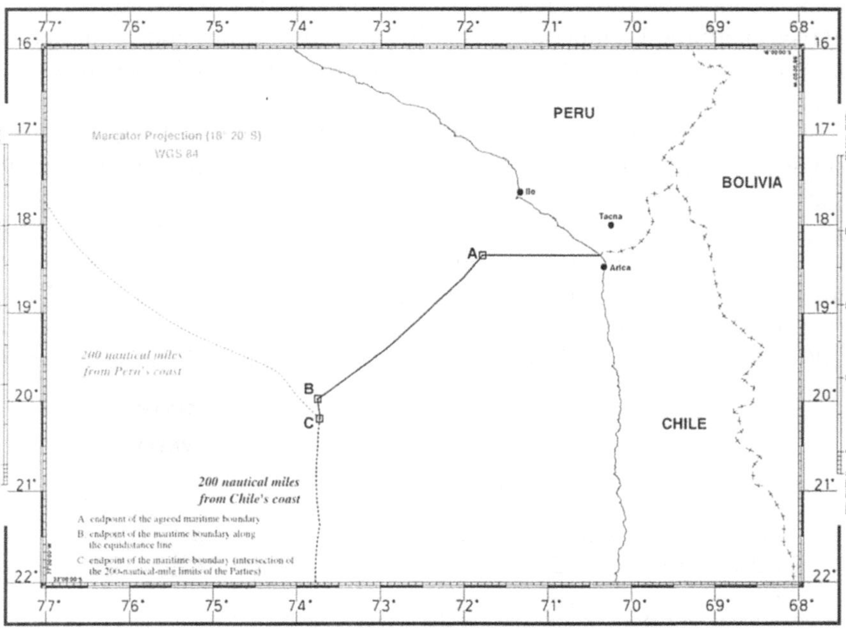

Map 20.2 Course of the maritime boundary

Another question submitted to the ICJ was to determine the starting point of the maritime boundary. By fifteen votes to one, the ICJ found that it is located at the intersection of the parallel of latitude passing through Boundary Marker No. 1 with the low-water line.[5]

The judgment is of interest for both law of treaties and law of the sea. The coasts of Peru and Chile along the area to be delimited 'are mostly uncomplicated and relatively smooth, with no distinct promontories or other distinguishing features'[6] and without islands that can influence the delimitation. This is a typical case where delimitation according to equidistance, corresponding approximately to the bisector of the angle formed by the respective coasts, would achieve an equitable solution. However, the dispute mostly related to whether Peru had previously agreed on a different delimitation, based on a parallel of latitude, which would have the effect of cutting off Peru from a large extent of waters and seabed that are closer to Peru than to Chile.

II The Santiago Declaration

Both parties agreed and the ICJ confirmed that the Santiago Declaration is an international treaty. It was subject to ratification by the three signatory states, open to accession by other states and registered according to Art. 102, para. 1, of the United Nations Charter.[7]

As to its main purpose, the Santiago Declaration represents a seminal turning point in international law of the sea. The three parties proclaimed that they possess exclusive sovereignty and jurisdiction over the sea and seabed along their coasts up to a minimum distance of 200 NM from these coasts (Arts II and III). It is the first international instrument that sets forth the concept of a 200 mile zone where the coastal states are entitled to exercise sovereign rights with regard to the exploitation of marine natural resources, both living and mineral. Adopted on the basis of the unilateral measures previously enacted by the two parties (the Chilean declaration of 23 June 1947 and the Peruvian decree of 1 August 1947), the Santiago Declaration is a direct precedent of what would become some decades later the 200 mile exclusive economic zone, as codified in the United Nations Convention on the Law of the Sea (Montego Bay, 1982; hereinafter 'UNCLOS').

5 This aspect of the judgment is not analyzed in this comment. The ICJ noted that it could be possible that the point where the land frontier starts does not coincide with the starting point of the maritime boundary (*Maritime Dispute, Judgment*, para. 175). In his declaration, judge Gaja remarked that this would create 'a situation in which, albeit for a limited stretch of the coast, the adjacent territorial sea would be under the sovereignty of a State other than the one to which the coast belongs.'

6 *Maritime Dispute, Judgment*, para. 16.

7 Ibid., para. 23.

The core of the Santiago Declaration is the outer 200 mile limit of the claimed maritime zone. In 1952, most states, and especially the great maritime powers, were not prepared to accept such an extended coastal state jurisdiction. Given the uncertain fate of its main objective, the major concern of the Declaration was to set forth the geographical, economic and political reasons that could justify and support such a claim, as it clearly appears from the preamble and several of its provisions.

The different question of the lateral delimitation of the maritime zones pertaining to the three parties, which could appear as a logical, although not indispensable, corollary to the 200 mile joint proclamation, was only partially considered in the Santiago Declaration. The parties did not include in it any provision on the delimitation of the maritime zones generated by their continental coasts. But they agreed on a provision relating to the delimitation of the maritime zones generated by their islands, making reference in this case to the parallel at the point where the land frontier reaches the sea:

> In the case of island territories, the zone of 200 nautical miles shall apply to the entire coast of the island or group of islands. If an island or group of islands belonging to one of the countries making the declaration is situated less than 200 nautical miles from the general maritime zone belonging to another of those countries, the maritime zone of the island or group of islands shall be limited by the parallel at the point at which the land frontier of the States concerned reaches the sea (Art. IV).[8]

There is no explanation as to why the parties focused on the exceptional case of islands and disregarded the continental coasts. The preparatory works do not shed light on this inconsistency. It only appears that a sort of lateral delimitation was implied in a Chilean proposal that was not adopted;[9] it also appears that the issue of islands was raised by the Ecuadorean delegate and that the other two delegates agreed on his proposal.[10] The argument put forward before the ICJ by Chile that Art. IV of the Santiago Declaration implied that delimitation according to the parallel would also be applicable for the maritime zones generated by the continental coasts is not devoid of some logical foundation, as any exception should be based on a rule. However, the ICJ relied on the ordinary meaning of the only existing provision and refrained from giving it another meaning that would imply another provision which did not exist:

> The ordinary meaning of paragraph IV reveals a particular interest in the maritime zones of islands which may be relevant even if a general

8 English translation provided in the judgment.
9 *Maritime Dispute, Judgment*, para. 67.
10 Ibid., para. 68.

> maritime zone has not yet been established. In effect, it appears that the States parties intended to resolve a specific issue which could obviously create possible future tension between them by agreeing that the parallel would limit insular zones.
>
> In light of the foregoing, the Court concludes that the ordinary meaning of paragraph IV, read in its context, goes no further than establishing the Parties' agreement concerning the limits between certain insular maritime zones and those zones generated by the continental coasts which abut such insular maritime zones.[11]

The ICJ consequently found that Chile and Peru did not, by adopting the Santiago Declaration, agree to the establishment of a lateral maritime boundary between them along the line of latitude running into the Pacific Ocean from the seaward terminus of their land boundary.[12] However, the ICJ made a remark that would be relevant for a question to be subsequently addressed:

> Nevertheless, various factors mentioned in the preceding paragraphs, such as the original Chilean proposal and the use of the parallel as the limit of the maritime zone of an island of one State party located less than 200 nautical miles from the general maritime zone of another State party, suggest that there might have been some sort of shared understanding among the States parties of a more general nature concerning their maritime boundaries. The Court will return to this matter later.[13]

There is, however, another inconsistency in Art. IV of the Santiago Declaration on which the ICJ did not elaborate. In fact, there are no islands that can justify the inclusion of Art. IV in the declaration. No 200 mile projections from islands can touch the parallel passing through the land frontier either at the border between Chile and Peru[14] or at the border between Ecuador and Peru. Here the Ecuadorean Puno island is located too close to the continental coast to have any such effect and the Ecuadorean Galapagos Archipelago is located too far away from the continental coast (at about 490 NM West) to produce any overlapping with the 200 mile limit from the coast. It thus appears that the parties agreed on the inclusion in the Santiago Declaration of a provision that is devoid of any practical meaning. In other words, Art. IV of the Santiago Declaration is most likely based on a

11 Ibid., paras. 61 and 62.
12 Ibid., para. 70.
13 Ibid., para. 69.
14 'The Court further considers that it is not necessary for it to address the existence of small islands located close to the coast in the region of the Peru-Chile land boundary. The case file demonstrates that the issue of insular zones in the context of the 1952 Santiago Declaration arose from a concern expressed by Ecuador. It is equally clear from the case file that the small islands do not appear to have been of concern to the Parties' (*Maritime Dispute, Judgment*, para. 64).

geographical error, since the parties delimited maritime areas that do not overlap and did not delimit areas that overlap.

III The Zone Agreement and the 'Tacit Agreement'

In 1954, the three states in question met again in Lima and adopted six treaties relevant to the Santiago Declaration, including the Agreement Relating to a Special Maritime Frontier Zone (hereinafter 'Zone Agreement'). It provides for the establishment of a special frontier zone where the presence of small fishing vessels cannot be considered as a violation of the waters of the parties' maritime zones. The preamble of the Zone Agreement seems to assume the existence of a maritime frontier between the state parties:

> Experience has shown that innocent and inadvertent violations of the maritime frontier [la frontera marítima] between adjacent States occur frequently because small vessels manned by crews with insufficient knowledge of navigation or not equipped with the necessary instruments have difficulty in determining accurately their position on the high seas.[15]

Art. 1 of the Zone Agreement specifies that there is a parallel that constitutes the maritime boundary between the countries concerned (Ecuador and Peru; Peru and Chile):

> A special zone is hereby established, at a distance of [a partir de] 12 nautical miles from the coast, extending to a breadth of 10 nautical miles on either side of the parallel which constitutes the maritime boundary [el límite marítimo] between the two countries.[16]

The ICJ drew the conclusion that the relevant provisions of the Zone Agreement constitute an acknowledgement of an already delimited boundary:

> the operative terms and purpose of the 1954 Special Maritime Frontier Zone Agreement are indeed narrow and specific. That is not however the matter under consideration by the Court at this stage. Rather, its focus is on one central issue, namely, the existence of a maritime boundary. On that issue the terms of the 1954 Special Maritime Frontier Zone Agreement, especially Article 1 read with the preambular paragraphs, are clear. They acknowledge in a binding international agreement that a maritime boundary already exists.[17]

15 English translation provided in the judgment.
16 English translation provided in the judgment.
17 *Maritime Dispute, Judgment*, para. 90.

The ICJ was unable to specify how and when the maritime boundary, which did not exist at the time of the Santiago Declaration,[18] was agreed upon by Chile and Peru (and Ecuador as well) in the short time between 1952 and 1954. It just found that a 'tacit agreement' was reached and that it was 'cemented' by the Zone Agreement:

> The 1954 Special Maritime Frontier Zone Agreement does not indicate when and by what means that boundary was agreed upon. The Parties' express acknowledgment of its existence can only reflect a tacit agreement which they had reached earlier. In this connection, the Court has already mentioned that certain elements of the 1947 Proclamations and the 1952 Santiago Declaration suggested an evolving understanding between the Parties concerning their maritime boundary ... In this case, the Court has before it an Agreement which makes clear that the maritime boundary along a parallel already existed between the Parties. The 1954 Agreement is decisive in this respect. That Agreement cements the tacit agreement.[19]

The assumption of the existence of a tacit agreement is a highly questionable aspect of the ICJ judgment. How can it be assumed that an agreement establishing a maritime boundary was concluded orally, at a time that cannot be specified, in a place that cannot be located, by plenipotentiaries that cannot be identified, according to preparatory works that are not documented, with no reference to such an agreement made by either of the two parties pleading before the ICJ? In fact, here the ICJ tells the parties that they had concluded an agreement that they themselves were unaware had been concluded. Is this believable or is this something closer to a miracle than to reality? According to a separate opinion, here the judgment proceeds 'in an almost Delphic manner':[20] in ancient Greece the answers given by the oracle of Apollo at Delphi were celebrated for both their wisdom and ambiguity.

In a previous judgment rendered in 2007, the ICJ held the stringent position that 'the establishment of a permanent maritime boundary is a matter of grave importance' and that 'evidence of a tacit legal agreement must be compelling.'[21] While quoting these passages in the 2014 judgment,[22] the ICJ completely departs from its own precedent, without any tangible explanation for such a change.

18 Supra, para. 2.
19 *Maritime Dispute, Judgment*, para. 91.
20 Opinion of Judge Owada, para. 8. See also the opinion of Judge Sebutinde and the joint opinion of Judges Xue, Gaja, Bhandari and Orrego Vicuña.
21 *Territorial and Maritime Dispute between Nicaragua and Honduras in the Caribbean Sea (Nicaragua v. Honduras), Judgment, I.C.J. Reports 2007*, p. 659, para. 253.
22 *Maritime Dispute, Judgment*, para. 91.

Even though the ICJ assumption of a tacit agreement is far from convincing, it is a matter of fact that the parties stated in the Zone Agreement that there is a parallel that constitutes the maritime boundary between them. Reference to the preparatory works for the Zone Agreement indicates that, strangely enough, the delegates of the three parties, after having discussed a proposal made by Ecuador, finally concurred that a lateral boundary between the maritime zones adjacent to their continental coasts and corresponding to the above-mentioned parallel had already been established under the Santiago Declaration. Nevertheless, and regardless of the perception of delegates, Art. IV of the Santiago Declaration did not delimit the maritime zones adjacent to the continental coasts of the parties and, insofar as it delimited the zones adjacent to their islands, is affected by an error of fact.[23] It thus appears that in 1954 the parties made a second error[24] in implicitly referring to a provision already affected by an error of fact. The double error is a further complication in a story already notable for inconsistencies and misunderstandings.

IV The outer limit under the Tacit Agreement (first segment of the boundary)

Once decided that a tacit agreement existed, the ICJ undertook a second step as questionable as the first one. The Court provided content to the tacit agreement. It found that the boundary line on the parallel extended up to a point (Point A) located at a distance of 80 NM from the starting point, approximately corresponding to the maximum distance reached by small fishing vessels in carrying out their activities at the considered time:

> On the basis of the fishing activities of the Parties at that time, which were conducted up to a distance of some 60 nautical miles from the main ports in the area, the relevant practice of other States and the work of the International Law Commission on the Law of the Sea, the Court considers that the evidence at its disposal does not allow it to conclude that the agreed maritime boundary along the parallel extended beyond 80 nautical miles from its starting-point.[25]

Here again the reasoning of the ICJ is far from convincing. As the ICJ elsewhere remarked,[26] the tacit agreement (if it ever existed) acknowledged in the Zone Agreement must be understood in the context of the national proclamations made by both countries in 1947 and the Santiago Declaration.

23 Insofar as the parties delimited areas that do not overlap (supra, para. 2).
24 Is it an error of law or an error of fact? The question is relevant for the application of Art. 48, para. 1, of the Vienna Convention on the Law of Treaties.
25 *Maritime Dispute, Judgment*, para. 117.
26 Ibid., para. 102.

All these instruments referred to a clearly stated 200 mile limit and were designed for the main purpose of preventing long-distance fishing and whaling vessels flying the flag of third countries from depleting the living resources located in such a broad maritime zone. There is no reason to believe that, when concluding the tacit agreement (if it ever existed) sometime between 1952 and 1954, the said states had small fishing vessels in mind and their inadvertent violations of fishing limits within 80 NM from the coast.[27] This question was to be specifically addressed by them only at a subsequent stage, i.e. in 1954 under the Zone Agreement, where no 80 mile limit was mentioned at all.

V The second and third segments of the boundary

After point A, which is the terminal point of what it found to be the boundary tacitly agreed upon by the parties, the ICJ turned to the delimitation of the remaining areas. It applied the usual three-stage methodology followed in several maritime boundary cases to provide more precise content to the rather vague 'equitable solution' rule provided for in Art. 74, para. 1, UNCLOS (delimitation of the exclusive economic zone between states with opposite or adjacent coasts).[28]

> The methodology which the Court usually employs in seeking an equitable solution involves three stages. In the first, it constructs a provisional equidistance line unless there are compelling reasons preventing that. At the second stage, it considers whether there are relevant circumstances which may call for an adjustment of that line to achieve an equitable result. At the third stage, the Court conducts a disproportionality test in which it assesses whether the effect of the line, as adjusted, is such that the Parties' respective shares of the relevant area are markedly disproportionate to the lengths of their relevant coasts.[29]

27 See the separate opinions of Judge Orrego Vicuña, para. 13, and Judge Owada, para. 24. The ICJ seems aware of the weakness of its own argument: 'The Court recalls that the all-purpose nature of the maritime boundary … means that evidence concerning fisheries activity, in itself, cannot be determinative of the extent of that boundary. Nevertheless, the fisheries activity provides some support for the view that the Parties, at the time when they acknowledged the existence of an agreed maritime boundary between them, were unlikely to have considered that it extended all the way to the 200-nautical-mile limit' (ibid., para. 111).

28 The 'equitable solution' rule is today generally considered as a customary rule of international law, which is applicable also to states, such as Peru, that are not parties to the UNCLOS. While Peru claims a 200 mile 'maritime domain,' and not an exclusive economic zone, Peru's agent formally declared before the ICJ on behalf of his government that the term 'maritime domain' used in Peru's Constitution is applied in a manner consistent with the maritime zones set out in the UNCLOS. The Court took note of this declaration 'which expresses a formal undertaking by Peru' (*Maritime Dispute, Judgment*, para. 178).

29 *Maritime Dispute, Judgment*, para. 180.

Despite the unusual character of a situation where the starting-point for the delimitation was located much further from the coast, precisely 80 NM from the closest point on the Chilean coast and about 45 NM from the closest point on the Peruvian coast,[30] the ICJ selected the appropriate basepoints on the respective coastlines to construct the provisional equidistance line.[31] The line so constructed runs almost straight in a south-west direction, reflecting the regular character of the two coasts, until it reaches the 200 NM limit measured from the Chilean baselines (Point B). Seaward of this point the 200 NM limit projections of the respective coasts no longer overlap.[32] The final short segment of the maritime boundary proceeds from Point B to Point C, where the 200 NM limits of the respective maritime entitlements intersect.[33]

As the provisional equidistance line avoided any excessive amputation of either state's maritime projections, the ICJ remarked that there was no basis for adjusting it.[34] The ICJ finally found that no significant disproportion between the lengths of the relevant coasts and the parties' shares of the relevant area was evident, such as would call into question the equitable nature of the delimitation.[35]

VI Moving from The Hague to Delphi

The dispute between Peru and Chile is unique among all maritime boundary cases. In no previous instance was the main question to be settled whether a whole boundary had ever been previously agreed upon by the parties. The treaties by which such a boundary could have been agreed upon, namely the 1952 Santiago Declaration and the 1954 Zone Agreement, did not have the specific purpose to delimit the maritime zones belonging to the parties. Moreover they were concluded at a time when the content of a rule of international law on maritime boundaries had yet to be defined and international practice on this matter was almost non-existent. It was only in 1958,

30 Ibid., para. 183.

31 'In view of the location of Point A at a distance of 80 nautical miles from the coast along the parallel, the nearest initial base point on the Chilean coast will be situated near the starting-point of the maritime boundary between Chile and Peru, and on the Peruvian coast at a point where the arc of a circle with an 80-nautical-mile radius from Point A intersects with the Peruvian coast. For the purpose of constructing a provisional equidistance line, only those points on the Peruvian coast which are more than 80 nautical miles from Point A can be matched with points at an equivalent distance on the Chilean coast' (*Maritime Dispute, Judgment*, para. 185).

32 Ibid., para. 186.

33 Ibid., para. 190.

34 Ibid., para. 191.

35 Ibid., para. 194. Given the unusual situation of an agreed line running for 80 NM along the parallel of latitude, the ICJ engaged in a broad assessment of disproportionality (ibid., para. 193).

following Art. 12, para. 1, of the Geneva Convention on the Territorial Sea and the Contiguous Zone and Art. 6 of the Geneva Contention on the Continental Shelf, that some guidance could be found on how delimitation should be made. Both provisions referred to the equidistance line and allowed for exceptions where special circumstances occurred. The several maritime boundary cases decided thereafter by the ICJ or arbitral tribunals showed all the complexities that the equidistance/special circumstances rules entails.

It thus appears that, when the delegates of Chile, Ecuador and Peru negotiated the Santiago Declaration and its collateral instruments, they could neither envisage nor seize the full legal and geometrical dimension of a maritime boundary issue. It was indeed too early. This may explain all the inconsistencies, misunderstandings and errors that the negotiations show. It is doubtful whether during the negotiations, sufficiently accurate maps were employed and cartographers were consulted to draw the relevant lines and to assess the results. It is likely, although unbelievable as it may seem today, that nobody at that time realized how illogical and inequitable a delimitation based on a parallel would represent in the geographical context of the maritime boundary between Peru and Chile.

In approaching such a unique case, the ICJ had two main choices. The first was to conclude that an agreement for a maritime boundary along the parallel already existed up to 200 NM from the coast and that Peru was consequently bound by a totally inequitable delimitation for the simple reason that it had put its signature under the words 'the parallel which constitutes the maritime boundary.' The second choice was to conclude that an agreement did not exist or, if it existed, was to be invalidated or terminated for some reasons contemplated by the law of treaties.[36] If so, a maritime boundary had to be drawn *ex novo* and, given the geographical circumstances, it could only be based on the equidistance line, despite the fact that Peru had put its signature under the words 'the parallel which constitutes the maritime boundary.' Either of the alternative choices would have meant to fully subscribe to the claim of one of the parties and to fully disregard the claim of the other.

A possible reading of the judgment is that the ICJ felt how inequitable either of the two main choices would have been, given the unique circumstances of the case. It chose a third option[37] and found a solution that, while

36 For the provisional nature of the boundary established by the Zone Agreement see the dissenting opinion of Judge Sebutinde, para. 8.

37 'The Judgment has adopted an unprecedented solution for effecting maritime delimitation in the context of the complex circumstances of this case. It appears to give satisfaction to one Party in following the parallel to the distance noted and to the other Party in continuing along an equidistance line, which were of course the two main approaches to this dispute, albeit with a different meaning and extent' (separate opinion of Judge Orrego Vicuña, para. 27).

departing from a solid legal logic, was in harmony with the inconsistencies, misunderstandings and errors that not only represented the main peculiarities of the case, but constituted its essence. This time, exceptionally, the ICJ moved its seat from The Hague to Delphi. Here a tacit agreement materialized, as a *deus ex machina*, giving the final touch in the same style of the picture initiated by the parties.

Two judges who voted with the majority explained why, despite their vote, they were not convinced about the content of the tacit agreement or its very existence. However, one of them remarked that 'the Parties' treatment of the extent of the agreed maritime boundary lacks the clarity that would have been expected in respect of a matter of that importance.[38] The other one emphasized that 'maritime disputes count, without doubt, amongst the most sensitive issues submitted by States to international adjudication,' hoping that the judgment 'will contribute to the maintenance of peaceful and friendly relations between Peru and Chile and, thereby, strengthen the public order of the oceans in Latin America.'[39]

These two remarks are at the core of this surprising case. However questionable the explanations provided by the ICJ may be, the judgment arrived at an equitable solution. As already noted, the answers given at Delphi were celebrated not only for their ambiguity, but also for their wisdom. Neither of the parties won the case, but both, building on this Delphic judgment, can now put an end to a long-standing dispute that presented serious political implications given the sensitivity of questions relating to national territory. In fact, it appears that the two countries are making all the technical steps to implement in good faith the ICJ judgment.

Bibliography

Agüero Colunga, M., *Consideraciones para la delimitación marítima del Perú*, Lima: Fondo Editorial del Congreso, 2001.

Bákula, J. M., *La imaginación creadora y el nuevo régimen jurídico del mar – Perú y Chile: ¿el desacuerdo es posible?*, Lima: Universidad del Pacífico, 2008.

Infante-Caffi, M. T., 'The Decision on the Maritime Boundary between Chile and Perú: International Law Revisited', in del Castillo, L. (ed.), *Law of the Sea, from Grotius to the International Tribunal for the Law of the Sea – Liber Amicorum Judge Hugo Caminos*, Leiden: Brill Nijhoff, 2015, pp. 529–44.

Jiménez de Aréchaga, E., 'Chile – Peru', in Charney, I. and Alexander, L. (eds), *International Maritime Boundaries*, Dordrecht: Martinus Nijhoff Publishers, 1993, I, p. 793.

Rodríguez Cuadros, M., *Delimitación marítima con equidad – El caso de Perú y Chile*, Lima: Peisa, 2007.

38 Declaration of Judge Sepúlveda-Amor, para. 18.

39 Declaration of Judge Skotnikov, para. 7.

Rodríguez Cuadros, M. (ed.), *Derecho internacional de la delimitación marítima*, Lima: Fondo Editorial Universidad Tecnológica del Perú, 2011 (see the contributions by Rodríguez Cuadros, Valle-Riestra and Horna).

Novak, F., and García-Corrochano, L., 'Presentación y anális general del fallo de la Corte Internacional de Justicia de la Haya sobre el diferendo marítimo entre el Perú y Chile', *Agenda Internacional* 21, 2014, 24–49.

Orrego Vicuña, F., 'International Law Issues in the Judgment of the International Court of Justice in the Peru-Chile Maritime Dispute Case', in del Castillo, L. (ed.), *Law of the Sea, from Grotius to the International Tribunal for the Law of the Sea – Liber Amicorum Judge Hugo Caminos*, Leiden: Brill Nijhoff, 2015, pp. 563–82.

21 *Territorial and Maritime Dispute between Nicaragua and Honduras in the Caribbean Sea (Nicaragua v. Honduras), 1999*

Chie Kojima

I Introduction

In the 2007 Judgment of the *Territorial and Maritime Dispute between Nicaragua and Honduras in the Caribbean Sea (Nicaragua v. Honduras)*,[1] the International Court of Justice (hereinafter the 'Court') determined the single maritime boundary dividing the territorial sea, continental shelf and exclusive economic zones of the parties by applying the angle-bisector method, which had not been used by the Court since the mid-1980s.[2] While the equidistance method provided in Article 15 of the United Nations Convention on the Law of the Sea (hereinafter 'UNCLOS')[3] is a general rule in the delimitation of the territorial sea,[4] the angle-bisector method proved to be 'a viable substitute method' when drawing an equidistance line is not possible or appropriate due to 'special circumstances'.[5] The Court referred to the unstable and changing nature of the mouth of the River Coco and stated that accretion caused by sediments carried to sea by the river 'might render any equidistance line so constructed today arbitrary and unreasonable in the near future'.[6] The Court thus clarified that it does not automatically apply the equidistance method to all cases.

Furthermore, *Nicaragua v. Honduras* involved the question of sovereignty over islands located in disputed maritime areas, which was an issue

1 *Territorial and Maritime Dispute between Nicaragua and Honduras in the Caribbean Sea (Nicaragua v. Honduras), Judgment, I.C.J. Reports 2007* (II), p. 659.

2 The cases in which a bisector line was drawn include: *Continental Shelf (Tunisia v. Libyan Arab Jamahiriya), Merits, Judgment, I.C.J. Reports 1982*, p. 18; *Delimitation of the Maritime Boundary in the Gulf of Maine Area, Judgment, I.C.J. Reports 1984*, p. 246; 'Delimitation of the Maritime Boundary between Guinea and Guinea Bissau, Award of 14 February 1985', *International Legal Materials* 25, 1986, 252.

3 *United Nations Convention on the Law of the Sea*, 10 December 1982, 1833 U.N.T.S. 3.

4 *Territorial and Maritime Dispute between Nicaragua and Honduras in the Caribbean Sea (Nicaragua v. Honduras)*, op. cit. n. 1, p. 745, para. 281.

5 Ibid., p. 746, para. 287.

6 Ibid., p. 742, para. 277.

raised by Nicaragua during the oral hearings. Based on evidence supporting post-colonial *effectivités,* the Court confirmed that the islands belonged to Honduras, and that it had a right to establish a territorial sea of 12 NM from the islands.[7] The Court delimited the maritime boundary by enclaving the islands in the Nicaraguan exclusive economic zone.

While some have argued that *Nicaragua v. Honduras* was a departure from earlier jurisprudence of the Court, others view the judgment as consistent with it.[8] The differences in opinion indicate that the Court's jurisprudence on maritime delimitation has not gained objectivity. However, *Nicaragua v. Honduras* helped clarify the Court's jurisprudence on methods of maritime delimitation. This chapter provides an overview of the 2007 Judgment and discusses its contribution to the development of international law as well as its implications to state practice in Latin America.

II Background and history

Both Nicaragua and Honduras obtained sovereignty over their respective territory when they became independent from Spain in 1821. However, the treaties recognizing the two states' independence and sovereignty over their respective territory comprising adjacent islands along their coasts did not specify the islands' names.[9] The two states made a series of attempts to agree on their boundary, which successfully led to the conclusion of the 1894 Gámez-Bonilla Treaty establishing the Mixed Boundary Commission.[10] Article II, paragraph 3, of the Treaty stipulated, in accordance with the principle of *uti possidetis juris,* that 'each Republic is the owner of the territory which at the date of independence constituted, respectively, the provinces of Honduras and Nicaragua'.[11] The commission succeeded in demarcating the boundary between the two states facing the Pacific Ocean, but failed to determine the disputed boundary in the western Caribbean Sea.

Subsequently, in accordance with Article III of the Treaty, Nicaragua and Honduras appointed King Alfonso XIII of Spain as the sole arbitrator to settle the dispute. The Arbitral Award handed down on 23 December 1906 drew a boundary from the mouth of the River Coco at Cape Gracias to Portillo de Teotecacinte, but the validity and binding character of the award were questioned by Nicaragua in 1912. Following the mediation by the

7 Ibid., p. 751, para. 302.

8 Judge Koroma wrote 'the Judgment is both consistent with and reflective of the jurisprudence on maritime delimitation, including the Court's case law', ibid., p. 774 (Separate Opinion of Judge Koroma).

9 Ibid., pp. 673–4, paras 34–5.

10 The translation of the treaty appears in the Judgment of 18 November 1960 in the *Case concerning the Arbitral Award made by the King of Spain on 23 December 1906* (*Honduras* v. *Nicaragua*), *I.C.J. Reports 1960*, pp. 199–202.

11 *Territorial and Maritime Dispute between Nicaragua and Honduras in the Caribbean Sea (Nicaragua v. Honduras)*, op. cit. n. 1, p. 674, para. 37.

Organization of the American States, Nicaragua submitted the dispute before the Court, which found that the award was valid and binding and that Nicaragua was under an obligation to give effect to it.[12] Since the two states could not agree on how to implement the award, the Inter-American Peace Committee upon Nicaragua's request established a Mixed Commission. In 1962, the Mixed Commission completed the demarcation of the boundary line and determined that the land boundary would begin at the mouth of the River Coco.

Honduras and Nicaragua maintained friendly relations from 1963 to 1979, but no progress was made in bilateral negotiations. The relations between the two states then deteriorated because of the capture and/or attack of fishing vessels by each state in the vicinity of the disputed maritime area. Furthermore, Honduras expressed its intention to ratify the 1986 Treaty on maritime delimitation with Colombia, in which the Honduran claim to a traditional boundary along the fifteenth parallel was recognized. Nicaragua sought a regional solution by instituting proceedings against Honduras before the Central American Court of Justice to suspend the procedure of ratification of the 1986 Treaty. Having ignored the Court's order to suspend the procedure of ratification, Honduras and Colombia exchanged instruments of ratification of the 1986 Treaty on 20 December 1999. The Court held, in its Judgment of 27 November 2001, that Honduras infringed a number of provisions of the Tegucigalpa Protocol to the Charter of the Organization of Central American States, in particular the concept of the 'territorial patrimony of Central America', included in the fundamental objectives and principles of the Central American Integration System.

Against this background, Nicaragua instituted proceedings against Honduras before the Court on 8 December 1999, requesting that it determines a single maritime boundary in areas of territorial sea, continental shelf and exclusive economic zone of the two states. During the oral proceedings, Nicaragua further raised the question of sovereignty over the islands located in the disputed maritime areas. On 8 October 2007, the Court rendered a judgment, to which Judges Ranjeva and Koroma appended separate opinions, Judge Parra-Aranguren and Judge *ad-hoc* Gaja declarations and Judge Torres Bernárdez a dissenting opinion.

III Judgment of 8 October 2007

A. Sovereignty over the islands

The Court's examination of sovereignty claims over the maritime features in the vicinity of the maritime area in dispute was divided into three phases: (1)

12 *Arbitral Award Made by the King of Spain on 23 December 1906 (Honduras v. Nicaragua), Judgment, I.C.J. Reports 1960*, p. 192.

the admissibility of the claim, (2) the type of maritime features whose sovereignty was to be determined by the Court, and (3) the legal basis of sovereignty over the islands specified by the Court.

First, the Court examined Nicaragua's claim and decided that the question of sovereignty was 'inherent in the original claim relating to the maritime delimitation' and, therefore, admissible.[13]

Second, the maritime features in the disputed maritime area included four main cays – Bobel Cay, Savanna Cay, Port Royal Cay and South Cay – and a number of smaller islets, cays and reefs. Both Nicaragua and Honduras did not identify all the islands and cays by name, but they did not dispute the fact that the four main cays remain above water at high tide. The Court, therefore, decided to consider sovereignty of these four cays as islands falling within the definition under Article 121, paragraph 1, of UNCLOS. It found that there was a lack of evidence concerning the legal status of other smaller islets, cays and reefs in the disputed area.[14]

Third, Nicaragua and Honduras argued different legal bases for sovereignty over the islands, while they agreed that none of the islands and cays in dispute, including the four main cays, were *terra nullius* when they gained independence from Spain in 1821. Nicaragua contended that it was impossible to establish the *uti possidetis juris* situation of 1821 and therefore held the original title over the cays under the principle of adjacency. Honduras asserted that its title over the cays was based on the doctrine of *uti possidetis juris*, which was evidenced by *effectivités*.

On this question, the Court stated that 'it is beyond doubt that the principle [of *uti possidetis juris*] is applicable'[15] if the Spanish Crown allocated the islands to one or the other of its colonial provinces. The Court also noted that both states agreed in Article II, paragraph 3, of the 1894 Gámez-Bonilla Treaty that 'each Republic [was] owner of the territory which at the date of independence constituted, respectively, the province of Honduras and Nicaragua'.[16] The Court, however, found that the application of the principle to the small islands would not settle the issue of sovereignty over them because of the following reasons. First, there was no evidence that Spain had allocated the islands to Nicaragua or Honduras.[17] Second, there was no legal

13 *Territorial and Maritime Dispute between Nicaragua and Honduras in the Caribbean Sea (Nicaragua v. Honduras)*, op. cit. n. 2, p. 697, para. 115.

14 The Court therefore left the question of sovereignty over those smaller islets, cays and reefs in the disputed area. L. A. de La Fayette wrote: '[p]resumably, they would belong to the State in whose waters they were located in accordance with the delimitation'. See, de La Fayette, L. A., 'Maritime Delimitation between Nicaragua and Honduras in the Caribbean Sea Case (Nicaragua v. Honduras)', *Max Planck Encyclopedia of Public International Law* VI, 2012, 1091–101.

15 *Territorial and Maritime Dispute between Nicaragua and Honduras in the Caribbean Sea (Nicaragua v. Honduras)*, op. cit. n. 2, pp. 706–7, para. 154.

16 Ibid.

17 Ibid., p. 708, para. 160.

or geographical reason that supports the adjacency argument put forward by Nicaragua.[18] Third, there was a lack of evidence to prove colonial *effectivités* in relation to the islands.[19]

Accordingly, the Court relied on post-colonial *effectivités* to determine sovereignty over the islands. Although the Court found some of the evidence submitted by Honduras unconvincing, it admitted that some other activities indeed constituted evidence of post-colonial *effectivités*, such as the application and enforcement of criminal and civil law, regulation of immigration and regulation of fishery activities. Since Nicaragua had not objected to these acts, the Court found that the evidence demonstrated an 'intention and will to act as sovereign' constituting a 'modest but real display of authority over the four islands' by Honduras.[20]

B. Delimitation of the maritime boundary

Nicaragua asserted that its maritime boundary with Honduras in the Caribbean Sea had not been delimited and proposed that the Court delimits it by means of the angle-bisector method. Honduras maintained the position that there existed a traditionally recognized boundary along the fifteenth parallel between the maritime spaces of Honduras and Nicaragua. Honduras argued that this traditional boundary originated from 'the principle of *uti possidetis juris*' and was founded on the practice of the parties and third states as well as the Spanish Crown's practice to use parallels and meridians as boundaries.

The Court found that Honduras could present no evidence of previous colonial power's practice, and therefore it could not delimit the maritime boundary based on the *uti possidetis juris* principle.[21] The Court further rejected Honduras' claim that neither the oil concessions, the 1986 Treaty between Colombia and Honduras nor the 1993 Treaty between Colombia and Jamaica were sufficient to prove that a tacit agreement establishing a boundary between the two states existed.[22] The Court applied a strict standard wherein evidence of a tacit agreement had to be compelling.[23]

Turning to the question of the methodology to be used for the maritime boundary delimitation, the Court emphasized that 'the equidistance method does not automatically have priority over other methods of delimitation and, in particular circumstances, there may be factors that make the application of the equidistance method inappropriate'.[24] The important factor in

18 Ibid., p. 709, para. 164.
19 Ibid., p. 710, para. 166.
20 Ibid., p. 721, para. 208.
21 Ibid., p. 729, para. 234.
22 Ibid., pp. 735–7, paras 254–8.
23 Ibid., p. 735, para. 253.
24 Ibid., p. 741, para. 272.

Nicaragua v. Honduras was that the mouth of the River Coco shifted considerably over time. Moreover, the parties agreed that the sediment carried to and deposited at sea by the river shaped its delta, as well as the coastline to the north and south of Cape Gracias a Dios, and exhibited a very active morpho-dynamism.[25] The Court, therefore, found that 'continued accretion at the Cape might render any equidistance line so constructed today arbitrary and unreasonable in the near future',[26] and concluded that it was impossible to identify base points and construct a provisional equidistance line.[27]

Envisaging a situation where the drawing of an equidistance line is not possible or appropriate, Article 15 of UNCLOS, which is identical to Article 12, paragraph 1, of the 1958 Convention on the Territorial Sea and the Contiguous Zone, stipulates that the equidistance method does not apply 'where it is necessary by reason of historic title or other special circumstances'. The Court stated that '[n]othing in the wording of Article 15 suggests that geomorphological problems are *per se* precluded from being "special circumstances" within the meaning of the exception, nor that such "special circumstances" may only be used as a corrective element to a line already drawn'.[28] Having considered the commentary of the International Law Commission and its own jurisprudence, the Court concluded that it faced 'special circumstances' in which it could not apply the equidistance principle.[29] Simultaneously, the Court reiterated that equidistance remained the general rule.[30]

Accordingly, the Court applied an alternative method, a line bisecting two lines drawn along the coastal fronts of the two states. The bisector method 'seeks to approximate the relevant coastal relationships, but does so on the basis of the macro-geography of a coastline as represented by a line drawn between two points on the coast', whereas the equidistance method does the same by considering the 'relationships between designated pairs of base points'.[31] The Court emphasized that the key elements for using the bisector method were the 'geographical configuration of the coast, and the geomorphological features of the area where the endpoint of the land boundary is located'.[32] In considering the relevant mainland coast, the Court rejected Nicaragua's argument that the entire Caribbean coast should be taken into account. Having considered different coastal fronts, the Court decided that the front extending from Punta Patuca to Wouhnta would avoid the problem of cutting off Honduran territory and simultaneously provide a coastal

25 Ibid., p. 742, para. 277.
26 Ibid.
27 Ibid., pp. 743–5, para. 280.
28 Ibid.
29 Ibid., p. 745, para. 281.
30 Ibid.
31 Ibid., p. 747, para. 289.
32 Ibid., p. 748, para. 292.

façade of sufficient length to account properly for the coastal configuration in the disputed area.[33] The bisector line, therefore, had an azimuth of 70° 14' 41.25" in the judgment.

With regard to the delimitation around the islands, which were confirmed to be under Honduras' sovereignty, the Court noted that Nicaragua did not dispute that the islands generated a territorial sea but argued that the Honduran islands should only have a 3-mile territorial sea because a 'full 12-mile territorial sea … would result in giving a disproportionate amount of the maritime areas in dispute to Honduras'.[34] Honduras insisted its claim of a 12-mile territorial sea for the four islands. On this question, the Court upheld the claim of Honduras by simply referring to Article 3 of UNCLOS.[35] Accordingly, Bobel Cay, Savanna Cay, Port Royal Cay and South Cay were accorded a territorial sea of 12 NM.[36]

In generating a 12-mile territorial sea from the islands, an overlap occurred between the territorial sea of Bobel Cay, Port Royal Cay and South Cay (Honduras) and the territorial sea of Edinburgh Cay (Nicaragua) both to the south and to the north of the fifteenth parallel. To solve this problem, the Court suggested to first provisionally draw an equidistance line and then consider whether the line should be adjusted in light of special circumstances, which was a method also used in the *Maritime Delimitation and Territorial Questions between Qatar and Bahrain (Qatar v. Bahrain)*.[37] Since the Court did not find any legally relevant 'special circumstances' in the area, it drew an equidistance line between the overlapping territorial seas of the three Honduran islands and Edinburgh Cay.[38]

Regarding the starting point of the delimitation, the Court upheld Honduras' submission and determined that the starting point should begin 3 miles out to sea from the point already identified by the Mixed Commission in 1962 along the azimuth of the bisector line it had drawn.[39] The Court held that the parties must negotiate in good faith to agree on a line that links the end of the land boundary as fixed by the 1906 Award and the starting point of the maritime boundary determined in the judgment.[40] Concerning the endpoint, the Court took into account certain interests of third states arising from bilateral treaties and consequently delimited the maritime boundary extending beyond the 82nd meridian without specifying a precise endpoint.[41]

33 Ibid., p. 749, para. 298.
34 Ibid., p. 751, para. 300.
35 Ibid., p. 751, para. 302.
36 Ibid.
37 Ibid., pp. 751–2, para. 303; *Maritime Delimitation and Territorial Questions between Qatar and Bahrain (Qatar v. Bahrain), Judgment, I.C.J. Reports 2001*, p. 94 para. 176.
38 Ibid., p. 752, paras 304–5.
39 Ibid., p. 756, para. 311.
40 Ibid., p.763.
41 Ibid., p. 759, para. 319.

IV Appraisals

A. Methods of maritime delimitation

UNCLOS comprises provisions concerning the delimitation of territorial sea (Art. 15), exclusive economic zone (Art. 74) and continental shelf (Art. 83). With regard to the delimitation of the territorial sea, Article 15 of UNCLOS stipulates the equidistance/special circumstances rule, which was equally embodied in Article 12, paragraph 1, of the 1958 Convention on the Territorial Sea and the Contiguous Zone.[42] However, Articles 74 and 83 do not specify any method of delimitation for the exclusive economic zone and the continental shelf. In the drafting process of these articles during the Third United Nations Conference on the Law of the Sea (UNCLOS III), one group of states that was in favour of the equidistance/special circumstances rule, and another group of states that was in favour of equitable principles conflicted with each other. Difficulties in delimiting a maritime boundary were already foreseen since the time of the UNCLOS III.[43] Without explicit references to the method of maritime delimitation of the exclusive economic zone and continental shelf under UNCLOS, jurisprudence of international courts and tribunals reflected certain trends in terms of choices of the method of maritime delimitation.

It has been observed that the most recent trend in the jurisprudence of maritime delimitation is the 'three-stage approach', which appeared after the two-stage approach widely employed before *Nicaragua v. Honduras*. The three-stage approach was first employed in the *Maritime Delimitation in the Black Sea (Romania v. Ukraine)*[44] in 2009, then in the *Territorial and Maritime Dispute (Nicaragua v. Colombia)*[45] in 2012 and in the *Delimitation of the Maritime Boundary between Bangladesh and Myanmar in the Bay of Bengal (Bangladesh/Myanmar)* in 2012.[46] The three-stage approach consists of: (1) constructing a provisional equidistance line, based on the geography of the parties' coasts and mathematical calculations, (2) determining whether there are any relevant circumstances requiring adjustment of the provisional equidistance line, and (3) checking whether the line, as adjusted, resulted in any significant disproportion between the ratio of the

42 *Convention on the Territorial Sea and the Contiguous Zone*, 29 April 1958, 516 U.N.T.S. 205.

43 Yanai, S., 'Can the UNCLOS Address Challenges of the 21st Century?', *German Yearbook of International Law* 57, 2014, 54.

44 *Maritime Delimitation in the Black Sea (Romania v. Ukraine), Judgment, I.C.J. Reports 2009*, pp. 101–3, paras 116–22.

45 *Territorial and Maritime Dispute (Nicaragua v. Colombia), Merits, Judgment, I.C.J. Reports 2012*, pp. 695–8, paras 190–9.

46 *Dispute concerning Delimitation of the Maritime Boundary between Bangladesh and Myanmar in the Bay of Bengal (Bangladesh/Myanmar), ITLOS Reports 2012*, p. 76, para. 240.

respective coastal lengths and the ratio of relevant maritime areas allocated to each party.[47] Despite this trend in maritime delimitation cases, it is generally understood that 'there is no single or mandatory delimitation method that is applicable to all delimitation cases, as each case arises under unique circumstances' as Judge Yanai writes.[48] In *Nicaragua v. Colombia*, the Court referred to *Nicaragua v. Honduras* and reiterated that the three-stage approach should not 'be applied in a mechanical fashion' and that it would not be 'appropriate in every case to begin with a provisional equidistance/median line'.[49]

Subsequent judgments of international courts and tribunals positively refer to *Nicaragua v. Honduras* by evaluating the angle-bisector method as secondary or alternative choice of delimitation method. In *Romania v. Ukraine*, the Court referred to *Nicaragua v. Honduras* in the following way: '[s]o far as delimitation between adjacent coasts is concerned, an equidistance line will be drawn unless there are compelling reasons that make this unfeasible in the particular case'.[50] The International Tribunal for the Law of the Sea similarly observed in *Bangladesh/Myanmar* that the angle-bisector method was 'an alternative to the equidistance/relevant circumstances method, where recourse to it has not been possible or appropriate'.[51]

Accordingly, *Nicaragua v. Honduras* contributed to clarify the international court's jurisprudence that the equidistance method, even though it is recognized as the general rule, does not automatically have priority over other methods of delimitation. A controversy, however, remains with regard to the criteria on admitting the deviation from the general rule. Some authors question whether the highly unstable nature of the mouth of a river can alone exclude the application of the equidistance method. Judge Ranjeva pointed out in his Separate Opinion that UNCLOS 'did not overlook the potential for extremely unstable coastlines'.[52] Article 7, paragraph 2, of UNCLOS stipulates a solution to identify basepoints for unstable coastlines:

> Where because of the presence of a delta and other natural conditions the coastline is highly unstable, the appropriate points may be selected along the furthest seaward extent of the low-water line and, notwithstanding subsequent regression of the low-water line, the straight

47 Ibid.

48 Yanai, S., 'Can the UNCLOS Address Challenges of the 21st Century?', op. cit. n. 43, p. 55.

49 *Territorial and Maritime Dispute (Nicaragua v. Colombia)*, op. cit. n. 45, p. 696, para. 194.

50 *Maritime Delimitation in the Black Sea (Romania v. Ukraine)*, op. cit. n. 44, p. 101, para. 116.

51 *Dispute concerning Delimitation of the Maritime Boundary between Bangladesh and Myanmar in the Bay of Bengal (Bangladesh/Myanmar)*, op. cit. n. 46, pp. 74–5 para. 234.

52 *Territorial and Maritime Dispute between Nicaragua and Honduras in the Caribbean Sea (Nicaragua v. Honduras)*, op. cit. n. 1, p. 767 (Separate Opinion of Judge Ranjeva).

baselines shall remain effective until changed by the coastal State in accordance with this Convention.

Judge Ranjeva criticized the Court's restrictive approach in *Nicaragua v. Honduras* that the geomorphological element was solely applied as a sufficient reason to justify the necessity of abandoning the general rule, instead of proving the impossibility of applying the general rule.[53]

Y. Tanaka pointed to the changing nature of the configuration of coastlines due to erosion, accretion and sea-level rise resulting from global warming.[54] He emphasized the stable nature of legal maritime boundaries by stating that maritime boundaries established by agreements and through international adjudication should not be affected by subsequent changes in the configuration of coasts.[55] Tanaka further argued that the shifting coastline alone would not automatically make the equidistance line inequitable since the Court was supposed to examine all relevant circumstances in adjusting the equidistance line in order to achieve an equitable result.[56]

It must be considered, however, that both a decision to follow the general rule and a decision not to follow the general rule are not free from subjective elements. One may recall that, in *Bangladesh/Myanmar*, the Tribunal stated that '[i]nternational courts and tribunals have developed a body of case law on maritime delimitation which has reduced the elements of subjectivity and uncertainty in the determination of maritime boundaries and in the choice of methods employed to that end'.[57] On this point, Judge Treves succinctly commented that the progress made towards reducing subjectivity and uncertainty is 'probably unavoidably, far from complete so that adjudication of disputes concerning delimitation of maritime areas remains an uncertain exercise where subjective elements are often decisive'.[58] Even though the equidistance principle seems to gain certain objectivity among scholars and legal practitioners, its application cannot be automatic for all maritime delimitation cases in principle.

B. Treatment of islands

In *Nicaragua v. Honduras*, the Court admitted 'the right to establish the breath of its territorial sea up to a limit of 12 nautical miles be that for its

53 Ibid.

54 Tanaka, Y., 'Reflections on Maritime Delimitation in the Nicaragua/Honduras Case', *Zeitschrift für ausländisches öffentliches Recht und Völkerrecht* 68, 2008, p. 925.

55 Ibid., pp. 925–6.

56 Ibid., p. 926.

57 *Dispute concerning Delimitation of the Maritime Boundary between Bangladesh and Myanmar in the Bay of Bengal (Bangladesh/Myanmar)*, op. cit. n. 47, p. 72, para. 226.

58 Treves, T., 'Maritime Delimitation and Offshore Features', in Jayakumar, S., Koh, T., and Beckman, R. (eds), *The South China Sea Disputes and Law of the Sea*, Cheltenham: Edward Elgar, 2014, p. 132.

mainland or for islands under its sovereignty'.[59] The right was also recognized in the judgments of *Nicaragua v. Colombia* in 2012[60] and *Bangladesh/Myanmar* in 2012.[61] The Court stated, in *Nicaragua v. Colombia*, that it had never restricted the right on the basis of an overlap with the continental shelf and exclusive economic zone of another state.[62]

With regard to the effect of islands in the delimitation of the exclusive economic zone or continental shelf boundary, the jurisprudence of international courts and tribunals provides no general rule.[63] The effect to be given to an island 'depends on the geographic realities and the circumstances of the specific case'.[64] In *Nicaragua v. Honduras*, the Court enclaved the islands with a 12 NM territorial sea within the Nicaraguan exclusive economic zone and drew an equidistance line between the overlapping territorial seas of the three Honduran islands and the Nicaraguan island.

C. State practice in Latin America

The implications of *Nicaragua v. Honduras* to state practice in Latin America are twofold: the recognition of principle of *uti possidetis* and the tradition of utilizing international institutions for the peaceful settlement of international disputes. First, the principle of *uti possidetis* was developed by Latin American states as early as the nineteenth century.[65] Although the Court could not apply the principle of *uti possidetis* in determining the maritime boundary due to the lack of evidence, it endorsed the applicability of the principle in the case.[66]

Second, Nicaragua and Honduras initially sought different peaceful means of dispute settlement such as negotiations, international commission and arbitral tribunal established under bilateral agreement, and mediation by a regional organization. Considering the tradition of Central American states as being the first ones to establish a regional international court in the world, it is not surprising that the parties positively utilized international institutions

59 *Territorial and Maritime Dispute between Nicaragua and Honduras in the Caribbean Sea (Nicaragua v. Honduras)*, op. cit. n. 1, p. 751, para. 302.
60 *Territorial and Maritime Dispute (Nicaragua v. Colombia)*, op. cit. n. 45, p. 690, para. 177.
61 *Dispute concerning Delimitation of the Maritime Boundary between Bangladesh and Myanmar in the Bay of Bengal (Bangladesh/Myanmar)* op. cit. n. 46, pp. 55–6, para. 169.
62 *Territorial and Maritime Dispute (Nicaragua v. Colombia)*, op. cit. n. 45, pp. 690–1, para. 178.
63 *Dispute concerning Delimitation of the Maritime Boundary between Bangladesh and Myanmar in the Bay of Bengal (Bangladesh/Myanmar)* op. cit. n. 46, p. 96, para. 317.
64 Ibid.
65 Infante Caffi, M. T., 'Boundary Disputes in Latin America', *Max Planck Encyclopedia of Public International Law* I, 2012, 1017–8.
66 *Territorial and Maritime Dispute between Nicaragua and Honduras in the Caribbean Sea (Nicaragua v. Honduras)*, op. cit. n. 1, pp. 706–7, para. 154.

for the common goal of resolving a maritime boundary dispute, except for one occasion when Honduras ignored the proceedings before the Central American Court of Justice brought forth by Nicaragua. In this historical context, it is significant that the Court finally played a role in resolving one of the long-lasting disputes among the Central American states with a judgment accepted by both parties.[67]

Bibliography

De La Fayette, L. A., 'Maritime Delimitation between Nicaragua and Honduras in the Caribbean Sea Case', *Max Planck Encyclopedia of Public International Law* VI, 2012, 1091–101.

Infante Caffi, M. T., 'Boundary Disputes in Latin America', *Max Planck Encyclopedia of Public International Law* I, 2012, 1017–28.

Kirk, E. A., 'Case Concerning Territorial and Maritime Dispute between Nicaragua and Honduras in the Caribbean Sea (Nicaragua v. Honduras), Judgment of 8 October 2007', *International and Comparative Law Quarterly* 57, 2008, pp. 701–9.

Lathrop, C. G., 'Territorial and Maritime Dispute between Nicaragua and Honduras in the Caribbean Sea (Nicaragua v. Honduras)', *American Journal of International Law* 102, 2008, pp. 113–9.

Tanaka, Y., 'Reflections on Maritime Delimitation in the Nicaragua/Honduras Case', *Zeitschrift für ausländisches öffentliches Recht und Völkerrecht* 68, 2008, 903–37.

Treves, T., 'Maritime Delimitation and Offshore Features', in Jayakumar, S., Koh, T., and Beckman, R. (eds), *The South China Sea Disputes and Law of the Sea*, Cheltenham: Edward Elgar, 2014, pp. 121–43.

Yanai, S., 'International Law concerning Maritime Boundary Delimitation', in D.J. Attard, M. Fitzmaurice and N.A. Martínez Gutiérrez (eds) *The IMLI Manual on International Maritime Law*, Volume I: The Law of the Sea, Oxford University Press, 2014, pp. 304–41.

Yanai, S., 'Can the UNCLOS Address Challenges of the 21st Century?', *German Yearbook of International Law* 57, 2014, 43–62.

67 According to the information received by the author from official sources, the Presidents of Nicaragua and Honduras met on the 8th of October in 2007 and confirmed that both states would accept the judgment and continue to negotiate for maintaining friendly relations between the two states.

22 *Case concerning the Arbitral Award made by the King of Spain on 23 December 1906 (Honduras v. Nicaragua), 1960*

Yoshifumi Tanaka

I Introduction

It is beyond serious argument that inter-state arbitration occupies an important place within the international law of peaceful settlement of international disputes. In his time-honoured publication, *Le droit des gens*, Emer de Vattel stressed the role of arbitration in international dispute settlement, stating that:

> Arbitration is a very reasonable mode, and one that is perfectly conformable to the law of nature, for the decision of every dispute which does not directly interest the safety of the nation.[1]

In Latin America, arbitration was accepted as a means of the settlement of disputes, including even territorial issues, in many treaties concluded between the mid-nineteenth and early twentieth centuries. In some cases, however, the validity of the arbitral award was disputed.[2] *Case concerning the Arbitral Award made by the King of Spain on 23 December 1906* (hereinafter '*King of Spain case*') is one such case where the validity of an arbitral award was at issue before the International Court of Justice (hereinafter 'ICJ' or 'the Court').[3] An essential issue to be considered in this case concerns reasons that may deprive the validity of an arbitral award. The *King of Spain*

1 de Vattel, Emmerich, *The Law of Nations; or Principles of the Law of Nature, Applied to the Conduct and Affairs of Nations and Sovereigns* (translated by Joseph Chitty, Philadelphia, T and J. W. Johnson and Co., Law Booksellers, 1853), 278, section. 329. For the original French text, see de Vattel, E., *Le droit des gens ou principes de la loi naturelle, Appliqués à la conduite et aux affaires des Nations et des Souverains* (The Classics of International Law Washington 1916).

2 See Dissenting Opinion of Judge Urrutia Holguín in *Case concerning the Arbitral Award made by the King of Spain on 23 December 1906 (Honduras v. Nicaragua)* (hereinafter the '*King of Spain case*'), *Judgment of 18 November 1960, I.C.J. Reports 1960*, pp. 223–6.

3 Another case concerning the nullity of an arbitral award before the ICJ is *Arbitral Award of 31 July 1989 (Guinea-Bissau v. Senegal), Judgment of 12 November 1991, I.C.J. Reports 1991*, p. 53.

case provides an important insight into the nullity of an arbitral award in relation to territorial disputes. Thus, this contribution will seek to succinctly examine the *King of Spain case*.

II Course of the litigation

On 7 October 1894, Honduras and Nicaragua concluded the Gámez-Bonilla Treaty in order to demarcate the boundary line between the two republics.[4] Although the Mixed Boundary Commission established by Article I of the Treaty succeeded in determining the boundary from the Pacific Coast to the *Portillo de Teotecacinte*, it was unable to agree on the boundary from that point to the Atlantic Coast.[5] With regard to the latter section of the boundary, the King of Spain, Alphonse XIII, handed down an arbitral award on 23 December 1906.[6] In a Note dated 19 March 1912, however, Nicaragua challenged the validity and binding character of the award. As a consequence, a dispute arose with regard to the award. Honduras and Nicaragua attempted to settle the dispute by direct negotiation or through the good offices or mediation of other states but to no avail. Thus, the Council of the Organization of American States (OAS), at Honduras's request, appointed an investigating committee and issued a report recommending the ICJ as an appropriate forum to settle the dispute. To this end, an *ad hoc* committee appointed by the OAS Council drafted a proposal for an agreement to submit the case before the ICJ. On 21 July 1957, the two parties reached an agreement to submit the dispute to the ICJ.[7]

On 1 July 1958, Honduras instituted proceedings against Nicaragua before the ICJ with regard to a dispute concerning the arbitral award made by the King of Spain.[8] By the application, Honduras requested the Court to declare, *inter alia*, that Nicaragua is under an obligation to give effect to the award.[9] By contrast, Nicaragua solicited the Court to adjudge and declare that the decision given by the King of Spain on 23 December 1906 did not possess the character of a binding arbitral award and that the arbitral decision was incapable of being executed by reason of its omissions, contradictions and obscurities.[10] In the *King of Spain case*, four issues in particular deserve

4 *King of Spain case, Judgment*, p. 199.

5 Ibid., p. 202,.

6 Ibid. See also *The Boundary Case between Honduras and Nicaragua*, 23 December 1906, *RIAA* 11, pp. 101–117.

7 *King of Spain case, Judgment*, p. 203. See also Schulte, C., *Compliance with Decisions of the International Court of Justice*, Oxford: Oxford University Press, 2004, p. 127.

8 *King of Spain case*, Judgment, pp. 194–5. Roberto Ago and Francisco Urrutia Holguín were respectively chosen as Judges *ad hoc* by the government of Honduras and the government of Nicaragua.

9 Ibid., p. 204. See also *I.C.J., Memorial submitted by the Government of the Republic of Honduras, King of Spain case*, p. 62.

10 *King of Spain case, Judgment*, pp. 204–5. See also *I.C.J., Counter-Memorial submitted by the Government of the Republic of Nicaragua, King of Spain case*, pp. 204–5.

consideration: (i) the validity of the designation of the King of Spain as arbitrator, (ii) the lapse of the Gámez-Bonilla Treaty, (iii) the conditions that nullify the arbitral award on 23 December 1906, and (iv) the incapability of execution of the arbitral award. The following sections will address these four issues, respectively.

III The validity of the designation of the King of Spain as arbitrator

The first issue to be examined pertains to the validity of the designation of the King of Spain as arbitrator. In this regard, Nicaragua argued that the requirements in relation to the composition of an arbitral tribunal set out in Articles III and V of the Gámez-Bonilla Treaty were not complied with in the designation process.[11] Under Article III of the Treaty, an arbitral tribunal shall be composed of three members: one representative for Honduras and another for Nicaragua and one Member of the foreign Diplomatic Corps accredited to Guatemala, the last to be elected by the first two or chosen by lot.[12] The record showed that on 2 December 1899, the two national arbitrators designated the Mexican Chargé d'affaires in Central America, Federico Gamboa, as third member of the arbitral tribunal. As he was recalled from Guatemala in April 1902, the two national arbitrators designated the Mexican Minister to Central America, Cayetano Romero, as third member of the tribunal on 21 August 1902. However, he left Guatemala soon after for health reasons. At a meeting on 2 October 1904, the two national arbitrators met with the Spanish Minister to Central America and, on this occasion, the King of Spain was designated as the sole arbitrator. In this regard, Nicaragua contended that before the two national arbitrators could proceed to this designation, it was necessary to exhaust the membership of the foreign Diplomatic Corps accredited to Guatemala and, thereafter, to come to an agreement on any other foreign or Central American public figure for the purpose of constituting a three-member arbitral tribunal.[13]

The Court was not persuaded by Nicaragua's argument. In the Court's view, it must be established by positive proof whether the procedure prescribed in the Treaty had not been complied with. Nonetheless, no such proof had been placed before the Court. According to the Court, it was within the power of the arbitrators to interpret and apply the articles in question in order to discharge their function of organising the arbitral tribunal. The Court therefore concluded that the requirements of the relevant articles of the Gámez-Bonilla Treaty, as interpreted by the two national arbitrators, had already been complied with when it was agreed by common consent that

11 Ibid., p. 168, para. 92 et seq.
12 *King of Spain case, Judgment*, p. 200.
13 Ibid., pp. 205–6.

the King of Spain be designated as arbitrator at the meeting of 2 October 1904.[14]

The Court's view appears to leave some room for discussion. Originally, the arbitral tribunal set out by the Gámez-Bonilla Treaty was designed to be composed of three arbitrators. Hence, some doubts might arise whether the establishment of the arbitral tribunal that consists of the sole arbitrator should be contrary to Article III of the Gámez-Bonilla Treaty. In considering this issue, the 1950 *Peace Treaties case* must be noted.[15] In this case, one of the questions submitted to the ICJ related to the composition of the commission established by the Peace Treaties with Bulgaria, Hungary and Romania of 1947.[16] In its advisory opinion of 1950, the Court held that the constitution of a two-member commission, rather than instituting a three-member commission according to the provisions of the treaties, is not the designated commission that caters for the Peace Treaties. Nor would the decisions of a commission of two members have the same degree of moral authority as those of a three-member commission.[17] It therefore found that the UN Secretary-General was unauthorised to appoint the third member of the commission upon the request of the other party to a dispute, should one party fail to appoint a representative to a Treaty Commission under the Peace Treaties.[18] In light of the *Peace Treaties* advisory opinion, there appears to be some scope to reconsider the question whether a one-member arbitral tribunal was not the intended arbitral tribunal that the Gámez-Bonilla Treaty envisioned.[19]

In any case it was only on 19 March 1912 when Nicaragua, for the first time, challenged the validity of the designation of the King of Spain as arbitrator. According to the Court, no question as to either the validity of his designation as arbitrator or his justification as such was at any time raised in the arbitral proceedings before the King. The Court thus held that it was unable to accept the Nicaraguan allegation that the designation of the King of Spain as arbitrator was invalid.[20]

IV The lapse of the Gámez-Bonilla Treaty

The second contention of Nicaragua was that the Gámez-Bonilla Treaty had lapsed before the King of Spain agreed to act as arbitrator.[21] Nicaragua

14 Ibid., p. 206.
15 *Interpretation of Peace Treaties with Bulgaria, Hungary and Romania, Advisory Opinion of 18 July 1950* (the Second Phase), *I.C.J. Reports 1950*, p. 221.
16 Ibid., pp. 223–4.
17 Ibid., p. 228.
18 Ibid., p. 230.
19 Separate Opinion of Sir Percy Spender, *King of Spain case, Judgment*, p. 219.
20 *King of Spain case, Judgment*, p. 207.
21 *I.C.J., Counter-Memorial submitted by the Government of Nicaragua, King of Spain case*, p. 175, para. 111 et seq.

argued that the Treaty took effect on 7 October 1894, the date on which it was signed, and that it lapsed ten years later, namely, on 7 October 1904, in accordance with Article XI. It was on 17 October 1904 when the King of Spain agreed to act as arbitrator. According to Nicaragua, his designation as arbitrator took effect ten days after the treaty had ceased to be in force. By contrast, Honduras argued that the treaty did not come into effect until the exchange of ratifications between the parties on 24 December 1896. According to Honduras, the ten-year period laid down in Article XI of the Treaty expired on 24 December 1906, and as a consequence, the award was granted during the currency of the treaty.[22]

In this regard, the Court took the position that the intention of the parties was that the treaty should come into force on the date of exchange of ratifications and that the ten-year period specified in Article XI should commence from that date, even though there was no expressed provision in the treaty with regard to the date on which it was to come into force.[23] In fact, on the very day the treaty expired, the president of Nicaragua stated in his telegram to the Spanish Minister to Central America that it would be satisfactory and an honour for Nicaragua if the King of Spain would accept his designation as arbitrator to settle the boundary dispute between Honduras and Nicaragua.[24] According to the Court, 'this furnishes a clear indication that Nicaragua did not regard the treaty as having lapsed on that day'.[25] In relation to this, the Court stressed the following facts: (i) the designation of the King of Spain as arbitrator was freely agreed to by Nicaragua; (ii) no objection was made by Nicaragua to the jurisdiction of the King of Spain as arbitrator either on the ground of irregularity in his designation as arbitrator or on the ground that the Gámez-Bonilla Treaty had lapsed even before the King of Spain had signified his acceptance of the office of arbitrator; and (iii) Nicaragua fully participated in the arbitral proceedings before the King. The Court therefore concluded that the Gámez-Bonilla Treaty was in force until 24 December 1906 and that the King's acceptance on 17 October 1904 of his designation as arbitrator was well within the currency of the treaty; and that it was no longer open to Nicaragua to rely on either of these contentions as furnishing a ground for the nullity of the award.[26] Furthermore, the Court held that Nicaragua, by expressed declaration and by conduct, recognised the award made by the King of Spain as valid and was no longer able to rescind upon that recognition and challenge the validity of the award.[27]

22 *King of Spain case, Judgment*, p. 208. See also *I.C.J., Reply submitted by the Government of Honduras, King of Spain case*, p. 522, para. 102.

23 In this regard, it may be relevant to note that Article VIII of the Gámez-Bonilla Treaty required 'constitutional ratifications'. Ibid., p. 522, para. 104.

24 *King of Spain case, Judgment*, p. 208.

25 Ibid., p. 209.

26 Ibid.

27 Ibid., p. 213.

It is of particular interest to note that the conduct of Nicaragua constituted the key element in reaching the conclusion. Here some consideration must be given to the doctrine of acquiescence and the related doctrine of estoppel. As the Chamber of the ICJ in the 1984 *Gulf of Maine case* stated, the concepts of acquiescence and estoppel both derive from the fundamental principles of good faith and equity.[28] According to the Chamber, 'acquiescence is equivalent to tacit recognition manifested by unilateral conduct which the other party may interpret as consent, while estoppel is linked to the idea of preclusion'.[29] While the distinction between acquiescence and estoppel is not necessarily clear-cut, it is argued that acquiescence presumes consent to have existed on the basis of the factual circumstances, whereas estoppel recognises the possibility that consent was non-existent in light of any fact within its actual or presumed knowledge.[30] It can also be considered that the passage of time is likely to be more important in the cases of acquiescence than in cases of estoppel.[31]

In relation to this, Bowett has enumerated three essential elements of estoppel: (i) the meaning of the statement must be clear and unambiguous; (ii) the statement or representation must be voluntary, unconditional and authorised; and (iii) reliance in good faith upon the representation of one party by the other party to his detriment.[32] These elements were confirmed by the ICJ in the 1969 *North Sea Continental Shelf* judgment.[33] Furthermore, according to Fitzmaurice, the essential condition for the operation of the rule of preclusion or estoppel is that the party invoking the rule must have 'relied upon' the statements or conduct of the other party, either to its own detriment or to the other's advantage.[34] In fact, the Chamber of the ICJ, in the *El Salvador/Honduras case*, also considered 'a statement or representation made by one party to another and reliance upon it by that other party to his detriment or to the advantage of the party making it' as an essential element of estoppel.[35]

28 *Delimitation of the Maritime Boundary in the Gulf of Maine Area (Canada/United States of America), Judgment of 12 October 1984, I.C.J. Reports 1984*, p. 305, para. 130.

29 Ibid.

30 Thirlway, H., *The Law and Procedure of the International Court of Justice: Fifty Years of Jurisprudence*, Vol. I: Oxford: Oxford University Press, 2013, pp. 40–41.

31 Ibid., p. 41.

32 Bowett, D. W., 'Estoppel before International Tribunals and Its Relation to Acquiescence', *British Yearbook of International Law* 33, 1957, pp. 188–94. This view was echoed by Kolb. See Kolb, R., *La bonne foi en droit international public: Contribution à l'étude des principes généraux de droit*, Paris: PUF, 2000, pp. 360 et seq.

33 *North Sea Continental Shelf (Federal Republic of Germany v. Denmark; Federal Republic of Germany v. Netherlands), Judgment of 20 February 1969, I.C.J. Reports 1969*, p. 26, para. 30.

34 *Temple of Preah Vihear (Cambodia v. Thailand), Separate Opinion of Sir Gerald Fitzmaurice, Judgment of 15 June 1962, Merits, I.C.J. Reports 1962*, p. 63.

35 *Land, Island and Maritime Frontier Dispute (El Salvador/Honduras: Nicaragua intervening), Judgment of 13 September 1990, Application by Nicaragua for Permission to Intervene, I.C.J. Reports 1990*, p. 118, para. 63.

In the *King of Spain case*, the ICJ did not make an explicit reference to either estoppel or acquiescence. In this regard, Judge *ad hoc* Urrutia Holguín maintained that the theory of estoppel could not be invoked against Nicaragua because it could not be said that Nicaragua's attitude between 1906 and 1912 was the cause of Honduras believing that the award was indeed accepted.[36] Considering that the Court did not discuss any reliance by Honduras on Nicaragua's conduct, the more acceptable view appears to be that the King of Spain's judgment was thought to rely essentially on the acquiescence of Nicaragua.[37] The *King of Spain case* is curious in the sense that more than 50 years elapsed between the arbitral award being made by the King of Spain and the matter being brought to the ICJ. However, it is relevant to recall that the validity of the King's designation was challenged five and a half years after the award. If the five and a half years were enough to support a finding of acquiescence in the circumstance, a considerable delay in instituting proceedings for settling the dispute is irrelevant.[38]

V Conditions to nullify the arbitral award made by the King of Spain

An important factor in the *King of Spain case* relates to conditions to nullify the arbitral award. In this regard, Nicaragua invoked three reasons that deprived the validity of the award made by the King of Spain: (i) excess of jurisdiction; (ii) essential error; and (iii) lack or inadequacy of reasons in support of the conclusions arrived at by the arbitrator.[39] Interestingly, similar grounds for invalidity of arbitral awards were already indicated by Vattel.[40] Excess of jurisdiction and essential error were also considered as reasons to deprive the validity of arbitral awards in the resolution of *Institut de droit international* of 1875.[41] However, the Court did not accept the allegations of Nicaragua for the following reasons.

36 Dissenting Opinion of Judge Urrutia Holguín in the *King of Spain case*, op. cit. n. 2, p. 236.

37 Thirlway, H., *The Law and Procedure of the International Court of Justice: Fifty Years of Jurisprudence*, op. cit. n. 30, p. 41.

38 Ibid., pp. 41–2.

39 *King of Spain case, Judgment*, p. 210. See also *I.C.J., Counter-Memorial submitted by the Government of Nicaragua, King of Spain case*, p. 184, para. 134 et seq; *I.C.J., Rejoinder submitted by the Government of Nicaragua, King of Spain case*, p. 776, para. 61 et seq.

40 Vattel, Emmerich de, *The Law of Nations; or Principles of the Law of Nature, Applied to the Conduct and Affairs of Nations and Sovereigns*, op. cit. n. 1, section 329. Concerning an analysis of Vattel's view on this subject, see Caflisch, L., 'Vattel and the Peaceful Settlement of International Disputes', in Chetail, V., and Haggenmacher, P. (eds), *Vattel's International Law in a XXIst Century Perspective*, Leiden: Nijhoff, 2011, p. 264.

41 Institut de droit international, Session de La Haye (1875), *Projet de règlement pour la procédure arbitrale internationale*, Article 27. Furthermore, under Article 35 of Model Rules of Arbitral Procedure, the validity of an award may be challenged by either party on one or more of the following grounds: (a) that the tribunal has exceeded its powers; (b)

First, whereas Nicaragua contended that the arbitrator breached the rules stipulated in paragraphs 3 and 4 of that Article that reflected the principle of *uti possidetis juris*,[42] the Court held that the complaint of Nicaragua was without foundation inasmuch as the decision of the arbitrator was based on historical and legal consideration in accordance with Article II (3) and (4) of the Gámez-Bonilla Treaty.[43] Nicaragua further argued that the arbitrator, in the award of 1906, exercised his discretion in granting compensation in order to establish a well-defined natural boundary line as provided for in Article II (6) of the Treaty and that this discretion was vested in the Mixed Boundary Commission and could not be exercised by the arbitrator. Nicaragua thus contended that the arbitrator exercised a power, which he did not possess.[44] According to the Court, however, the rules specified in Article II were intended not only for the guidance of the Mixed Commission to which they expressly referred, but were also intended to furnish guidance for the arbitration. There existed no convincing reasons in support of Nicaragua's view. The Court therefore ruled that the King of Spain did not exceed the authority conferred upon him.[45] Second, the Court considered that there was no precise indication of essential error within the Nicaraguan argument that would have the effect of nullifying the award.[46] Third, the Court held that the ground of inadequacy of reasons alleged by Nicaragua was also without foundation for the award dealt in logical order and in some detail with all relevant considerations and contained ample reasoning and explanations in support of the arbitrator's conclusions.[47]

In this regard, an issue arises to what extent the Court should examine the substance of the arbitral award made by the King of Spain. As the Court rightly observed, the award was not subject to appeal. The Court was called upon to only decide whether the award was proved to be null and with having no effect, not to pronounce on whether the award should be right or wrong.[48] Hence the Court took a formalistic approach while examining the

that there was corruption on the part of a member of the tribunal; (c) that there has been a failure to state the reasons for the award or a serious departure from a fundamental rule of procedure; and (d) that the undertaking to arbitrate or the *compromis* is a nullity. *Yearbook of the International Law Commission* 2, 1958, p. 86.

42 *I.C.J., Counter-Memorial submitted by the Government of Nicaragua, King of Spain case*, p. 182, para. 129 et seq; *Case concerning the Arbitral Award made by the King of Spain on 23 December 1906 (Honduras v. Nicaragua), Declaration of Judge Moreno Quintana, I.C.J. Reports 1960*, pp. 217–8. See also *Dissenting Opinion of Judge Urrutia Holguín*, op. cit. n. 3, p. 227.

43 *King of Spain case, Judgment*, p. 215.

44 Ibid.

45 Ibid.

46 Ibid.

47 Ibid., p. 216. However, Judge *ad hoc* Urrutia Holguín stated that the King of Spain committed essential errors related to the exceeding of powers in the application of the *uti possidetis juris* rule. *Dissenting Opinion of Judge Urrutia Holguín*, op. cit. n. 2, pp 233–4.

48 *King of Spain case, Judgment*, p. 214.

question whether the award relied on 'historical and legal considerations' or whether it dealt with all relevant considerations 'in logical order'. For the Court, the *manner* of the examination of relevant materials is of utmost importance in the award. Here it is noteworthy that 'convincing reason' or 'precise indication' was referred to by the Court as a criterion for determining the excess of jurisdiction and essential error.[49] It appears that the ICJ set out a high standard on this matter. According to the standard, it would be difficult to prove that the arbitrator committed an essential error.

VI Incapability of execution of the arbitral award

Finally, Nicaragua contended that the award was incapable of execution by reason of its omissions, contradictions and obscurities for the mouth of a river was not a fixed point and cannot serve as a common boundary between two states.[50] However, the Court considered that the determination of the boundary in this section should give rise to no difficulty.[51] Nicaragua further argued that the delimitation in the operative clause left a gap of a few kilometres between the point of departure of the frontier line from the junction of the Poteca or Bodega with the Guineo or Namaslí up to the *Portillo de Teotecacinte*. In the Court's view, however, the award did not contain any gap in reference to the drawing of the frontier line between the junction of the Poteca or Bodega with Guineo or Namaslí and the *Portillo de Teotecacinte*. The Court did not consider that the award was incapable of execution by reason of any omissions, contradictions or obscurities.[52]

In conclusion, the ICJ ruled, by fourteen votes to one, that the award made by the King of Spain on 23 December 1906 was valid and binding and that Nicaragua was under an obligation to honour its effect.[53] Nonetheless, the implementation of the judgment was not easy as Nicaragua had to withdraw from an inhabited area where it had occupied for several decades. Faced with practical difficulties in implementing the judgment, the Inter-American Peace Committee offered assistance to the parties. In particular, it settled the issues of demarcation and undertook a final inspection of the boundary markers. In July 1963, nearly three years after the Court's judgment, the dispute over the arbitral award was eventually settled.[54]

49 Ibid., p. 215.
50 Ibid., p. 210.
51 Ibid., p. 216.
52 Ibid., pp. 216–7.
53 Ibid., p. 217.
54 Schulte, C., *Compliance with Decisions of the International Court of Justice*, op. cit. n. 7, pp. 129–31; Merrills, J. G., 'The International Court of Justice and the Adjudication of Territorial and Boundary Disputes', *Leiden Journal of International Law* 13, 2000, p. 899.

Conclusions

On the basis of the above considerations, three points in particular merit highlighting.

First, in the *King of Spain case*, the ICJ stressed the conduct of Nicaragua as evidence that Nicaragua accepted the validity of the arbitral award made by the King of Spain. According to the Court's approach, the conduct of parties to a dispute constitutes a key element when considering the validity of arbitral awards.

Second, the ICJ is not a Court of Appeal of arbitration and its task was limited to examine whether the award issued on 23 December 1906 was proved to be a nullity, thereby having no effect. In examining the nullity of arbitral awards, however, one should evaluate the substance of the award. In this regard, the Court took a formalistic approach in the present case by focusing on the *manner* of the examination of relevant materials in the award.

Third, it is notable that the Court referred to 'convincing reason' or 'precise indication of essential error' as the criterion for determining the nullity of arbitral award. However, the criterion is not an objective one. The existence of 'convincing reason' or 'precise indication' is subject to a case-by-case appreciation of the Court. In light of the limited number of case laws in this matter, whether the criterion can be generalized needs careful consideration.

Fourth, the role of international institutions in pre- and post-adjudicative phases in the *King of Spain* dispute merits particular attention. At the pre-adjudicative phase, the OAS can be said to have performed a valuable role in encouraging the parties to refer the dispute to the ICJ. At the post-adjudicative phase, the Inter-American Peace Committee performed an important role in facilitating cooperation between the disputing parties in the implementation of the ICJ judgment. Thus, the *King of Spain case* provides an interesting example with regard to the interaction between international adjudication and regional international institutions.[55]

Bibliography

Antunes, N. S. M., 'Acquiescence', in Wolfrum, R. (ed.), *Max Planck Encyclopedia of Public International Law* (online edition).

Bowett, D. W., 'Estoppel before International Tribunals and Its Relation to Acquiescence', *British Yearbook of International Law* 33, 1957, 176–202.

Caflisch, L., 'Vattel and the Peaceful Settlement of International Disputes', in Chetail, V. and Haggenmacher, P. (eds), *Vattel's International Law in a XXIst Century Perspective*, Leiden: Nijhoff, 2011, pp. 257–66.

55 J.G. Merrills, *International Dispute Settlement*, 5th edn, Cambridge: Cambridge University Press, 2011, pp. 274 et seq.

Cottier, T. and Müller, J. P., 'Estoppel', in Wolfrum R. (ed.), *Max Planck Encyclopedia of Public International Law* (online edition).

Das, H., 'L'estoppel et l'acquiescement: assimilations pragmatiques et divergences conceptuelles', *Revue Belge de Droit International* 2, 1997, 207–634.

Kolb, R., *La bonne foi en droit international public: Contribution à l'étude des principes généraux de droit*, Paris: Presses Universitaire de France, 2000.

MacGibbon, I. C., 'The Scope of Acquiescence in International Law', *British Yearbook of International Law* 31, 1954, 143–86.

Merrills, J. G., 'The International Court of Justice and the Adjudication of Territorial and Boundary Disputes', *Leiden Journal of International Law* 13, 2000, 873–901.

Merrills, J. G., *International Dispute Settlement*, Cambridge: Cambridge University Press, 5th edn, 2011.

Oliver, C. T., 'Case Concerning the Arbitral Award Made by the King of Spain on 23 December 1906 (Honduras v. Nicaragua). ICJ Reports, 1960, p. 192', *AJIL* 55, 1961, 478–89.

Schulte, C., *Compliance with Decisions of the International Court of Justice*, Oxford: Oxford University Press, 2004.

Thirlway, H., *The Law and Procedure of the International Court of Justice: Fifty Years of Jurisprudence*, Vol. I, Oxford: Oxford University Press, 2013.

de Vattel, Emmerich, *Le droit des gens out principes de la loi naturelle appliqués à la conduit et aux affaires des nations et des souverains*, Nouvelle edition, Paris: Guillaumin et C(ie), Libraires, 1863 (reprint, Elibron Classics. 2005). *The Law of Nations; or Principles of the Law of Nature, Applied to the Conduct and Affairs of Nations and Sovereigns* (translated by Joseph Chitty), Philadelphia, T and J. W. Johnson and Co., Law Booksellers, 1853.

23 *Territorial and Maritime Dispute (El Salvador/Honduras), 2002*

Antonio Remiro Brotóns

I Basis of jurisdiction

The jurisdiction of the International Court of Justice (hereinafter 'the Court' or 'ICJ') was based on a Special Agreement signed on 24 May 1986, which amounts to a compromise that settled a *pactum de contrahendo* duly included in the General Treaty of Peace of 30 October 1980 between El Salvador and Honduras.[1] Both countries had foregoing declarations accepting the Court's jurisdiction according to Article 36.2 of the Statute and were also parties to the Pact of Bogotá.[2] In order to prevent that any basis of jurisdiction might enable Honduras to unilaterally demand El Salvador before the ICJ, the latter proceeded to denounce the Pact on 24 November 1973 and then, two days later, to modify its declaration accepting the Court's jurisdiction according to Article 36.2 of the Statute, thereby including drastic limitations that affected all sorts of territorial disputes.

II The object of the dispute

The complexity of this case is due to its three dimensions: first, the land boundary disputed over six different sectors; second, the sovereignty over the islands of the Gulf of Fonseca; and, third, the determination of the legal situation of maritime zones in and out the referred Gulf in the Pacific Ocean.

III A chamber instead of the full court

This was the first case in which Latin American countries requested the Court to constitute an *ad hoc* chamber to adjudicate a dispute. The sole

1 General Peace Treaty between the Republics of El Salvador and Honduras (Lima, 30 October 1980, 21856 UNTS).

2 American Treaty on Pacific Settlement (Pact of Bogotá) (Bogotá, 30 April 1948, 30 UNTS 449). Honduras had registered its declaration of acceptance of the Court jurisdiction in 1948, with no conditions, renewing it successively. It had also ratified the Pact in 1950. El Salvador, having ratified the Pact that same year, had formulated its declaration in 1930 with a reservation (under the Permanent Court of International Justice [PCIJ]) for an indefinite lapse of time.

precedent was that of the delimitation of the maritime frontier in the area of the Gulf of Maine (Canada/United States).[3] It is only by means of an agreement between the parties that a sitting of the full Court can be avoided. Such an agreement may occur after a unilateral application,[4] but it is logical – and also confirmed in the practice of states – that the cases introduced by means of a Special Agreement constitute the natural ground for the *ad hoc* chambers.

The parties stated in their Special Agreement (Art. 1)[5] their intention to strictly control not only the number of sitting members but also the composition of this special chamber. The parties made clear that the three members of the Court that had to compose the chamber should be those mentioned in their joint proposal, together with the *ad hoc* judges, being one appointed by each party – this agreement 'being essential for the formation of the Chamber'. However, this statement appeared hardly compatible with both the Statute (Art. 26.2) and Rules of Court (Art. 17.2), since the Statute requires only the approval of the parties as to the number of judges indicated by the Court, and the Rules determine that the President of the Court will ascertain the 'views' of the parties regarding the composition of the chamber and shall report to the Court accordingly. Notwithstanding, the Court remains sovereign in its decision and may exercise its prerogative in the manner desired by the parties, thus keeping up appearances. This assumption is what appeared to have actually occurred.

The chamber, with only one Latin American (Brazilian) member, did not initially include a Spanish-speaking member, and it was only after the death of the *ad hoc* judge chosen by Honduras that this situation was modified. The inclusion in this case of one sole Spanish-speaking judge was, in the author's opinion, decisive for the drafting of the judgment, especially in relation to the disputed sectors of the land boundary.

IV The judgment on the merits

The judgment rendered by the chamber is extremely long and difficult to read and digest.[6] This is a consequence of both the multiple dimensions of the dispute submitted to the chamber and its very nature, especially with reference to the land boundary, which constitutes two-thirds of the decision. The chamber then ruled according to each of the points submitted by the

3 *Delimitation of the Maritime Boundary in the Gulf of Maine Area (Canada/United States of America), Judgment, I.C.J. Reports 1984*, p. 246.

4 *Elettronica Sicula S.p.A (ELSI) (United States of America v. Italy), Judgment, I.C.J. Reports 1989*, p. 15.

5 Special Agreement to submit to the decision of the International Court of Justice the terrestrial, insular and maritime border dispute existing between the two countries (Esquipulas, 24 May 1986, 24358 UNTS).

6 *Land, Island and Maritime Frontier Dispute (El Salvador/Honduras: Nicaragua intervening), Judgment, I.C.J. Reports 1992*, p. 351.

parties. First, it considered the six sectors in dispute in the continent (the *bolsones*); second, the legal situation of the islands of the Gulf; and, finally, the legal situation of the maritime zones in and out of the Gulf. The judges voted on fifteen points (six in relation to the land boundary; six regarding the legal situation of the islands; three of the determination of the maritime areas being in and out of the Gulf), eight of these were decided unanimously and the remaining seven by a four-to-one majority.[7] Judge Oda appended a declaration and a dissenting opinion regarding the legal situation of the waters inside and outside of the Gulf.[8] Judges *ad hoc* Valticos[9] and Torres Bernárdez[10] appended separate opinions, the latter being overly long and very detailed.

A. The land boundary

The first task entrusted to the chamber was 'to delimit the boundary line in the zones or sections not described in Article 16 of the General Treaty of Peace of 30 October 1980' (Art. 1.1 of the Special Agreement). This task referred to six disputed zones (named *bolsones*, due to their shape) that amounted to approximately four hundred square kilometres.

The chamber had to decide 'taking into account the rules of International Law applicable between the Parties, including, where pertinent, the provisions of the General Treaty of Peace' (Art. 5 of the Special Agreement). The principle of *uti possidetis juris* of 1821, the year in which Central America gained independence from the Spanish Crown was neither mentioned in the 1986 Special Agreement nor in the General Treaty of Peace of 1980. Nevertheless both parties agreed that the *uti possidetis* was the first, and, in the case of Honduras, the sole principle to be applied. El Salvador added the *effectivités* and what it called 'arguments of a human nature', under the auspices of Article 26 of the General Treaty of Peace, expressly mentioned by Article 5 of the Special Agreement when dealing with the applicable law.

The judgment's most noteworthy paragraphs concerning the sectors of the disputed land boundary are those contained in its introduction.[11] There, the chamber considers the scope and the importance conferred to *uti possidetis juris*, which is considered as an elusive principle, hidden within a jungle of pronounced titles of Indian law, including civil, military and ecclesiastic, also compromised by changes in the jurisdiction of provinces and municipalities (*alcald jurisdicti*) over the centuries. The chamber had also to address:

7 Ibid., paras 430–32.

8 *I.C.J., Dissenting Opinion of Judge Oda, Land, Island and Maritime Frontier Dispute (El Salvador/Honduras: Nicaragua intervening).*

9 *I.C.J., Separate Opinion of Judge Valticos, Land, Island and Maritime Frontier Dispute (El Salvador/Honduras: Nicaragua intervening).*

10 *I.C.J., Separate Opinion of Judge Torres Bernárdez, Land, Island and Maritime Frontier Dispute (El Salvador/Honduras: Nicaragua intervening).*

11 Ibid., para. 40–67.

(i) the notion and pertinence of certain dubious titles, particularly the *ejidales* (translated as 'formal title-deeds to commons'); (ii) the role of the *effectivités* and of the private property titles subsequent to independence (the 'republican titles'); (iii) the relationship between titles and *effectivités*; (iv) the acquiescence and the *reconnaissance* as capable of modifying situations protected by *uti possidetis*; (v) the demographic factor; (vi) the localization of resources; (vii) the consideration of the topographic features for drawing the limits; (viii) the bond of the *bolsones* with the adjacent sectors in which the boundary had been agreed; (ix) the proof elements; and, finally, (x) the relevance of the negotiations between the parties, among other issues.

El Salvador placed trust above all in the *ejidales* titles as the best means to prove the *uti possidetis juris.* This trust was misplaced as the chamber rejected its pretension of transforming the limits of the Indian settlements, based on these titles, in international boundaries, according to a peculiar interpretation of the *uti possidetis* principle that, as contented by El Salvador, had been adopted by Article 26 of the General Treaty of Peace. The chamber stated that 'It was the administrative boundaries between Spanish colonial administrative units, not the boundaries between Indian settlements as such, which were transformed, by the operation of the *uti possidetis juris,* into international boundaries in 1821.'[12] The extensive and documented considerations of Judge Torres Bernárdez when outlining the wisdom of this criterion suggest he exercised particular influence over the course of the chamber's deliberation.[13]

The chamber then presented a detailed application of the principles and criteria stated below, sector by sector, leaving room for pragmatic solutions, stretching concepts that melted like butter in the hands and thus invoking an equity *infra legem.*[14] The judicial truth may stifle the historical truth when interpreting former institutions out of their own context. The *uti possidetis* had finally became a convenient tool. The judges confessed their frustration: '*uti possidetis juris* is essentially a retrospective principle, investing as international boundaries administrative limits intended originally for quite other purposes.'[15] And emphasized further on: 'It is rather as if the disputed boundaries must be constructed like a jig-saw puzzle from certain already cut pieces so that the extent and location of the resulting boundary depends upon the size and shape of the fitting piece.'[16]

The judges, like adventurers, had to deeply investigate an unexplored region. Judge Torres Bernárdez, the only Spanish-speaking member of the chamber, apparently assumed the lead.[17]

12 Ibid., para. 50.

13 *I.C.J., Separate Opinion of Judge Torres Bernárdez,* op. cit., paras 28–37.

14 *Land, Island and Maritime Frontier Dispute (El Salvador/Honduras: Nicaragua intervening), Judgment,* op. cit., para. 68–322.

15 Ibid., para. 43.

16 Ibid., para. 44.

17 *I.C.J., Separate Opinion of Judge Torres Bernárdez,* op. cit., paras 8–37.

The *bolsones* were awarded by unanimous decision, with the exception of the fourth, being the largest of all sectors, with one dissenting vote, particularly from the *ad hoc* Judge Valticos, proposed by El Salvador. On the other hand, there was a singular reasoning as regards the sixth sector, because the Spanish Crown's administrative limit had been established by the course of the Goascorán River, which flowed into the Gulf. According to El Salvador, the Goascorán had abruptly changed its course due to a supposed *avulsion*. El Salvador considered that the date in which this abrupt change had occurred was not decisive provided that it was subsequent to the date in which the Goascorán was adopted as an administrative limit. Had it been previous to 1821 independence, the Spanish colonial law was to be applied. Had it been after 1821, international law was to be applied. El Salvador maintained that in either case, the rule was the same: the frontier had to remain on the old course and not transition *in tandem* with the river.

The Court rejected El Salvador's view considering it new and incompatible with the history of the dispute, as it had only appeared in 1972. The chamber understood that the *avulsion* had not been proved. After consulting both the chart of the Gulf of Fonseca, which was drawn up at the end of the eighteenth century, and a report of the expedition, which was provided by Honduras, the chamber arrived at the conclusion that in 1821 the Goascorán was already running along its present course. It observed that both documents failed to provide proof as regards the boundary, but they did present proof of an actual geographical fact.[18] The delimitation became more difficult due to the multiplicity of the river mouths that flow into the Gulf and because of the importance afforded to the parties' behaviour in previous negotiations (this being mentioned as corroborative data by the chamber). The approach decided by the chamber related to this *bolsón*, following Honduras' claim, was the fact that motivated El Salvador's request for the revision of the judgment some ten years later.

B. The legal situation of the islands

The second task entrusted to the chamber was 'to determine the legal situation of the islands' (Art. 2.2 of the Special Agreement). Again, for Honduras this determination had to be exacted on the exclusive basis of the *uti possidetis juris* of 1821. Conversely for El Salvador, the *effective possession* had also to be taken into account. El Salvador claimed the islands of the Gulf, except for Zacate Grande, which pertained to Honduras, and the Farallones of Cosigüina, to Nicaragua.

The chamber noted that the parties endowed it with jurisdiction to determine the legal situation of *all* the islands. The chamber understood, however, that the said jurisdiction should be exercised only in relation to

18 *Land, Island and Maritime Frontier Dispute (El Salvador/Honduras: Nicaragua intervening), Judgment*, op. cit., para. 316.

those islands (Meanguera and Meanguerita) in dispute at the date of the signature of the Special Agreement. The chamber finally included the isle of El Tigre.[19]

The chamber clearly asserted that the starting point to determine the islands' sovereignty should be in accordance to the 1821 *uti possidetis juris*, taking into account the 'colonial effectivités.' Nevertheless, as long as it proved impossible to ascertain a clear and categorical answer as to which administrative entity the islands pertained, the chamber considered it particularly appropriate to look at the conduct of the new states during the period immediately following independence as elements of probable acquiescence. This reasoning finds support in Article 26 of the 1980 General Peace Treaty. It however does not imply that effective possession prevails over *uti possidetis juris*. As a consequence, the chamber considered that 'where the relevant administrative boundary was ill-defined or its position disputed ... the behaviour of the two newly independent States in the years following independence may well serve as a guide to where the boundary was, either in their shared view, or in the view acted on by one and acquiesced in by the other.'[20] The chamber then insisted that the islands were not *terra nullius* and, thus, they were not susceptible of being acquired through occupation. Nevertheless, it held that 'the effective possession by one of the Gulf States of any island of the Gulf could constitute an *effectivité*, though a post-colonial one, throwing light on the contemporary appreciation of the legal situation. Possession backed by the exercise of sovereignty may be taken as evidence confirming the *uti possidetis juris* title.'[21]

The chamber rightly declared, as it could not be otherwise, that the isle of El Tigre pertains to Honduras. With regards to Meanguera, the Salvadorian historical titles were less clear than even El Salvador had anticipated. However, the fragmentary and ambiguous characteristics of the said titles were compensated with its effective possession by El Salvador, since Honduras did not immediately react. Indeed, the conduct of Honduras *vis-à-vis* earlier Salvadorian *effectivités* revealed 'an admission, recognition, acquiescence or other form of tacit consent to the situation.'[22] Meanguera was then allocated to El Salvador.[23] Meanguerita, as an exiguous and

19 The dissenting judge Torres Bernárdez considered that this Honduran isle was not in dispute in 1986 and, consequently, did not require to be judicially adjudicated.

20 *Land, Island and Maritime Frontier Dispute (El Salvador/Honduras: Nicaragua intervening), Judgment,* op. cit., para. 345.

21 *Land, Island and Maritime Frontier Dispute (El Salvador/Honduras: Nicaragua intervening), Judgment,* op. cit., para. 347. See *I.C.J., Separate Opinion of Judge Torres Bernárdez,* op. cit., paras 78–90, 91–5 and 96–104.

22 *Land, Island and Maritime Frontier Dispute (El Salvador/Honduras: Nicaragua intervening), Judgment,* op. cit., para. 364.

23 Judge Torres Bernárdez agreed with this allocation, but did not share the reasoning of the chamber to that effect. See *I.C.J., Separate Opinion of Judge Torres Bernárdez,* op. cit., paras 105–59.

uninhabited isle in close proximity to nearby Meanguera, followed it as a dependency.[24]

From a certain viewpoint, Honduras was compensated as the chamber decided to consider both islands 'invisible' or irrelevant when projecting the Honduran continental coast over the closing line of the Gulf, thereby facilitating the oceanic condition of the country. A different solution would otherwise have created an insurmountable barrier with regards to Meanguera and Meanguerita. Considering the fact that the Salvadorian *ad hoc* judge, Valticos, was Greek, it might be expected that he would be especially sensitive with regards to the legal effects of the physical reality of the Salvadorian islands; but the latter justified his position by stating that 'the very peculiar character of the Gulf of Fonseca as a historic bay with three riparians' and that: 'the conclusions drawn by the Chamber are a consequence of that particular situation and cannot, of course, be given a more general scope in circumstances of a different kind'.[25]

C. The legal situation of the maritime spaces

The third and last task required from the chamber was 'to determine the legal situation of the ... maritime spaces' (Art. 2.2 of the Special Agreement).[26]

The parties differed as to the interpretation of the jurisdiction granted to the Court. The chamber would be competent to identify such spaces, their number and nature, but would that also be the case when it came to the process of delimiting the said spaces between the parties? El Salvador answered with a categorical 'no'; while Honduras said 'yes'. The chamber then had to decide.

The chamber attempted to find the common ground between the parties by interpreting the text of Article 2.2 of the Special Agreement according to its ordinary meaning in its own context: 'the object of the verb "determine" is not the maritime spaces themselves but the legal situation of these spaces. No indication of a common intention to obtain delimitation by the Chamber can therefore be derived from this text as it stands.'[27] The chamber further adds that this conclusion is also confirmed should the phrase be interpreted in its broader context, by taking into account the Special Agreement as a whole, and the 1980 General Treaty of Peace.[28]

24 Judge Torres Bernárdez, however, disagreed with this since there were neither Salvadorian *effectivité* nor Honduran acquiescence in Meanguerita (*I.C.J., Separate Opinion of Judge Torres Bernárdez*, op. cit., paras 169–176).

25 *I.C.J. Separate Opinion of Judge Valticos*, op. cit., p. 281.

26 Ibid., paras 372–420.

27 Ibid., para. 373.

28 Judge Torres Bernárdez maintained a different approach. See: *I.C.J., Separate Opinion of Judge Torres Bernárdez*, op. cit., para. 184–207.

Starting from this basis, the chamber deals successively with the condition of the Gulf of Fonseca and the maritime projection of riparian states outside the Gulf.[29]

1. The condition of the Gulf of Fonseca

El Salvador championed the *condominium* (co-ownership) of the three riparian states within the Gulf. The judgment of the Central American Court of Justice (hereinafter 'CACJ') dated 9 March 1917[30] served to reinforce its position. El Salvador has indeed incorporated this judgment into its Constitution and then requested the Court to confirm the *dicta* of the CACJ. Honduras, however, not a party in those proceedings, protested by rejecting the co-ownership; it argued instead the existence of a 'community of interests' going beyond the first three nautical miles from the coast of each riparian state. Moreover, it maintained the delimitation agreement with Nicaragua, subscribed in June 1900. This 'community of interests' between the riparian states produced, as stated by Honduras: 'a perfect equality of rights which has nevertheless never been transformed by the same States into a condominium.' But, as the chamber countered, 'it seems odd ... to postulate a community of interest regime as an argument against a condominium regime, for a condominium is almost an ideal juridical embodiment of the community of interest's requirements.'[31]

It is here that the chamber pays extraordinary attention to the 1917 judgment of the CACJ 'as an important part of the Gulf's history.'[32] With Judge Oda dissenting,[33] the chamber decided that the Gulf was a historical bay conjointly inherited from Spain by the three riparian republics: 'there seems no reason in principle why a succession should not create a joint sovereignty where a single and undivided maritime area passes to two or more new States.'[34] The chamber understood this to be the meaning given to the Gulf *condominium* or co-ownership referred to by the CACJ in 1917. The chamber then adopted the same criterion from that of the 1917

29 Judge Oda however voted against, 'as a result of his understanding of the law of the sea, greatly at variance with the views underlying the Judgment' (*I.C.J., Dissenting Opinion of Judge Oda, Land, Island and Maritime Frontier Dispute (El Salvador/Honduras: Nicaragua intervening)*, para. 1).

30 *El Salvador v. Nicaragua*, CACJ, Judgment of 9 March 1917, 11 *Am. J. Int'l L.* 674 (1917).

31 *Land, Island and Maritime Frontier Dispute (El Salvador/Honduras: Nicaragua intervening), Judgment,* op. cit., para. 407.

32 Ibid., para. 387.

33 Judge Oda is very critical with the Chamber's approach which abstained to dissect the 1917 judgment, thereby granting it an authority that it objectively lacked and thus consolidating its mistakes (See *I.C.J., Dissenting Opinion of Judge Oda, Land, Island and Maritime Frontier Dispute (El Salvador/Honduras: Nicaragua intervening)*, p. 732.).

34 *Land, Island and Maritime Frontier Dispute (El Salvador/Honduras: Nicaragua intervening), Judgment,* op. cit., para. 399.

judgment, taking the latter into account 'as a relevant precedent decision of a competent court and as a, in the words of Article 38 of the Court's Statute, "a subsidiary means for determination of rules of law".'[35] The joint ownership remained valid, except for the exclusive three nautical mile sovereignty belt from the coastal zone of each state and with the reservation of the delimitation agreement signed in June 1900 between Honduras and Nicaragua, which thus was accepted by El Salvador. It must be noted that the belt refers to the continental coast, and in principle as such, the islands would not have a belt of exclusive waters, irrespective of their width, in the absence of a riparians' agreement. Otherwise, the right of innocent passage would be recognized within the waters of the Gulf, as they are internal waters subject to a particular historic regime.

2. The closing line of the Gulf

The Court assumed that the closing line of the Gulf was drawn between Punta Amapala (El Salvador) and Punta Cosigüina (Nicaragua). This was mentioned in the 1917 judgment of the Central American Court and was recognized, in practice, by the three riparian states in the Gulf. The chamber considered that the closing line of a historical bay where waters are internal constitutes – as Honduras asserted – a base line of the territorial sea and title of maritime zones away from the Gulf. The rights over the waters in the central portion of said line were assigned to the three riparian states as a result of the common appurtenance of the Gulf, this as long as they do not mutually agree on a delimitation. This central portion is 13.75 nautical miles long, excluding the first three nautical miles from Punta Amapala (El Salvador) at one end, and from Punta Cosiguïna (Nicaragua) at the other.

3. The sovereign access of Honduras to the Pacific Ocean

Following the judgment's delimitation of the six disputed sectors (the *bolsones*), the Court upheld Honduran sovereign rights and jurisdiction over the maritime spaces beyond the Gulf's closing line. The length of the line, which amounts to 19.75 maritime miles, would, *per se*, have been enough to keep Honduras locked in the Gulf had an orthodox interpretation of the law of the sea been upheld by the chamber. The Gulf-locked condition of Honduras should have been strengthened by the screen effect of Salvadorian Meanguera and Meanguerita. In this sense, El Salvador did nothing to favour its interests by insisting on the thesis of the Gulf *condominium*. Its fidelity to the concept was indeed fatal. This was not the case of Honduras when exploiting the benefits of the Salvadorian plea for the *condominium* via the

35 Ibid., para. 403.

notion of 'community of interests', which, to a certain extent, was a masked version of the same idea.

El Salvador contested the Court's jurisdiction to rule on this matter, and also the recognition of any of Honduras' rights beyond the Gulf. The chamber considered that the Special Agreement subscribed by the parties when requested 'to determine the legal situation of the maritime spaces' empowered it to establish whether Honduras had oceanic rights, without delimiting them with respect to its neighbours. According to the chamber, the closing line of the Gulf was the base line of the territorial sea of the three riparian states, thus attributing all to a shared central zone with rights to a territorial sea, an exclusive economic zone and a continental shelf out of the Gulf. Any such limits had yet to be established and would require mutual agreement, according to the rules of international law. The chamber sustained that it was equitable to not to confine Honduras to the bottom of the Gulf since it 'has by far the longest coastline of the Gulf and only Gulf coastline facing the Ocean'.[36]

Maybe the Court should have refrained from deciding this point, by applying the 'indispensable third party' principle, irrespective of the interpretation given to the Special Agreement between El Salvador and Honduras.

V The sequels of the procedure: the Nicaraguan intervention

The maritime dimension of the dispute, thereby dealing with the legal condition of the Gulf and rights of the riparian states in the Pacific Ocean also involved Nicaragua as a third state. This is the reason why it requested to intervene in the present procedure. The terms according to which the said intervention was accepted by the Court were established in the judgment rendered on 13 September 1990.

As to the 1992 judgment, it referred briefly to the effect on Nicaragua in its final paragraphs.[37] The chamber considered that 'it is correct that a State permitted to intervene under Article 62 of the Statute, but which does not acquire the status of party to the case, is not bound by the Judgment given in the proceedings in which it has intervened.'[38] The chamber concluded that 'in the circumstances of the present case, this Judgment is not *res judicata* for Nicaragua.'[39] This point was not submitted to a vote, but both Judge

36 Ibid., para. 418. Judge Oda dissented from the majority and considered that the findings of the chamber were 'hardly tenable in the light of any rule, traditional or contemporary of the law of the sea' (*I.C.J., Dissenting Opinion of Judge Oda, Land, Island and Maritime Frontier Dispute (El Salvador/Honduras: Nicaragua intervening)*, para. 53).

37 Ibid., paras 421–4 and 369–71.

38 Ibid., para. 423.

39 Ibid., para. 424.

Oda, in his declaration, and Judge Torres Bernárdez, in his separate opinion,[40] observed that irrespective of the *res judicata*, Nicaragua would be bound by this judgment as it relates to the legal position of the maritime zones within the Gulf.

This assumption was based on the idea that what is true for two shall also be so for three. The Gulf of Fonseca cannot be considered as a historical bay only for the parties involved in the case, this would be also applicable to the third riparian state, alien to the case. Nevertheless, it may be argued whether it was pertinent for the Court to exercise jurisdiction on this aspect of the dispute which directly affected the legitimate interests of a third state for which the relative effect of the judgment might not provide enough protection. It can only be suggested, as a mitigating circumstance, that Nicaragua being the third riparian state in the Gulf, was obliged to adhere to the 1917 judgment of the CACJ which, as previously noted, influenced the chamber's decision. Certainly, the chamber refrained from any pronouncement on this point by observing that Nicaragua was just a participant and not a party in the proceedings. According to the chamber: 'It does not appear, therefore, that this Chamber is at present required now to pronounce upon whether the 1917 judgment is *res judicata* between the States parties to it.'[41] Notwithstanding, it could be suggested that this circumstance certainly influenced the mood and mindset of the judges.

In conclusion, Nicaragua is not formally bound by the judgment of the chamber, but this judgment as far as the maritime dimension is concerned, introduced additional factors of instability and uncertainty that perhaps can only be overcome via a three-sided negotiation or, eventually, through arbitration or judicial remedy.

Bibliography

Decaux, E. (1992) "Le différend frontalier terrestre, insulaire et maritime (El Salvador/Honduras), arrêt de la Chambre de la CIJ du 11 septembre 1992. La délimitation terrestre", *AFDI*, 38: 393–426.

Evans, M. (1992) 'Case concerning the Land, Island and Maritime Frontier Dispute (El Salvador/Honduras): The Nicaraguan Intervention, *The International and Comparative Law Quarterly*, Vol. 41, No. 4, 896–906.

Kohen, M. G. (1993) "L'*uti possidetis* revisité: l'arrêt du 11 septembre 1992 dans l'affaire El Salvador/Honduras", *RGDIP*, 97: 939–973.

Lucchini, L. (1992) "Le différend entre le Honduras et El Salvador devant La CIJ. Aspects insulaires et maritimes", *AFDI*, 38: 427–459.

Ratner, S. R. (1991) 'Land, Island and Maritime Frontier Dispute (El Salvador/Honduras), Application to Intervene, *The American Journal of International Law*, Vol. 85, No. 4, pp. 680–686.

40 *I.C.J.*, *Separate Opinion of Judge Torres Bernárdez*, op. cit., para. 208.

41 *Land, Island and Maritime Frontier Dispute (El Salvador/Honduras: Nicaragua intervening), Judgment*, op. cit., para. 402.

Rottem, G. (1993) 'Land, Island and Maritime Frontier Dispute, *The American Journal of International Law*, Vol. 87, No. 4, pp. 618–626.

Shaw, M. N. (1993) 'Case concerning the Land, Island and Maritime Frontier Dispute (El Salvador/Honduras: Nicaragua Intervening), Judgment of 11 September 1992, *The International and Comparative Law Quarterly*, Vol. 42, No. 4, pp. 929–937.

Part VI

Material contribution of Latin American cases to the development of international law

Environmental law and activities carried in the border area

24 *Construction of a Road in Costa Rica along the San Juan River (Nicaragua v. Costa Rica), 2011* and *Certain activities carried out by Nicaragua in the border area (Costa Rica v. Nicaragua), 2010*

Ximena Fuentes and Mariana Durney

This chapter is concerned with two cases between Costa Rica and Nicaragua before the International Court of Justice (hereinafter ICJ or 'the Court'). This is not the first time that these two countries have been before the Court. Indeed, Nicaragua and Costa Rica are two frequent visitors to the ICJ.[1] These two proceedings were joined in accordance with Article 47 of the Rules of the Court. By its order of 17 April 2013 the Court decided to join the two cases because, in its view, this allowed 'to address simultaneously the totality of the various interrelated issues raised by the Parties'.[2] Indeed, the two disputes revolve around activities conducted in the border area that might have potential transboundary effects: the two states involved claim that the other has violated its territorial sovereignty and the two cases deal with allegations of transboundary environmental harm. Therefore, it is quite clear that these cases are closely linked.

The purpose of this chapter is to discuss the main issues addressed by the recent Judgment that the ICJ handed down on 16 December 2015. The two

1 Nicaragua and Costa Rica have been involved in four cases before the Court including these two cases. The other two cases are: a) Armed Activities and b) Navigational Rights. In 1986, in the context of the Central American crisis, Nicaragua accused Costa Rica of assisting the *Contras* in their armed incursions in Nicaragua. The case was discontinued since Nicaragua and Costa Rica and three other Central American states reached an agreement to find a negotiated solution to their problems (The Esquipulas II process of negotiations). The case is nevertheless important because Nicaragua invoked as a basis of the Court's jurisdiction the Pact of Bogotá, this being the second time in which the 1948 Pact of Bogotá had been invoked after 40 years of being a dormant treaty. The second case is Navigational Rights in the River San Juan, in which Costa Rica demanded respect for its right to navigate the San Juan for the purposes of commerce. The ICJ handed down its decision on 13 July 2009.

2 *Construction of a Road in Costa Rica along the San Juan River (Nicaragua* v. *Costa Rica), Joinder of Proceedings, Order of 17 April 2013, I.C.J. Reports 2013*, p. 184.

principal questions that the parties submitted to the Court are: (i) the delimitation of the boundary in the Atlantic sector of the San Juan River; and (ii) the question whether or not Costa Rica complied with international law when authorizing and undertaking the construction of a road along the San Juan River. The following are some of the interesting problems involved in these two cases: a) the interpretation of old boundary treaties and international awards in a context in which, over the years, the geography of the region has been subject to dramatic change; b) the evolution of the rules of environmental protection and their impact on the interpretation and application of nineteenth-century treaties; and, c) questions of evidence and the role of experts. During the proceedings both parties additionally requested the Court to issue provisional measures. A number of interesting issues were raised during those incidental proceedings and for that reason a section of this chapter will be devoted to provisional measures.

I The delimitation of the boundary in the relevant area

The delimitation of the boundary between Costa Rica and Nicaragua in the San Juan River was effected by the 1858 Jerez-Cañas Treaty. In this treaty the parties accorded that the boundary runs along the right bank of the San Juan River until its outlet in the Caribbean Sea.[3] In addition, they agreed to grant in perpetuity rights of navigation in the waters of the river belonging to Nicaragua to Costa Rica for the purposes of commerce.[4] Years after the conclusion of the 1858 Treaty, Nicaragua aired discontent with the Treaty and claimed that it was null. The dispute was resolved by President Cleveland of the United States.[5] The Cleveland Award of 1888 declared that the 1858 Treaty was valid. The parties also asked the arbitrator to interpret some points of doubtful interpretation that had been raised by Nicaragua. With regard to the relevant area, the Cleveland Award stated that:

> The boundary line between the Republics of Costa Rica and Nicaragua, on the Atlantic side, begins at the extremity of Punta de Castilla, at the mouth of the San Juan de Nicaragua River, as they both existed on the 15th day of April 1858. The ownership of any accretion to said Punta de Castilla is to be governed by the laws applicable to that subject.[6]

3 Article 2 of the Jerez-Cañas Treaty 1858. Text in: *Certain Activities in the Border Area (Costa Rica v. Nicaragua), Memorial of Costa Rica*, Volume II, Annex I.

4 Article 5 of the Jerez-Cañas Treaty. With regard to the rights of navigation, Costa Rica submitted an application against Nicaragua and the Court handed down its Judgment on 13 July 2009.

5 Award of the President of the United States in regard to the validity of the Treaty of Limits between Costa Rica and Nicaragua of 15 July 1858, Decision of 22 March of 1888, in: legal.un.org/riaa/cases/vol_XXVIII/189-236.pdf (accessed on 3 April 2016).

6 Ibid., para. 3 (1), p. 209.

In 1896, the parties agreed to establish a commission in charge of the demarcation of the boundary. The commission included an engineer appointed by the President of the United States, General Alexander, who was mandated to resolve the differences between the parties that might emerge during the demarcation process. Indeed, these differences did emerge and were resolved by Alexander. By that time, changes had already occurred in the geography of the area. The original Punta Castilla had been swept over by the sea and, therefore, Alexander decided that:

> I have accordingly made personal inspection of this ground, and declare the initial line of the boundary to run as follows, to wit:
>
> > Its direction shall be due northeast and southwest, across the bank of sand, from the Caribbean Sea into the waters of Harbor Head Lagoon. It shall pass, at its nearest point, 300 feet on the northwest side from the small hut now standing in that vicinity. On reaching the waters of Harbor Head Lagoon the boundary line shall turn to the left, or southeastward, and shall follow the water's edge around the harbor until it reaches the river proper by the first channel met. Up this channel, and up the river proper, the line shall continue to ascend as directed in the treaty.[7]

The interpretation of General Alexander's decision is central for the resolution of the *Certain Activities in the Border Area* dispute. Nicaragua claimed that the 'first channel met' referred to by Alexander is the *caño* or channel that existed in 1896 and that it had attempted in recent years to clear this old *caño*. Costa Rica contradicted Nicaragua and claimed that Nicaragua had commenced to construct a new channel across the northern tip of Isla Portillos in an illegal attempt to annex what is now Costa Rican territory in violation of the 1858 Treaty, the 1888 Cleveland Award and the 1897–1900 Alexander Awards, respectively. In plain words, Costa Rica asserted that Nicaragua had attempted to reopen a dispute that was already settled and that it had created an artificial non-existent *ex post facto* dispute to justify the occupation of Costa Rican territory.[8]

This type of dispute is far from being a novelty in the context of Latin American boundaries. Despite the fact that Latin America has been inclined to solve disputes through direct negotiations or arbitration, the truth is that

7 First Award of the Engineer-Umpire, under the Convention between Costa Rica and Nicaragua of 8 April 1896 for the Demarcation of the Boundary between the two Republics, Decision of 30 September 1897 in: legal.un.org/riaa/cases/vol_XXVIII/189-236.pdf, p. 208 (accessed on 3 April 2016).

8 *Certain Activities in the Border Area (Costa Rica v. Nicaragua), Oral Pleadings*, speech by Mr Marcelo Kohen, Public sitting held on Tuesday 14 April 2015, 10 am sitting, pp. 40 ff.

in the region there is a tendency to re-open old disputes.[9] The poor geographical knowledge at the time of the conclusion of boundary treaties or of the drafting of arbitral awards might account as one of the reasons that explains this tendency to re-open disputes in Central and South America. When faced with disputes such as this, courts can expect that the arguments of the parties will likely revolve around the following aspects: the *res judicata* character of arbitral awards; the interpretation of the arbitrators' intention; the relevance of the conduct of the parties (if any); the value accorded to maps; the importance of the principle of stability of boundaries; and, finally, the impact that contemporary geographical knowledge might bear upon the interpretation of old instruments or awards.

The Court could have taken the opportunity to deal with the various interesting aspects of the territorial dispute in hand; however, the final decision was mainly based on two elements: (i) the factual evidence submitted by the parties and (ii) the interpretation of the purpose of the 1858 Treaty.

With regard to the impact of the geographical changes alleged by Nicaragua, in particular, the existence of an old *caño* that was later covered by the vegetation of the area, the Court weighed the evidence submitted by the parties and rejected the arguments of Nicaragua. It decided that the evidence did not allow the existence of an old *caño* to be established.[10] This factual conclusion was reinforced with a legal argument that emphasized that the purpose of the 1858 Treaty was to grant to Costa Rica perpetual rights of navigation in the San Juan River, which were linked to Costa Ricas's sovereignty of the right bank of the river. Nicaragua's claim with regard to the existence of an old *caño* would have deprived Costa Rica of sovereignty over the right bank of the river.[11]

With regard to the *effectivités* invoked by the parties in the disputed territory, the Court stated that they could not affect the title over territory resulting from the 1858 Treaty and the Cleveland and Alexander Awards.[12] The Court may be criticized in that it did not stop to examine all the implications of the conduct of the parties. The conduct of the parties can modify a pre-existing title to territory, whether by operation of a tacit agreement or estoppel, as is illustrated in the *Temple of Preah Vihear case*.[13]

9 Fuentes, X., 'Latin American States and the ICJ', in Klein, N. (ed.), *Litigating International Law Disputes. Weighing the Options*, Cambridge: CUP, 2014, p. 81.

10 *Certain Activities in the Border Area (Costa Rica v. Nicaragua), Construction of a Road in Costa Rica Along the San Juan River, (Nicaragua v. Costa Rica), Judgment, I.C.J. Reports, 2015*, para. 90.

11 Ibid., para. 91.

12 Ibid., para. 89.

13 *The Temple of Preah Vihear (Cambodia v. Thailand), Judgment, I.CJ. Reports 1962.*

II The environmental disputes: the environmental impact of the dredging activities carried out by Nicaragua and of the construction of a road by Costa Rica

Costa Rica claimed that the construction of a new channel in the river (which Nicaragua describes as the cleaning of an existing one) and the dredging activities conducted by Nicaragua in the border caused significant harm to Costa Rica. In addition, it claimed that Nicaragua breached its obligations under general international law since it neither notified nor consulted about these activities with Costa Rica, and because it did not conduct a proper transboundary Environmental Impact Assessment (EIA). All these obligations are part of general international law.

With regard to the obligation not to cause significant harm, the ICJ stated in the *Pulp Mills case* that:

> A State is thus obliged to use all the means at its disposal in order to avoid activities which take place in its territory, or in any area under its jurisdiction, causing significant damage to the environment of another State. This Court has established that this obligation 'is now part of the corpus of international law relating to the environment.'
> (Legality of the Threat or Use of Nuclear Weapons, Advisory Opinion, I.C.J. Reports 1996 (I), p. 242, para. 29)[14]

With regard to the obligation to notify and consult, there is a significant amount of state practice to assert that this obligation has become customary international law.[15] And, in relation to the international EIA, the Court stated in the *Pulp Mills case* that this is 'a requirement under general international law where there is a risk that the proposed activity may have a significant adverse impact in a transboundary context.'[16]

Nicaragua and Costa Rica agree that these are obligations pursuant to general international law. They claimed, each in its turn, however, that there is a *lex specialis* applicable to their own particular situation. In this line, with regard to the works it undertook in the San Juan River, Nicaragua claimed, on the one hand, that the dredging activities and the cleaning of the *caños* are subject only to the 1858 Treaty, which is the exclusive applicable law to the exclusion of general international law. As expected from a nineteenth-century instrument, the 1858 Treaty contains very few substantive and procedural obligations in relation to the protection of the environment.

On the other hand, Costa Rica claims that Nicaragua's attempt to isolate the 1858 Treaty from the rest of international law would be against the

14 *Pulp Mills (Argentina v. Uruguay), Judgment, I.C.J. Reports 2010*, para. 101.

15 Rieu-Clarke, A., 'The obligation to notify and consult on planned measures concerning international watercourses - learning lessons from recent international case-law', *Yearbook of International Environmental Law* 2013, p. 102.

16 *Pulp Mills (Argentina v. Uruguay), Judgment, I.C.J. Reports 2010*, para 204.

principles of treaty interpretation. According to Costa Rica, treaties must be interpreted in the broader context of other rules of general international law applicable between the parties. The recent award in the *Kishenganga Arbitration* is apposite here:

> It is established that principles of international environmental law must be taken into account even when (unlike the present case) interpreting treaties concluded before the development of that body of law. The *Iron Rhine* Tribunal applied concepts of customary international environmental law to treaties dating back to the mid-nineteenth century, when principles of environmental protection were rarely if ever considered in international agreements and did not form any part of customary international law. Similarly, the International Court of Justice in *Gabcikovo-Nagymaros* ruled that, whenever necessary for the application of a treaty, 'new norms have to be taken into consideration, and … new standards given proper weight.' It is therefore incumbent upon this Court to interpret and apply this 1960 Treaty in light of the customary international principles for the protection of the environment in force today.[17]

With regard to the construction of a road along the San Juan River by Costa Rica, the *lex specialis* argument adopts the form of an exemption to the obligation to conduct an EIA in view of an emergency situation. As stated by Costa Rica during the oral hearings:

> The road was an appropriate infrastructure measure to the threats faced by Costa Rica at this time, not least in view of the need to mobilize its police force, its emergency services, and, where the circumstances might necessitate, mobilize the riparian population along the river.[18]

According to Costa Rican counsel, '[i]nternational law provides an emergency exemption to the obligation to carry out an EIA'.[19] Moreover, Costa Rica claims that in this case the construction of the road does not pose a risk of significant environmental damage and, therefore, that 'the threshold for preparing the EIA has not been reached'. It is clear that the obligation to carry out an international EIA when certain activities pose a risk of significant environmental harm is part of customary international law. However,

17 *Indus Waters Kishenganga Arbitration (Pakistan v. India), Partial Award, 18 February 2013*, para. 452.

18 *Construction of a Road in Costa Rica along the San Juan River (Nicaragua v. Costa Rica), Oral Pleadings*, speech by Mr Arnoldo Brenes, Public sitting held on Thursday 23 April 2015, para. 30.

19 Ibid., speech by Mr Marcelo Kohen, Public sitting held on Thursday 23 April 2015, para. 39.

international law lacks an adequate regulation of EIAs. Yet this was recognized by the Court in the *Pulp Mills case* when it stated that the content of an EIA depends on what is required by domestic law.[20] Another fundamental problem in the operation of EIA concerns the assessment of the risk that some might claim that cannot depend on the sole appreciation of the state responsible for planning the activity, as Costa Rica appears to assert.

The Court's decision is important because it clarified that this obligation to carry out an EIA exists only where there is a significant risk of environmental harm and that this risk should be assessed by the State responsible for planning the activity.[21] The immediate question should be: how do we know that the assessment of the risk conducted by the State concerned has been adequately performed? The obligation to notify and consult could play an important role in this regard, however, the Court concluded that the obligation to notify and consult also emerges only in case of a risk of significant harm.[22] The Court's decision might have contributed to make environmental protection in the transboundary context somehow more difficult.

III Questions of evidence

Evidence is a prominent subject in cases concerning environmental issues. In the *Pulp Mills case*, it was not only an important matter but also a controversial one. In that case, experts of both sides appeared as counsel acting on behalf of the parties. This was criticized as there was no opportunity to cross-examine the experts and, therefore, the Court did not benefit from the opportunity to confront points of view during the oral pleadings.[23] Apart from that, the Court itself was criticized for not making use of its power to appoint independent experts under Article 50 of the Statute and Article 67 of the Rules of Court, having preferred instead to rely on the advice of 'ghost experts' that were unknown to both the parties and to the general public.[24]

Yet, contentious cases that have come before the Court after the *Pulp Mills case* indicate that the practice of the Court in this field has changed. Indeed, in the *Whaling in the Antarctic case*[25] and in these two cases between Costa Rica and Nicaragua, experts have appeared as party-appointed experts and they have been subject to cross-examination. In its turn, the judges of the Court have been actively addressing questions to the experts.

20 *Pulp Mills on the River Uruguay (Argentina v. Uruguay), Judgment, I.C.J. Reports 2010*, para. 205.

21 *Certain Activities on the Border Area (Costa Rica v. Nicaragua), Construction of a Road in Costa Rica Along the San Juan River, Judgment, I.C.J, Reports, 2015*, para. 104.

22 Ibid. 105.

23 *Pulp Mills on the River Uruguay (Argentina v. Uruguay), Joint Dissenting Opinion Judges Al-Khasawneh and Simma*, para. 6 and 7.

24 Ibid., para. 14.

25 *Whaling in the Antarctic (Australia v. Japan: New Zealand intervening), Judgment, I.C.J. Reports 2014.*

In the *Certain Activities on the Border Area (Costa Rica v. Nicaragua)* there was also an additional interesting development concerning evidence. Nicaragua requested the Court to exercise its power under Article 62 of the Rules of Court and requested Costa Rica to produce, before the opening of the oral proceedings, studies relating to impact of the dredging of the San Juan River on the flow of the Colorado River.[26]

IV The requests for provisional measures

In total, the parties filed five requests for provisional measures during the proceedings in the two cases under analysis. Both requests were based on 'rights to sovereignty and territorial integrity'. In the first order on provisional measures of 8 March 2011, the aspect of territorial sovereignty appeared more prominent than the environmental element. The Court made clear that the 'rights at issue in these proceedings derived from sovereignty claimed by the parties over the same territory'.[27] The Court was solely relying on the manner in which the parties understood and portrayed their respective claims.

In this first order, the Court declined one of Costa Rica's most important requests, namely an order that Nicaragua should cease its dredging operations. The Court found that the evidence provided by Costa Rica was insufficient to confirm a risk of irreparable prejudice to Costa Rica's environment and the imminence of such risk. The Court also declined to order provisional measures with regard to the alleged construction of a new channel on the river, to the felling of trees and the dumping of sediment, all this based on Nicaragua's assertion that it had already finished the works in the area of the *caño*. One might wonder whether this finding – along with a subsequent order by which the presence of private persons in the area was not prohibited[28] paved the way for the events that followed.

In its order of 22 November 2013, the Court upheld Costa Rica's request on the basis of the evidence –which included satellite imagery – in relation to the construction of two new *canos* (and a trench that almost reached the sea) in the disputed area and the presence of Nicaraguan personnel, infrastructure and equipment, including an army encampment, on the beach. In view of all these facts, the Court ordered additional measures.[29] It prohibited

26 The request was submitted on 4 January 2011 and Costa Rica produced the requested document at its own initiative on 6 January 2011. See *Certain Activities carried out by Nicaragua in the Border Areas (Costa Rica v. Nicaragua), Provisional Measures, Order of 8 March 2011, I.C.J. Reports 2011*, p. 12.

27 Ibid., p. 19.

28 *Certain Activities Carried Out by Nicaragua in the Border Area (Costa Rica v. Nicaragua); Construction of a Road in Costa Rica along the San Juan River (Nicaragua v. Costa Rica), Order of 16 July 2013, Provisional Measures, I.C.J. Reports 2013*, p. 235.

29 *Certain Activities Carried Out by Nicaragua in the Border Area (Costa Rica v. Nicaragua); Construction of a Road in Costa Rica along the San Juan River (Nicaragua v. Costa Rica), Provisional Measures, Order of 22 November 2013, I.C.J. Reports 2013*, p. 369.

any dredging and other activities in the disputed territory (in particular, works of any kind on the two new *canos*); it ordered Nicaragua to fill the trench on the beach (within two weeks) and to immediately inform the Court of the completion of that work and submit a report (in one week); it ordered the removal of all Nicaraguan personnel (civilian, police or security) from the disputed territory and of any private persons under Nicaraguan jurisdiction or control; and, finally, it imposed upon the parties the duty to provide regular information on compliance with the measures decreed (at three-month intervals). The reasons given by the Court to justify the prescribed provisional measures suggest that the Court was more concerned with the protection of the territorial sovereignty of Costa Rica than with the protection of the environment itself.[30]

On the other hand, the provisional measures sought by *Nicaragua in the Construction of a Road in Costa Rica along the San Juan River (Nicaragua v. Costa Rica) case* were more environmentally motivated. In its request for provisional measures Nicaragua sought that Costa Rica immediately and unconditionally provided an EIA and other technical reports. Nicaragua also requested the implementation of various urgent measures that could contribute to eliminate or significantly reduce the sedimentation of the river due to the construction of the road. Nicaragua further requested an order from the Court to ban Costa Rica from reassuming all construction activities on the road while the Court was seized of the case.[31]

On 13 December 2013, the Court decided that the circumstances, as presented to it, were not such to require the exercise of its power to indicate provisional measures. Firstly, regarding the EIA the Court deemed that it could not decide on this issue at this stage and that this was a decision for the merits.[32] Secondly, the Court rejected the other provisional measures sought by Nicaragua (all of them linked to the right to be free from trans-boundary harm), based on the inconclusiveness of the evidence.[33] Ironically, it was precisely the lack of an EIA that made the proof of any existing risk more difficult to scrutinize.

Another interesting feature of the proceedings on provisional measures is the monitoring of their implementation. On the two occasions in which provisional measures were granted, the Court imposed on the parties the obligation to report on their compliance. Although the possibility of requesting information from the parties on any matter connected with the

30 When assessing the plausibility of the rights invoked and the linking element, the Court stated that any environmental harm would infringe Costa Rica's alleged territorial rights (ibid., p.360) and that the construction of the *caños* could affect Costa Rica's rights of sovereignty, as well as 'environmental rights connected thereto' (ibid., p. 361).

31 *Construction of a Road in Costa Rica along the San Juan River (Nicaragua v. Costa Rica); Certain Activities Carried Out by Nicaragua in the Border Area (Costa Rica v. Nicaragua), Provisional Measures, Order of 13 December 2013, I.C.J. Reports 2013*, p. 401.

32 Ibid., p. 404.

33 Ibid., p. 407.

implementation of provisional measures is explicitly set forth in Article 78 of the Rules of Court, this power has only recently been put into practice by the Court.[34]

The order of 22 November 2013 is quite remarkable in this respect. First, it imposed the duty for Nicaragua to inform as soon as it completed the filling works that were prescribed. Nicaragua was granted specific and very short time to discharge this duty. Secondly, the order imposed a duty upon both parties to regularly inform the Court, at three-month intervals, as to their compliance with the provisional measures.[35]

Finally and inextricably connected with the above, there is the issue of the consequences that should follow non-compliance with provisional measures. The Court reiterated in its orders the binding nature of the provisional measures and that the question of compliance by both parties may be a matter to be considered in its findings on the merits. In its final submissions, Costa Rica requested the Court to adjudge and declare that Nicaragua should 'pay all the costs and expenses incurred by Costa Rica in requesting and obtaining the Order on Provisional Measures of 22 November 2013 ... on a full indemnity basis'.[36] Costa Rica also included a general request for reparations in the following terms: to 'provide satisfaction so to achieve full reparation of the injuries caused to Costa Rica in a manner to be determined by the Court'.

In its judgment of 15 December 2015, the Court addressed the issue of the infringement of provisional measures and the consequential remedies. First, the Court declared by unanimity, that Nicaragua had breached the Order of March 2011 by excavating two *caños* in 2013 and by establishing a military presence in the disputed territory. Regarding reparation for such a breach, the Court took the view that the aforesaid declaratory finding sufficed in this respect and that it should be considered as 'adequate satisfaction' for the non-material injury.[37]

In a rather puzzling way, the Court continued to reject the specific submission made by Costa Rica with respect to the costs and expenses which she incurred by asking new provisional orders due to the lack of compliance

34 The first time that the Court requested an *Implementation Report* was in the *Application of the International Convention on the Elimination of all Forms of Racial Discrimination (Georgia v. Russian Federation), Provisional Measures, Order of 15 October 2008, I.C.J. Reports 2008*, p. 399.

35 *Certain Activities Carried Out by Nicaragua in the Border Area (Costa Rica v. Nicaragua); Construction of a Road in Costa Rica along the San Juan River (Nicaragua v. Costa Rica), Provisional Measures, Order of 22 November 2013, I.C.J. Reports 2013*, p. 370.

36 *I.C.J. Pleadings, Certain Activities carried out by Nicaragua in the Border Area (Costa Rica v. Nicaragua); Construction of a Road in Costa Rica along the San Juan River (Nicaragua v. Costa Rica),CR 2015/14, Final Submissions of Costa Rica, (f and d)*, p. 70.

37 *Certain Activities carried out by Nicaragua in the Border Areas (Costa Rica v. Nicaragua), Construction of a Road in Costa Rica along the San Juan River (Nicaragua v. Costa Rica), Judgment, I.C.J. Reports 2015*, p. 54.

by Nicaragua with the first ones. The Court began by conceding that Costa Rica was entitled to receive compensation for the material damage caused by Nicaragua's failure to comply with its obligations (including the breach of the provisional measures, as it is clear from para. 142). Then, the Court recalls that it has the power to set aside the rule enshrined in Article 64 of the Statute on the payment of the costs. Further, the Court acknowledges that Costa Rica had to, as a matter of necessity, engage in new proceedings on provisional measures on account of Nicaragua's breach. However, after all this reasoning, the Court reaches the conclusion that 'an award of costs to Costa Rica ... would not be appropriate'. There is no further elaboration. The joint Declaration of Judges Tomka, Greenwood, Sebutinde and *ad hoc* Dugard regrets that the Court failed to develop a satisfactory reasoning in its decision not to award costs to Costa Rica.

Bibliography

X. Fuentes, 'Latin American States and the ICJ', in N. Klein (ed.), *Litigating International Law Disputes: Weighing the Options*, (Cambridge University Press) 2014; pp. 79–105.

A. Rieu-Clarke, 'The obligation to notify and consult on planned measures concerning international watercourses: learning lessons from recent international case-law', *Yearbook of International Environmental Law* 2013; pp. 102–130.

25 *Dispute regarding Navigational and Related Rights (Costa Rica v. Nicaragua), 2005*

Martin Dawidowicz

I Introduction

Costa Rica and Nicaragua both gained independence from Spanish colonial rule in 1821. Shortly after independence, Costa Rica and Nicaragua, together with El Salvador, Guatemala and Honduras decided to establish the Federal Republic of Central America. The Federal Republic of Central America would be a short-lived union: it soon encountered insurmountable problems and was dissolved in 1839 after Costa Rica and Nicaragua had seceded as independent states in the previous year. A period of instability followed in Nicaragua. During the mid-1850s, war broke out between Costa Rica and Nicaragua. At the termination of hostilities, the two countries resolved to settle outstanding bilateral matters, relating, *inter alia*, to their common boundary, to the navigational regime on the San Juan River and to the possibility of building an inter-oceanic canal across the Central American isthmus. After two failed attempts in 1857, Costa Rica and Nicaragua, through the mediation of the Salvadoran Foreign Minister, finally concluded an agreement by entering into the 1858 Treaty of Territorial Limits (Treaty of Limits).[1] Thus, the Treaty of Limits established the present boundary between Costa Rica and Nicaragua.

The Treaty of Limits is not an ordinary boundary treaty. It is an example *par excellence* of a treaty establishing a boundary regime.[2] Article II of the Treaty of Limits fixed the course of the boundary between Costa Rica and Nicaragua from the Pacific Ocean to the Caribbean Sea.[3] Part of the border between the two countries runs along the right bank (i.e. the Costa Rican side) of the San Juan River; namely, from a point three miles below Castillo

1 See generally *Dispute regarding Navigational and Related Rights (Costa Rica v. Nicaragua), Judgment, I.C.J. Reports 2009*, p. 213 at 226–31, paras 15–28; Treaty of Territorial Limits, Costa Rica and Nicaragua, San José, 15 April 1858, reproduced in ibid., Costa Rica's Application, p. 20 (English translation).

2 Compare *Dispute regarding Navigational and Related Rights (Costa Rica v. Nicaragua)*, op. cit., n. 1, Verbatim Record, 9 March 2009, CR 2009/6, p. 15, para. 30 (Mr Crawford on behalf of Costa Rica), pp. 38–9, paras 7–8 (Mr Caflisch on behalf of Costa Rica).

3 Treaty of Territorial Limits, op. cit., n. 1 (Art. II).

Viejo, a small town in Nicaragua, to the end of Punta de Castilla, where the river enters the Caribbean Sea. Article VI of the Treaty of Limits contained the essential bargain of the territorial settlement: it conferred on Nicaragua full and exclusive sovereignty over the entire course of the San Juan River, in exchange for which Costa Rica was granted a perpetual right of free navigation '*con objetos de comercio*' on that approximately 87 mile-long stretch of the river, which constitutes the border between the two countries.[4] Treaties involving territorial settlements are by their nature characterized by permanence and the objective of achieving finality and stability. The Treaty of Limits is no exception.[5] And yet the boundary regime established between Costa Rica and Nicaragua in the Treaty of Limits appears to be one of the most litigated bilateral treaties in the history of the International Court of Justice (hereinafter 'the ICJ' or 'the Court').

During the late-nineteenth century, Nicaragua had on various occasions challenged the validity of the Treaty of Limits and, as a result, the parties submitted the question to arbitration by US President Cleveland. It was additionally agreed that, if the Treaty of Limits was found to be valid, President Cleveland should also decide whether Costa Rica could navigate the San Juan River with vessels of war or of the revenue service. In the Cleveland Award of 22 March 1888, President Cleveland held that the Treaty of Limits was valid. With reference to Article VI of the Treaty of Limits, President Cleveland further stated that Costa Rica did not have the right of navigation on the San Juan River with vessels of war, although it could navigate with such revenue service vessels as may be connected to navigation 'for the purposes of commerce'.[6]

On 5 August 1914, Nicaragua signed a treaty with the US (the Chamorro-Bryan Treaty)[7] that granted it perpetual ownership rights for the construction and maintenance of an inter-oceanic canal through the San Juan River. On 24 March 1916, Costa Rica filed a case against Nicaragua before the Central American Court of Justice[8] (CACJ) alleging that Nicaragua had breached its obligation to consult with it prior to entering into any canalization project pursuant to Article VIII of the Treaty of Limits. On 30 September 1916, the CACJ ruled that, by not consulting Costa Rica, Nicaragua had violated rights guaranteed to it under the Treaty of Limits and

4 *Dispute regarding Navigational and Related Rights (Costa Rica v. Nicaragua)*, op. cit., n. 1, p. 229, para. 19.

5 Ibid., pp. 243–4, paras 68–9.

6 Ibid., pp. 229–30, para. 20.

7 *Convention for the Construction of a Canal by the River San Juan*, United States and Nicaragua, Washington, DC, 5 August 1914, 220 *CTS* 215.

8 The Central American Court of Justice was created by the *General Treaty of Peace and Amity*, Costa Rica, Guatemala, Honduras, Nicaragua and El Salvador, Washington, DC, 20 December 1907, 206 *CTS* 63. See further Part I, Chapter III (for a discussion of the Central American Court of Justice).

the Cleveland Award.[9] These early disputes in relation to the boundary regime in the Treaty of Limits have not settled matters. The validity of the Treaty of Limits may no longer be in question, but it appears that its interpretation nevertheless remains in doubt, as it is still the subject of litigation before the Court.

On 29 September 2005, Costa Rica instituted proceedings before the ICJ against Nicaragua with respect to a dispute concerning navigational and related rights of Costa Rica on the San Juan River.[10] The Court, in its 2009 judgment in the *Navigational and Related Rights case*, upheld many of Costa Rica's submissions on the scope and extent of its navigational and related rights on the San Juan River whilst also recognising Nicaragua's broad power of regulation of navigation. But matters did not end there. The *Navigational and Related Rights case* has triggered a decade of litigation before the Court with two additional cases (later joined) – *Certain Activities carried out by Nicaragua in the Border Area* and *Construction of a Road in Costa Rica along the San Juan River*[11] – at least in part involving alleged additional breaches of the Treaty of Limits and alleged non-compliance with the Court's judgment in *Navigational and Related Rights.*[12] And still further disputes about the Treaty of Limits (including with respect to its Article IV regarding the common Bays of San Juan del Norte and Salinas) cannot be excluded.[13]

The preamble to the Treaty of Limits between Costa Rica and Nicaragua explains that it was designed to 'put an end to all the differences which have obstructed the perfect understanding and harmony that must prevail among

9 See *Dispute regarding Navigational and Related Rights (Costa Rica v. Nicaragua)*, op. cit., n. 1, p. 230, para. 22. Further: 'Costa Rica v. Nicaragua, Central American Court of Justice, Opinion and Decision of the Court, 30 September 1916', *American Journal of International Law* 11, 1917, 181 (English translation).

10 *Dispute regarding Navigational and Related Rights (Costa Rica v. Nicaragua)*, op. cit., n. 1, Costa Rica's Application, 29 September 2005.

11 See *Certain Activities carried out by Nicaragua in the Border Area (Costa Rica v. Nicaragua)*, Costa Rica's Application, 18 November 2010; *Construction of a Road in Costa Rica along the San Juan River (Nicaragua v. Costa Rica)*, Nicaragua's Application, 22 December 2011; *Certain Activities carried out by Nicaragua in the Border Area, Joinder of Proceedings, Order of 17 April 2013, I.C.J. Reports 2013*, p. 166; *Construction of a Road in Costa Rica along the San Juan River, Joinder of Proceedings, Order of 17 April 2013, I.C.J. Reports 2013*, p. 184; *Certain Activities carried out by Nicaragua in the Border Area (Costa Rica v. Nicaragua)*; *Construction of a Road in Costa Rica along the San Juan River (Nicaragua v. Costa Rica)*, ICJ, Judgment of 16 December 2015. See further Part I, Chapter III.

12 See e.g. *Certain Activities carried out by Nicaragua in the Border Area (Costa Rica v. Nicaragua)*, op. cit., n. 11, Costa Rica's Application, p. 2; *Construction of a Road in Costa Rica along the San Juan River*, op. cit., n. 11, Nicaragua's Application, p. 30.

13 See *Dispute regarding Navigational and Related Rights*, op. cit., n. 1, Verbatim Record, 5 March 2009, CR 2009/4, p. 17, paras 35–6 (Mr Argüello on behalf of Nicaragua); Verbatim Record, 9 March 2009, CR/2009/6, p. 66, para. 7 (Mr Ugalde on behalf of Costa Rica).

them for their safety and prosperity'.[14] This laudable aspiration is yet to be fully achieved. The waters of the San Juan River may be shallow, but the dispute between Costa Rica and Nicaragua over the boundary regime in the Treaty of Limits evidently runs deep.

II Dispute regarding navigational and related rights

With these introductory remarks in mind, the present chapter will now briefly examine the *Navigational and Related Rights case*. This section will in turn consider the facts of the dispute before the Court in *Navigational and Related Rights*; the Court's reasoning in arriving at its judgment in the case; the Court's judgment and its contribution to international law; and the alleged non-compliance with the Court's judgment.

A. The dispute before the Court

As already observed, on 29 September 2005, Costa Rica brought ICJ proceedings against Nicaragua with respect to a dispute concerning navigational and related rights of Costa Rica on the San Juan River. Costa Rica notably claimed that Nicaragua had breached its obligation under Article VI of the Treaty of Limits to allow all Costa Rican vessels and their passengers to navigate freely on the San Juan for purposes of commerce, including communication and the transportation of passengers and tourism on that stretch of the San Juan where the border lies on the Costa Rican bank of the river. Costa Rica further claimed that Nicaragua had breached the Treaty of Limits by imposing various regulations of navigation (e.g. with respect to departure clearance certificates, visas and tourist cards and the charging of multiple associated fees), which infringed its right of free navigation on the San Juan. Finally, Costa Rica claimed that Nicaragua had violated a bilateral custom permitting riparians of the Costa Rican bank to fish in the San Juan for subsistence purposes.[15]

The key issue before the Court concerned the effect of the passage of time on the interpretation of Article VI of the 1858 Treaty of Limits – a provision that it may be recalled granted Nicaragua sovereignty over the San Juan River in exchange for which Costa Rica received a perpetual right of free navigation *con objetos de comercio* on that part of the San Juan, which constitutes the border between them. The parties' disagreement was greatest on the meaning of the Spanish words *con objetos de comercio* in Article VI of the Treaty of Limits – the Spanish version of the treaty being the only authoritative one. For its part, Costa Rica claimed that the Spanish words *con objetos de comercio* in Article VI of the Treaty of Limits should be translated into

14 Treaty of Territorial Limits, op. cit., n. 1 (preamble).
15 *Dispute regarding Navigational and Related Rights (Costa Rica v. Nicaragua)*, op. cit., n. 1, pp. 221–6, paras 12–14 (for the parties' submissions).

English as meaning 'for purposes of commerce'; as a minimum, its navigational rights on the San Juan River were therefore not only limited to the transport of goods but also encompassed passengers, including tourists. For Nicaragua, the relevant words should be translated into English as meaning 'with articles of trade'; in other words, Costa Rica's navigational rights were limited to the commercial transport of goods.[16]

B. The Court's reasoning

The Court's analysis proceeded in several steps. It first determined the extent of Costa Rica's right of free navigation on the San Juan River. It then considered, within the scope of the Costa Rican right thus defined, to what extent Nicaragua had the power to regulate navigation by Costa Rican boats and whether the specific regulations it had enacted in relation thereto were compatible with Costa Rica's rights. Finally, it considered whether inhabitants of the Costa Rican bank of the San Juan River had a customary right to engage in subsistence fishing.[17]

The Court turned first to the meaning of the Spanish phrase *con objetos de comercio* in Article VI of the 1858 Treaty of Limits in order to determine whether it should mean 'for the purposes of commerce' – as Costa Rica alleged – or 'with articles of trade' – as alleged by Nicaragua. Having determined by way of treaty interpretation that the parties must have understood the phrase *con objetos de comercio* to mean 'for purposes of commerce',[18] the Court next turned to the meaning of *comercio* or 'commerce' in order to determine the precise extent of Costa Rica's right of free navigation. For Costa Rica, the word 'commerce' covered 'any activity in pursuit of commercial purposes, and includes, *inter alia*, the transport of passengers, tourists among them, as well as of goods'.[19] Costa Rica further alleged that 'commerce' was a broad concept that extended even beyond for-profit activities to cover 'communication and dealings of some persons or peoples with others'.[20] It followed on this view that the word 'commerce' included 'movement and contact between inhabitants of the villages on the Costa Rican bank of the San Juan River, and the use of the river for purposes of navigation by Costa Rican public officials providing the local population with essential services, such as health, education and security'.[21] For Nicaragua, by contrast, 'commerce' covered 'solely the purchase and sale of merchandise, of physical goods, and excludes all services, such as passenger transport'.[22] In

16 Ibid., p. 236, para. 45.
17 Ibid., pp. 231–2, para. 29.
18 Ibid., pp. 238–40, paras 50–6.
19 Ibid., p. 241, para. 59.
20 Ibid.
21 Ibid.
22 Ibid., p. 240, para. 58.

any event, the word 'commerce' should be given the meaning it had when the Treaty of Limits was entered into in 1858, which 'necessarily meant trade in goods and did not extend to services, the inclusion of services being a very recent development'.[23]

The Court subscribed neither to the broad interpretation of the meaning of 'commerce' advanced by Costa Rica nor the narrow one put forward by Nicaragua. After reaffirming the cardinal principle that 'the terms used in a treaty must be interpreted in light of what is determined to have been the parties' common intention, which is, by definition, contemporaneous with the treaty's conclusion', the Court in relevant part stated:

> [T]here are situations in which the parties' intent upon conclusion of the treaty was, or may be presumed to have been, to give the terms used – or some of them – a meaning or content capable of evolving, not one fixed once and for all, so as to make allowance for, among other things, developments in international law. In such instances, it is indeed in order to respect the parties' common intention at the time the treaty was concluded, not to depart from it, that account should be taken of the meaning acquired by the terms in question upon each occasion on which the treaty is to be applied.[24]

The Court, relying on its earlier case law, continued:

> [W]here the parties have used generic terms in a treaty, the parties necessarily having been aware that the meaning of the terms was likely to evolve over time, and where the treaty has been entered into for a very long period or is "of continuing duration", the parties must be presumed, as a general rule, to have intended those rules to have an evolving meaning.[25]

The Court concluded that this was so in respect of the term *comercio* as used in Article VI of the Treaty of Limits. First, *comercio* was a generic term. Second, the Treaty of Limits was entered into for an unlimited duration; from the outset it was intended to create a boundary regime characterized by its perpetuity. Indeed the Court noted that the object of the Treaty of Limits was 'to achieve a permanent settlement between the parties of their territorial disputes'.[26] And the perpetual right of free navigation guaranteed to Costa Rica under Article VI of the Treaty of Limits formed an integral part of that territorial settlement.[27] Thus the Court concluded that the term

23 Ibid.
24 Ibid., p. 242, paras 63–4.
25 Ibid., pp. 242–3, paras 65–6.
26 Ibid., p. 243, para. 68.
27 Ibid., p. 244, para. 69.

'*comercio*' had to be given an evolutive interpretation.[28] The Court on this basis found that Costa Rica's right of navigation under Article VI of the Treaty of Limits applied to the transport of persons as well as the transport of goods, as the activity of transporting persons (including tourists) could nowadays be commercial in nature provided it was carried out for profit-making purposes.[29]

The Court's reasoning meant that non-commercial navigation of private vessels belonging to the inhabitants of the villages of the Costa Rican bank of the San Juan River in order to meet the basic requirements of everyday life, such as taking children to school or in order to give or receive medical treatment, was not protected by Article VI of the Treaty of Limits. The Court nonetheless held that such a minimal right of navigation – which applied equally to certain Costa Rican official vessels used solely for the purpose of providing the local population with assistance in order to meet the necessities of daily life – could be inferred from the provisions of the Treaty of Limits as a whole and, in particular, the manner in which the boundary was fixed along the river bank.[30]

Having determined the extent of Costa Rica's navigational rights on the San Juan River, the Court next considered Nicaragua's power to regulate that navigation. The Court concluded that Nicaragua had the power to regulate navigation under the Treaty of Limits but that power was not unlimited, being tempered by the rights and obligations of the parties. In balancing the interests of the parties, the Court explained that a regulation required certain characteristics based essentially on the principles of reasonableness, non-arbitrariness and non-discrimination. The Court applied these criteria to the impugned regulations and found that Nicaragua had duly exercised its powers of regulation regarding certain matters but had breached the Treaty of Limits in implementing measures requiring visas and tourist cards and the payment of charges in respect of Costa Rican vessels, boat operators and their passengers exercising the freedom of navigation.[31] Finally, the Court found that the inhabitants of the Costa Rican bank of the San Juan River were entitled to engage in fishing for subsistence purposes as a customary right.[32]

28 Ibid., p. 244, para. 70. For criticism see ibid., pp. 283–5 (Sep. Op. Judge Skotnikov), pp. 293–9 (Dec. Judge *ad hoc* Guillaume).

29 Ibid., p. 244, para. 71.

30 Ibid., pp., 244–8, paras 72–84. For criticism see ibid., pp. 300–1 (Dec. Judge *ad hoc* Guillaume).

31 Ibid., pp. 248–63, paras 85–133. For criticism see ibid., pp. 273–7 (Sep. Op. Judge Sepúlveda-Amor), pp. 286–9 (Sep. Op. Judge Skotnikov), pp. 301–2 (Dec. Judge *ad hoc* Guillaume).

32 Ibid., pp. 263–6, paras 134–44. For criticism see ibid., pp. 278–82 (Sep. Op. Judge Sepúlveda-Amor).

C. The Court's judgment and its contribution to international law

The Court's main legal contribution in *Navigational and Related Rights* was to reaffirm the principle of evolutionary treaty interpretation – a principle well-established (albeit not well-developed) in its jurisprudence.[33] As Judge *ad hoc* Guillaume observed in *Navigational and Related Rights*, the question of the effect of the passage of time on treaty interpretation has been the subject of lively debate between proponents of 'contemporaneous' interpretation and advocates of 'evolutionary' interpretation. Judge *ad hoc* Guillaume rightly concluded that evolutionary interpretation does not appear to be a separate method of interpretation but rather the result of the proper application of ordinary means of interpretation as codified in Article 31 Vienna Convention on the Law of Treaties.[34] Put differently, there appears to be nothing special about evolutionary treaty interpretation – a point not always well understood.[35]

Evolutionary treaty interpretation has become a characteristic feature of the jurisprudence of the European Court of Human Rights; however, it is not at all clear that such an interpretation is more suited for some types of treaty (notably in the field of human rights) than it is for others. The Court's reliance on evolutionary treaty interpretation in *Navigational and Related Rights* – albeit seemingly unnecessary in the actual circumstances of the case[36] – underlines this point. In essence, evolutionary treaty interpretation is but an application of a basic principle – the intention of the parties – the determination of which is the main task in the work of interpretation.[37] Since the Court's 2009 judgment in *Navigational and Related Rights*, Special Rapporteur Nolte has explicitly relied on the judgment in support of the

33 For an assessment see further Bjorge, E., 'International Court of Justice, Case Concerning the Dispute regarding Navigational and Related Rights (Costa Rica v. Nicaragua) Judgment of 13 July 2009', *International and Comparative Law Quarterly* 60, 2009, pp. 271–9; Lathrop, C. G., 'Dispute regarding Navigational and Related Rights (Costa Rica v. Nicaragua)', *American Journal of International Law* 104, 2010, pp. 454–61; Dawidowicz, M., 'The Effect of the Passage of Time on the Interpretation of Treaties: Some Reflections on Costa Rica v. Nicaragua', *Leiden Journal of International Law* 24, 2011, pp. 201–22. Generally: Bjorge, E., *The Evolutionary Interpretation of Treaties*, Oxford: Oxford University Press, 2014; Djeffal, C., *Static and Evolutive Treaty Interpretation: A Functional Reconstruction*, Cambridge: Cambridge University Press, 2016.

34 *Dispute regarding Navigational and Related Rights (Costa Rica v. Nicaragua)*, op. cit., n. 1, p. 294 (Dec. Judge *ad hoc* Guillaume).

35 To the same effect see Bjorge E., 'International Court of Justice, Case Concerning the Dispute regarding Navigational and Related Rights (Costa Rica v. Nicaragua)', op. cit., n. 34, p. 279 (with further references).

36 See generally Dawidowicz, 'The Effect of the Passage of Time on the Interpretation of Treaties: Some Reflections on Costa Rica v. Nicaragua', op. cit., n. 33.

37 See e.g. *Dispute regarding Navigational and Related Rights (Costa Rica v. Nicaragua)*, op. cit., n. 1, p. 242, para. 64. Also: Bjorge, *The Evolutionary Interpretation of Treaties*, op. cit., n. 33, p. 2.

same conclusion in the ILC's work on subsequent agreements and subsequent practice in relation to treaty interpretation.[38]

D. The alleged non-compliance with the Court's judgment

In his declaration appended to the Court's judgment in *Navigational and Related Rights*, Judge *ad hoc* Guillaume 'express[ed] the hope that it will enable the two countries to overcome their past difficulties in respect of the river'.[39] In a similar vein, during the oral proceedings, Costa Rica's Agent explained: 'It is our hope that the decision of the Court will finally settle the dispute so that both Costa Rica and Nicaragua can return to fraternal and friendly relations on all fronts.'[40] This laudable aspiration has evidently not yet been fulfilled. In *Certain Activities carried out by Nicaragua in the Border Area*, Costa Rica has, among other alleged breaches of the Treaty of Limits and international law,[41] claimed that Nicaragua has continued to impair Costa Rica's navigational rights under the Treaty of Limits in clear violation of the Court's judgment in *Navigational and Related Rights.* Indeed Costa Rica in this case specifically asked the Court to adjudge and declare that Nicaragua had violated its judgment in *Navigational and Related Rights.* In its judgment, the Court found that Nicaragua has breached Costa Rica's rights of navigation on the San Juan River pursuant to the Treaty of Limits.[42]

In *Construction of a Road in Costa Rica along the San Juan River*, Costa Rica also invoked Nicaragua's alleged non-compliance with the Court's judgment as a circumstance (among several others) leading to the urgent construction of a road on its territory along the San Juan River.[43] Nicaragua brought ICJ proceedings against Costa Rica in this case, *inter alia*, on the basis that Costa Rica's construction of the said road allegedly prejudiced Nicaraguan territory contrary to the Treaty of Limits. As part of its request for relief, Nicaragua initially asked the Court for a declaration allowing it by

38 See Nolte, G., 'First report on subsequent agreements and subsequent practice in relation to treaty interpretation', UN Doc. A/CN.4/660, pp. 23–7; and also ILC Report (2013), UN Doc. A/68/10, p. 12 (Draft Conclusion 3 – 'Interpretation of treaty terms as capable of evolving over time').

39 *Dispute regarding Navigational and Related Rights* (*Costa Rica v. Nicaragua),* op. cit., n. 1, p. 302 (Dec. Judge *ad hoc* Guillaume).

40 Ibid., Verbatim Record, 2 March 2009, CR 2009/2, p. 14, para. 10 (Mr Ugalde on behalf of Costa Rica).

41 See *Certain Activities carried out by Nicaragua in the Border Area (Costa Rica v. Nicaragua)*, op. cit., n. 11, Costa Rica's Application, p. 2.

42 Ibid., Costa Rica's Memorial, pp. 290–4, 303; Verbatim Record, 15 April 2015, CR 2015/4, pp. 10–22, paras 1–40 (Ms Del Mar on behalf of Costa Rica); *Certain Activities carried out by Nicaragua in the Border Area (Costa Rica v. Nicaragua)*, op. cit., n. 11, para. 229(4).

43 See *Construction of a Road in Costa Rica along the San Juan River (Nicaragua v. Costa Rica)*, op. cit., n. 11, Costa Rica's Counter-Memorial, pp. 22–5.

way of countermeasures to suspend Costa Rica's right of navigation on the San Juan River under the Treaty of Limits.[44] The request was, however, subsequently withdrawn as being unnecessary since Nicaragua would not require the Court's permission to take such action. In any event, Nicaragua explained that such a countermeasure was 'not envisaged at the present time'.[45]

Costa Rica had earlier in the proceedings stated that the Nicaraguan claim that prejudice to its territory could affect the navigational rights granted to it under the Treaty of Limits appeared to be 'an ill-disguised and impermissible attempt to re-litigate the Court's 2009 judgment in Navigational Rights'.[46] Costa Rica immediately added: 'It is in reality another in a series of attempts by Nicaragua to eliminate Costa Rica's navigational rights on the San Juan River.'[47] The matter may ultimately have become moot in *Construction of a Road in Costa Rica along the San Juan River*, but the fact that Nicaragua appears to have reserved its right to suspend Costa Rica's navigational rights under the Treaty of Limits by way of countermeasures reinforces the regrettable notion that the Court's judgment in *Navigational and Related Rights* has so far not enabled the two countries to overcome their past difficulties in respect of the San Juan River.

III Conclusion

The 1858 Treaty of Limits appears to be one of the most litigated bilateral treaties in the history of the Court. This is somewhat surprising given that the object of the Treaty of Limits was to achieve a permanent settlement between the parties of their territorial disputes. Treaties involving territorial settlements are by nature characterized by permanence and the objective of achieving finality and stability. The Court in its 2009 judgment in *Navigational and Related Rights* upheld many of Costa Rica's submissions on the scope and extent of its navigational and related rights on the San Juan River – chiefly by reaffirming the principle of evolutionary treaty interpretation – while also recognising Nicaragua's broad power of regulation of navigation. But disputes between the parties remain.

The *Navigational and Related Rights* case has triggered additional litigation before the Court. This litigation has at least in part involved alleged additional breaches of the Treaty of Limits and alleged non-compliance with the Court's judgment in *Navigational and Related Rights*. In *Certain Activities carried out by Nicaragua in the Border Area*, the Court found that Nicaragua has breached Costa Rica's navigational rights under the Treaty of

44 Ibid., Nicaragua's Memorial, pp. 244–9, 252.

45 Ibid., Verbatim Record, 30 April 2015, CR 2015/16, p. 57, para. 17 (Mr Pellet on behalf of Nicaragua).

46 Ibid., Costa Rica's Counter-Memorial, p. 136.

47 Ibid.

Limits, as affirmed by its judgment in *Navigational and Related Rights*. And it appears that still further disputes in relation to the Treaty of Limits may arise. The preamble to the Treaty of Limits between Costa Rica and Nicaragua explains that it was designed to 'put an end to all the differences which have obstructed the perfect understanding and harmony that must prevail among them for their safety and prosperity'.[48] The aftermath to the Court's judgment in *Navigational and Related Rights* suggests that this aspiration is yet to be fully achieved.

Bibliography

Bjorge, E., 'International Court of Justice, Case Concerning the Dispute regarding Navigational and Related Rights (Costa Rica v. Nicaragua) Judgment of 13 July 2009', *International and Comparative Law Quarterly* 60, 2009, 271–9.

Bjorge, E., *The Evolutionary Interpretation of Treaties*, Oxford: Oxford University Press, 2014.

Dawidowicz, M., 'The Effect of the Passage of Time on the Interpretation of Treaties: Some Reflections on Costa Rica v. Nicaragua', *Leiden Journal of International Law* 24, 2011, 201–22.

Djeffal, C., *Static and Evolutive Treaty Interpretation: A Functional Reconstruction*, Cambridge: Cambridge University Press, 2016.

Lathrop, C. G., 'Dispute regarding Navigational and Related Rights (Costa Rica v. Nicaragua)', *American Journal of International Law* 104, 2010, 454–61.

Nolte, G., 'First report on subsequent agreements and subsequent practice in relation to treaty interpretation', UN Doc. A/CN.4/660, 23–7.

48 Treaty of Territorial Limits, op. cit., n. 1 (preamble).

26 *Pulp Mills on the River Uruguay (Argentina v. Uruguay)*, 2006

Sandrine Maljean-Dubois and Vanessa Richard

On 20 April 2010, the International Court of Justice (hereinafter 'ICJ' or 'the Court') pronounced its ruling regarding the dispute opposing Argentina and Uruguay in the *Pulp Mills on the River Uruguay* case.[1] The dispute originates from the construction of two pulp mills on the left bank of the Uruguay River, near the Uruguayan city of Fray Bentos and opposite the Argentinian region of Gualeguaychú. The construction works of the first mill, called CMB (ENCE),[2] started in November 2005. However, the Spanish sponsors of the project announced they were definitively abandoning it in September 2006. The second pulp mill, Orion (Botnia),[3] was constructed a few kilometres downstream of the site planned for CMB (ENCE). The project was launched in 2003 and the Orion (Botnia) mill has been in operation since 9 November 2007.

As emphasized by the ICJ in 2006,[4] this is the biggest foreign investment that Uruguay has ever hosted; thus the stakes were quite high for Uruguay. From the very beginning, on the other bank of the river, the project has stirred up vehement protests from the Argentine authorities as well as from the population who feared pollution damage in a region where the main economic activities are tourism, fishing and agriculture. Since December 2005, the bridges between the two countries have occasionally been blocked by gatherings of Argentine citizens.

In this tense context, several judicial and non-judicial procedures at the domestic, regional and international levels have been triggered, including procedures based on the social and environmental responsibility of

1 *Pulp Mills on the River Uruguay (Argentina v. Uruguay), Judgment, I.C.J. Reports 2010*, p. 14.

2 CMB stands for *Celulosas de M'Bopicuá S.A.*, and ENCE for *Empresa Nacional de Celulosas de España*.

3 The Orion mill is undertaken by Botnia S.A. and Botnia Fray Bentos S.A., two Uruguayan companies created in 2003 by the Finnish company Oy Metsä-Botnia AB.

4 *Pulp Mills on the River Uruguay (Argentina v. Uruguay), Request for the Indication of Provisional Measures, Order of 13 July 2006, I.C.J. Reports 2006*, para. 48.

enterprises.[5] The two states agreed in 2005 on the creation of a High-Level Technical Group, the GTAN,[6] which aimed to find a negotiated solution. The King of Spain also offered mediation. None of these initiatives resulted in reaching a settlement. On 4 May 2006, Argentina brought the dispute before the ICJ, grounding the referral on Article 60, paragraph 1 of a bilateral treaty in force between the two countries: the Statute of the River Uruguay signed on 26 February 1975 (the Statute).[7] Its object and purpose are to promote the 'optimum and rational utilization of the River Uruguay' (Article 1). It frames the river's uses and provides for obligations of consultation, information and cooperation between the parties, as well as the prevention of environmental harm.

In the following months, the Court examined two requests for the indication of provisional measures. First, on 13 July 2006, it dismissed an Argentine request submitted at the same time as the main complaint, which demanded the suspension of the construction of Orion (Botnia).[8] Second, on 23 January 2007, the ICJ dismissed another request submitted by Uruguay, the aim of which was to obtain an emergency removal of the blockades that Argentine opponents to the project had organized on the bridges and roads connecting the two countries.[9] A few months before this request, Uruguay had already referred this matter to the Mercosur arbitral tribunal. In an award of 6 September 2006, the latter had refused to command Argentina to remove the blockades but had stated that Argentina had failed to comply with its due diligence obligations.[10]

Regarding the substance of the case, Argentina alleged before the ICJ that Uruguay had breached the Statute by unilaterally authorizing the construction of two pulp mills and by allowing the commissioning of Orion (Botnia). According to Argentina, such facilities would cause significant harm to the environment of the Uruguay River and its area of influence. It added that 'the process envisaged by the CMB and Orion projects ... is inherently polluting' and that '90 per cent of fish production in the Argentina-Uruguay section of the river ... is located within the areas affected by the mills, which are also a breeding area for the river's migratory fish stocks,' not to mention

5 M.-P. Lanfranchi, 'L'affaire des usines de pâte à papier, un état des lieux', *L'Observateur des Nations Unies* 24, 2010, p. 38.

6 For *Grupo Técnico de Alto Nivel.*

7 UNTS, vol. 1295, no. I-21425, p. 340.

8 *Order of 13 July 2006*, op. cit.

9 *Pulp Mills on the River Uruguay (Argentina v. Uruguay), Request for the Indication of Provisional Measures, Order of 23 January 2007, I.C.J. Reports 2007.*

10 *Laudo del Tribunal arbitral 'ad hoc' de MERCOSUR sobre 'Omisión del Estado Argentino en Adoptar Medidas Apropiadas para Prevenir y/o Hacer Cesar los Impedimentos a la Libre Circulación Derivados de los Cortes en Territorio Argentino de vías de Acceso a los Puentes Internacionales General San Martín y General Artigas'.* Online. Available HTTP: www.mercosur.int/msweb/portal%20intermediario/es/controversias/laudo.html (accessed 18 February 2016).

the fact that the pulp mills would be very near to the surrounding towns.[11] Argentina requested the ICJ to find that these were 'internationally wrongful acts by which Uruguay engaged its responsibility,' to impose that Uruguay 're-establish on the ground and in legal terms the situation that existed before these internationally wrongful acts were committed' and, to declare that Uruguay had to 'pay compensation to the Argentine Republic for the damage caused.'[12] Uruguay requested the ICJ to determine that the construction was lawful and declare 'Uruguay's right to continue operating the Botnia plant in conformity with the provisions of the 1975 Statute.'[13]

The judgment on the merits of the case was eagerly awaited by both parties, whose relationship had been badly damaged by the dispute. However, the case raised expectations beyond the parties as it is emblematic of sustainable development issues that balance economic development against the protection of natural resources. From the stage of provisional measures, the Court said it was aware of the stakes: 'account must be taken of the need to safeguard the continued conservation of the river environment and the rights of economic development of the riparian States.'[14] In its 2010 judgment, the Court nevertheless reaches only an unstable balance between these considerations. On the one hand, it considers that Uruguay did not comply with its procedural obligations under Articles 7 to 12 of the Statute. On the other, it dismisses all allegations concerning the breach of substantive prevention obligations, and it states that the declaration by the Court on the breach of procedural obligations constitutes appropriate satisfaction. Such a result reflects the difficulties the Court faced in solving a dispute on the basis of pieces of evidence that were highly technical. The conclusions it reached in this regard are quite traditional, not to say excessively cautious as regards the determination of both the burden and content of proof.[15] The Court adopted a rather restrictive interpretation of the obligations to prevent environmental harm. Grounding its reasoning on a strict distinction between procedural and substantive obligations, the Court comes to specify the scope of the obligation to prevent transboundary harm.

I A watertight distinction between procedural and substantive obligations

The judgment rests on a *summa divisio* between procedural and substantive obligations. Such a distinction does not stand up to analysis: first because the frontier between the two categories of obligation is ill-defined; second because the results they lead to are hardly satisfactory. According to the

11 *Order of 13 July 2006*, op. cit., para. 8.
12 *Judgment of 20 April 2010*, op. cit., para. 24.
13 Ibid., para. 23.
14 *Order of 13 July 2006*, op. cit., para. 80.
15 On these issues see Kerbrat and Maljean-Dubois, op. cit., 39–75.

Court, the process of authorizing, constructing and commissioning the pulp mill must be analyzed under the angle of the procedural obligations of information, notification and cooperation provided for by Articles 7 to 12 of the Statute. After the mill's commissioning, the question is rather to determine whether Uruguay has complied with its substantive obligations to prevent harm.[16] Such a chronological distinction proves to be artificial, since procedural and substantive obligations are not called for to operate successively. Some of the obligations considered as substantive by the Court should be applied *before* the commissioning of the pulp mill, for example, the obligation to conduct an environmental impact assessment (EIA). The EIA is an obligatory stage of the procedure and is generally described as a procedural obligation. It is thus unsurprising that the Court has trouble in deciding which category it belongs to and therefore treats it under both angles.

Procedural and substantive obligations are inextricably intertwined: compliance with obligations to prevent environmental harm – which the Court views as substantive – is inseparable from compliance with procedural requirements. As compliance with procedural obligations is the only way to prevent environmental harm, procedural obligations are as important as substantive obligations. Being non-operational, this distinction entails questionable consequences.[17] In the Court's opinion, non-compliance with procedural obligations is independent from compliance or non-compliance with substantive obligations.[18] The Court however recognizes that there is a 'functional link' between the two categories of obligations because of their common purpose: the prevention of environmental harm,[19] but it considers that this link 'does not prevent the States parties from being required to answer for those obligations separately, according to their specific content, and to assume, if necessary, the responsibility resulting from the breach of them, according to the circumstances.'[20] The Court considers that as Uruguay's breaches of procedural obligations occurred in the past and have come to an end, there is no cause to order their cessation.[21] It leads the Court to deny any material compensation to Argentina. Since 'the procedural obligations under the 1975 Statute did not entail any ensuing prohibition on Uruguay's building of the Orion (Botnia) mill, failing consent by Argentina, after the expiration of the period for negotiation' and since 'the operation of the Orion (Botnia) mill has not resulted in the breach of substantive obligations laid down in the 1975 Statute,' 'ordering the dismantling of the mill would not, in the view of the Court, constitute an appropriate remedy for the breach of procedural obligations.'[22] Having ruled out the *restitutio in*

16 *Judgment*, op. cit., para. 46.
17 *Joint dissenting opinion of Judges Al-Khasawneh and Simma*, op. cit., para. 26.
18 *Judgment*, op. cit., para. 68.
19 Ibid., para. 79.
20 Ibid.
21 Ibid., para. 269.
22 Ibid., para. 275.

integrum, the Court refuses 'for the same reasons' the Argentine claim for monetary compensation for alleged injuries suffered in various economic sectors, specifically tourism and agriculture.[23] In short, the Court does not draw any consequence from the fact that procedural obligations have been breached.[24] And yet, by breaching its procedural obligations, hasn't Uruguay also breached some substantive obligations? Conversely, could compliance with its procedural obligations have led Uruguay to revise its decision to construct the pulp mill, or to reconsider its location or possibly the technology used at the mill?

In the end, it appears the Court had in advance determined the result it desired to obtain – to not question the huge investment that constitutes the construction of the pulp mill in Uruguay – and then determined the course of reasoning that led to the desired result. It also spares the Court of finding itself in a delicate position. In July 2006, it had decided that a provisional suspension of the construction works was not necessary.[25] Recalling its *Passage through the Great Belt* decision to refuse to order provisional measures, the Court had then merely mentioned that its decision on the merits could *in fine* result in Uruguay having to decommission the facilities.[26]

II The scope of obligations to prevent harm in a transboundary context

The Court upholds a narrow conception of its role that leads it to rule that Uruguay was not prevented to construct the pulp mills awaiting a decision on the merits. It then asserts or re-asserts some fundamental principles related to cooperation and prevention of harm in a transboundary context. Finally, it does not find that Uruguay has breached substantive obligations.

A. The scope of Article 12 of the Statute: Right to have a say or right of veto?

Articles 7 to 12 of the Statute define the different stages of a detailed procedure. If one of the parties 'plans to construct new channels, substantially modify or alter existing ones or carry out any other works which are liable to affect navigation, the *régime* of the river or the quality of its waters,'[27] it has to inform the joint river commission, the CARU. The CARU determines *prima facie* whether the plan might cause significant damage to the other party. Should the Commission find this to be the case or a decision cannot be reached, the party concerned shall notify the other of the plan through

23 Ibid., para. 276.
24 *Dissenting opinion of Judge ad hoc Vinuesa*, para. 6.
25 *Order of 13 July 2006*, op. cit.
26 Ibid., paras. 70–1.
27 Article 7.

the CARU. The notified party can object, request further information or suggest changes. If parties cannot reach an agreement within 180 days, Article 12 provides that the dispute settlement clause of Article 60 – which enables 'either Party to submit to the Court any dispute concerning the interpretation or application of the Statute which cannot be settled by direct negotiations' – shall apply. The purpose of this procedure is to support cooperation between parties, each one having a 'right to have a say' on the projects the other party might have. Can this 'right to have a say' that Article 12 bestows amount to a right of veto? Nothing points in this direction. Article 12 however enables the parties to ask the judge to settle a potential dispute in this respect. This opportunity could have radically changed the role of the Court.[28] It could indeed have resulted in Uruguay being forbidden to authorize the pulp mills' construction until the Court reached a decision. Such an interpretation would have been in line with the spirit of the procedure established by Articles 7 to 12. The Court did not follow this line of reasoning and its position is criticized with reason by several judges.[29]

B. Cooperation and prevention: Customary and treaty-based obligations

Since the Court is referred to on the basis of the dispute settlement clause of the Statute and the Statute's provisions are rather detailed, the ICJ refers little to customary law. Therefore, observers can hardly draw lessons beyond the case-specific. The Court nevertheless reasserts or asserts on several occasions the customary nature of certain obligations and in the end clarifies their content and scope along the lines of its previous case law.

1. The importance of procedural obligations in the management of shared natural resources

The Court rightly points out that 'These [procedural] obligations are all the more vital when a shared resource is at issue ... which can only be protected through close and continuous co-operation between the riparian States.'[30] The cooperation obligation is inferred from the Statute: it is a treaty-based obligation. The Court states that the CARU is the masterpiece of the scheme.[31] In line with its order on provisional measures,[32] it underlines the critical role that joint river commissions play in the sharing of information and, more broadly, in cooperation. According to the Court, the CARU is an international organization: 'far from being merely a transmission mechanism

28 *Joint dissenting opinion of Judges Al-Khasawneh and Simma*, op. cit., para. 25.
29 *Judgment*, op. cit., para. 154; *Declaration of Judge Skotnikov*, para. 3.
30 Ibid., para. 81.
31 *Judgment*, op. cit., para. 93.
32 *Order of 13 July 2006*, op. cit., para. 81.

between the parties, CARU has a permanent existence of its own.'[33] Such an approach perpetuates the negotiation framework since 'neither of [the parties] may depart from that framework unilaterally.'[34]

2. *The customary nature of the obligation to prevent transboundary environmental harm*

The Court reminds that:

> [t]he principle of prevention, as a customary rule, has its origins in the due diligence that is required of a State in its territory. It is 'every State's obligation not to allow knowingly its territory to be used for acts contrary to the rights of other States' *(Corfu Channel (United Kingdom v. Albania), Merits, Judgment, I.C.J. Reports 1949,* p. 22.) A State is thus obliged to use all the means at its disposal in order to avoid activities which take place in its territory, or in any area under its jurisdiction, causing significant damage to the environment of another State. This Court has established that this obligation 'is now part of the corpus of international law relating to the environment.'[35]

The Court here refers to the expression ('is now part of') employed in its 1996 Advisory Opinion on the *Legality of the Threat or Use of Nuclear Weapons*[36] and repeated in its 1997 *Gabčíkovo-Nagymaros* judgment.[37] The phrase is inspired by Principle 21 of the 1972 Stockholm Declaration.[38] Its scope had however remained vague in 1996 and 1997, probably voluntarily. It is henceforth clearly established as a customary obligation. On the merits, this obligation is also clear. From then on, 'A State is thus obliged to use all the means at its disposal in order to avoid activities which take place in its territory, or in any area under its jurisdiction, causing significant damage to the environment of another State.'[39] It commands to apply 'diligence' and 'use all the means at [the state's] disposal.' However, whereas states had to respect the environment of other states in 1997, in 2010 they must not cause

33 *Judgment*, op. cit., para. 87.

34 Ibid., para. 90. See also the critic of Judge Torres Bernardez, *Pulp Mills on the River Uruguay (Argentina v. Uruguay), Separate opinion of Judge ad hoc Torres Bernárdez*, para. 11.

35 Ibid., para. 101.

36 *Legality of the Threat or Use of Nuclear Weapons, Advisory Opinion, 8 July 1996, I.C.J. Reports 1996*, para. 29.

37 *Gabčíkovo-Nagymaros Project (Hungary v. Slovakia), Judgment, 25 September 1997, I.C.J. Reports 1997*, para. 53.

38 *Declaration of the United Nations Conference on the Human Environment*, 16 June 1972. See also Principle 2 of the *Rio Declaration on Environment and Development*, 13 June 1992, A/CONF.151/26.

39 *Judgment*, op. cit., para. 101.

any 'significant damage'. First, the positive obligation to respect the environment becomes an obligation to refrain from causing harm. Second, the Court creates a threshold where none existed. This notion of 'significant damage' remains to be clarified. In any case, by referring to the *Corfu Channel case* and to the notion of due diligence, the Court is probably attempting to better define the consequences of this principle in the procedural field, including beyond the case-specific.

The Court then analyzes Uruguay's compliance with its substantive prevention obligations. It states that 'the attainment of optimum and rational utilization requires a balance between the Parties' rights and needs to use the river for economic and commercial activities on the one hand, and the obligation to protect it from any damage to the environment that may be caused by such activities, on the other,'[40] and mentions the 'interconnectedness between equitable and reasonable utilization of a shared resource and the balance between economic development and environmental protection that is the essence of sustainable development.'[41] A twofold balance must be reached: on the one hand, between the rights and needs of the parties and; on the other hand, between the different uses of the river and the protection of the environment.

From this prevention obligation, the Court infers positive obligations incumbent upon the parties.[42] As in the *Gabčíkovo-Nagymaros* case, the Court reminds that 'in the field of environmental protection, vigilance and prevention are required on account of the often irreversible character of damage to the environment and of the limitations inherent in the very mechanism of reparation of this type of damage.'[43] It also endeavours to give full effect to the scientific 'ecological balance' concept and specifies that 'the negative impact of human activities on the waters of the river may affect other components of the ecosystem of the watercourse such as its flora, fauna, and soil.'[44] Likewise, regarding the obligation to prevent pollution and to preserve the aquatic environment (Article 41), the Court interprets the Statute as subjecting the parties to a positive obligation – once again an obligation of means – which is rather cumbersome.[45] Since then, the Seabed Disputes Chamber of the International Tribunal for the Law of the Sea (ITLOS) has built on these contributions of the ICJ and has deepened them.[46]

40 *Judgment*, op. cit., para. 75. See also para. 174.
41 Ibid., para. 177.
42 Ibid., para. 185.
43 Ibid. *Gabcíkovo-Nagymaros Project*, op. cit., para. 140.
44 *Judgment*, op. cit., para. 188.
45 Ibid., para. 223.
46 See ITLOS, *Responsibilities and obligations of States sponsoring persons and entities with respect to activities in the Area (Request for Advisory Opinion submitted to the Seabed Disputes Chamber), Advisory Opinion in Case no. 17, 1 February 2011*; see also ITLOS, *Request for an advisory opinion submitted by the Sub-Regional Fisheries Commission (SRFC), Advisory Opinion in Case no. 21, 2 April 2015.*

3. The recognition of the customary nature of the EIAs obligation

The Court recognizes the customary nature of the obligation 'to undertake an environmental impact assessment where there is a risk that the proposed industrial activity may have a significant adverse impact in a transboundary context, in particular, on a shared resource.'[47] In line with the parties' agreement on the existence of an EIA obligation – and consistently with the Statute read as a whole – the Court interprets the Statute in an evolutive manner by referring again to general international law.[48] The Court quotes its judgment in the *Dispute Regarding Navigational and Related Rights case* and considers that the parties' will was to give some of the Statute's provisions 'a meaning or content capable of evolving, not one fixed once and for all, so as to make allowance for, among other things, developments in international law.'[49] The ICJ infers such EIA obligations from a 'practice, which in recent years has gained so much acceptance among States' but also from 'due diligence, and the duty of vigilance and prevention which it implies,'[50] although it does not specify whether it is a customary obligation or an obligation stemming from Article 41 of the Statute. In any case, it is this connection with due diligence that makes the Court affirm that 'if a party planning works liable to affect the regime of the river or the quality of its waters did not undertake an environmental impact assessment on the potential effects of such works' it will not have exercised due diligence. The Court adds that once the project is commissioned, a continuous monitoring of the project's impacts must be implemented throughout the whole life-cycle of it.[51]

C. Uruguay's compliance and non-compliance

Regarding the procedural obligations of information and notification, the Court is severe *vis-à-vis* Uruguay: it finds that Uruguay did not comply with the whole of the cooperation mechanism provided for by Article 7 to 12.[52] By requiring that the EIA be passed on together with the notification, the Court's interpretation goes even further than the 1991 Espoo Convention,[53] which does not include the EIA in the documents that must be passed on.[54] It is all the easier for the Court to sentence Uruguay for non-compliance

47 *Judgment*, op. cit., para. 204.

48 Ibid.

49 Ibid.; *Dispute Regarding Navigational and Related Rights (Costa Rica v. Nicaragua), Judgment, 13 July 2009, I.C.J. Reports 2009*, para. 64.

50 *Judgment*, op. cit., para. 204.

51 Ibid., para. 205.

52 Ibid., para. 149.

53 *Convention on Environmental Impact Assessment in a Transboundary Context*, Espoo, Finland, 25 February 1991, UNTS, vol. 1989, p. 309.

54 Article 3 of the Espoo Convention.

with its procedural obligations then it does not entail any consequence in terms of compensation.

Equitable and reasonable use of a shared natural resource, balance between the rights and needs of the parties, need to combine economic development with environmental protection, 'real community of interests and rights,'[55] positive obligations of states, customary obligation to conduct an EIA... The Court has taken a number of steps forward. The Court however finds that Uruguay did not breach its obligations on the merits. It progressively rejects all Argentina's allegations. The Court's dismissals are sometimes based on a restrictive interpretation of substantive obligations, and sometimes on a lack of sufficient evidence. Such a restrictive approach applies for example to EIAs. The Court only goes part of the way when it recognizes the obligation to conduct an EIA as customary. Indeed, it considers that neither the Statute nor general international law specify 'the scope and content of an [EIA].' Consequently, 'it is for each State to determine in its domestic legislation or in the authorization process for the project, the specific content ... required in each case, having regard to the nature and magnitude of the proposed development and its likely adverse impact on the environment as well as to the need to exercise due diligence in conducting such an assessment.'[56]

III Conclusion

At the end of its judgment, the Court emphasizes that the Statute 'places the Parties under a duty to co-operate with each other' and that it 'encompasses ongoing monitoring of ... the Orion (Botnia) mill.'[57] One could indeed fear that the parties would not be able to agree and that the situation that followed the *Gabčíkovo-Nagymaros* judgment would repeat itself: a few months after the judgment, Slovakia had requested a complementary judgment on the grounds that it hadn't managed to reach an agreement with Hungary on the means to execute it. The case is still pending and has trouble in reaching a solution.

Regarding the *Pulp mills* case, the parties demonstrated a will to turn the page during the first months. The two heads of state reached an agreement in July 2010 on the establishment of a joint committee composed of two Argentine scientists and two Uruguayan scientists, entrusted with assessing the environmental impact of Orion.[58] Soon however, after the scientists began their mission in early October 2010,[59] the release of an alarming report

55 *Judgment*, op. cit., para. 281.

56 Ibid., para. 205.

57 Ibid., para. 281. See also *Separate opinion of Judge Greenwood*, op. cit.

58 Notice of Agence France Presse, 28 July 2010.

59 'Cientificos argentinos y uruguayos realizan el primer control en la pastera,' *La Nación*, 6 October 2010.

on the air pollution allegedly emitted by the pulp mill – written under the auspices of the Argentine Ministry of the Environment – sparked off a new crisis. This issue seems to be outside of the scope of the monitoring the parties agreed on. Likewise, it was considered to be outside of the remit of the ICJ.[60] In 2013, tensions reached a new level due to a unilateral Uruguayan decision to considerably increase the pulp mill's production. Argentina threatened to refer the matter to the Court again.[61] Uruguay denied planning the construction of another pulp mill. Tension has temporarily decreased but the battle might not be over.

Bibliography

Journal articles

'Cientificos argentinos y uruguayos realizan el primer control en la pastera,' *La Nación*, 6 October 2010.

'Cruce con Uruguay por un informe sobre contaminación,' *Clarín*, 14 October 2010.

'Uruguay's first lady praises Cristina Fernandez and blast opposition,' 29 August 2013. Online. Available http://en.mercopress.com/2013/08/29/uruguay-s-first-lady-praises-cristina-fernandez-and-blast-opposition (accessed 29 May 2015).

'Argentina threatens legal action over UPM pulp mill', AFP, 3 October 2013.

Y. Kerbrat, S. Maljean-Dubois, 'La Cour internationale de Justice face aux enjeux de protection de l'environnement: réflexions critiques sur l'arrêt du 20 avril 2010, *Usines de pâte à papier sur le fleuve Uruguay* (Argentine c. Uruguay),' *Revue Générale de Droit International Public* 1, vol. CXV, 2001, 39–75.

M.-P. Lanfranchi,'L'affaire des usines de pâte à papier, un état des lieux,' *L'Observateur des Nations Unies* 24, 2010, 37–63.

V. Richard, E. Truilhé-Marengo, 'La coopération sur un fleuve partagé, l'anticipation des risques environnementaux et la CIJ: un pas en avant, deux pas en arrière?,' *Bulletin du Droit de l'Environnement Industriel* 28, 2010, 17–21.

Jurisprudence

Legality of the Threat or Use of Nuclear Weapons, Advisory Opinion, 8 July 1996, I.C.J. Reports 1996, p. 226.

Gabčíkovo-Nagymaros Project (Hungary v. Slovakia), Judgment, 25 September 1997, I.C.J. Reports 1997, p. 7.

Pulp Mills on the River Uruguay (Argentina v. Uruguay), Request for the Indication of Provisional Measures, Order of 13 July 2006, I.C.J. Reports 2006, p. 113.

60 'Cruce con Uruguay por un informe sobre contaminación,' *Clarín*, 14 October 2010.

61 'Argentina threatens legal action over UPM pulp mill,' AFP, 3 October 2013. See also 'Uruguay's first lady praises Cristina Fernandez and blast opposition,' 29 August 2013. Online. Available http://en.mercopress.com/2013/08/29/uruguay-s-first-lady-praises-cristina-fernandez-and-blast-opposition (accessed 29 May 2015).

Dispute Regarding Navigational and Related Rights (Costa Rica v. Nicaragua), Judgment, 13 July 2009, I.C.J. Reports 2009, p. 213.
Pulp Mills on the River Uruguay (Argentina v. Uruguay), Judgment, I.C.J. Reports 2010, p. 14.
Responsibilities and obligations of States with respect to activities in the Area, Advisory Opinion, 1 February 2011, ITLOS Reports 2011, p. 10.
Request for an advisory opinion submitted by the Sub-Regional Fisheries Commission (SRFC), Advisory Opinion in Case no. 21, 2 April 2015, ITLOS Reports (not published). Available at: https://www.itlos.org/fileadmin/itlos/documents/cases/case_no.21/advisory_opinion/C21_AdvOp_02.04.pdf (accessed 24 July 2016).

Treaties and other International Instruments

Declaration of the United Nations Conference on the Human Environment, 16 June 1972, A/CONF48/14/Rev/1.
Resolution 3129 (XXVIII) of the United Nations General Assembly, 'Cooperation in the field of the environment concerning natural resources shared by two or more States,' 13 December 1973.
Convention on Environmental Impact Assessment in a Transboundary Context, Espoo, Finland, 25 February 1991, UNTS, vol. 1989, p. 309.
Rio Declaration on Environment and Development, 13 June 1992, A/CONF.151/26.

Index

For Product Safety Concerns and Information please contact our EU representative GPSR@taylorandfrancis.com Taylor & Francis Verlag GmbH, Kaufingerstraße 24, 80331 München, Germany